DIGITAL FILTERS
Analysis, Design, and Applications

McGraw-Hill Series in Electrical and Computer Engineering

Senior Consulting Editor

Stephen W. Director, Carnegie Mellon University

Circuits and Systems
Communications and Signal Processing
Computer Engineering
Control Theory
Electromagnetics
Electronics and VLSI Circuits
Introductory
Power and Energy
Radar and Antennas

Previous Consulting Editors

Ronald N. Bracewell, Colin Cherry, James F. Gibbons, Willis W. Harman, Hubert Heffner, Edward W. Herold, John G. Linvill, Simon Ramo, Ronald A. Rohrer, Anthony E. Siegman, Charles Susskind, Frederick E. Terman, John G. Truxal, Ernest Weber, and John R. Whinnery

Communications and Signal Processing

Senior Consulting Editor

Stephen W. Director, Carnegie Mellon University

Antoniou: *Digital Filters: Analysis, Design, and Applications*
Candy: *Signal Processing: The Model-Based Approach*
Candy: *Signal Processing: The Modern Approach*
Carlson: *Communications Systems: An Introduction to Signals and Noise in Electrical Communication*
Cherin: *An Introduction to Optical Fibers*
Collin: *Antennas and Radiowave Propagation*
Collin: *Foundations for Microwave Engineering*
Cooper and McGillem: *Modern Communications and Spread Spectrum*
Davenport: *Probability and Random Processes: An Introduction for Applied Scientists and Engineers*
Drake: *Fundamentals of Applied Probability Theory*
Huelsman and Allen: *Introduction to the Theory and Design of Active Filters*
Jong: *Method of Discrete Signal and System Analysis*
Keiser: *Local Area Networks*
Keiser: *Optical Fiber Communications*
Kraus: *Antennas*
Kuc: *Introduction to Digital Signal Processing*
Papoulis: *Probability, Random Variables, and Stochastic Processes*
Papoulis: *Signal Analysis*
Papoulis: *The Fourier Integral and Its Applications*
Peebles: *Probability, Random Variables, and Random Signal Principles*
Proakis: *Digital Communications*
Schwartz: *Information Transmission, Modulation, and Noise*
Schwartz and Shaw: *Signal Processing*
Smith: *Modern Communication Circuits*
Taub and Schilling: *Principles of Communication Systems*

Also available from McGraw-Hill

Schaum's Outline Series in Electronics & Electrical Engineering

Most outlines include basic theory, definitions, and hundreds of example problems solved in step-by-step detail, and supplementary problems with answers.

Titles on the current list include:

Acoustics
Analog & Digital Communications
Basic Circuit Analysis, 2d edition
Basic Electrical Engineering
Basic Electricity
Basic Equations of Engineering Science
Basic Mathematics for Electricity & Electronics, 2d Edition
Digital Principles, 2d edition
Electric Circuits, 2d edition
Electric Machines & Electromechanics
Electric Power Systems
Electromagnetics, 2d Edition
Electronic Circuits
Electronic Communication
Electronic Devices & Circuits
Electronics Technology
Feedback & Control Systems, 2d edition
Introduction to Digital Systems
Microprocessor Fundamentals, 2d edition
State Space & Linear Systems

Schaum's Solved Problems Books

Each title in this series is a complete and expert source of solved problems containing thousands of problems with worked out solutions.

Related titles on the current list include:

3000 Solved Problems in Calculus
2500 Solved Problems in Differential Equations
3000 Solved Problems in Electric Circuits
2000 Solved Problems in Electromagnetics
2000 Solved Problems in Electronics
3000 Solved Problems in Linear Algebra
2000 Solved Problems in Numerical Analysis
3000 Solved Problems in Physics

Available at most college bookstores, or for a complete list of titles and prices write to: Schaum Division
McGraw-Hill, Inc.
Princeton Road, S-1
Hightstown, NJ 08520

DIGITAL FILTERS
Analysis,
Design,
and Applications

Second Edition

Andreas Antoniou
University of Victoria

McGraw-Hill, Inc.
New York St. Louis San Francisco Auckland Bogotá
Caracas Lisbon London Madrid Mexico Milan Montreal
New Delhi Paris San Juan Singapore Sydney Tokyo Toronto

DIGITAL FILTERS : Analysis, Design and Applications
International Editions 1993

Exclusive rights by McGraw-Hill Book Co. - Singapore for manufacture and export. This book cannot be re-exported from the country to which it is consigned by McGraw-Hill.

Copyright © 1993 by McGraw-Hill, Inc. All rights reserved. Previously published under the title of *Digital Filters: Analysis and Design*. Copyright © 1979 by McGraw-Hill, Inc. All rights reserved. Except as permitted under the United States Copyright Act of 1976, no part of this publication may be reproduced or distributed in any form or by any means, or stored in a data base or retrieval system, without the prior written permission of the publisher.

1 2 3 4 5 6 7 8 9 0 CMO PMP 9 8 7 6 5 4 3

This book was set in Times Roman by Electronic Technical Publishing Services.
The editors were Anne T. Brown and John M. Morriss;
the production supervisor was Louise Karam.
The cover was designed by Carla Bauer.
Project supervision was done by Electronic Technical Publishing Services.

Library of Congress Cataloging-in-Publication Data

Antoniou, Andreas, (date).
　　Digital filters: analysis, design, and applications / Andreas Antoniou. - 2nd ed.
　　　　p.　　cm. - (McGraw-Hill series in electrical and computer engineering.
　　Communications and signal processing)
　　　　Includes bibliographical references and index.
　　　　ISBN 0-07-002121-X
　　　　1. Electric filters, Digital.　　I. Title.　　II. Series.
TK7872.F5A64　　　　1993
621.3815'324-dc20　　　　　　　　　　　　　　　　　　　　92-35793

When ordering this title, use ISBN 0-07-112600-7

Printed in Singapore

ABOUT THE AUTHOR

Andreas Antoniou served as the founding chairman of the Department of Electrical and Computer Engineering at the University of Victoria, British Columbia, from 1983 to 1990 and is now Professor and Coordinator, Victoria Micronet Centre, Networks of Centres of Excellence Program, in the same department. He taught in the Department of Electrical and Computer Engineering at Concordia University in Montreal from 1970 to 1983 and was chairman of the department from 1977 to 1983. He received the B.Sc. and Ph.D. degrees in electrical engineering from the University of London, England. Nonacademic positions before 1970 include Member of Scientific Staff, Hirst Research Centre, GEC Ltd., London; Senior Scientific Officer, Post Office Research Department, London; and Member of the Scientific Staff, R & D Labs, Northern Electric Co. (now Bell-Northern Research), Ottawa, Ontario. He served as Associate Editor, IEEE Transactions on Circuits and Systems, from June 1983 to May 1985 and as Editor from June 1985 to May 1987. Dr. Antoniou was elected Fellow of the Institute of Electrical and Electronics Engineers for contributions to active and digital filters, and to electrical engineering education; he is also a Fellow of the Institution of Electrical Engineers and a member of the Association of Professional Engineers and Geophysicists of B.C. He was awarded the Ambrose Flemming Premium by the Institution of Electrical Engineers for his well-known paper on gyrator circuits and has published extensively on the design and applications of active and digital filters and on digital signal processing.

To my mother and father

CONTENTS

Preface to the Second Edition xvii
Preface to the First Edition xxi
Introduction xxv

1 Elementary Analysis 1
   1.1 Introduction 1
   1.2 Types of Discrete-Time Signals 1
   1.3 The Digital Filter as a System 2
   1.4 Characterization of Digital Filters 8
   1.5 Digital-Filter Networks 9
   1.6 Introduction to Time-Domain Analysis 15
   1.7 Convolution Summation 21
   1.8 Stability 25
   1.9 State-Space Analysis 26
       References 31
       Problems 31

2 The z Transform 43
   2.1 Introduction 43
   2.2 Review of Complex Analysis 44
   2.3 Definition of z Transform 47
   2.4 z-Transform Theorems 48
   2.5 Inverse z Transform 53
   2.6 Complex Convolution 59
       References 62
       Problems 62

3 The Application of the z Transform 66
   3.1 Introduction 66
   3.2 The Discrete-Time Transfer Function 67
   3.3 Stability 69

3.4	Time-Domain Analysis	78
3.5	Frequency-Domain Analysis	80
3.6	Amplitude and Delay Distortion	85
3.7	Introduction to the Design Process	87
	References	89
	Problems	89

4 Realization 97

4.1	Introduction	97
4.2	Direct Realization	98
4.3	Direct Canonic Realization	102
4.4	State-Space Realization	104
4.5	Ladder Realization	104
4.6	Lattice Realization	109
4.7	Cascade Realization	113
4.8	Parallel Realization	114
4.9	Implementation	115
4.10	Topological Properties	119
	References	129
	Additional References	129
	Problems	130

5 Analog-Filter Approximations 138

5.1	Introduction	138
5.2	Basic Concepts	139
5.3	Butterworth Approximation	143
5.4	Chebyshev Approximation	146
5.5	Elliptic Approximation	153
5.6	Bessel Approximation	168
5.7	Transformations	169
	References	172
	Problems	172

6 Continuous-Time, Sampled, and Discrete-Time Signals 177

6.1	Introduction	177
6.2	The Fourier Transform	177
6.3	Generalized Functions	182
6.4	Fourier Series	197
6.5	Poisson's Summation Formula	198
6.6	Sampled Signals	199
6.7	The Sampling Theorem	203
6.8	Interrelations	205
6.9	The Processing of Continuous-Time Signals	207
	References	213
	Problems	215

7 Approximations for Recursive Filters 220

7.1	Introduction	220
7.2	Realizability Constraints	221
7.3	Invariant-Impulse-Response Method	221

	7.4	Modified Invariant-Impulse-Response Method	224
	7.5	Matched-z-Transformation Method	228
	7.6	Bilinear-Transformation Method	231
	7.7	Digital-Filter Transformations	238
		References	244
		Problems	244
8	**Recursive Filters Satisfying Prescribed Specifications**		**249**
	8.1	Introduction	249
	8.2	Design Procedure	250
	8.3	Design Formulas	251
	8.4	Design Using the Formulas and Tables	262
	8.5	Constant Group Delay	267
	8.6	Amplitude Equalization	270
		Reference	271
		Problems	271
9	**Design of Nonrecursive Filters**		**274**
	9.1	Introduction	274
	9.2	Properties of Nonrecursive Filters	275
	9.3	Design Using the Fourier Series	280
	9.4	Use of Window Functions	282
	9.5	Design Based on Numerical-Analysis Formulas	301
	9.6	Comparison between Recursive and Nonrecursive Designs	305
		References	306
		Additional References	306
		Problems	306
10	**Random Signals**		**310**
	10.1	Introduction	310
	10.2	Random Variables	310
	10.3	Random Processes	314
	10.4	First- and Second-Order Statistics	316
	10.5	Moments and Autocorrelation	318
	10.6	Stationary Processes	319
	10.7	Frequency-Domain Representation	319
	10.8	Discrete-Time Random Processes	323
	10.9	Filtering of Discrete-Time Random Signals	325
		References	327
		Problems	327
11	**Effects of Finite Word Length in Digital Filters**		**330**
	11.1	Introduction	330
	11.2	Number Representation	331
	11.3	Coefficient Quantization	339
	11.4	Low-Sensitivity Structures	344
	11.5	Product Quantization	350
	11.6	Signal Scaling	352
	11.7	Minimization of Output Roundoff Noise	360
	11.8	Application of Error-Spectrum Shaping	364

11.9	Limit-Cycle Oscillations	367
	References	380
	Additional References	382
	Problems	383

12 Wave Digital Filters 388

12.1	Introduction	388
12.2	Sensitivity Considerations	389
12.3	Wave Network Characterization	390
12.4	Element Realizations	391
12.5	Lattice Wave Digital Filters	405
12.6	Ladder Wave Digital Filters	412
12.7	Filters Satisfying Prescribed Specifications	416
12.8	Frequency-Domain Analysis	419
12.9	Scaling	422
12.10	Elimination of Limit-Cycle Oscillations	423
12.11	Related Synthesis Methods	425
12.12	A Cascade Synthesis Based on the Wave Characterization	426
12.13	Choice of Structure	434
	References	436
	Additional References	437
	Problems	438

13 The Discrete Fourier Transform 444

13.1	Introduction	444
13.2	Definition	444
13.3	Inverse DFT	446
13.4	Properties	446
13.5	Interrelation between the DFT and the z Transform	449
13.6	Interrelation between the DFT and the CFT	454
13.7	Interrelation between the DFT and the Fourier Series	456
13.8	Nonrecursive Approximations through the Use of the DFT	458
13.9	Simplified Notation	462
13.10	Periodic Convolutions	463
13.11	Fast-Fourier-Transform Algorithms	465
13.12	Digital-Filter Implementation	477
	References	482
	Additional References	483
	Problems	483

14 Design of Recursive Filters Using Optimization Methods 489

14.1	Introduction	489
14.2	Problem Formulation	490
14.3	Newton's Method	492
14.4	Quasi-Newton Algorithms	496
14.5	Minimax Algorithms	509
14.6	Improved Minimax Algorithms	513
14.7	Design of Recursive Filters	517

	14.8	Design of Recursive Delay Equalizers	521
		References	535
		Additional References	538
		Problems	538

15 Design of Nonrecursive Filters Using Optimization Methods 544

	15.1	Introduction	544
	15.2	Problem Formulation	545
	15.3	Remez Exchange Algorithm	549
	15.4	Improved Search Methods	554
	15.5	Efficient Remez Exchange Algorithm	562
	15.6	Gradient Information	566
	15.7	Prescribed Specifications	571
	15.8	Generalization	574
	15.9	Digital Differentiators	578
	15.10	Arbitrary Amplitude Responses	583
	15.11	Multiband Filters	583
		References	587
		Additional References	587
		Problems	588

16 Digital Signal Processing Applications 592

	16.1	Introduction	592
	16.2	Sampling-Frequency Conversion	593
	16.3	Quadrature-Mirror-Filter Banks	602
	16.4	Hilbert Transformers	613
	16.5	Adaptive Digital Filters	625
	16.6	Two-Dimensional Digital Filters	637
		References	644
		Additional References	646
		Problems	646

Appendix A Elliptic Functions 653

	A.1	Introduction	653
	A.2	Elliptic Integral of the First Kind	653
	A.3	Elliptic Functions	656
	A.4	Imaginary Argument	656
	A.5	Formulas	659
	A.6	Periodicity	660
	A.7	Transformation	662
	A.8	Series Representation	663
		References	665

Index 667

PREFACE TO THE SECOND EDITION

The purpose of this edition, like that of the first edition, is to introduce the reader to theories, techniques, methods, and procedures which can be used to analyze, design, and implement digital filters. Emphasis is placed on methods that are simple, efficient, and robust. The revisions made can be classified into two categories, namely, revisions aimed at rationalizing or improving the exposition of known methods and revisions that involve deleting dated or obsolete material, adding new material, and possibly updating known principles in the light of recent developments in the field.

Chapters 1 to 10 of the first edition are concerned with fundamentals and the changes made are largely of the first category. The state-space approach of Chap. 1 has been replaced by one based on the flow-graph approach, which is more intuitive and easier to understand. The new derivation avoids the need for the digital filter to be time-invariant and, therefore, highlights the important fact that state-space methods can be used to represent time-variable filters and systems. Chapter 2 now begins with an introduction to complex analysis and deals exclusively with the two-side z transform. The treatment of the one-side z transform as a distinct mathematical entity has been dropped. In this way, the possibility of having a discrete-time function that has different one-sided and two-sided z transforms [e.g., if $x(nT) \neq 0$ for $n < 0$] is avoided. The one-sided z transform is simply a special case of the two-sided one. The penalty of this change is that we are now obliged to show that the inverse z transforms of some elementary z transforms are zero for $n < 0$, which is not too difficult. Chapter 2 now includes a convergence theorem which illustrates clearly how certain bounds imposed on the discrete-time function are interrelated with corresponding bounds in the z domain that guarantee the convergence of the z transform. Chapter 3 includes a more detailed exposition of the various stability criteria. The Schur-Cohn and Schur-Cohn-Fujisawa criteria, which form the link between the Routh-Hurwitz and the Jury-Marden criterion, have been added. The later criterion was referred to as the Jury criterion in the first edition, but the name of Marden has been added in recognition of his involvement in the derivation of some of the relevant formulas. Chapter 3 concludes with an introduction to the design process.

Chapter 4 has been restructured and a lattice realization method due to Gray and Markel has been included. Also the realization of digital filters in terms of systolic structures, which are suitable for VLSI implementation, has been considered to some extent. Chapter 5 now includes the inverse-Chebyshev analog-filter approximation method. Note that Chebyshev is now referred to by the North American version of his name throughout the volume. In Chap. 6 the section dealing with generalized functions has been rationalized further. Specifically, the impulse function is treated with more rigor and precision and the relation between a theoretical and a practical impulse funciton is clarified. Along with the interrelations among the various transforms found in the first edition, Chap. 6 also includes the interrelation between the Fourier series and Fourier transform; in addition, it provides a derivation of Poisson's summation formula, which is probably the most important result in establishing the various interrelations in Chaps. 6 and 13. Chapters 7 to 10 and 13 are approximately the same as before except that Chap. 8 now includes formulas for the design of inverse-Chebyshev digital filters, Chap. 9 includes the Dolph-Chebyshev window function, and Chap. 13 includes the overlap-and-save method as an alternative of the overlap-and-add method for the efficient implementation of nonrecursive filters.

Chapter 11 includes new methods for obtaining low-sensitivity and low-noise digital-filter structures and considers the issue of limit-cycle oscillation in greater detail. In the first edition, several pages are dedicated to the derivation of upper bounds on the amplitudes of limit-cycle oscillations. This material is no longer of considerable importance and has been replaced by methods that deal with the elimination of limit-cycle oscillations. Chapter 12 has been expanded somewhat and now includes the design of lattice wave digital filters; these offer some attractive solutions in the design of quadrature-mirror filters.

Chapter 14 of the first edition deals with the hardware implementation of digital filters. Nowadays, this task is largely accomplished by using general-purpose digital signal-processing chips which can be designed by engineers who need not be experts on the subject of digital-filter design. For this reason, old Chap. 14 has been dropped. It has been replaced by new Chaps. 14 and 15, which deal with the solution of the approximation problem in recursive and nonrecursive filters, respectively, using optimization methods. These methods are very useful when filters with arbitrary amplitude and phase responses are required.

Chapter 16, which is also a new chapter, deals with some of the numerous applications of digital filters to digital signal processing. The applications considered include downsampling and upsampling using decimators and interpolators, the design of quadrature-mirror-image filters and their application in time-division to frequency-division multiplex translation, Hilbert transformers and their application in single-sideband modulation, adaptive filters, and two-dimensional digital filters.

In the first edition, a set of computer programs was included in Appendix B. This proved useful to a number of readers who happened to know the computer language used, namely, HPL. However, eventually HPL became obsolete and these programs ceased to be useful. In the second edition, no computer programs are included to avoid the problem of obsolescence. Instead, algorithms are presented in a pseudo-code format that is easy to program in any one of the current computer languages.

The book can serve as a text for a sequence of two one-semester courses on digital filters for senior undergraduate or first-year graduate students. For students with an adequate background of discrete-time system theory, analog filters, the continuous and discrete Fourier transforms, and the theory of random variables and processes, a one-semester course could be offered comprising Chaps. 6 to 9, 11, 12, 14, 15 and 16. With the strengthening of Chap. 6 and the addition of Chap. 16, the book can also be used for courses in digital signal processing.

The book is supported by a detailed solutions manual. Furthermore, a set of transparencies for classroom use is currently being prepared and will be made available to instructors adopting the book upon the payment of a nominal handling fee.

I would like to thank Pan Agathoklis, Chris Charalambous, Paulo S. R. Diniz, Eliahu I. Jury, Wu-Sheng Lu, S. Douglas Peters, Dale Shpak, and P. P. Vaidyanathan for reading parts of the manuscript and for suggesting improvements; Randy K. Howell for constructing the plots in Figs. 14.12 and 14.13; Catherine Chang for help provided in cutting and pasting the illustrations, and copying the manuscript; Anne Brown for her encouragement and involvement as the McGraw-Hill editor in charge of the project; Peter H. Bauer, University of Notre Dame, Behrouz Peikari, Southern Methodist University, Russell E. Trahan, University of New Orleans, and Gregory H. Wakefield, University of Michigan, for their reviews and useful comments; the staff at Electronic Technical Publishing Services, in particular, Don Heard, Production Manager, Richard Hayes, Composition Manager, Lee Smith, Art Manager, and Candy Lafrenz, Prepress Manager, for putting together such an attractive volume; Lynne Barrett for helping with the proofreading of the manuscript; the University of Victoria for granting me the sabbatical year that enabled the completion of the second edition; Micronet, Networks of Centres of Excellence Program, and the Natural Sciences and Engineering Research Council of Canada for supporting the research that led to some of the new results in Chapters 14 and 15. Last but not least, I would like to thank my wife Rosemary for her sacrifices and understanding.

Andreas Antoniou

PREFACE TO THE FIRST EDITION

This book attempts to put together theories, techniques, and procedures which can be used to analyze, design, and implement digital filters. The digital filter is viewed as a system which can be implemented by means of software or hardware. In other words, the book is concerned both with the construction of algorithms (or computer programs) that can be used to filter recorded signals and also with the design of dedicated digital hardware that can be used to perform real-time filtering tasks like the many required in communications systems.

The prerequisite knowledge is a typical undergraduate mathematics background of calculus, complex variables, and simple differential equations. Section 5.5 entails a basic understanding of elliptic functions. Since this topic is normally excluded from undergraduate curricula, Appendix A provides a brief but adequate treatment of these functions. Chapter 11 requires a basic understanding of random variables and processes, which are reviewed in Chap. 10.

Chapter 1 introduces the digital filter as a discrete-time system which can be linear or nonlinear, causal or noncausal, and so on. Time-domain analysis is then introduced at an elementary level. The analysis is accomplished by solving the difference equation of the filter by induction. Although this technique is unlikely to be employed in practice, it is of pedagogical value as it provides the newcomer with a solid grasp of the physical nature of a digital filter. The chapter concludes with an alternative and more advanced time-domain analysis based on a state-space characterization.

The basic mathematical tool for the analysis of digital filters, the z transform, is described in Chap. 2. Its application in the time-domain and frequency-domain analysis of linear, time-invariant filters is considered in Chap. 3.

The realization of digital filters, namely the process of translating the transfer function into a digital-filter network, is discussed in Chap. 4. Later in this chapter a very useful topological theorem, known as Tellegen's theorem, is described. It is then used to develop the concepts of reciprocity, interreciprocity, and transposition. The chapter concludes with a convenient sensitivity analysis based on Tellegen's theorem.

Transfer-function approximations for digital filters of the recursive type are almost invariably obtained indirectly by using analog-filter approximations (e.g., Tschebyscheff, elliptic, etc.). Such approximations are considered in Chap. 5. The elliptic approximation is treated in detail because this is the most efficient of the available approximations. Although the derivation of this approximation turns out to be quite involved, a simple and easy-to-apply formulation is possible, as demonstrated at the end of Sec. 5.5.

Up to Chapter 5, the digital filter is treated as a distinct entity with its own methods of analysis. In Chap. 6, a theoretical link is established between discrete-time and continuous-time signals, which leads to a direct interrelation between the z transform and the Fourier transform. By this means, the enormous wealth of analog techniques can be applied in the analysis and design of digital filters; in addition, digital filters can be applied in the processing of analog signals.

Chapter 7 deals with the approximation problem for recursive filters. Methods are described by which a given continuous-time transfer function can be transformed into a corresponding discrete-time transfer function, e.g., impulse-invariant-response method and bilinear transformation method. A detailed procedure which can be used to design Butterworth, Tschebyscheff, and elliptic filters satisfying prescribed specifications is found in Chap. 8. This can be readily programmed.

Nonrecursive filters and their approximations are considered in Chap. 9. The use of window functions is described in detail. Emphasis is placed on Kaiser's window function, which has proved very versatile. In Chap. 10 the concept of a random process is introduced as a means of representing random signals. Such signals arise in digital filters because of the inevitable signal quantization. The effects of finite word length in digital filters along with appropriate, up-to-date methods of analysis are discussed in Chap. 11. The topics considered include coefficient quantization, product quantization, scaling, and limit-cycle bounds.

A relatively recent innovation in the domain of digital filtering has been the introduction of wave digital filters. Filters of this type can have certain attractive features, e.g., low sensitivity, and are considered in Chap. 12. The chapter includes step-by-step procedures by which filters satisfying prescribed specifications can be designed. The chapter concludes with a list of guidelines that can be used in the choice of a digital-filter network (or structure).

Chapter 13 presents the discrete Fourier transform and the associated fast Fourier-transform method as mathematical tools in the software implementation of digital filters. Much effort is spent in clearly conveying the exact interrelations between the discrete Fourier transform and (1) the z transform, (2) the continuous Fourier transform, and (3) the Fourier series because if these interrelations are not thoroughly understood, the user of the fast Fourier-transform method is likely to finish with masses of meaningless numbers.

Chapter 14, which concludes the book, deals with the hardware implementation of digital filters. It starts with a brief review of boolean algebra and then proceeds to a detailed description of the various types of combinational and sequential circuits that may be used as components in the implementation. Later the main integrated-circuit families are discussed and their features compared. Subsequently, in Sec. 14.7, three specific approaches to the implementation are described and their merits and demerits discussed. The chapter concludes with a brief outline of past, present, and future trends in the domain of digital filtering.

Most of the techniques are illustrated by examples and selected sets of problems are included throughout the book. Also 17 useful computer programs are included in Appendix B, which can be used by the filter designer in real-life designs and by the student in developing projects and in solving some of the challenging problems included.

The book can serve as a text for a sequence of two one-semester courses on digital filters for senior undergraduate or first-year graduate students. It could also serve for a one-semester course for groups of students with an adequate background of discrete-time system theory, analog filters, the Fourier transform, and the theory of random variables and processes. Such a course could comprise the following:

1. Review of Chap. 6
2. Brief discussion of analog-filter approximations (Chap. 5) with the emphasis placed on application rather than derivation
3. Chapters 7 to 9
4. Chapter 11, possibly in conjunction with Sec. 4.7
5. A choice for Chaps. 12 to 14, depending on the course orientation desired

The book should also be of interest to the filter designer, in particular Appendix B, and also to the scientist who has his own discrete-time signal to process.

The book is supported by a comprehensive Solutions Manual and also by a Tape Cartridge, which contains the programs given in Appendix B. The latter will be made available to institutions adopting the book upon the payment of a nominal handling fee.

I wish to thank V. Bhargava, A. G. Constantinides, R. Crochiere, L. B. Jackson, O. Monkewich, A. Papoulis, W. Saraga, and L. Weinberg for reading parts of the manuscript and for suggesting numerous improvements; S. W. Director, R. F. J. Filipowsky, and S. K. Mitra for their detailed reviews and suggestions; E. I. Jury for suggesting a number of improvements for the second printing of the book; J. C. Callaghan and M. N. S. Swamy for encouraging and supporting this work; June Anderson for typing the manuscript; Concordia University for supporting this work; and the National Research Council of Canada for supporting the research that led to many of the new results presented. Last but not least, I wish to thank my wife Rosemary and also Anthony, David, Constantine, and Helen for their sacrifices and understanding.

Andreas Antoniou

INTRODUCTION

Signals arise in almost every field of science and engineering, e.g., in acoustics, biomedical engineering, communications, control systems, radar, physics, seismology, and telemetry. Two general classes of signals can be identified, namely, continuous-time and discrete-time signals.

A *continuous-time signal* is one that is defined at each and every instant of time. Typical examples are a voltage waveform and the velocity of a space vehicle as a function of time. A *discrete-time signal*, on the other hand, is one that is defined at discrete instants of time, perhaps every millisecond, second, or day. Examples of this type of signal are the closing price of a particular commodity on the Stock Exchange and the daily precipitation as functions of time.

A discrete-time signal, like a continuous-time signal, can be represented by a unique function of frequency referred to as the *frequency spectrum* of the signal. This is a description of the frequency content of the signal.

Filtering is a process by which the frequency spectrum of a signal can be modified, reshaped, or manipulated according to some desired specification. It may entail amplifying or attenuating a range of frequency components, rejecting or isolating one specific frequency component, etc. The uses of filtering are manifold, e.g., to eliminate signal contamination such as noise, to remove signal distortion brought about by an imperfect transmission channel or by inaccuracies in measurement, to separate two or more distinct signals which were purposely mixed in order to maximize channel utilization, to resolve signals into their frequency components, to demodulate signals, to convert discrete-time signals into continuous-time signals, and to bandlimit signals.

The digital filter is a digital system that can be used to filter discrete-time signals. It can be implemented by means of software (computer programs), dedicated hardware, or a combination of software and hardware. Software digital filters may be implemented using a low-level language on a general-purpose digital signal-processing chip or in terms of a high-level language on a personal computer or workstation. At the other extreme, hardware digital filters can be designed using a number of highly specialized interconnected VLSI chips. Both software and hardware digital filters can be used to process real-time or non-real-time (recorded) signals, except that the latter are usually much faster and can process signals whose frequency spectra extend to much higher frequencies.

Software digital filters made their appearance along with the first digital computer in the late forties, although the name digital filter did not emerge until the midsixties. Early in the history of the digital computer many of the classical numerical analysis formulas of Newton, Stirling, Everett, and others were used to carry out interpolation, differentiation, and integration of functions (signals) represented by means of sequences of numbers (discrete-time signals). Since interpolation, differentiation, or integration of a signal represents a manipulation of the frequency spectrum of the signal, the subroutines or programs constructed to carry out these operations were essentially digital filters. In subsequent years, many complex and highly sophisticated algorithms and programs were developed to perform a variety of filtering tasks in numerous applications, e.g., data smoothing and prediction, pattern recognition, electrocardiogram processing, and spectrum analysis, speech coding, and speech synthesis. More recently, two-dimensional digital filters have emerged that can be used for the processing of images and two-dimensional geophysical signals. Work is currently in progress to extend the principles involved to multidimensional digital filters.

A bandlimited continuous-time signal can be transformed into a discrete-time signal by means of sampling. Conversely, the discrete-time signal so generated can be used to regenerate the original continuous-time signal by means of interpolation, by virtue of Shannon's sampling theorem. As a consequence, hardware digital filters can be used to perform real-time filtering tasks which in the not too distant past were performed almost exclusively by analog filters. The advantages to be gained are the traditional advantages associated with digital systems in general:

1. Component tolerances are uncritical.
2. Component drift and spurious environmental signals have no influence on the system performance.
3. Accuracy is high.
4. Physical size is small.
5. Reliability is high.

A very important additional advantage of digital filters is the ease with which their characteristics can be changed or adapted by simply changing the contents of a finite number of registers. Owing to this feature, digital filters are naturally suited for the design of programmable filters that can be used to perform a multiplicity of filtering tasks, and for the design of adaptive filters that can be used in a diverse range of applications, such as system identification, channel equalization, signal enhancement, and signal prediction.

DIGITAL FILTERS
Analysis, Design, and Applications

CHAPTER 1

ELEMENTARY ANALYSIS

1.1 INTRODUCTION

A digital filter, like an analog filter, can be represented by a network which comprises a collection of interconnected elements. *Analysis* of a digital filter is the process of determining the response of the filter network to a given excitation. *Design* of a digital filter, on the other hand, is the process of synthesizing and implementing a filter network so that a set of prescribed excitations results in a set of desired responses.

This chapter is an introduction to the analysis of digital filters. First, the fundamental concepts of time invariance, causality, etc., as applied to digital filters are discussed. Then an elementary time-domain analysis is described. The chapter concludes with a more advanced time-domain analysis based on a state-space approach.

1.2 TYPES OF DISCRETE-TIME SIGNALS

A *continuous-time* signal can be represented by a function $x(t)$ whose domain is a range of numbers (t_1, t_2), where $-\infty \leq t_1$ and $t_2 \leq \infty$. Similarly, a *discrete-time* signal can be represented by a function $x(nT)$, where T is a constant and n is an integer in the range (n_1, n_2) such that $-\infty \leq n_1$ and $n_2 \leq \infty$. Alternatively, a discrete-time signal can be represented by $x(n)$ or x_n. In this book we shall be using the first two notations, namely, $x(nT)$ and $x(n)$.

Continuous-time and discrete-time signals can be nonquantized or quantized. A *nonquantized* signal is one that can assume any value within a specific range, whereas

2 DIGITAL FILTERS: ANALYSIS, DESIGN, AND APPLICATIONS

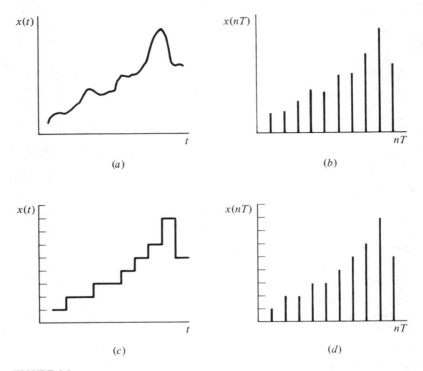

FIGURE 1.1
Types of signals: (*a*) nonquantized continuous-time signal, (*b*) nonquantized discrete-time signal, (*c*) quantized continuous-time signal, (*d*) quantized discrete-time signal.

a *quantized* signal is one that can assume only a finite number of discrete values. The ambient temperature as a function of time is a nonquantized signal. The ambient temperature, however, as measured by a digital thermometer is a quantized signal. The various types of signals are illustrated in Fig. 1.1.

1.3 THE DIGITAL FILTER AS A SYSTEM

A digital filter can be represented by the block diagram of Fig. 1.2. Input $x(nT)$ and output $y(nT)$ are the excitation and response of the filter, respectively. The response is related to the excitation by some rule of correspondence. We can indicate this fact

FIGURE 1.2
Digital filter.

notationally as

$$y(nT) = \mathcal{R}x(nT)$$

where $\mathcal{R}$ is an operator.

Like other systems, digital filters can be classified as time-invariant or time-dependent, causal or noncausal, and linear or nonlinear [1].

1.3.1 Time Invariance

A digital filter is said to be *time-invariant* if its response to an arbitrary excitation does not depend on the time of application of the excitation. As in other types of systems, the response of a digital filter depends on a number of internal system parameters. In a time-invariant digital filter, these parameters do not change with time.

Before we describe a test that can be used to check digital filters for time-invariance, the notion of a relaxed filter must be explained. Digital filters, like other types of systems, have internal storage or memory elements that can store values of signals. Such elements can serve as sources of internal signals and, consequently, a nonzero filter response may be produced in a case where the excitation is zero. If all the memory elements of a digital filter are empty or their contents are set to zero, the filter is said to be *relaxed*. The response of such a filter is zero for all n if the excitation is zero for all n.

Formally, an initially relaxed digital filter with excitation $x(nT)$ and response $y(nT)$, such that $x(nT) = y(nT) = 0$ for $n < 0$, is said to be time-invariant if and only if

$$\mathcal{R}x(nT - kT) = y(nT - kT)$$

for all possible excitations $x(nT)$ and all integers k. In other words, in a time-invariant digital filter, the response produced if the excitation $x(nT)$ is delayed by a period kT is numerically equal to the original response $y(nT)$ delayed by a period kT. This must be the case, if the internal parameters of the filter do not change with time. The behavior of a time-invariant filter is illustrated in Fig. 1.3. As can be seen, the response of the filter to the delayed excitation shown in Fig. 1.3b is equal to the response shown in Fig. 1.3a delayed by kT.

A filter that does not satisfy the above requirement for time-invariance is said to be *time-dependent*.

Example 1.1. (*a*) A digital filter is characterized by the equation

$$y(nT) = \mathcal{R}x(nT) = 2nTx(nT)$$

Check the filter for time invariance. (*b*) Repeat part (*a*) if

$$y(nT) = \mathcal{R}x(nT) = 12x(nT - T) + 11x(nT - 2T)$$

[†]Numbered references will be found at the end of each chapter.

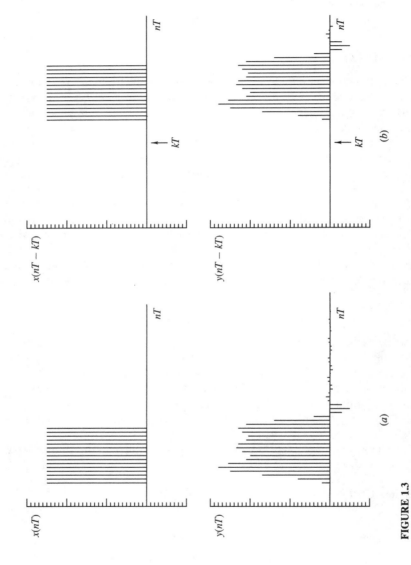

FIGURE 1.3
Time invariance: (a) response to an excitation $x(nT)$, (b) response to a delayed excitation $x(nT - kT)$.

Solution. (a) The response to a delayed excitation is
$$\mathcal{R}x(nT - kT) = 2nTx(nT - kT)$$
The delayed response is
$$y(nT - kT) = 2(nT - kT)x(nT - kT)$$
Clearly, for $k \neq 0$
$$\mathcal{R}x(nT - kT) \neq y(nT - kT)$$
and, therefore, the filter is time-dependent.

(b) In this case
$$\mathcal{R}x(nT - kT) = 12x[(n - k)T - T] + 11x[(n - k)T - 2T] = y(nT - kT)$$
for all possible $x(nT)$ and all integers k, and so the filter is time-invariant.

1.3.2 Causality

A *causal* digital filter is one whose response at a specific instant is independent of subsequent values of the excitation. More precisely, an initially relaxed digital filter in which $x(nT) = y(nT) = 0$ for $n < 0$ is said to be causal if and only if
$$\mathcal{R}x_1(nT) = \mathcal{R}x_2(nT) \quad \text{for } n \leq k$$
for all possible distinct excitations $x_1(nT)$ and $x_2(nT)$ such that
$$x_1(nT) = x_2(nT) \quad \text{for } n \leq k \tag{1.1}$$
Conversely, if
$$\mathcal{R}x_1(nT) \neq \mathcal{R}x_2(nT) \quad \text{for } n \leq k$$
for at least one pair of distinct excitations $x_1(nT)$ and $x_2(nT)$ such that
$$x_1(nT) = x_2(nT) \quad \text{for } n \leq k$$
then the filter is *noncausal*.

The above causality test can be easily justified. If all possible pairs of excitations $x_1(nT)$ and $x_2(nT)$ that satisfy Eq. (1.1) produce responses that are equal at instants $nT \leq kT$, then the filter response must depend only on values of the excitation at instants prior to nT, where $x_1(nT)$ and $x_2(nT)$ are assumed to be equal, and the filter is causal. This possibility is illustrated in Fig. 1.4. On the other hand, if at least two *distinct* excitations $x_1(nT)$ and $x_2(nT)$ that satisfy Eq. (1.1) produce responses that are not equal at instants $nT \leq kT$, then the filter response must depend on values of the excitation at instants subsequent to nT, since the differences between $x_1(nT)$ and $x_2(nT)$ occur after nT, and the filter is noncausal.

Example 1.2. (a) A digital filter is represented by
$$y(nT) = \mathcal{R}x(nT) = 3x(nT - 2T) + 3x(nT + 2T)$$

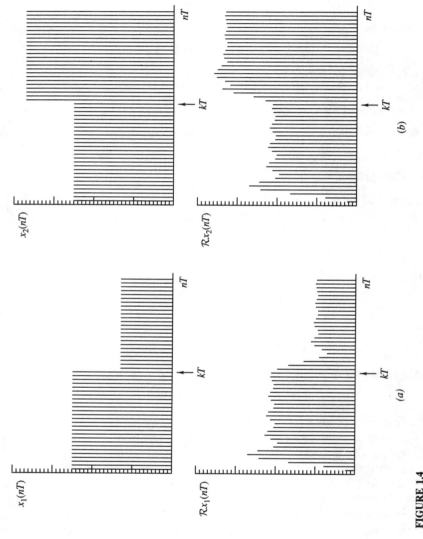

FIGURE 1.4
Causality: (*a*) response to $x_1(nT)$, (*b*) response to $x_2(nT)$.

Check the filter for causality. (b) Repeat part (a) if
$$y(nT) = \mathcal{R}x(nT) = 3x(nT - T) - 3x(nT - 2T)$$

Solution. (a) Let $x_1(nT)$ and $x_2(nT)$ be distinct excitations that satisfy Eq. (1.1) and assume that
$$x_1(nT) \neq x_2(nT) \qquad \text{for } n > k$$
For $n = k$
$$\mathcal{R}x_1(nT)|_{n=k} = 3x_1(kT - 2T) + 3x_1(kT + 2T)$$
$$\mathcal{R}x_2(nT)|_{n=k} = 3x_2(kT - 2T) + 3x_2(kT + 2T)$$
but since
$$3x_1(kT + 2T) \neq 3x_2(kT + 2T)$$
it follows that
$$\mathcal{R}x_1(nT) \neq \mathcal{R}x_2(nT) \qquad \text{for } n = k$$
i.e., the filter is noncausal.

(b) For this case
$$\mathcal{R}x_1(nT) = 3x_1(nT - T) - 3x_1(nT - 2T) \qquad \text{and}$$
$$\mathcal{R}x_2(nT) = 3x_2(nT - T) - 3x_2(nT - 2T)$$
If $n \leq k$, then $n - 1, n - 2 < k$ and so
$$x_1(nT - T) = x_2(nT - T) \qquad x_1(nT - 2T) = x_2(nT - 2T)$$
for $n \leq k$ or
$$\mathcal{R}x_1(nT) = \mathcal{R}x_2(nT) \qquad \text{for } n \leq k$$
i.e., the filter is causal.

1.3.3 Linearity

A digital filter is *linear* if and only if it satisfies the conditions
$$\mathcal{R}\alpha x(nT) = \alpha \mathcal{R}x(nT)$$
$$\mathcal{R}[x_1(nT) + x_2(nT)] = \mathcal{R}x_1(nT) + \mathcal{R}x_2(nT)$$
for all possible values of α and all possible excitations $x_1(nT)$ and $x_2(nT)$. These conditions are sometimes referred to as the *homogeneity* and *additivity* conditions, respectively [1].

The response of a linear filter to an excitation $\alpha x_1(nT) + \beta x_2(nT)$, where α and β are arbitrary constants, can be expressed as
$$y(nT) = \mathcal{R}[\alpha x_1(nT) + \beta x_2(nT)] = \mathcal{R}\alpha x_1(nT) + \mathcal{R}\beta x_2(nT)$$
$$= \alpha \mathcal{R}x_1(nT) + \beta \mathcal{R}x_2(nT)$$

Therefore, the above two conditions can be combined into one as
$$\mathcal{R}[\alpha x_1(nT) + \beta x_2(nT)] = \alpha \mathcal{R} x_1(nT) + \beta \mathcal{R} x_2(nT)$$
If this condition is violated for any pair of excitations or any constant α or β, then the filter is *nonlinear*.

Example 1.3. (*a*) The response of a digital filter is of the form
$$y(nT) = \mathcal{R} x(nT) = 7x^2(nT - T)$$
Check the filter for linearity. (*b*) Repeat part (*a*) if
$$y(nT) = \mathcal{R} x(nT) = (nT)^2 x(nT + 2T)$$

Solution. (*a*) For a constant α
$$\mathcal{R} \alpha x(nT) = 7\alpha^2 x^2(nT - T)$$
whereas
$$\alpha \mathcal{R} x(nT) = 7\alpha x^2(nT - T)$$
Clearly if $\alpha \neq 1$ then
$$\mathcal{R} \alpha x(nT) \neq \alpha \mathcal{R} x(nT)$$
and therefore the filter is nonlinear.

(*b*) For this case
$$\mathcal{R}[\alpha x_1(nT) + \beta x_2(nT)] = (nT)^2 [\alpha x_1(nT + 2T) + \beta x_2(nT + 2T)]$$
$$= \alpha (nT)^2 x_1(nT + 2T) + \beta (nT)^2 x_2(nT + 2T)$$
$$= \alpha \mathcal{R} x_1(nT) + \beta \mathcal{R} x_2(nT)$$
i.e., the filter is linear.

In this book we shall be concerned almost exclusively with linear, causal, and time-invariant filters.

1.4 CHARACTERIZATION OF DIGITAL FILTERS

Analog filters are characterized in terms of differential equations. Digital filters, on the other hand, are characterized in terms of *difference* equations. Two types of digital filters can be identified, *nonrecursive* and *recursive* filters.

1.4.1 Nonrecursive Filters

The response of a nonrecursive filter at instant nT is of the form
$$y(nT) = f\{\ldots, x(nT - T), x(nT), x(nT + T), \ldots\}$$
If we assume linearity and time invariance, $y(nT)$ can be expressed as
$$y(nT) = \sum_{i=-\infty}^{\infty} a_i x(nT - iT) \qquad (1.2)$$

where a_i represents constants. Now on assuming that the filter is causal and noting that $x(nT+T), x(nT+2T), \ldots$ are subsequent values of the excitation with respect to instant nT, we must have
$$a_i = 0 \quad \text{for } i \leq -1$$
and so
$$y(nT) = \sum_{i=0}^{\infty} a_i x(nT - iT)$$
If, in addition, $x(nT) = 0$ for $n < 0$ and $a_i = 0$ for $i > N$,
$$\begin{aligned} y(nT) &= \sum_{i=0}^{n} a_i x(nT - iT) + \sum_{i=n+1}^{\infty} a_i x(nT - iT) \\ &= \sum_{i=0}^{N} a_i x(nT - iT) + \sum_{i=N+1}^{n} a_i x(nT - iT) \\ &= \sum_{i=0}^{N} a_i x(nT - iT) \end{aligned} \quad (1.3)$$

Therefore, a linear, time-invariant, causal, nonrecursive filter can be represented by an Nth-order linear difference equation. N is the *order* of the filter.

1.4.2 Recursive Filters

The response of a recursive filter is a function of elements in the excitation as well as the response sequence. In the case of a linear, time-invariant, causal filter
$$y(nT) = \sum_{i=0}^{N} a_i x(nT - iT) - \sum_{i=1}^{N} b_i y(nT - iT) \quad (1.4)$$
i.e., if instant nT is taken to be the present, the present response is a function of the present and past N values of the excitation as well as the past N values of the response. Note that Eq. (1.4) simplifies to Eq. (1.3) if $b_i = 0$, and essentially the nonrecursive filter is a special case of the recursive one.

1.5 DIGITAL-FILTER NETWORKS

The basic elements of digital filters are the *unit delay*, the *adder*, and the *multiplier*. Their characterizations and symbols are given in Table 1.1. Ideally, the adder will produce the sum of its inputs and the multiplier will multiply its input by a constant instantaneously. The unit delay, on the other hand, will record its input at instant nT and deliver its previous input to the output. In effect, the unit delay is a memory element. The implementation of the digital-filter elements can assume various forms, depending on the representation of the signals to be processed. If the signals are sequences of binary numbers, the adder and multiplier will be combinational or sequential digital circuits, and the unit delay will be in the form of some type of

Table 1.1 Digital-filter elements

Element	Symbol	Equation
Unit delay	$x(nT) \to \boxed{T} \to y(nT)$	$y(nT) = x(nT - T)$
Adder	$x_1(nT), x_2(nT), \ldots, x_k(nT) \to \oplus \to y(nT)$	$y(nT) = \sum_{i=1}^{k} x_i(nT)$
Multiplier	$x(nT) \to \otimes \to y(nT)$, with m	$y(nT) = mx(nT)$

register. Collections of unit delays, adders, and multipliers can be interconnected to form *digital-filter networks*.

1.5.1 Network Analysis

The *analysis* of a digital-filter network, which is the process of deriving the difference equation characterizing the network, can be carried out by using the element equations given in Table 1.1. Network analysis can often be simplified by using the *shift operator* $\mathcal{E}$, which is defined by

$$\mathcal{E}^r f(nT) = f(nT + rT)$$

The shift operator is one of the basic operators of numerical analysis and its main properties are as follows:

1. Since

$$\mathcal{E}^r [a_1 f_1(nT) + a_2 f_2(nT)] = a_1 f_1(nT + rT) + a_2 f_2(nT + rT)$$
$$= a_1 \mathcal{E}^r f_1(nT) + a_2 \mathcal{E}^r f_2(nT)$$

we conclude that $\mathcal{E}$ is a linear operator which *distributes* with respect to a sum of functions of nT.

2. Since
$$\mathcal{E}^r\mathcal{E}^p f(nT) = \mathcal{E}^r f(nT+pT) = f(nT+rT+pT)$$
$$= \mathcal{E}^{r+p} f(nT)$$
the $\mathcal{E}$ operator obeys the usual *law of exponents*.

3. If $g(nT) = \mathcal{E}^r f(nT)$ then $\mathcal{E}^{-r} g(nT) = f(nT)$ for all $f(nT)$, and if $f(nT) = \mathcal{E}^{-r} g(nT)$ then $g(nT) = \mathcal{E}^r f(nT)$ for all $g(nT)$. Therefore, $\mathcal{E}^{-r}$ is the inverse of $\mathcal{E}^r$ and vice versa, i.e.,
$$\mathcal{E}^{-r}\mathcal{E}^r = \mathcal{E}^r\mathcal{E}^{-r} = 1$$

4. A linear combination of powers of $\mathcal{E}$ defines a meaningful operator, e.g., if
$$f(\mathcal{E}) = 1 + a_1\mathcal{E} + a_2\mathcal{E}^2$$
then
$$f(\mathcal{E})f(nT) = (1 + a_1\mathcal{E} + a_2\mathcal{E}^2)f(nT)$$
$$= f(nT) + a_1 f(nT+T) + a_2 f(nT+2T)$$

Further, given an operator $f(\mathcal{E})$ of the above type, an *inverse* operator $f(\mathcal{E})^{-1}$ may be defined such that
$$f(\mathcal{E})^{-1} f(\mathcal{E}) = f(\mathcal{E}) f(\mathcal{E})^{-1} = 1$$

5. If $f_1(\mathcal{E})$, $f_2(\mathcal{E})$, and $f_3(\mathcal{E})$ are operators that comprise linear combinations of powers of $\mathcal{E}$, then they satisfy the *distributive, commutative,* and *associative* laws of algebra, i.e.,
$$f_1(\mathcal{E})[f_2(\mathcal{E}) + f_3(\mathcal{E})] = f_1(\mathcal{E})f_2(\mathcal{E}) + f_1(\mathcal{E})f_3(\mathcal{E})$$
$$f_1(\mathcal{E})f_2(\mathcal{E}) = f_2(\mathcal{E})f_1(\mathcal{E})$$
$$f_1(\mathcal{E})[f_2(\mathcal{E})f_3(\mathcal{E})] = [f_1(\mathcal{E})f_2(\mathcal{E})]f_3(\mathcal{E})$$

Operators of the preceding type can be used to construct more complicated operators of the form
$$F(\mathcal{E}) = f_1(\mathcal{E})f_2(\mathcal{E})^{-1} = f_2(\mathcal{E})^{-1}f_1(\mathcal{E})$$
which may also be expressed without danger of ambiguity as
$$F(\mathcal{E}) = \frac{f_1(\mathcal{E})}{f_2(\mathcal{E})}$$

Owing to the above properties, the shift operator $\mathcal{E}$ can to a large extent be treated like an ordinary algebraic quantity [2], and operators that are linear combinations of powers of $\mathcal{E}$ can be treated as polynomials. For example, the difference equation of a recursive filter given in Eq. (1.4) can be expressed as
$$y(nT) = \left(\sum_{i=0}^{N} a_i \mathcal{E}^{-i}\right) x(nT) - \left(\sum_{i=1}^{N} b_i \mathcal{E}^{-i}\right) y(nT)$$

or

$$y(nT) = \left(\frac{\sum_{i=0}^{N} a_i \mathcal{E}^{-i}}{1 + \sum_{i=1}^{N} b_i \mathcal{E}^{-i}} \right) x(nT)$$

The application of the above principles in the analysis of digital-filter networks is illustrated in Example 1.4b.

Example 1.4. (a) Analyze the network of Fig. 1.5a. (b) Repeat part (a) for the network of Fig. 1.5b.

Solution. (a) The signals at nodes A and B are $y(nT-T)$ and $e^\alpha y(nT-T)$, respectively. Thus,

$$y(nT) = x(nT) + e^\alpha y(nT - T)$$

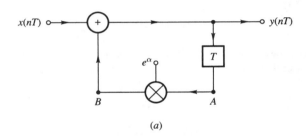

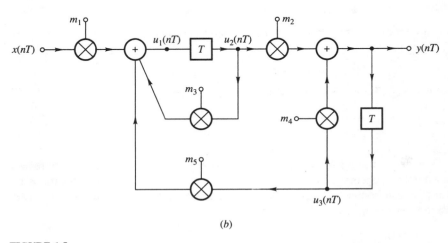

FIGURE 1.5
Digital-filter networks (Example 1.4): (a) first-order filter, (b) second-order filter.

(b) From Fig. 1.5b

$$u_1(nT) = m_1 x(nT) + m_3 u_2(nT) + m_5 u_3(nT)$$

$$u_2(nT) = \mathcal{E}^{-1} u_1(nT) \qquad u_3(nT) = \mathcal{E}^{-1} y(nT)$$

$$y(nT) = m_2 u_2(nT) + m_4 u_3(nT)$$

and on eliminating $u_2(nT)$ and $u_3(nT)$ in $u_1(nT)$ and $y(nT)$, we have

$$(1 - m_3 \mathcal{E}^{-1}) u_1(nT) = m_1 x(nT) + m_5 \mathcal{E}^{-1} y(nT)$$

and

$$(1 - m_4 \mathcal{E}^{-1}) y(nT) = m_2 \mathcal{E}^{-1} u_1(nT)$$

The latter equation can be expressed as

$$(1 - m_3 \mathcal{E}^{-1})(1 - m_4 \mathcal{E}^{-1}) y(nT) = m_2 \mathcal{E}^{-1} (1 - m_3 \mathcal{E}^{-1}) u_1(nT)$$

and on eliminating $(1 - m_3 \mathcal{E}^{-1}) u_1(nT)$, we obtain

$$[1 - (m_3 + m_4) \mathcal{E}^{-1} + m_3 m_4 \mathcal{E}^{-2}] y(nT) = m_1 m_2 \mathcal{E}^{-1} x(nT)$$
$$+ m_2 m_5 \mathcal{E}^{-2} y(nT)$$

Therefore

$$y(nT) = a_1 x(nT - T) + b_1 y(nT - T) + b_2 y(nT - 2T)$$

where $\quad a_1 = m_1 m_2 \qquad b_1 = m_3 + m_4 \qquad b_2 = m_2 m_5 - m_3 m_4$

1.5.2 Signal Flow-Graph Analysis

Given a digital-filter network, a corresponding signal flow graph can readily be deduced by replacing (1) each multiplier by a directed branch with transmittance equal to the constant of the multiplier, (2) each unit delay by a directed branch with transmittance equal to the operator $\mathcal{E}^{-1}$, and (3) each adder by a node with one outgoing branch and as many incoming branches as there are inputs in the adder. For example, the network of Fig. 1.6a can be represented by the flow graph of Fig. 1.6b.

In view of the equivalence alluded to in the preceding paragraph, the analysis of digital-filter networks can also be carried out by using *flow-graph* techniques [3–6]. The analysis can be accomplished by reducing the flow graph of the network to a single transmittance between nodes $x(nT)$ and $y(nT)$ through node elimination [5]. The approach is straightforward and reliable but can be time-consuming if the number of nodes is large.

An alternative approach, which will be used in Sec. 1.9.1 in the derivation of a state-space representation of digital filters, is based on *Mason's gain formula* [5, 6]. In this approach, the response at node j of a digital filter to an excitation applied at node i is given by

$$y_j(nT) = \left(\frac{1}{\Delta} \sum_k T_k \Delta_k \right) x_i(nT)$$

14 DIGITAL FILTERS: ANALYSIS, DESIGN, AND APPLICATIONS

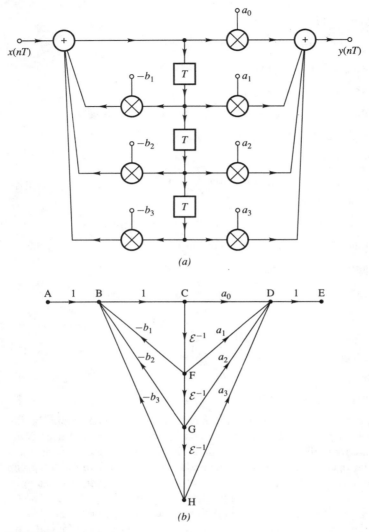

FIGURE 1.6
(a) Digital-filter structure, (b) signal flow graph.

where T_k is the transmittance of the kth direct path between nodes i and j, Δ is the determinant of the flow graph, and Δ_k is the determinant of the subgraph that does not touch (has no nodes or branches in common with) the kth direct path between nodes i and j.

The graph determinant Δ is given by

$$\Delta = 1 - \sum_u L_{u1} + \sum_v P_{v2} - \sum_w P_{w3} + \ldots$$

where L_{u1} is the loop transmittance of the uth loop, P_{v2} is the product of the loop transmittances of the vth pair of nontouching loops (loops that have neither nodes nor branches in common), P_{w3} is the product of loop transmittances of the wth triplet of nontouching loops, and so on.

The subgraph determinant Δ_k can be determined by applying the formula for Δ to the subgraph that does not touch the kth direct path between nodes i and j.

The derivation of Mason's formula can be found in [6]. Its application is illustrated by the following example.

Example 1.5. Analyze the network of Fig. 1.6a using Mason's gain formula.

Solution. From Fig. 1.6b, the direct paths of the flow graph are ABCDE, ABCFDE, ABCFGDE, and ABCFGHDE and hence

$$T_1 = a_0, \quad T_2 = a_1\mathcal{E}^{-1}, \quad T_3 = a_2\mathcal{E}^{-2}, \quad T_4 = a_3\mathcal{E}^{-3}$$

The loops of the graph are BCFB, BCFGB, and BCFGHB and hence

$$L_{11} = -b_1\mathcal{E}^{-1}, \quad L_{21} = -b_2\mathcal{E}^{-2}, \quad L_{31} = -b_3\mathcal{E}^{-3}$$

All loops are touching, since branch BC is common to all of them, and so

$$P_{v2} = P_{w3} = \cdots = 0$$

Hence

$$\Delta = 1 + b_1\mathcal{E}^{-1} + b_2\mathcal{E}^{-2} + b_3\mathcal{E}^{-3}$$

The determinants of the subgraphs Δ_k, $k = 1, 2, 3$, and 4, can similarly be determined by identifying each subgraph that does not touch the kth direct path. As can be seen in Fig. 1.6b, branch BC is common to all direct paths between input and output, and therefore it does not appear in any of the subgraphs. Consequently, no loops are present in the k subgraphs and so

$$\Delta_1 = \Delta_2 = \Delta_3 = \Delta_4 = 1$$

Using Mason's formula, we obtain

$$y(nT) = \left(\frac{\sum_{i=0}^{3} a_i \mathcal{E}^{-i}}{1 + \sum_{i=1}^{3} b_i \mathcal{E}^{-i}} \right) x(nT)$$

or

$$y(nT) = \left(\sum_{i=0}^{3} a_i \mathcal{E}^{-i} \right) x(nT) - \left(\sum_{i=1}^{3} b_i \mathcal{E}^{-i} \right) y(nT)$$

1.6 INTRODUCTION TO TIME-DOMAIN ANALYSIS

The *time-domain* analysis of analog systems is facilitated by using several *elementary functions* such as the unit impulse, the unit step, etc. Similar functions can be used for digital filters. Such functions are defined in Table 1.2. The discrete-time *unit step,*

Table 1.2 Discrete-time elementary functions

Function	Definition	Waveform
Unit impulse	$\delta(nT) = \begin{cases} 1 & n = 0 \\ 0 & n \neq 0 \end{cases}$	
Unit step	$u(nT) = \begin{cases} 1 & n \geq 0 \\ 0 & n < 0 \end{cases}$	
Unit ramp	$r(nT) = \begin{cases} nT & n \geq 0 \\ 0 & n < 0 \end{cases}$	
Exponential	e^{anT}	
Sinusoid	$\sin \omega n T$	

unit ramp, *exponential*, and *sinusoid* are generated by letting $t = nT$ in the corresponding continuous-time functions. The discrete-time *unit impulse* $\delta(nT)$, however, is generated by letting $t = nT$ in the pulse function $p_{t_0}(t)$ given by

$$p_{t_0}(t) = \begin{cases} 1 & \text{for } |t| \leq t_0 < T \\ 0 & \text{otherwise} \end{cases}$$

The role of $\delta(nT)$ in discrete-time systems is analogous to the role of the continuous-time unit impulse $\delta(t)$ in analog systems. Nevertheless, the two functions are defined differently, and as a consequence $\delta(nT)$ cannot be generated by letting $t = nT$ in $\delta(t)$. The interrelation between $\delta(nT)$ and $\delta(t)$ is discussed in the footnote of p. 206.

The time-domain response of simple digital filters can be determined by solving the difference equation directly using induction. Although this approach is somewhat primitive, it merits consideration since it demonstrates the mode by which digital filters operate. The approach is best illustrated by the following examples.

Example 1.6. (*a*) Find the impulse response of the filter in Fig. 1.5a. The filter is initially relaxed, that is, $y(nT) = 0$ for $n < 0$. (*b*) Find the unit-step response of the filter.

Solution. (*a*) From Example 1.4a

$$y(nT) = x(nT) + e^\alpha y(nT - T) \qquad (1.5)$$

With $x(nT) = \delta(nT)$, we can write

$$y(0) = 1 + e^\alpha y(-T) = 1$$
$$y(T) = 0 + e^\alpha y(0) = e^\alpha$$
$$y(2T) = 0 + e^\alpha y(T) = e^{2\alpha}$$

$$\dotsb\dotsb\dotsb\dotsb\dotsb\dotsb$$

$$y(nT) = u(nT)e^{n\alpha}$$

The response is plotted in Fig. 1.7 for $\alpha < 0$, $\alpha = 0$, and $\alpha > 0$.

(*b*) With $x(nT) = u(nT)$

$$y(0) = 1 + e^\alpha y(-T) = 1$$
$$y(T) = 1 + e^\alpha y(0) = 1 + e^\alpha$$
$$y(2T) = 1 + e^\alpha y(T) = 1 + e^\alpha + e^{2\alpha}$$

$$\dotsb\dotsb\dotsb\dotsb\dotsb\dotsb$$

$$y(nT) = u(nT) \sum_{k=0}^{n} e^{k\alpha}$$

This is a geometric series in powers of e^α and hence

$$y(nT) - e^\alpha y(nT) = u(nT)(1 - e^{(n+1)\alpha})$$

or

$$y(nT) = u(nT) \frac{1 - e^{(n+1)\alpha}}{1 - e^\alpha}$$

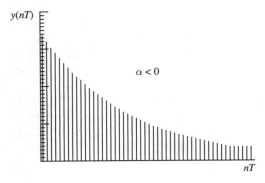

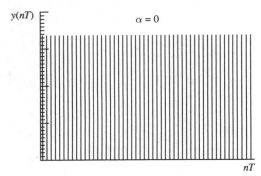

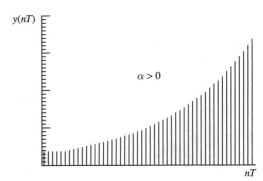

FIGURE 1.7
Impulse response of first-order filter (Example 1.6a).

For $\alpha < 0$, the steady-state value of the response is obtained by evaluating $y(nT)$ for $n \to \infty$, i.e.,

$$\lim_{n \to \infty} y(nT) = \frac{1}{1 - e^{\alpha}}$$

For $\alpha = 0$, using l'Hospital's rule, we get

$$y(nT) = \lim_{\alpha \to 0} \frac{d(1 - e^{(n+1)\alpha})/d\alpha}{d(1 - e^{\alpha})/d\alpha} = n + 1$$

and hence $y(nT) \to \infty$ as $n \to \infty$. For $\alpha > 0$

$$\lim_{n \to \infty} y(nT) \approx \frac{e^{n\alpha}}{e^{\alpha} - 1} \to \infty$$

(see Fig. 1.8).

Example 1.7. Find the response of the filter in Fig. 1.5a if $x(nT) = u(nT) \sin \omega nT$. Assume that $\alpha < 0$.

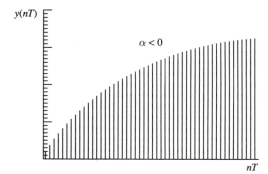

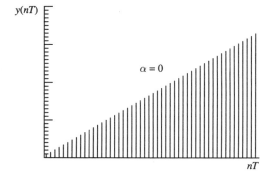

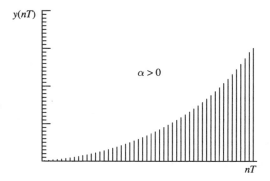

FIGURE 1.8
Unit-step response of first-order filter (Example 1.6b).

Solution. The filter is linear and so

$$y(nT) = \mathcal{R}u(nT)\sin\omega nT = \mathcal{R}u(nT)\frac{1}{2j}(e^{j\omega nT} - e^{-j\omega nT})$$

$$= \frac{1}{2j}[\mathcal{R}u(nT)e^{j\omega nT} - \mathcal{R}u(nT)e^{-j\omega nT}]$$

$$= \frac{1}{2j}[y_1(nT) - y_2(nT)] \tag{1.6}$$

where

$$y_1(nT) = \mathcal{R}u(nT)e^{j\omega nT} \quad \text{and} \quad y_2(nT) = \mathcal{R}u(nT)e^{-j\omega nT}$$

With the filter initially relaxed, the use of Eq. (1.5) gives

$$y_1(0) = e^0 + e^\alpha y_1(-T) = 1$$

$$y_1(T) = e^{j\omega T} + e^\alpha y_1(0) = e^{j\omega T} + e^\alpha$$

$$y_1(2T) = e^{j2\omega T} + e^\alpha y_1(T) = e^{j2\omega T} + e^{\alpha+j\omega T} + e^{2\alpha}$$

$$\cdots\cdots\cdots\cdots\cdots\cdots\cdots\cdots\cdots\cdots\cdots\cdots\cdots$$

$$y_1(nT) = u(nT)(e^{j\omega nT} + e^{\alpha+j\omega(n-1)T} + \cdots + e^{(n-1)\alpha+j\omega T} + e^{n\alpha})$$

$$= u(nT)e^{j\omega nT}(1 + e^{\alpha-j\omega T} + \cdots + e^{n(\alpha-j\omega T)})$$

$$= u(nT)e^{j\omega nT}\sum_{k=0}^{n}e^{k(\alpha-j\omega T)}$$

and, as in part (b) of Example 1.6, we have

$$y_1(nT) = u(nT)\frac{e^{j\omega nT} - e^{(n+1)\alpha-j\omega T}}{1 - e^{\alpha-j\omega T}}$$

Now consider the function

$$H(e^{j\omega T}) = \frac{e^{j\omega T}}{e^{j\omega T} - e^\alpha} = \frac{1}{1 - e^{\alpha-j\omega T}}$$

and let

$$H(e^{j\omega T}) = M(\omega)e^{j\theta(\omega)}$$

where

$$M(\omega) = |H(e^{j\omega T})| = \frac{1}{\sqrt{1 + e^{2\alpha} - 2e^\alpha \cos\omega T}} \tag{1.7}$$

and

$$\theta(\omega) = \arg H(e^{j\omega T}) = \omega T - \tan^{-1}\frac{\sin\omega T}{\cos\omega T - e^\alpha} \tag{1.8}$$

Using these relations, $y_1(\omega T)$ can be expressed as

$$y_1(nT) = u(nT)M(\omega)(e^{j[\theta(\omega)+\omega nT]} - e^{(n+1)\alpha+j[\theta(\omega)-\omega T]}) \tag{1.9}$$

and on replacing ω by $-\omega$ in $y_1(nT)$, we get

$$y_2(nT) = u(nT)M(-\omega)(e^{j[\theta(-\omega)-\omega nT]} - e^{(n+1)\alpha+j[\theta(-\omega)+\omega T]}) \quad (1.10)$$

From Eqs. (1.7) and (1.8), we note that $M(\omega)$ is an even function and $\theta(\omega)$ is an odd function of ω, i.e.,

$$M(-\omega) = M(\omega) \quad \text{and} \quad \theta(-\omega) = -\theta(\omega)$$

Hence Eqs. (1.6), (1.9), and (1.10) yield

$$y(nT) = u(nT)\frac{M(\omega)}{2j}(e^{j[\theta(\omega)+\omega nT]} - e^{-j[\theta(\omega)+\omega nT]})$$

$$- u(nT)\frac{M(\omega)}{2j}e^{(n+1)\alpha}(e^{j[\theta(\omega)-\omega T]} - e^{-j[\theta(\omega)-\omega T]})$$

$$= u(nT)M(\omega)\sin[\omega nT + \theta(\omega)]$$

$$- u(nT)M(\omega)e^{(n+1)\alpha}\sin[\theta(\omega) - \omega T]$$

Evidently, the response of the filter consists of two components. The second term represents a transient sinusoidal component whose amplitude reduces to zero as $n \to 0$ since $\alpha < 0$. Therefore, we can write

$$\tilde{y}(nT) = \lim_{n \to \infty} y(nT) = M(\omega)\sin[\omega nT + \theta(\omega)]$$

In effect, the steady-state response of the filter to a sinusoid with an amplitude of unity and zero displacement relative to the nT axis is a sinusoid with amplitude $M(\omega)$ displaced by a phase angle $\theta(\omega)$, as illustrated in Fig. 1.9. Functions $M(\omega)$ and $\theta(\omega)$ are called the *amplitude response* and *phase response*, respectively, and together they describe the performance of the filter as a function of frequency. From Eqs. (1.7) and (1.8), we note that these functions are the magnitude and angle of $H(e^{j\omega T})$. This function is called the *frequency response* of the filter.

Steady-state sinusoidal analysis is very important in practice since it allows us to find the steady-state response of the filter to an arbitrary linear combination of sinusoids. This is possible by virtue of the linearity of the filter.

If α were greater than zero in the above analysis, then the amplitude of the transient component would, in principle, increase indefinitely. In such a case, the filter is said to be *unstable* (see Sec. 1.8).

General methods for the time-domain analysis of digital filters are based on the application of the z transform and will be considered in detail in Chap. 3.

1.7 CONVOLUTION SUMMATION

The response of a digital filter to an arbitrary excitation can be expressed in terms of the impulse response of the filter.

An excitation $x(nT)$ can be written as

$$x(nT) = \sum_{k=-\infty}^{\infty} x_k(nT)$$

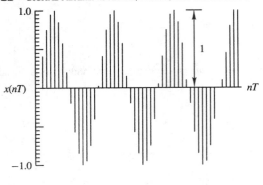

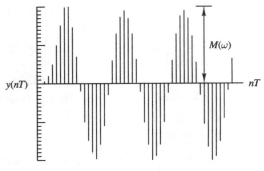

FIGURE 1.9
Steady-state sinusoidal response of first-order filter (Example 1.7).

where
$$x_k(nT) = \begin{cases} x(kT) & \text{for } n = k \\ 0 & \text{otherwise} \end{cases}$$

Alternatively
$$x_k(nT) = x(kT)\delta(nT - kT)$$

and hence
$$x(nT) = \sum_{k=-\infty}^{\infty} x(kT)\delta(nT - kT) \quad (1.11)$$

Now consider a linear time-invariant filter in which
$$\mathcal{R}\delta(nT) = h(nT) \quad \text{and} \quad y(nT) = \mathcal{R}x(nT)$$

From Eq. (1.11), we have
$$y(nT) = \mathcal{R} \sum_{k=-\infty}^{\infty} x(kT)\delta(nT - kT) = \sum_{k=-\infty}^{\infty} x(kT)\mathcal{R}\delta(nT - kT)$$
$$= \sum_{k=-\infty}^{\infty} x(kT)h(nT - kT) = \sum_{k=-\infty}^{\infty} h(kT)x(nT - kT) \quad (1.12)$$

where the second form at the right is deduced by a simple change of variable. This relation is of considerable importance in the characterization as well as analysis of digital filters and is known as the *convolution summation*.

Two special cases of the convolution summation are of particular interest. If the filter is causal, $h(nT) = 0$ for $n < 0$, and so

$$y(nT) = \sum_{k=-\infty}^{\infty} x(kT)h(nT - kT) = \sum_{k=0}^{\infty} h(kT)x(nT - kT) \qquad (1.13a)$$

If, in addition, $x(nT) = 0$ for $n < 0$, we have

$$y(nT) = \sum_{k=0}^{n} x(kT)h(nT - kT) = \sum_{k=0}^{n} h(kT)x(nT - kT) \qquad (1.13b)$$

1.7.1 Graphical Interpretation

The convolution summation is illustrated in Fig. 1.10. The impulse response $h(kT)$ is first folded about the y axis, as in Fig. 1.10c, and is then shifted to the right by a time interval nT, as in Fig. 1.10d, to yield $h(nT - kT)$. Then $x(kT)$ is multiplied by $h(nT - kT)$, as in Fig. 1.10e. The sum of all values in Fig. 1.10e is the response of the filter at instant nT.

Example 1.8. (*a*) Using the convolution summation, find the unit-step response of the filter in Fig. 1.5a. (*b*) Hence find the response to an excitation

$$x(nT) = \begin{cases} 1 & \text{for } 0 \leq n \leq 4 \\ 0 & \text{otherwise} \end{cases}$$

Solution. (*a*) From part (*a*) of Example 1.6

$$h(nT) = u(nT)e^{n\alpha}$$

Hence from Eq. (1.13b)

$$y(nT) = \mathcal{R}u(nT) = \sum_{k=0}^{n} u(kT)e^{k\alpha}u(nT - kT)$$

$$= \sum_{k=0}^{n} u(nT - kT)e^{k\alpha} = u(nT)\sum_{k=0}^{n} e^{k\alpha}$$

and as in part (*b*) of Example 1.6

$$y(nT) = u(nT)\frac{1 - e^{(n+1)\alpha}}{1 - e^{\alpha}}$$

(*b*) For this part, we observe that

$$x(nT) = u(nT) - u(nT - 5T)$$

and so

$$\mathcal{R}x(nT) = \mathcal{R}u(nT) - \mathcal{R}u(nT - 5T)$$

24 DIGITAL FILTERS: ANALYSIS, DESIGN, AND APPLICATIONS

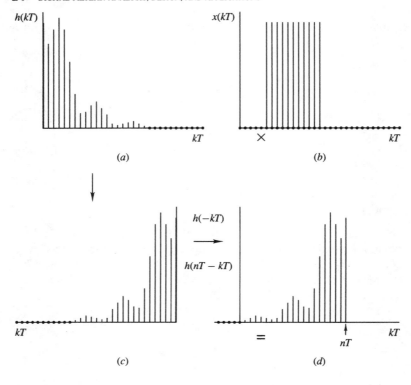

FIGURE 1.10
Convolution summation.

Thus

$$y(nT) = \begin{cases} u(nT)\dfrac{1 - e^{(n+1)\alpha}}{1 - e^{\alpha}} & \text{for } n \leq 4 \\ \dfrac{e^{(n-4)\alpha} - e^{(n+1)\alpha}}{1 - e^{\alpha}} & \text{for } n > 4 \end{cases}$$

1.7.2 Alternative Classification of Digital Filters

If the impulse response of a digital filter is of finite duration such that $h(nT) = 0$ for $n > N$, then Eq. (1.13b) gives

$$y(nT) = \sum_{k=0}^{N} h(kT)x(nT - kT)$$

This equation is of the same form as Eq. (1.3) with $h(0) = a_0, h(T) = a_1, \ldots, h(NT) = a_N$ and, in effect, the filter is nonrecursive. Conversely, it readily follows from Eq. (1.3) that if the filter is nonrecursive, the impulse response is of finite duration. In recursive filters, the impulse response is usually, but not always, of infinite duration, but if the impulse response is of inifinite duration, then the filter is always recursive. Evidently, one may classify digital filters as *finite-duration impulse response* (FIR) filters and *infinite-duration impulse response* (IIR) filters. In this book, we shall be referring to digital filters as nonrecursive and recursive although the alternative classification is equally valid and widely used.

1.8 STABILITY

A digital filter is said to be *stable* if and only if any bounded excitation results in a bounded response, i.e., if

$$|x(nT)| < \infty \qquad \text{for all } n$$

implies that

$$|y(nT)| < \infty \qquad \text{for all } n$$

For a linear and time-invariant filter, Eq. (1.12) gives

$$|y(nT)| \leq \sum_{k=-\infty}^{\infty} |h(kT)| \cdot |x(nT - kT)|$$

and if

$$|x(nT)| \leq M < \infty \qquad \text{for all } n$$

we have

$$|y(nT)| \leq M \sum_{k=-\infty}^{\infty} |h(kT)|$$

Clearly if

$$\sum_{k=-\infty}^{\infty} |h(kT)| < \infty \qquad (1.14)$$

then

$$|y(nT)| < \infty \qquad \text{for all } n$$

and, therefore, Eq. (1.14) constitutes a sufficient condition for stability.

A filter can be classified as stable only if its response is bounded for all possible bounded excitations. Consider the bounded excitation

$$x(nT - kT) = \begin{cases} M & \text{if } h(kT) \geq 0 \\ -M & \text{if } h(kT) < 0 \end{cases}$$

where M is a positive constant. From Eq. (1.12)

$$y(nT) = |y(nT)| = \sum_{k=-\infty}^{\infty} M|h(kT)|$$

or

$$|y(nT)| = M \sum_{k=-\infty}^{\infty} |h(kT)|$$

Evidently, the response will be bounded if and only if Eq. (1.14) holds and, therefore, Eq. (1.14) constitutes a necessary and sufficient condition for stability. Under these circumstances, the filter is said to be *bounded-input, bounded-output* (or BIBO) stable.

In nonrecursive filters, the impulse response is of finite duration and hence Eq. (1.14) is always satisfied. Consequently, these filters are always stable.

Example 1.9. Check the filter of Fig. 1.5a for stability.

Solution. From part (*a*) of Example 1.6

$$h(nT) = u(nT)e^{n\alpha}$$

and hence

$$\sum_{k=-\infty}^{\infty} |h(kT)| = 1 + |e^{\alpha}| + \cdots + |e^{k\alpha}| + \cdots$$

The series converges if

$$\left| \frac{e^{(k+1)\alpha}}{e^{k\alpha}} \right| < 1 \quad \text{or} \quad e^{\alpha} < 1$$

Therefore, for stability

$$\alpha < 0$$

1.9 STATE-SPACE ANALYSIS

The analysis of analog systems is often simplified by using *state-space* techniques. Similar techniques can be developed for digital filters, as will now be shown.

1.9.1 Characterization

Consider an arbitrary digital-filter network containing N unit delays and assume that each and every loop in the network includes at least one unit delay, i.e., the network has no delay-free loops. The flow graph of such a network is computable (i.e., there

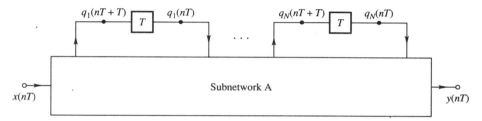

FIGURE 1.11
Arbitrary digital-filter network.

are no variables that depend on themselves) and is, therefore, realizable. The network can be drawn as shown in Fig. 1.11 where subnetwork A consists exclusively of interconnected adders and multipliers. Subnetwork A cannot contain any loops since each of the loops in the network is broken if the unit delays in Fig. 1.11 are removed.

Now let $q_i(nT)$, for $i = 1, 2, \ldots, N$, be variables at the outputs of unit delays. These are stored quantities and can thus be referred to as *state variables*. The signals at the inputs of the unit delays can obviously be represented by corresponding variables $q_i(nT + T)$. If all the state variables are assumed to be zero and the input $x(nT)$ is assumed to be nonzero, then Mason's gain formula gives the response at the input of the ith unit delay as

$$q_i(nT + T) = \left(\frac{1}{\Delta} \sum_k T_k \Delta_k\right) x(nT)$$

where T_k is the transmittance of the kth direct path between input and node i, Δ is the determinant of the flow graph, and Δ_k is the determinant of the subgraph that does not touch the kth direct path between input and node i (see Sec. 1.5.2). Since there are no loops in the graph of subnetwork A, we have

$$\Delta = \Delta_k = 1$$

Furthermore, since there are no unit delays in subnetwork A, the transmittances T_k are independent of the shift operator $\mathcal{E}^{-1}$ and we conclude that

$$q_i(nT + T) = b_i x(nT) \tag{1.15}$$

for $i = 1, 2, \ldots, N$ where $b_1, b_2, \ldots, b_N$ are constants independent of nT for a time-invariant digital filter.

Similarly, if input $x(nT)$ and all the state variables except the jth state variable are zero, we have

$$q_i(nT + T) = a_{ij} q_j(nT) \tag{1.16}$$

for $i = 1, 2, \ldots, N$ where $a_{1j}, a_{2j}, \ldots, a_{Nj}$ are constants independent of nT for a time-invariant digital filter.

Now if the filter is linear, the response at the input of the ith unit delay is obtained from Eqs. (1.15) and (1.16) as

$$q_i(nT + T) = \sum_{j=1}^{N} a_{ij} q_j(nT) + b_i x(nT) \tag{1.17}$$

by applying the principle of superposition. The response of the filter $y(nT)$ due to input excitation $x(nT)$ and state variables $q_j(nT)$ can similarly be deduced as

$$y(nT) = \sum_{j=1}^{N} c_j q_j(nT) + d_0 x(nT) \tag{1.18}$$

Therefore, from Eqs. (1.17) and (1.18), the digital filter can be characterized by the system

$$\mathbf{q}(nT + T) = \mathbf{A}\mathbf{q}(nT) + \mathbf{b}x(nT) \tag{1.19}$$

$$y(nT) = \mathbf{c}^T \mathbf{q}(nT) + dx(nT) \tag{1.20}$$

where

$$\mathbf{A} = \begin{bmatrix} a_{11} & a_{12} & \cdots & a_{1N} \\ a_{21} & a_{22} & \cdots & a_{2N} \\ \vdots & \vdots & \vdots & \vdots \\ a_{N1} & a_{N2} & \cdots & a_{NN} \end{bmatrix}, \quad \mathbf{b} = \begin{bmatrix} b_1 \\ b_2 \\ \vdots \\ b_N \end{bmatrix}$$

$$\mathbf{c}^T = [c_1 \quad c_2 \quad \cdots \quad c_N], \quad d = d_0$$

and

$$\mathbf{q}(nT) = [q_1(nT) \quad q_2(nT) \quad \cdots \quad q_N(nT)]^T$$

is a column vector whose elements are the state variables of the digital-filter network. This is referred to as a *state-space* characterization and is analogous to that used in continuous-time systems.

Note that the choice of state variables for the representation of a digital filter is not unique. In fact, any other set $\tilde{\mathbf{q}}(n) = \mathbf{T}\mathbf{q}(n)$, where $\mathbf{T}$ is an $N \times N$ matrix, can be used (see Sec. 11.7).

Example 1.10. Obtain a state-space characterization for the filter of Fig. 1.6a.

Solution. State variables can be assigned to the flow graph of the filter as depicted in Fig. 1.12, where

$$q_1(nT + T) = q_2(nT)$$

$$q_2(nT + T) = q_3(nT)$$

$$q_3(nT + T) = -b_3 q_1(nT) - b_2 q_2(nT) - b_1 q_3(nT) + x(nT)$$

The output of the filter can be expressed as

$$y(nT) = a_3 q_1(nT) + a_2 q_2(nT) + a_1 q_3(nT) + a_0 q_3(nT + T)$$

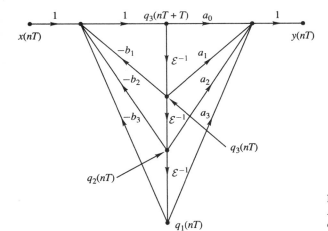

FIGURE 1.12
Assignment of state variables for digital filter of Fig. 1.6a.

and on eliminating $q_3(nT + T)$, we obtain

$$y(nT) = (a_3 - a_0 b_3)q_1(nT) + (a_2 - a_0 b_2)q_2(nT)$$
$$+ (a_1 - a_0 b_1)q_3(nT) + a_0 x(nT)$$

Hence the filter can be represented by Eqs. (1.19) and (1.20) with

$$\mathbf{A} = \begin{bmatrix} 0 & 1 & 0 \\ 0 & 0 & 1 \\ -b_3 & -b_2 & -b_1 \end{bmatrix}, \quad \mathbf{b} = \begin{bmatrix} 0 \\ 0 \\ 1 \end{bmatrix}$$

$$\mathbf{c}^T = [\,(a_3 - a_0 b_3) \quad (a_2 - a_0 b_2) \quad (a_1 - a_0 b_1)\,], \quad d = a_0$$

and

$$\mathbf{q}(nT) = [\,q_1(nT) \quad q_2(nT) \quad q_3(nT)\,]^T$$

1.9.2 Time-Domain Analysis

The preceding state-space characterization leads directly to a relatively simple *time-domain* analysis.

For $n = 0, 1, \ldots$, Eq. (1.19) gives

$$\mathbf{q}(T) = \mathbf{A}\mathbf{q}(0) + \mathbf{b}x(0)$$
$$\mathbf{q}(2T) = \mathbf{A}\mathbf{q}(T) + \mathbf{b}x(T)$$
$$\mathbf{q}(3T) = \mathbf{A}\mathbf{q}(2T) + \mathbf{b}x(2T)$$
$$\ldots\ldots\ldots\ldots\ldots\ldots\ldots\ldots\ldots\ldots$$

Hence

$$\mathbf{q}(2T) = \mathbf{A}^2\mathbf{q}(0) + \mathbf{A}\mathbf{b}x(0) + \mathbf{b}x(T)$$
$$\mathbf{q}(3T) = \mathbf{A}^3\mathbf{q}(0) + \mathbf{A}^2\mathbf{b}x(0) + \mathbf{A}\mathbf{b}x(T) + \mathbf{b}x(2T)$$

and in general

$$q(nT) = A^n q(0) + \sum_{k=0}^{n-1} A^{(n-1-k)} b x(kT)$$

where A^0 is the $N \times N$ unity matrix. Therefore, from Eq. (1.20), we obtain

$$y(nT) = c^T A^n q(0) + c^T \sum_{k=0}^{n-1} A^{(n-1-k)} b x(kT) + d x(nT)$$

If $x(nT) = 0$ for $n < 0$, Eq. (1.15) yields $q(0) = 0$. Thus for an initially relaxed filter

$$y(nT) = c^T \sum_{k=0}^{n-1} A^{(n-1-k)} b x(kT) + d x(nT)$$

The *impulse* response $h(nT)$ of the filter is

$$h(nT) = c^T \sum_{k=0}^{n-1} A^{(n-1-k)} b \delta(kT) + d \delta(nT)$$

For $n = 0$

$$h(0) = d\delta(nT) = d_0$$

and for $n > 0$

$$h(nT) = c^T A^{(n-1)} b \delta(0) + c^T A^{(n-2)} b \delta(T) + \cdots + d\delta(nT)$$

Therefore

$$h(nT) = \begin{cases} d_0 & \text{for } n = 0 \\ c^T A^{(n-1)} b & \text{for } n > 0 \end{cases} \quad (1.21)$$

Similarly, the *unit-step* response of the filter is

$$y(nT) = c^T \sum_{k=0}^{n-1} A^{(n-1-k)} b u(kT) + d u(nT)$$

Hence, for $n \geq 0$

$$y(nT) = c^T \sum_{k=0}^{n-1} A^{(n-1-k)} b + d$$

Example 1.11. An initially relaxed digital filter can be represented by the matrices

$$A = \begin{bmatrix} 0 & 1 \\ \frac{1}{4} & -\frac{1}{2} \end{bmatrix} \quad b = \begin{bmatrix} 0 \\ 1 \end{bmatrix} \quad c^T = \begin{bmatrix} \frac{7}{8} & \frac{5}{4} \end{bmatrix} \quad d = \frac{3}{2}$$

Find $h(17T)$.

Solution. From Eq. (1.21)

$$h(17T) = \mathbf{c}^T \mathbf{A}^{16} \mathbf{b}$$

By forming $\mathbf{A}^2, \mathbf{A}^4, \mathbf{A}^8$, and then $\mathbf{A}^{16}$, we have

$$h(17T) = \begin{bmatrix} \frac{7}{8} & \frac{5}{4} \end{bmatrix} \begin{bmatrix} \frac{610}{65{,}536} & -\frac{987}{32{,}768} \\ -\frac{987}{131{,}072} & \frac{1597}{65{,}536} \end{bmatrix} \begin{bmatrix} 0 \\ 1 \end{bmatrix} = \frac{1076}{262{,}144}$$

1.9.3 Applications of State-Space Method

The state-space method offers the advantage that filters can be analyzed through the manipulation of matrices which can be carried out very efficiently using array or vector processors. Another important advantage of this method is that it can be used to characterize and analyze time-dependent filters, that is, filters in which one or more of the elements of $\mathbf{A}$, $\mathbf{b}$, and $\mathbf{c}^T$ and possibly constant d depend on nT. This advantage follows from the fact that it is unnecessary to assume time-invariance in the derivation of the state-space equations.

Time-varying filters like adaptive filters are now used quite extensively in a variety of communications applications. The state-space method can also be used to realize digital filters that have certain important advantages, e.g., increased signal-to-noise ratio (see Sec. 11.7).

REFERENCES

1. R. J. Schwarz and B. Friedland, *Linear Systems*, McGraw-Hill, New York, 1965.
2. R. Butler and E. Kerr, *An Introduction to Numerical Methods*, Pitman, London, 1962.
3. J. R. Abrahams and G. P. Coverley, *Signal Flow Analysis*, Pergamon, New York, 1965.
4. B. C. Kuo, *Automatic Control Systems*, Prentice-Hall, Englewood Cliffs, N.J., 1962.
5. N. Balabanian and T. A. Bickart, *Electrical Network Theory*, Wiley, New York, 1969.
6. S. J. Mason, "Feedback Theory—Further Properties of Signal-Flow Graphs," *Proc. IRE*, vol. 44, pp. 920–926, July 1956.

PROBLEMS

1.1. By using appropriate tests, check the filters characterized by the following equations for time invariance, causality, and linearity:
  (a) $\mathcal{R}x(nT) = 2x(nT - gT)$ where $g > 0$
  (b) $\mathcal{R}x(nT) = \begin{cases} 6x(nT - 5T) & \text{for } x(nT) \leq 6 \\ 7x(nT - 5T) & \text{for } x(nT) > 6 \end{cases}$
  (c) $\mathcal{R}x(nT) = (nT + 3T)x(nT - 3T)$

1.2. Repeat Prob. 1.1 for the filters characterized by the following equations:
  (a) $\mathcal{R}x(nT) = 5nTx^2(nT)$
  (b) $\mathcal{R}x(nT) = 3x(nT + 3T)$
  (c) $\mathcal{R}x(nT) = x(nT)\sin\omega nT$

1.3. Repeat Prob. 1.1 for the filters characterized by the following equations:
(a) $\mathcal{R}x(nT) = K_1 \Delta x(nT)$ where $\Delta x(nT) = x(nT + T) - x(nT)$
(b) $\mathcal{R}x(nT) = K_2 \nabla x(nT)$ where $\nabla x(nT) = x(nT) - x(nT - T)$
(c) $\mathcal{R}x(nT) = x(nT + T)e^{-nT}$
(d) $\mathcal{R}x(nT) = x^2(nT + T)e^{-nT} \sin \omega nT$

1.4. Analyze the filter networks shown in Fig. P1.4a and b.

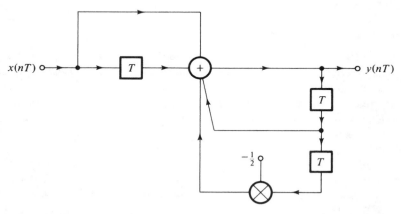

FIGURE P1.4a

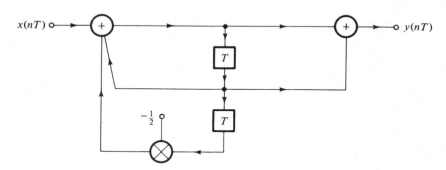

FIGURE P1.4b

1.5. Analyze the filter networks shown in Fig. P1.5a and b.
1.6. Two second-order filter sections of the type shown in Fig. P1.5a are connected in cascade as in Fig. P1.6. The parameters of the two sections are $a_{11}, a_{21}, -b_{11}, -b_{21}$ and $a_{12}, a_{22}, -b_{12}, -b_{22}$, respectively. Deduce the characterization of the combined filter.
1.7. The two sections in Prob. 1.6 are connected in parallel as in Fig. P1.7. Obtain the difference equation of the combined filter.

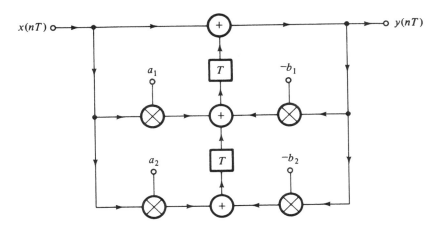

FIGURE P1.5a

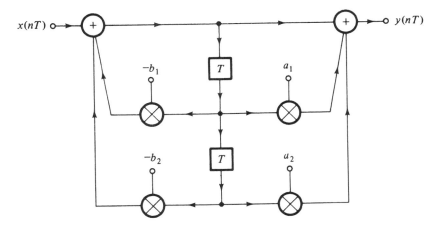

FIGURE P1.5b

FIGURE P1.6

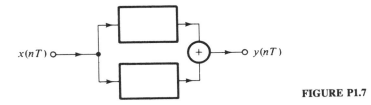

FIGURE P1.7

1.8. Fig. P1.8 shows a network with three inputs and three outputs. Derive a set of equations characterizing the network.

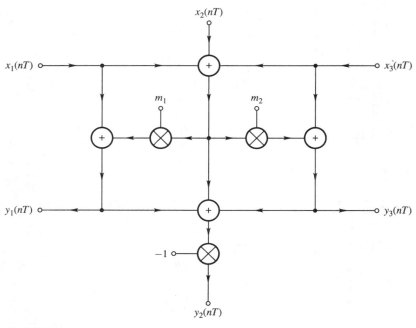

FIGURE P1.8

1.9. The network of Fig. P1.9 can be characterized by the equation

$$\mathbf{b} = \mathbf{Ca}$$

where $\mathbf{b} = [b_1 \quad b_2 \quad b_3]^T$ and $\mathbf{a} = [a_1 \quad a_2 \quad a_3]^T$ are column vectors and $\mathbf{C}$ is a 3×3 matrix. Obtain $\mathbf{C}$.

1.10. By using appropriate tests, check the filters of Fig. P1.10 for time invariance, linearity, and causality.

(a) The filter of Fig. P1.10a uses a device N whose response is given by

$$\mathcal{R}x(nT) = |x(nT)|$$

(b) The filter of Fig. P1.10b uses a multiplier M whose parameter is given by

$$m = 0.1x(nT)$$

(c) The filter of Fig. P1.10c uses a multiplier M whose parameter is given by

$$m = 0.1v(nT)$$

where $v(nT)$ is an independent signal.

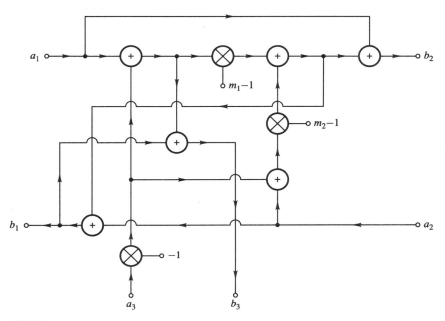

FIGURE P1.9

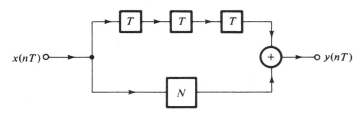

FIGURE P1.10a

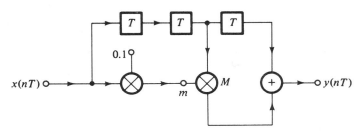

FIGURE P1.10b

1. An initially relaxed digital filter employs a device D, as shown in Fig. P1.11, which is characterized by the equation

$$w(nT) = 2(-1)^n |v(nT)|$$

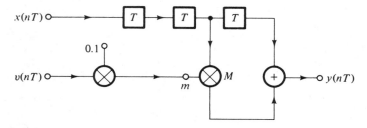

FIGURE P1.10c

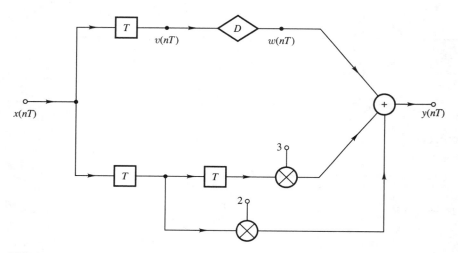

FIGURE P1.11

(a) Deduce the difference equation of the filter.

(b) By using appropriate tests determine whether the filter is linear or nonlinear, time-invariant or time-dependent, and causal or noncausal.

(c) Evaluate the time-domain response for the period 0 to $10T$ if the input signal is given by

$$x(nT) = u(nT) - 2u(nT - 4T)$$

where $u(nT)$ is the unit step.

(d) What is the order of the filter?

1.12. The digital filter of Fig. P1.12 uses a device D whose response to an input $w(nT)$ is $d_0 + d_1 w(nT)$, where d_0 and d_1 are nonzero constants. By using appropriate tests, check the filter for time invariance, linearity, and stability.

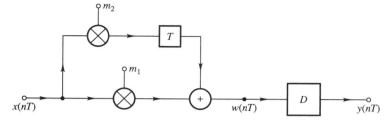

FIGURE P1.12

1.13. Show that

$$r(nT) = \mu_2(nT - T) \qquad u(nT) = \mu_1(nT) \qquad \delta(nT) = T\mu_0(nT)$$

where

$$\mu_{i-1}(nT) = \frac{1}{T}\nabla\mu_i(nT)$$

and

$$\mu_2(nT) = \begin{cases} nT + T & \text{for } n \geq 0 \\ 0 & \text{otherwise} \end{cases}$$

1.14. The filter of Fig. P1.14 is initially relaxed. Find the time-domain response for $n = 0, 1, \ldots, 6$ if

$$x(nT) = \begin{cases} \sin \omega nT & \text{for } n \geq 0 \\ 0 & \text{otherwise} \end{cases}$$

where $\omega = \pi/6T$ and $T = 1$.

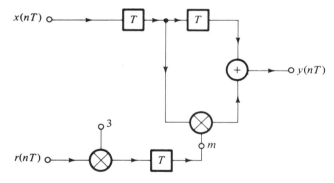

FIGURE P1.14

1.15. (*a*) Show that

$$r(nT) = \begin{cases} 0 & \text{for } n \leq 0 \\ T\sum_{k=1}^{n} u(nT - kT) & \text{otherwise} \end{cases}$$

(*b*) By using this relation obtain the unit-ramp response of the filter shown in Fig. 1.5*a* in closed form. The filter is initially relaxed.

(*c*) Sketch the response for $\alpha > 0$, $\alpha = 0$, and $\alpha < 0$.

1.16. The excitation in Fig. 1.5a is

$$x(nT) = \begin{cases} 1 & \text{for } 0 \le n \le 4 \\ 2 & \text{for } n > 4 \\ 0 & \text{for } n < 0 \end{cases}$$

Find the response in closed form.

1.17. Repeat Prob. 1.16 for an excitation

$$x(nT) = \begin{cases} 1 & \text{for } n = 0 \\ 0 & \text{for } n < 0, n = 1, 2, 3, 4 \\ 1 & \text{for } n > 4 \end{cases}$$

1.18. Fig. P1.18 shows a second-order recursive filter. Compute the unit-step response for $0 \le n \le 15$ if
(a) $\alpha = 1$ $\quad \beta = -\frac{1}{2}$
(b) $\alpha = \frac{1}{2}$ $\quad \beta = -\frac{1}{8}$
(c) $\alpha = \frac{5}{4}$ $\quad \beta = -\frac{25}{32}$

Compare the three responses and determine the frequency of ringing in terms of T, where possible.

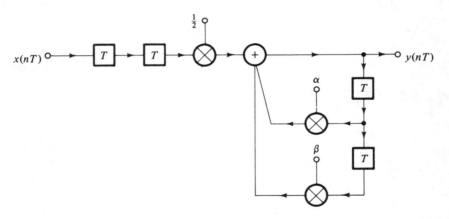

FIGURE P1.18

1.19. Fig. P1.19 shows a filter comprising a cascade of two first-order sections. The input signal is

$$x(nT) = \begin{cases} \sin \omega nT & \text{for } n \ge 0 \\ 0 & \text{otherwise} \end{cases}$$

and $T = 1$ ms. Compute the steady-state amplitude and phase angle of $y(nT)$ for a frequency $f = 10$ Hz. Repeat for $f = 100$ Hz.

1.20. Fig. P1.20 shows a linear first-order filter.
(a) Assuming a sinusoidal excitation, derive an expression for the steady-state gain of the filter.
(b) Plot the gain in decibels (dB) versus $\log f$ for $f = 0$ to 1.0 kHz if $T = 1$ ms.

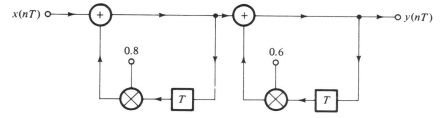

FIGURE P1.19

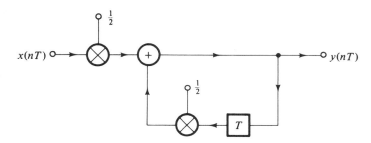

FIGURE P1.20

(c) Determine the lowest frequency at which the gain is reduced by 3 dB relative to the gain at zero frequency.

1.21. Two first-order filters of the type shown in Fig. 1.5a are connected in parallel as in Fig. P1.7. The multiplier constants for the two filters are $m_1 = e^{0.6}$ and $m_2 = e^{0.7}$. Find the unit-step response of the combined network in closed form.

1.22. The unit-step response of a filter is

$$y(nT) = \begin{cases} nT & \text{for } n \geq 0 \\ 0 & \text{for } n < 0 \end{cases}$$

(a) Using the convolution summation, find the unit-ramp response.

(b) Check the filter for stability.

1.23. A nonrecursive filter has an impulse response

$$h(nT) = \begin{cases} nT & \text{for } 0 \leq n \leq 4 \\ (8-n)T & \text{for } 5 \leq n \leq 8 \\ 0 & \text{otherwise} \end{cases}$$

The sampling frequency is 2π rad/s.

(a) Deduce the network of the filter.

(b) By using the convolution summation determine the response of the filter $y(nT)$ at $nT = 4T$ if the input signal is given by

$$x(nT) = u(nT - T)e^{-nT}$$

(c) Illustrate the solution in part (b) by a graphical construction.

1.24. An initially relaxed causal filter was tested with an input signal

$$x(nT) = u(nT) + u(nT - 2T)$$

and the response for $n = 0, 1, 2, 3, 4, 5, \ldots, 100, \ldots$ was found to be $y(nT) = 3, 5, 9, 11, 12, 12, \ldots, 12, \ldots$.

(a) Find the impulse response of the filter for $0 \leq n \leq 5$.

(b) Find the response for $0 \leq n \leq 5$ if the input is changed to
$$x(nT) = u(nT) - u(nT - 2T)$$

1.25. An initially relaxed filter was tested with an input signal
$$x(nT) = 2u(nT)$$

and found to have a response

n	0	1	2	3	4	5	$\cdots$	100	$\cdots$
$y(nT)$	2	6	12	20	30	30	$\cdots$	30	$\cdots$

(a) Deduce the difference equation of the filter.

(b) Construct a possible network for the filter.

1.26. (a) A digital filter has an impulse response
$$h(nT) = u(nT - T)\frac{1}{n}$$

By using an appropriate test, check the filter for stability.

(b) Repeat part (a) for the filter characterized by
$$h(nT) = u(nT - T)\frac{1}{n!}$$

1.27. Repeat Prob. 1.26 for the filters represented by the following impulse responses:

(a) $h(nT) = \dfrac{u(nT)n}{2^n}$

(b) $h(nT) = \dfrac{u(nT)n}{n+1}$

(c) $h(nT) = u(nT - T)\dfrac{(n+1)}{n^2}$

1.28. (a) Check the filter of Fig. P1.28a for stability.

(b) Repeat part (a) for the filter of Fig. P1.28b.

1.29. Derive state-space representations for the filters of Fig. P1.4a and b.

1.30. Derive state-space representations for the filters of Fig. P1.5a and b.

1.31. Derive a state-space representation for the filter of Fig. P1.31.

1.32. An initially relaxed digital filter is characterized by the state-space equations with
$$\mathbf{A} = \begin{bmatrix} 0 & 1 \\ -\dfrac{5}{16} & -1 \end{bmatrix} \quad \mathbf{b} = \begin{bmatrix} 0 \\ 1 \end{bmatrix} \quad \mathbf{c}^T = \begin{bmatrix} \dfrac{11}{8} & 2 \end{bmatrix} \quad d = 2$$

(a) Calculate the impulse response for $n = 0, 1, 2, \ldots, 5$, and for $n = 17$ using the state-space method.

(b) Deduce the difference equation of the filter.

(c) Repeat part (a) using the difference equation.

(d) Calculate the unit-step response for $n = 5$.

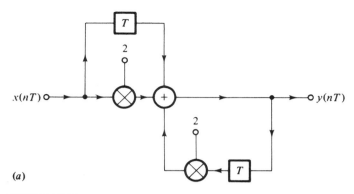

FIGURE P1.28a

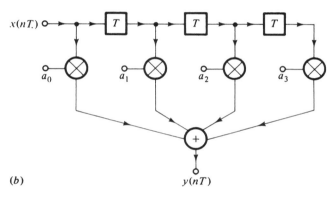

FIGURE P1.28b

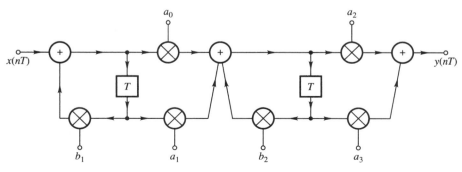

FIGURE P1.31

1.33. A digital filter is characterized by the state-space equations with

$$\mathbf{A} = \begin{bmatrix} 0 & 1 \\ -\frac{1}{4} & \frac{1}{2} \end{bmatrix} \quad \mathbf{b} = \begin{bmatrix} 0 \\ 1 \end{bmatrix} \quad \mathbf{c}^T = \begin{bmatrix} -\frac{1}{4} & \frac{3}{2} \end{bmatrix} \quad d = 1$$

(a) Assuming that $y(nT) = 0$ for $n < 0$, find $y(nT)$ for $n = 0$ to 5 if $x(nT) = \delta(nT)$.
(b) Repeat part (a) if $x(nT) = u(nT)$.
(c) Derive a network for the filter.

1.34. A signal

$$x(nT) = 3u(nT)\cos \omega nT$$

is applied at the input of the filter in Prob. 1.32. Find the response at instant 5T if $\omega = 1/10T$ by using the convolution summation.

1.35. Find the response of the filter in Prob. 1.32 at $n = 5$ if the excitation is

$$x(nT) = u(nT - T)e^{-nT}$$

1.36. Find the response of the filter in Prob. 1.33 at $n = 5$ if the excitation is

$$x(nT) = u(nT) + u(nT - 2T)$$

CHAPTER 2

THE z TRANSFORM

2.1 INTRODUCTION

The analysis of linear, time-invariant digital filters is almost invariably carried out by using the *z transform*. The principal reason for this is that upon the application of the z transform, the difference equations characterizing digital filters are transformed into algebraic equations which are usually much easier to solve.

The z transform, like the Laplace and Fourier transforms, is useful because it has an inverse, namely, the inverse z transform. The application of the z transform to a discrete-time signal $x(nT)$ yields a representation of the signal in terms of a rational function $X(z)$ where z is a complex variable. If the signal is to be processed by a digital filter, then the required processing can be carried out in the z domain through algebraic manipulations. In this way, a transformed version of $X(z)$, say $Y(z)$, is obtained and by applying the inverse z transform to $Y(z)$, the processed signal is obtained.

This chapter begins with a brief review of complex analysis. The subjects considered include the differentiability and analyticity of functions of a complex variable and their representation in terms of power series like the Laurent series. The z transform and its inverse are then defined in terms of the Laurent series. The convergence and salient properties of the z transform are then presented in terms of a series of theorems. The application of the z transform to digital filters is postponed to Chap. 3.

2.2 REVIEW OF COMPLEX ANALYSIS

A complex variable z can be expressed as $z = x + jy$ where x and y are real variables called the real and imaginary parts of z. A complex variable W may similarly be defined which is a function of z. Such a relation can be expressed as

$$W = F(z)$$

and if U and V are the real and imaginary parts of W, we have

$$W = F(z) = U(x, y) + jV(x, y)$$

2.2.1 Limit

A function $F(z)$ is said to have a *limit* F_0 as z approaches z_0, if (*a*) $F(z)$ is defined in a neighborhood of z_0 (except perhaps at point z_0) and (*b*) for every positive real number ϵ there exists a positive real number δ such that $|F(z) - F_0| < \epsilon$ for all values of $z \neq z_0$ in the disk $|z - z_0| < \delta$. Limit F_0 can be expressed as

$$F_0 = \lim_{z \to z_0} F(z)$$

A function $F(z)$ is said to be *continuous* at point $z = z_0$ if $F(z_0)$ is defined and is given by

$$F(z_0) = \lim_{z \to z_0} F(z) = F_0$$

Extending this concept somewhat, a *continuous function* is one that is continuous at all the points where it is defined.

2.2.2 Differentiability and Analyticity

The concept of limit leads readily to the definition of *differentiability* of a complex function.

Definition 2.1 Differentiability. A function $F(z)$ is said to be differentiable at a point $z = z_0$ if the limit

$$F'(z_0) = \lim_{\Delta z \to 0} \frac{F(z_0 + \Delta z) - F(z_0)}{\Delta z} \tag{2.1}$$

exists. This limit is called the derivative of $F(z)$ at the point $z = z_0$.

If we let $z_0 + \Delta z = z$ in Eq. (2.1), we obtain

$$F'(z_0) = \lim_{z \to z_0} \frac{F(z) - F(z_0)}{z - z_0} \tag{2.2}$$

Hence the derivative exists if and only if the quotient in Eq. (2.2) approaches a unique value independently of the path that z may take to approach z_0.

A closely related property to differentiability is the *analyticity* of a complex function.

Definition 2.2 Analyticity. A function $F(z)$ is said to be *analytic* at a point $z = z_0$ if it is defined and has a derivative at every point in some neighborhood of z_0. A function $F(z)$ is said to be analytic[†] in a domain D if it is analytic at every point in D.

Differentiability is a crucial requirement in practice and, consequently, the importance of analyticity cannot be overstated. Indeed, complex analysis is concerned exclusively with analytic functions. Two important equations that pertain to the analyticity of a function are the *Cauchy-Riemann equations* which are given by

$$\frac{\partial U}{\partial x} = \frac{\partial V}{\partial y} \quad \text{and} \quad \frac{\partial U}{\partial y} = -\frac{\partial V}{\partial x}$$

These equations are necessary and sufficient for a function to be analytic; that is, if the real and imaginary parts of a function satisfy the Cauchy-Riemann equations in domain D, then the function is analytic in D, and conversely.

2.2.3 Zeros and Singularities

If a function $F(z)$ is analytic in a domain D and is zero at a point z_0, then the function is said to have a *zero* at z_0. If in addition to $F(z)$, the derivatives $F'(z), \ldots, F^{(n-1)}(z)$ are also zero, then the function is said to have a zero of *order n* at point z_0. An analytic function $F(z)$ is said to have an nth-order zero at infinity if $F(1/z)$ has an nth-order zero at $z = 0$.

A point z_∞ at which a function $F(z)$ ceases to be analytic is referred to as a *singular point* of the function; alternatively, the function is said to have a singularity at z_∞. Singularities can be *isolated* or *nonisolated*. In the first case, the singularity has a neighborhood which contains no other singular points. For example, the function $1/z^5$ has an isolated singularity at $z = 0$. On the other hand, the function $\tan 1/z$ has a nonisolated singularity at $z = 0$ since the function is not analytic at an infinite number of points clustered in any neighborhood of $z = 0$, namely, at $z = \pm 2/\pi, \pm 2/3\pi, \ldots$. A function $F(z)$ is singular at infinity if $F(1/z)$ is singular at $z = 0$.

2.2.4 Laurent Series

The z transform is closely linked with the *Laurent*[‡] *series*, as will be demonstrated below. This very important series which includes the well-known Taylor series as a special case is given by the following theorem:

Theorem 2.1 Laurent theorem. (*a*) If $F(z)$ is an analytic and single-valued function on two concentric circles C_1 and C_2 with center a and in the annulus between them, as

---

[†]The terms holomorphic and regular have also been used in the literature.

[‡]Pierre Alphonse Laurent (1813–1854) was a French mathematician.

depicted in Fig. 2.1, then it can be represented by the Laurent series

$$F(z) = \sum_{n=-\infty}^{\infty} a_n(z-a)^{-n} \qquad (2.3)$$

where

$$a_n = \frac{1}{2\pi j} \oint_\Gamma F(z)(z-a)^{n-1} dz \qquad (2.4)$$

The contour of integration Γ is a closed contour in the counterclockwise sense, lying in the annulus between circles C_1 and C_2 and encircling the inner circle.

(b) The Laurent series converges and represents $F(z)$ in the open annulus obtained by continuously increasing the radius of C_2 and decreasing the radius of C_1 until each of C_1 and C_2 reaches a point where $F(z)$ is singular.

(c) The Laurent series of $F(z)$ in an annulus of convergence is unique. However, $F(z)$ may have different Laurent series in different annuli about the same center.

The Laurent series can also be written as

$$F(z) = \sum_{n=0}^{\infty} b_n(z-a)^n + \sum_{n=1}^{\infty} \frac{c_n}{(z-a)^n}$$

where the second summation is called the *principal part* of $F(z)$. If the principal part comprises a finite number of terms, then $F(z)$ has a singularity at point $z = a$ which is called a *pole*; if all $c_n = 0$ for $n > m$, then $F(z)$ has a pole of order m; and if the principal part of $F(z)$ has an infinite number of terms, then $F(z)$ is said to have an *essential singularity* at point a. For example, the function $1/z$ has an isolated singularity at $z = 0$ which is a simple pole; the function $1/z^5$ has an isolated singularity at $z = 0$ which is a pole of order 5; function $e^{1/z} = 1 + 1/z + 1/2z^2 + \cdots$ has an isolated essential singularity at $z = 0$; and function $\tan 1/z$ has a nonisolated essential singularity at $z = 0$.

Coefficient c_1, which is referred to as the *residue of $F(z)$ at point $z = a$*, is of considerable importance as will be demonstrated below. It can be represented by the

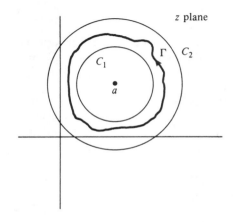

FIGURE 2.1
Domain of $G(z)$.

notation

$$c_1 = \text{res}_{z \to a} F(z)$$

The Laurent series of an arbitrary analytic function can be obtained by evaluating the contour integral in Eq. (2.4) for $n = \ldots, -2, -1, 0, 1, 2, \ldots$. This can be readily done by using the so-called *residue theorem* which is as follows:

Theorem 2.2 Residue theorem. If $F(z)$ is an analytic function on a simple contour Γ and inside Γ, except for a finite number of singular points $a_1, a_2, \ldots, a_M$, then

$$\frac{1}{2\pi j} \oint_\Gamma F(z) dz = \sum_{i=1}^{M} \text{res}_{z \to a_i} F(z) \tag{2.5}$$

where the integral is taken in the counterclockwise sense.

Formulas for the residues for the case where the singularities of $F(z)$ in the finite z plane are poles[†] are given in Sec. 2.5 [see Eqs. (2.9a) and (2.9b)].

The above review of complex analysis should prove more than adequate for our purposes. The keen reader who would like to check the proofs is referred to the work of Kreyszig [1].

2.3 DEFINITION OF z TRANSFORM

The z *transform* of a discrete-time function $f(nT)$ is defined as

$$F(z) = \sum_{n=-\infty}^{\infty} f(nT) z^{-n} \tag{2.6}$$

for all z for which $F(z)$ converges (see Theorems 2.3 and 2.4). It is sometimes represented by the simplified notation

$$F(z) = \mathcal{Z} f(nT)$$

A comparison of Eqs. (2.3) and (2.6) readily shows that the right-hand side of Eq. (2.6) is a Laurent series for $F(z)$ about the origin and consequently Theorem 2.1 applies.

With digital filters, we need only concern ourselves with z transforms whose singularities are poles. According to Theorem 2.1c, functions of this class may have more than one Laurent series. For example, the function represented by the zero-pole plot of Fig. 2.2a has three Laurent series about the origin, one for each of the annuli identified in Fig. 2.2b.

[†]Functions of this type are sometimes called *meromorphic*.

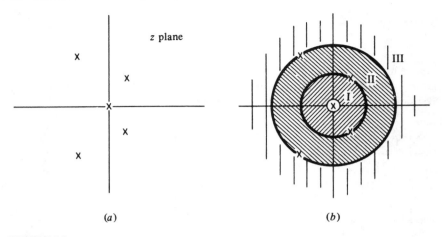

FIGURE 2.2
A function with three Laurent series about the origin: (*a*) zero-pole plot, (*b*) annuli of convergence.

2.4 z-TRANSFORM THEOREMS

The properties of the *z* transform can be conveniently described by means of a number of theorems. A more detailed discussion on the subject can be found in the work of Jury [2]. We start below with two theorems that deal with the convergence of the *z* transform.

Theorem 2.3 Absolute convergence. If

(i) $\quad\quad\quad\quad f(nT) = 0 \quad\quad \text{for } n < -N_1$

(ii) $\quad\quad\quad\quad |f(nT)| \leq K_1 \quad\quad \text{for } -N_1 \leq n < N_2$

(iii) $\quad\quad\quad\quad |f(nT)| \leq K_2 r^n \quad\quad \text{for } n \geq N_2$

where N_1 and N_2 are positive constants, then the *z* transform as defined in Eq. (2.6) exists and converges absolutely if and only if

$$\rho = |z| > r$$

Proof. If $z = \rho e^{j\theta}$, conditions (i) to (iii) give

$$\sum_{n=-\infty}^{\infty} |f(nT)z^{-n}| = \sum_{n=-\infty}^{\infty} |f(nT)| \cdot |z^{-n}|$$

$$\leq \sum_{n=-N_1}^{N_2-1} K_1 \rho^{-n} + \sum_{n=N_2}^{\infty} K_2 \left(\frac{r}{\rho}\right)^n$$

$$\leq K_1 \sum_{n=-N_1}^{N_2-1} \rho^{-n} + K_2 \sum_{n=N_2}^{\infty} \left(\frac{r}{\rho}\right)^n$$

Now if $\rho > r$, the first summation at the right-hand side is finite and is given by

$$\sum_{n=-N_1}^{N_2-1} \rho^{-n} = \frac{\rho^{N_1} - \rho^{-N_2}}{1 - \rho^{-1}}$$

while the second summation converges. Consequently,

$$\sum_{n=-\infty}^{\infty} |f(nT)z^{-n}| \leq K_0$$

where K_0 is finite and thus for $\rho > r$, $F(z)$ converges absolutely. If $\rho \leq r$, then the second summation does not converge and, therefore, we conclude that $F(z)$ converges absolutely if and only if

$$\rho = |z| > r$$

In other words, if $f(nT)$ is bounded by the dashed lines in Fig. 2.3a, then its z transform converges absolutely in the shaded region in Fig. 2.3b.

Since $F(z)$ converges for $\rho > r$ and diverges for $\rho < r$, $\rho = \rho_c = r$ is the *radius of convergence* of $F(z)$.

From Theorems 2.1 and 2.3, we conclude that the z transform as defined in Eq. (2.6) is the unique Laurent series of $F(z)$ about the origin that converges in the open annulus

$$\rho_c \leq |z| \leq R_2 \tag{2.7}$$

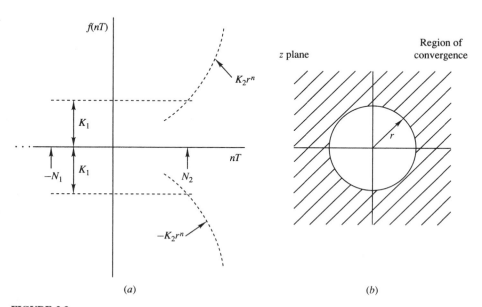

(a) (b)

FIGURE 2.3
Convergence of z transform: (a) bounds in time domain, (b) region of convergence in z domain.

where $|z| = \rho_c$ is the circle passing through the most distant pole of $F(z)$ from the origin and $R_2 \to \infty$.

Theorem 2.4 Uniform convergence. If $F(z)$ converges absolutely for $|z| > \rho_c$, then $F(z)$ converges uniformly and is analytic in the region $|z| > \rho_c$.

Proof. If $F(z)$ converges absolutely for $|z| > \rho_c$, then it converges absolutely for any neighborhood of z in the region $|z| > \rho_c$. Furthermore, the derivative of $F(z)$ exists at every point in any neighborhood of z in the region $|z| > \rho_c$. Therefore, $F(z)$ converges uniformly and is analytic in this region.

Some additional theorems that are often very useful in the analysis and design of digital filters are described below. We assume that

$$\mathcal{Z}f(nT) = F(z) \quad \text{and} \quad \mathcal{Z}g(nT) = G(z)$$

and that a and b are constants.

Theorem 2.5 Linearity

$$\mathcal{Z}[af(nT) + bg(nT)] = aF(z) + bG(z)$$

Proof

$$\mathcal{Z}[af(nT) + bg(nT)] = \sum_{n=-\infty}^{\infty} [af(nT) + bg(nT)]z^{-n}$$

$$= a \sum_{n=-\infty}^{\infty} f(nT)z^{-n} + b \sum_{n=-\infty}^{\infty} g(nT)z^{-n}$$

$$= aF(z) + bG(z)$$

Theorem 2.6 Translation

$$\mathcal{Z}f(nT + mT) = z^m F(z)$$

Proof

$$\mathcal{Z}f(nT + mT) = \sum_{n=-\infty}^{\infty} f(nT + mT)z^{-n}$$

$$= z^m \sum_{n=-\infty}^{\infty} f[(n+m)T]z^{-(n+m)}$$

$$= z^m \sum_{n=-\infty}^{\infty} f(nT)z^{-n} = z^m F(z)$$

Theorem 2.7 Complex scale change

$$\mathcal{Z}[w^{-n}f(nT)] = F(wz)$$

Proof

$$\mathcal{Z}[w^{-n}f(nT)] = \sum_{n=-\infty}^{\infty} f(nT)w^{-n}z^{-n}$$

$$= \sum_{n=-\infty}^{\infty} f(nT)(wz)^{-n}$$

$$= F(wz)$$

Theorem 2.8 Complex differentiation

$$\mathcal{Z}[nTf(nT)] = -Tz\frac{dF(z)}{dz}$$

Proof

$$\mathcal{Z}[nTf(nT)] = \sum_{n=-\infty}^{\infty} nTf(nT)z^{-n} = -Tz\sum_{n=-\infty}^{\infty} f(nT)(-n)z^{-n-1}$$

$$= -Tz\sum_{n=-\infty}^{\infty} f(nT)\frac{d}{dz}(z^{-n})$$

$$= -Tz\frac{d}{dz}\left[\sum_{n=-\infty}^{\infty} f(nT)z^{-n}\right] = -Tz\frac{dF(z)}{dz}$$

Theorem 2.9 Real convolution

$$\mathcal{Z}\sum_{k=-\infty}^{\infty} f(kT)g(nT-kT) = \mathcal{Z}\sum_{k=-\infty}^{\infty} f(nT-kT)g(kT) = F(z)G(z)$$

Proof

$$\mathcal{Z}\sum_{k=-\infty}^{\infty} f(kT)g(nT-kT) = \sum_{n=-\infty}^{\infty}\sum_{k=-\infty}^{\infty} f(kT)g(nT-kT)z^{-n}$$

$$= \sum_{k=-\infty}^{\infty}\sum_{n=-\infty}^{\infty} f(kT)g(nT-kT)z^{-n}$$

$$= \sum_{k=-\infty}^{\infty} f(kT)z^{-k}\sum_{n=-\infty}^{\infty} g(nT-kT)z^{-(n-k)}$$

$$= \sum_{n=-\infty}^{\infty} f(nT)z^{-n}\sum_{n=-\infty}^{\infty} g(nT)z^{-n}$$

$$= F(z)G(z)$$

Example 2.1. Find the z transforms of (a) $\delta(nT)$, (b) $u(nT)$, (c) $u(nT-T)K$, (d) $u(nT)Kw^n$ (e) $u(nT)e^{-\alpha nT}$, (f) $r(nT)$ (see Table 1.2), and (g) $u(nT)\sin \omega nT$.

Solution

(a)
$$\mathcal{Z}\delta(nT) = \delta(0) + \delta(T)z^{-1} + \cdots$$
$$= 1$$

(b)
$$\mathcal{Z}u(nT) = u(0) + u(T)z^{-1} + u(2T)z^{-2} + \cdots$$
$$= 1 + z^{-1} + z^{-2} + \cdots$$
$$= (1 - z^{-1})^{-1} = \frac{z}{z-1}$$

(c) From Theorem 2.6
$$\mathcal{Z}u(nT-T)K = Kz^{-1}\mathcal{Z}u(nT) = \frac{K}{z-1}$$

(d) From Theorem 2.7
$$\mathcal{Z}[u(nT)Kw^n] = K\mathcal{Z}\left[\left(\frac{1}{w}\right)^{-n} u(nT)\right] = K\mathcal{Z}u(nT)|_{z\to z/w} = \frac{Kz}{z-w}$$

(e) By letting $K = 1$ and $w = e^{-\alpha T}$ in the preceding example we have
$$\mathcal{Z}[u(nT)e^{-\alpha nT}] = \frac{z}{z - e^{-\alpha T}}$$

(f) From Theorem 2.8
$$\mathcal{Z}r(nT) = \mathcal{Z}[nTu(nT)] = -Tz\frac{d}{dz}[\mathcal{Z}u(nT)]$$
$$= \frac{Tz}{(z-1)^2}$$

(g) From part (e)
$$\mathcal{Z}[u(nT)\sin \omega nT] = \mathcal{Z}\left[\frac{u(nT)}{2j}(e^{j\omega nT} - e^{-j\omega nT})\right]$$
$$= \frac{1}{2j}\mathcal{Z}[u(nT)e^{j\omega nT}] - \frac{1}{2j}\mathcal{Z}[u(nT)e^{-j\omega nT}]$$
$$= \frac{1}{2j}\left(\frac{z}{z - e^{j\omega T}} - \frac{z}{z - e^{-j\omega T}}\right)$$
$$= \frac{z \sin \omega T}{z^2 - 2z \cos \omega T + 1}$$

A list of the common z transforms is given in Table 2.1. An extensive list can be found in the work of Jury [2].

TABLE 2.1
Standard z transforms

$f(nT)$	$F(z)$
$\delta(nT)$	1
$u(nT)$	$\dfrac{z}{z-1}$
$u(nT-T)K$	$\dfrac{K}{z-1}$
$u(nT)Kw^n$	$\dfrac{Kz}{z-w}$
$u(nT-T)Kw^{n-1}$	$\dfrac{K}{z-w}$
$u(nT)e^{-\alpha nT}$	$\dfrac{z}{z-e^{-\alpha T}}$
$r(nT)$	$\dfrac{Tz}{(z-1)^2}$
$r(nT)e^{-\alpha nT}$	$\dfrac{Te^{-\alpha T}z}{(z-e^{-\alpha T})^2}$
$u(nT)\sin\omega nT$	$\dfrac{z\sin\omega T}{z^2 - 2z\cos\omega T + 1}$
$u(nT)\cos\omega nT$	$\dfrac{z(z-\cos\omega T)}{z^2 - 2z\cos\omega T + 1}$
$u(nT)e^{-\alpha nT}\sin\omega nT$	$\dfrac{ze^{-\alpha T}\sin\omega T}{z^2 - 2ze^{-\alpha T}\cos\omega T + e^{-2\alpha T}}$
$u(nT)e^{-\alpha nT}\cos\omega nT$	$\dfrac{z(z - e^{-\alpha T}\cos\omega T)}{z^2 - 2ze^{-\alpha T}\cos\omega T + e^{-2\alpha T}}$

2.5 INVERSE z TRANSFORM

If $F(z)$ converges in some open annulus of the type defined by Eq. (2.7), then $f(nT)$ can be uniquely determined as

$$f(nT) = \frac{1}{2\pi j} \oint_\Gamma F(z) z^{n-1} dz \qquad (2.8)$$

by virtue of Theorem 2.1a and 2.1c, where Γ is a contour in the counterclockwise sense enclosing all the singularities (poles) of $F(z)z^{n-1}$. Function $f(nT)$ is said to be the inverse z transform of $F(z)$. Eq. (2.8) is often represented by the simplified notation

$$f(nT) = \mathcal{Z}^{-1} F(z)$$

If

$$F(z)z^{n-1} = F_0(z) = \frac{N(z)}{\prod_{i=1}^{M}(z-p_i)^{m_i}}$$

where M and m_i are positive integers, then by using Theorem 2.2, we have

$$f(nT) = \sum_{i=1}^{M} \text{res}_{z=p_i} F_0(z)$$

where

$$\text{res}_{z=p_i} F_0(z) = \frac{1}{(m_i-1)!} \lim_{z \to p_i} \frac{d^{m_i-1}}{dz^{m_i-1}}[(z-p_i)^{m_i} F_0(z)] \qquad (2.9a)$$

for a pole of order m_i, and

$$\text{res}_{z=p_i} F_0(z) = \lim_{z \to p_i}[(z-p_i) F_0(z)] \qquad (2.9b)$$

for a simple pole.

Note that $F_0(z)$ may have a simple pole at the origin when $n = 0$ and possibly higher-order poles for $n < 0$. This fact must be taken into account in the determination of $f(0), f(-T), f(-2T), \ldots$.

Example 2.2. Find the inverse z transforms of

(a) $$F(z) = \frac{Kz}{z-q}$$

(b) $$F(z) = \frac{(2z-1)z}{2(z-1)(z+0.5)}$$

(c) $$F(z) = \frac{1}{2(z-1)(z+0.5)}$$

Solution

(a) For $n \geq 0$

$$f(nT) = \text{res}_{z=q}\left[F(z)z^{n-1}\right] = \text{res}_{z=q}\left(\frac{Kz^n}{z-q}\right) = Kq^n$$

For $n = -1$, $F_0(z)$ has an additional simple pole at the origin and hence

$$f(nT) = \text{res}_{z=0}\left[\frac{K}{z(z-q)}\right] + \text{res}_{z=q}\left[\frac{K}{z(z-q)}\right] = -\frac{K}{q} + \frac{K}{q} = 0$$

For $n = -k$ where $k > 1$, $F_0(z)$ has an additional higher-order pole at the origin and so

$$f(nT) = \text{res}_{z=0}\left[\frac{K}{z^k(z-q)}\right] + \text{res}_{z=q}\left[\frac{K}{z^k(z-q)}\right]$$

Using the general formula for residues, we obtain

$$\text{res}_{z=0}\left[\frac{K}{z^k(z-q)}\right] = \frac{K}{(k-1)!}\lim_{z\to 0}\frac{d^{k-1}}{dz^{k-1}}(z-q)^{-1}$$

$$= \frac{K}{(k-1)!}\lim_{z\to 0}\frac{d^{k-2}}{dz^{k-2}}(-1)(z-q)^{-2}$$

$$= \frac{K}{(k-1)!}\lim_{z\to 0}\frac{d^{k-3}}{dz^{k-3}}(-1)(-2)(z-q)^{-3}$$

$$\cdots\cdots\cdots\cdots\cdots\cdots\cdots$$

$$= \frac{K}{(k-1)!}\lim_{z\to 0}(-1)(-2)\ldots[-(k-1)](z-q)^{-k}$$

$$= \frac{K}{(k-1)!}(-1)^{k-1}(k-1)!(-q)^{-k} = -\frac{K}{q^k}$$

On the other hand

$$\text{res}_{z=q}\left[\frac{K}{z^k(z-q)}\right] = \frac{K}{q^k}$$

and hence $f(nT) = 0$ for $n < 0$. Therefore

$$f(nT) = u(nT)Kq^n$$

(b) For $n \geq 0$

$$f(nT) = \text{res}_{z=1}[F(z)z^{n-1}] + \text{res}_{z=-0.5}[F(z)z^{n-1}]$$

$$= \left.\frac{(2z-1)z^n}{2(z+0.5)}\right|_{z=1} + \left.\frac{(2z-1)z^n}{2(z-1)}\right|_{z=-0.5}$$

$$= \frac{1}{3} + \frac{2}{3}\left(-\frac{1}{2}\right)^n$$

In order to find $f(nT)$ for $n < 0$, we note that $F(z)$ can be expressed as

$$F(z) = \frac{K_1 z}{z-1} + \frac{K_2 z}{z+0.5}$$

where $K_1 = 1/3$ and $K_2 = 2/3$; hence from Example 2.2(a), we obtain $f(nT) = 0$. Therefore, for any value of n, we have

$$f(nT) = u(nT)\left[\frac{1}{3} + \frac{2}{3}\left(-\frac{1}{2}\right)^n\right]$$

(c) In this case, $F(z)z^{n-1}$ has a pole at the origin if $n = 0$ and so $f(0)$ must be obtained individually, i.e.,

$$f(0) = \left.\frac{1}{2(z-1)(z+0.5)}\right|_{z=0} + \left.\frac{1}{2(z+0.5)z}\right|_{z=1} + \left.\frac{1}{2(z-1)z}\right|_{z=-0.5}$$

$$= -1 + \frac{1}{3} + \frac{2}{3} = 0$$

Similarly, for $n < 0$ we have $f(nT) = 0$. On the other hand, for $n > 0$

$$f(nT) = \left.\frac{z^{n-1}}{2(z + 0.5)}\right|_{z=1} + \left.\frac{z^{n-1}}{2(z - 1)}\right|_{z=-0.5}$$

$$= \tfrac{1}{3} - \tfrac{1}{3}\left(-\tfrac{1}{2}\right)^{n-1}$$

Alternatively, for any value of n

$$f(nT) = u(nT - T)\left[\tfrac{1}{3} - \tfrac{1}{3}\left(-\tfrac{1}{2}\right)^{n-1}\right]$$

Owing to the uniqueness of the Laurent series in a given annulus of convergence, the inverse z transform can also be obtained in other ways: for example, by equating coefficients, by performing long division, by expressing $F(z)$ in terms of binomial series or a partial-fraction expansion, or by applying the convolution theorem. Some of these techniques are illustrated in the examples that follow.

2.5.1 Use of Binomial Series

A function $(1 + u)^r$, where r is a positive or negative integer, can be expressed in terms of a *binomial series* as

$$(1 + u)^r = \binom{r}{0} + \binom{r}{1}u + \binom{r}{2}u^2 + \cdots + \binom{r}{s}u^s + \cdots$$

for $|u| \leq 1$ where

$$\binom{r}{s} = \frac{r(r-1)\ldots(r-s+1)}{s!}$$

and $0! = 1$.

Example 2.3. Find the inverse z transform of

$$F(z) = \frac{Kz^m}{(z - w)^k}$$

where m and k are integers, and K and w are constants.

Solution. $F(z)$ can be expressed as

$$F(z) = Kz^{m-k}[1 + (-wz^{-1})]^{-k}$$

$$= Kz^{m-k}\left[\binom{-k}{0} + \binom{-k}{1}(-wz^{-1}) + \binom{-k}{2}(-wz^{-1})^2 + \cdots \right.$$

$$\left. + \binom{-k}{n}(-wz^{-1})^n + \cdots\right]$$

$$= \sum_{n=-\infty}^{\infty} Ku(nT)\frac{(-k)(-k-1)\cdots(-k-n+1)(-w)^n z^{-n+m-k}}{n!}$$

If we let $n = n' + m - k$ and then replace n' by n, we have

$$F(z) = \sum_{n=-\infty}^{\infty} \left\{ Ku[(n+m-k)T] \right.$$

$$\left. \times \frac{(-k)(-k-1)\cdots(-n-m+1)(-w)^{n+m-k}}{(n+m-k)!} \right\} z^{-n}$$

Hence

$$f(nT) = \mathcal{Z}^{-1}\left[\frac{Kz^m}{(z-w)^k}\right]$$

$$= Ku[(n+m-k)T]\frac{(-k)(-k-1)\cdots(-n-m+1)(-w)^{n+m-k}}{(n+m-k)!}$$

2.5.2 Use of Partial Fractions

The z transform can be expressed in terms of *partial fractions* as

$$F(z) = \sum_{i=1}^{K} F_i(z)$$

where $F_1(z), F_2(z), \ldots$ are simple one-pole z transforms. Thus

$$\mathcal{Z}^{-1}F(z) = \sum_{i=1}^{K} \mathcal{Z}^{-1} F_i(z)$$

where each inverse z transform on the right-hand side can be obtained from Table 2.1.

Example 2.4. Find the inverse z transforms of

(a) $$F(z) = \frac{z}{\left(z - \frac{1}{2}\right)\left(z - \frac{1}{4}\right)}$$

(b) $$F(z) = \frac{z}{z^2 + z + \frac{1}{2}}$$

Solution. (a) $F(z)$ can be expressed as

$$F(z) = \frac{z}{\left(z - \frac{1}{2}\right)\left(z - \frac{1}{4}\right)} = \frac{2}{z - \frac{1}{2}} - \frac{1}{z - \frac{1}{4}}$$

and from Table 2.1

$$f(nT) = 2u(nT - T)\left(\tfrac{1}{2}\right)^{n-1} - u(nT - T)\left(\tfrac{1}{4}\right)^{n-1}$$

$$= 4u(nT - T)\left[\left(\tfrac{1}{2}\right)^n - \left(\tfrac{1}{4}\right)^n\right]$$

(b) By expanding $F(z)/z$ we get

$$\frac{F(z)}{z} = \frac{1}{z^2 + z + \frac{1}{2}} = \frac{A}{z - p_1} + \frac{B}{z - p_2}$$

where $\quad p_1 = \dfrac{e^{j3\pi/4}}{\sqrt{2}} \quad p_2 = \dfrac{e^{-j3\pi/4}}{\sqrt{2}} \quad A = -j \quad B = j$

and so $\quad F(z) = \dfrac{-jz}{z - p_1} + \dfrac{jz}{z - p_2}$

From Table 2.1
$$f(nT) = u(nT)(-jp_1^n + jp_2^n)$$
$$= \left(\tfrac{1}{2}\right)^{n/2} u(nT) \dfrac{1}{j} (e^{j3\pi n/4} - e^{-j3\pi n/4})$$
$$= 2\left(\tfrac{1}{2}\right)^{n/2} u(nT) \sin \dfrac{3\pi n}{4}$$

Alternatively, one could expand $F(z)$ into partial fractions, as in part (a).

2.5.3 Use of Convolution Theorem

From Theorem 2.9
$$\mathcal{Z}^{-1}[F(z)G(z)] = \sum_{k=-\infty}^{\infty} f(kT)g(nT - kT)$$

Thus, if a z transform can be factored into a product of two z transforms whose inverses are available, performing the convolution summation will yield the desired inverse.

Example 2.5. Find the inverse z transforms of

(a) $\qquad Y(z) = \dfrac{z}{(z - 1)^2}$

(b) $\qquad Y(z) = \dfrac{z}{(z - 1)^3}$

Solution
(a) Let
$$F(z) = \dfrac{z}{z - 1} \quad \text{and} \quad G(z) = \dfrac{1}{z - 1}$$

From Table 2.1
$$f(nT) = u(nT) \quad \text{and} \quad g(nT) = u(nT - T)$$

and hence for $n < 0$, the convolution summation yields $y(nT) = 0$. For $n \geq 0$
$$y(nT) = \sum_{k=0}^{n} u(kT)u(nT - T - kT)$$
$$= u(0)u(nT - T) + u(T)u(nT - 2T) + \cdots$$
$$+ u(nT - T)u(0) + u(nT)u(-T)$$
$$= 1 + 1 + \cdots + 1 + 0 = n$$

Therefore $\qquad y(nT) = nu(nT)$

(b) For this example we can write

$$F(z) = \frac{z}{(z-1)^2} \quad \text{and} \quad G(z) = \frac{1}{z-1}$$

in which case

$$f(nT) = nu(nT)$$
$$g(nT) = u(nT - T)$$

Thus for $n < 0$, $y(nT) = 0$. For $n \geq 0$

$$y(nT) = \sum_{k=0}^{n} ku(kT)u(nT - T - kT)$$
$$= 0 \cdot [u(nT - T)] + 1 \cdot [u(nT - 2T)] + \cdots + (n-1)u(0) + nu(-T)$$
$$= 0 + 1 + 2 + \cdots + n - 1 + 0$$
$$= \sum_{k=1}^{n-1} k$$

Now by writing the series $1, 2, \ldots, n-1$ first in the forward order and then in the reverse order, and subsequently adding the two series number by number, a series of $(n-1)$ n's is obtained. As a result, twice the above sum is equal to $(n-1) \times n$[†] and therefore we obtain

$$y(nT) = \sum_{k=1}^{n-1} k = \tfrac{1}{2}n(n-1)$$

Therefore
$$y(nT) = \frac{1}{2}n(n-1)u(nT)$$

2.6 COMPLEX CONVOLUTION

The inverse z transform of a product of z transforms can be formed by using the real convolution. By analogy, the z transform of a product of two time-domain functions can be formed by using the complex convolution.

Theorem 2.10 Complex convolution. If

$$\mathcal{Z}f(nT) = F(z) = \sum_{n=-\infty}^{\infty} f(nT)z^{-n}$$

$$\mathcal{Z}g(nT) = G(z) = \sum_{n=-\infty}^{\infty} g(nT)z^{-n}$$

[†]Gauss is reputed to have astonished his mathematics teacher by obtaining the sum of the numbers 1 to 100 in just a few seconds by using this technique.

then
$$Y(z) = \mathcal{Z}[f(nT)g(nT)] = \frac{1}{2\pi j} \oint_{\Gamma_1} F(v) G\left(\frac{z}{v}\right) v^{-1} dv$$
$$= \frac{1}{2\pi j} \oint_{\Gamma_2} F\left(\frac{z}{v}\right) G(v) v^{-1} dv$$

where Γ_1 (or Γ_2) is a contour in the common region of convergence of $F(v)$ and $G(z/v)$ [or $F(z/v)$ and $G(v)$].

Proof. From Eq. (2.8)

$$Y(z) = \sum_{n=-\infty}^{\infty} [f(nT)g(nT)] z^{-n}$$
$$= \sum_{n=-\infty}^{\infty} f(nT) \left[\frac{1}{2\pi j} \oint_{\Gamma_2} G(v) v^{n-1} dv \right] z^{-n}$$
$$= \frac{1}{2\pi j} \oint_{\Gamma_2} \left[\sum_{n=-\infty}^{\infty} f(nT) \left(\frac{z}{v}\right)^{-n} \right] G(v) v^{-1} dv$$
$$= \frac{1}{2\pi j} \oint_{\Gamma_2} F\left(\frac{z}{v}\right) G(v) v^{-1} dv$$

If Γ_2 is a circle in the region of convergence and $v = \rho e^{j\theta}$ and $z = r e^{j\phi}$, then the above integral can be put in the form

$$Y(re^{j\phi}) = \frac{1}{2\pi} \int_0^{2\pi} F\left[\frac{r}{\rho} e^{j(\phi-\theta)}\right] G(\rho e^{j\theta}) d\theta$$

which is recognized as a convolution integral.

The complex convolution will be used later in the design of nonrecursive filters.

Example 2.6. Find the z transform of
$$y(nT) = u(nT) e^{-\alpha nT} \sin \omega nT$$

Solution. Let
$$f(nT) = u(nT) e^{-\alpha nT} \quad \text{and} \quad g(nT) = u(nT) \sin \omega nT$$

From Table 2.1
$$F(z) = \frac{z}{z - e^{-\alpha T}} \quad \text{and} \quad G(z) = \frac{z \sin \omega T}{(z - e^{j\omega T})(z - e^{-j\omega T})}$$

and hence
$$Y(z) = \frac{1}{2\pi j} \oint_{\Gamma_2} F\left(\frac{z}{v}\right) G(v) v^{-1} dv$$
$$= \frac{1}{2\pi j} \oint_{\Gamma_2} \frac{-z e^{\alpha T} \sin \omega T}{(v - z e^{\alpha T})(v - e^{j\omega T})(v - e^{-j\omega T})} dv$$

$F(z/v)$ and $G(v)$ converge in the regions

$$|v| < |ze^{\alpha T}| \quad \text{and} \quad |v| > 1$$

respectively, as illustrated in Fig. 2.4. By evaluating the residues at $v = e^{+j\omega T}$ and $e^{-j\omega T}$ we have

$$Y(z) = \frac{ze^{-\alpha T} \sin \omega T}{z^2 - 2ze^{-\alpha T} \cos \omega T + e^{-2\alpha T}}$$

A relation that is often used in the design of digital filters is stated in terms of the following theorem:

Theorem 2.11   Parseval's discrete-time formula

$$\sum_{n=-\infty}^{\infty} |f(nT)|^2 = \frac{1}{2\pi} \int_0^{\omega_s} |F(e^{j\omega T})|^2 d\omega$$

Proof. Although function $f(nT)$ in Eq. (2.6) has been implicitly assumed to be real, the z transform can be applied to complex signals as long as $F(z)$ converges. If we assume that $g(nT)$ is the complex conjugate of $f(nT)$, i.e., $g(nT) = f^*(nT)$, then

$$G(z) = \sum_{n=-\infty}^{\infty} f^*(nT) z^{-n} = \left[\sum_{n=-\infty}^{\infty} f(nT)(z^{-1})^n \right]^*$$

$$= F^*(z^{-1})$$

and if we let $z = 1$ in Theorem 2.10, we obtain

$$\sum_{n=-\infty}^{\infty} f(nT) f^*(nT) = \frac{1}{2\pi j} \oint_\Gamma F(v) F^*(v^{-1}) v^{-1} dv$$

Now if we let $v = e^{j\omega T}$, Parseval's relation is obtained.

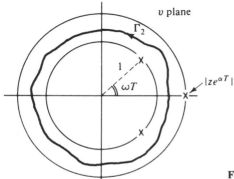

FIGURE 2.4
Region of convergence of $F(z/v)$ and $G(v)$ (Example 2.6).

REFERENCES

1. E. Kreyszig, *Advanced Engineering Mathematics*, Wiley, New York, 1972.
2. E. I. Jury, *Theory and Application of the z-Transform Method*, Wiley, New York, 1964.

PROBLEMS

2.1. Check the following functions for analyticity:
  (a) $F(z) = 2z + 3$
  (b) $F(z) = z^3 + 3z^2$
  (c) $F(z) = z + \dfrac{1}{z}$
  (d) $F(z) = \dfrac{z + 1}{z - 1}$

2.2. Check the following functions for analyticity:
  (a) $F(z) = |z|^2$
  (b) $F(z) = e^x \cos y$
  (c) $F(z) = (z^2 + 3z)^2$
  (d) $F(z) = \dfrac{1}{z^4}$

2.3. (a) Find all Laurent series of
$$F(z) = \frac{1}{1 - z^2}$$
with center $z = 1$.

  (b) Find all Laurent series of
$$F(z) = \frac{7z^2 + 9z - 18}{z^3 - 9z}$$
with center $z = 0$.

  (c) Illustrate the region of convergence for each of the above series.

2.4. (a) Find all Laurent series of
$$F(z) = \frac{4z - 1}{z^4 - 1}$$
with center $z = 0$.

  (b) Find all Laurent series of
$$F(z) = \frac{e^z}{(z - 1)^2}$$
with center $z = 1$.

  (c) Illustrate the region of convergence in each case.

2.5. Find the locations and orders of the zeros of the following functions:
  (a) $F(z) = (z - 1)^4$
  (b) $F(z) = z^{-3} \cos^3 \pi z$
  (c) $F(z) = \dfrac{z + j}{z^2 + 3}$
  (d) $F(z) = \dfrac{(z^2 + 4)^2}{(z + 1)^4}$

2.6. Find the locations and types of the singularities of the following functions:
(a) $F(z) = z^2 + z^{-1}$
(b) $F(z) = \dfrac{z^7}{(z^2+1)^3}$
(c) $F(z) = \dfrac{1}{z^3 + 6z^2 + 11z + 6}$

2.7. Find the locations and types of the singularities of the following functions:
(a) $F(z) = \cosh \dfrac{1}{z-\pi}$
(b) $F(z) = \cot \dfrac{1}{z}$
(c) $F(z) = \tan^2 \dfrac{1}{z-1}$

2.8. Find the z transforms of the following functions:
(a) $u(nT)(2 + 3e^{-2nT})$
(b) $u(nT) - \delta(nT)$
(c) $\mathcal{E}^{-k} u(nT)$

2.9. Find the z transforms of the following functions:
(a) $\nabla u(nT)$
(b) $u(nT)[1 + (-1)^n] e^{-nT}$
(c) $u(nT) \sin(\omega nT + \psi)$
(d) $u(nT) \cosh \alpha nT$

2.10. Find the z transforms of the following functions:
(a) $u(nT) nT$
(b) $u(nT)(nT)^2$
(c) $u(nT)(nT)^3$
(d) $u(nT) nT e^{-4nT}$

2.11. Find the z transforms of the following discrete-time signals:
(a) $x(nT) = \begin{cases} 1 & \text{for } 0 \le n \le k \\ 0 & \text{otherwise} \end{cases}$
(b) $x(nT) = \begin{cases} nT & \text{for } 0 \le n \le 5 \\ 0 & \text{otherwise} \end{cases}$

2.12. Find the z transforms of the following discrete-time signals:
(a) $x(nT) = \begin{cases} 0 & \text{for } n < 0 \\ 1 & \text{for } 0 \le n \le 5 \\ 2 & \text{for } 5 < n \le 10 \\ 3 & \text{for } n > 10 \end{cases}$
(b) $x(nT) = \begin{cases} 0 & \text{for } n < 0 \\ nT & \text{for } 0 \le n < 5 \\ (n-5)T & \text{for } 5 \le n < 10 \\ (n-10)T & \text{for } n \ge 10 \end{cases}$

2.13. Find the z transforms of the following discrete-time signals:
(a) $x(nT) = \begin{cases} 0 & \text{for } n < 0 \\ 2 + nT & \text{for } n \ge 0 \end{cases}$
(b) $x(nT) = u(nT) \sinh \alpha nT$

2.14. Find the z transforms of the following discrete-time signals:
(a) $x(nT) = u(nT) nT(1 + e^{-\alpha nT})$
(b) $x(nT) = \begin{cases} 0 & \text{for } n \le 0 \\ \dfrac{e^{-\alpha nT}}{nT} & \text{for } n > 0 \end{cases}$ $\left(\text{Note that } \ln \dfrac{1}{1-y} = \sum_{k=1}^{\infty} \dfrac{y^k}{k}\right)$

2.15. By using Theorem 2.9, form the z transforms of
(a) $y(nT) = \sum_{k=0}^{n} r(nT - kT)u(kT)$
(b) $y(nT) = \sum_{k=0}^{n} u(nT - kT)u(kT)e^{-\alpha kT}$

2.16. Prove that
$$\mathcal{Z} \sum_{k=0}^{n} f(kT) = \frac{z}{z-1} \mathcal{Z} f(nT)$$

2.17. Assuming that $f(nT) = 0$ for $n < 0$, show that
(a) $f(0) = \lim_{z \to \infty} F(z)$
(b) $f(\infty) = \lim_{z \to \infty} (z - 1)F(z)$

2.18. Find $f(0)$ and $f(\infty)$ for the following z transforms:
(a) $F(z) = \dfrac{2z - 1}{z - 1}$
(b) $F(z) = \dfrac{(e^{-\alpha T} - 1)z}{z^2 - (1 + e^{-\alpha T})z + e^{-\alpha T}}$
(c) $F(z) = \dfrac{Tze^{-4T}}{(z - e^{-4T})^2}$

2.19. Find the z transforms of
(a) $y(nT) = \sum_{i=0}^{N} a_i x(nT - iT)$
(b) $y(nT) = \sum_{i=0}^{N} a_i x(nT - iT) - \sum_{i=1}^{N} b_i y(nT - iT)$

2.20. Form the z transform of
$$x(nT) = [u(nT) - u(nT - NT)]W^{kn}$$

2.21. Find the inverse z transforms of
(a) $F(z) = \dfrac{2}{2z - 1}$
(b) $F(z) = \dfrac{5}{z - e^{-T}}$
(c) $F(z) = \dfrac{3z}{3z + 2}$
(d) $F(z) = \dfrac{2z}{z^2 - 2z + 1}$

2.22. Find the inverse z transforms of
(a) $F(z) = \dfrac{z + 2}{z^2 - 0.25}$
(b) $F(z) = \dfrac{z^2}{(z - 0.5)^5}$
(c) $F(z) = \dfrac{3z^4 - z^3 - z^2}{z - 1}$

2.23. The z transform
$$F(z) = \frac{N(z)}{D(z)}$$
has a simple pole at $z = p_i$. Show that
$$\operatorname{res}_{z = p_i} F(z) = \lim_{z \to p_i} (z - p_i)F(z) = \frac{N(p_i)}{D'(p_i)} \qquad \text{where } D'(z) = \frac{dD(z)}{dz}$$

2.24. Find the inverse z transforms of the following by using Eq. (2.8):
(a) $F(z) = \dfrac{z^2}{z^2 + 1}$
(b) $F(z) = \dfrac{2z^2}{2z^2 - 2z + 1}$

2.25. Find the inverse z transforms of the following by using Eq. (2.8):

(a) $F(z) = \dfrac{1}{(z-0.8)^4}$

(b) $F(z) = \dfrac{6z}{(2z^2+2z+1)(3z-1)}$

2.26. Find the inverse z transforms of the following by using the partial-fraction method:

(a) $F(z) = \dfrac{(z-1)^2}{z^2-0.1z-0.56}$

(b) $F(z) = \dfrac{4z^3}{(2z+1)(2z^2-2z+1)}$

2.27. Find the inverse z transforms of the following by using Theorem 2.9:

(a) $F(z) = \dfrac{z^2}{z^2-2z+1}$

(b) $F(z) = \dfrac{z^2}{(z-e^{-T})(z-1)}$

2.28. Find the inverse z transform of the following by equating coefficients of equal powers of z:

$$F(z) = \dfrac{z(z+1)}{(z-1)^3}$$

2.29. Find the inverse z transform of the following by means of long division:

$$F(z) = \dfrac{z(z^2+4z+1)}{(z-1)^4}$$

2.30. Find the z transform of

$$x(nT) = u(nT)e^{-\alpha nT} \sin(\omega nT + \psi)$$

by using the complex-convolution theorem.

CHAPTER 3

THE APPLICATION OF THE z TRANSFORM

3.1 INTRODUCTION

Through the use of the z transform, a digital filter can be characterized by a discrete-time transfer function, which plays the same key role as the continuous-time transfer function in an analog filter. In this chapter, the discrete-time transfer function is first defined, its properties are then examined, and finally its application in time-domain and frequency-domain analysis is described.

In Sec. 3.2, it is shown that the transfer function is a ratio of polynomials in complex variable z and, as a result, a digital filter can be represented by a set of zeros and poles. In Sec. 3.3, it is shown that the stability of a filter is closely linked with the location of its poles. Several stability criteria that enable one to determine whether a filter is stable or not with minimal computational effort are then presented. Sections 3.4 and 3.5 deal with general time-domain and frequency-domain methods, respectively, that can be used to analyze filters of arbitrary order and complexity. The chapter concludes with a general introduction to the design process. To be specific, the design process is broken into distinct steps and the tasks involved in each step are delineated.

3.2 THE DISCRETE-TIME TRANSFER FUNCTION

The *transfer function* of a digital filter is defined as the ratio of the z transform of the response to the z transform of the excitation.

Consider a linear, time-invariant digital filter, and let $x(nT)$, $y(nT)$, and $h(nT)$ be the excitation, response, and impulse response, respectively. By using the convolution summation in Eq. (1.12) we have

$$y(nT) = \sum_{k=-\infty}^{\infty} x(kT)h(nT - kT)$$

and therefore, from Theorem 2.9,

$$\mathcal{Z}y(nT) = \mathcal{Z}h(nT)\mathcal{Z}x(nT)$$

or
$$Y(z) = H(z)X(z) \qquad (3.1)$$

In effect, the transfer function of a digital filter is the z transform of the impulse response.

In later chapters we shall be dealing with analog and digital filters at the same time. To avoid confusion, we refer to the transfer function of an analog filter as *continuous-time* and that of a digital filter as *discrete-time*.

3.2.1 Derivation of $H(z)$

The exact form of $H(z)$ can be derived (i) from the difference equation characterizing the filter, (ii) from the filter network, and (iii) from a state-space characterization, if one is available.

(i) For causal, recursive filters

$$y(nT) = \sum_{i=0}^{N} a_i x(nT - iT) - \sum_{i=1}^{N} b_i y(nT - iT)$$

and hence

$$\mathcal{Z}y(nT) = \sum_{i=0}^{N} a_i z^{-i} \mathcal{Z}x(nT) - \sum_{i=1}^{N} b_i z^{-i} \mathcal{Z}y(nT)$$

or
$$\frac{Y(z)}{X(z)} = H(z) = \frac{\sum_{i=0}^{N} a_i z^{N-i}}{z^N + \sum_{i=1}^{N} b_i z^{N-i}} \qquad (3.2)$$

By factoring the numerator and denominator polynomials, $H(z)$ can be put in the form

$$H(z) = \frac{N(z)}{D(z)} = \frac{H_0 \prod_{i=1}^{N}(z - z_i)}{\prod_{i=1}^{N}(z - p_i)^{m_i}} \qquad (3.3)$$

where $z_1, z_2, \ldots, z_N$ are the zeros and $p_1, p_2, \ldots, p_N$ are the poles of $H(z)$, and m_i is the order of pole p_i. Thus digital filters, like analog filters, can be represented by zero-pole plots like the one in Fig. 3.1.

In nonrecursive filters, $b_i = 0$ for $i = 1, 2, \ldots, N$, and so the poles in these filters are all located at the origin of the z plane.

(ii) The z-domain characterizations of the unit delay, the adder, and the multiplier are obtained from Table 1.1 as

$$Y(z) = z^{-1}X(z) \qquad Y(z) = \sum_{i=1}^{K} X_i(z) \qquad Y(z) = mX(z)$$

By using these relations, $H(z)$ can be derived directly from the filter network.

Example 3.1. Find the transfer function of the filter shown in Fig. 3.2.

Solution. We can write

$$U(z) = X(z) + \frac{1}{2}z^{-1}U(z) - \frac{1}{4}z^{-2}U(z)$$

$$Y(z) = U(z) + z^{-1}U(z)$$

Hence
$$U(z) = \frac{X(z)}{1 - \frac{1}{2}z^{-1} + \frac{1}{4}z^{-2}} \qquad Y(z) = (1 + z^{-1})U(z)$$

and
$$H(z) = \frac{z(z+1)}{z^2 - \frac{1}{2}z + \frac{1}{4}}$$

(iii) Alternatively, $H(z)$ can be deduced from the state-space equations (see Sec. 1.9)

$$\mathbf{q}(nT + T) = \mathbf{A}\mathbf{q}(nT) + \mathbf{b}x(nT) \tag{3.4a}$$

$$y(nT) = \mathbf{c}^T \mathbf{q}(nT) + dx(nT) \tag{3.4b}$$

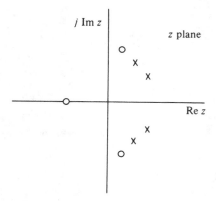

FIGURE 3.1
Typical zero-pole plot for $H(z)$.

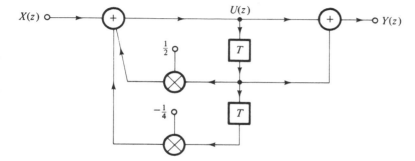

FIGURE 3.2
Second-order recursive filter (Example 3.1).

By applying the z transform

$$\mathcal{Z}\mathbf{q}(nT+T) = \mathbf{A}\mathcal{Z}\mathbf{q}(nT) + \mathbf{b}\mathcal{Z}x(nT) = \mathbf{A}\mathbf{Q}(z) + \mathbf{b}X(z) \tag{3.5}$$

Also

$$\mathcal{Z}\mathbf{q}(nT+T) = z\mathcal{Z}\mathbf{q}(nT) = z\mathbf{Q}(z) \tag{3.6}$$

and from Eqs. (3.5) and (3.6)

$$z\mathbf{Q}(z) = \mathbf{A}\mathbf{Q}(z) + \mathbf{b}X(z) \quad \text{or} \quad \mathbf{Q}(z) = (z\mathbf{I} - \mathbf{A})^{-1}\mathbf{b}X(z)$$

where $\mathbf{I}$ is the $N \times N$ unity matrix. Now from Eq. (3.4 b)

$$Y(z) = \mathbf{c}^T\mathbf{Q}(z) + dX(z)$$

and on eliminating $\mathbf{Q}(z)$ we have

$$H(z) = \frac{N(z)}{D(z)} = \mathbf{c}^T(z\mathbf{I} - \mathbf{A})^{-1}\mathbf{b} + d \tag{3.7}$$

where $N(z)$ and $D(z)$ are polynomials in z.

3.3 STABILITY

As can be seen in Eq. (3.2), the discrete-time transfer function is a rational function of z with real coefficients, and for causal filters the degree of the numerator polynomial is equal to or less than that of the denominator polynomial. We shall now show that the poles of the transfer function or, alternatively, the eigenvalues of matrix $\mathbf{A}$ in a state-space characterization determine whether the filter is stable or unstable, as in analog systems.

3.3.1 Constraint on Poles

Consider an Nth-order causal filter characterized by the transfer function of Eq. (3.3). The impulse response of such a filter is given by Eq. (2.8) as

$$h(nT) = \mathcal{Z}^{-1}H(z) = \frac{1}{2\pi j} \oint_\Gamma H(z)z^{n-1}dz$$

For $n = 0$

$$h(0) = R_0 + \sum_{i=1}^{N} \text{res}_{z=p_i} \left[\frac{H(z)}{z}\right]$$

where

$$R_0 = \text{res}_{z=0} \left[\frac{H(z)}{z}\right]$$

if $H(z)$ has no zeros or poles at $z = 0$ and $R_0 = 0$ in all other cases. On the other hand, for $n > 0$

$$h(nT) = \sum_{i=1}^{N} \text{res}_{z=p_i}[H(z)z^{n-1}] = \sum_{i=1}^{N} p_i^{n-1} \text{res}_{z=p_i} H(z)$$

With

$$p_i = r_i e^{j\psi_i}$$

such that

$$r_i \leq r_m < 1 \qquad \text{for } i = 1, 2, \ldots, N \tag{3.8}$$

we can write

$$\sum_{n=0}^{\infty} |h(nT)| = |h(0)| + \sum_{n=1}^{\infty} \left|\sum_{i=1}^{N} r_i^{n-1} e^{j(n-1)\psi_i} \text{res}_{z=p_i} H(z)\right|$$

$$\leq |h(0)| + \sum_{n=1}^{\infty} \sum_{i=1}^{N} r_i^{n-1} |\text{res}_{z=p_i} H(z)| \tag{3.9}$$

Now if p_k is a pole of order m_k of some function $F(z)$, then function $(z - p_k)^{m_k} F(z)$ is analytic; hence, from Eq. (2.9a), the residue of $F(z)$ at $z = p_k$ is finite (see Sec. 2.2). Therefore, $h(0)$ as well as the residues of $H(z)$ are finite and as a result

$$|\text{res}_{z=p_i} H(z)| \leq R_m \qquad \text{for } i = 1, 2, \ldots, N$$

where R_m is a positive constant. Thus from Eqs. (3.8) and (3.9)

$$\sum_{n=0}^{\infty} |h(nT)| \leq |h(0)| + N R_m \sum_{n=1}^{\infty} r_m^{n-1} < \infty$$

since the summation on the right-hand side converges. Consequently, the necessary and sufficient condition for stability given in Eq. (1.14) is satisfied. In other words, if the condition in Eq. (3.8) is satisfied the filter is stable.

If a single pole of $H(z)$, say pole p_k, is located on or outside the unit circle $|z| = 1$, then as $n \to \infty$

$$h(nT) \approx r_k^{n-1} e^{j(n-1)\psi_k} \operatorname{res}_{z=p_k} H(z)$$

since $r_i^{n-1} \to 0$ for $i \neq k$. Hence

$$\sum_{n=0}^{\infty} |h(nT)| \approx |\operatorname{res}_{z=p_k} H(z)| \lim_{Q \to \infty} \sum_{n=0}^{Q} r_k^{n-1} \to \infty$$

i.e., the condition for stability given in Eq. (1.14) is violated in this case and the filter is unstable. Therefore, a digital filter is *stable* if and only if its poles satisfy the condition

$$|p_i| < 1 \qquad \text{for } i = 1, 2, \ldots, N$$

The permissible region for the location of poles is illustrated in Fig. 3.3.

In nonrecursive filters the poles are located at the origin of the z plane; i.e., these filters are always stable.

Example 3.2. Check the filter of Fig. 3.4 for stability.

Solution. The transfer function of the filter is

$$H(z) = \frac{z^2 - z + 1}{z^2 - z + \frac{1}{2}} = \frac{z^2 - z + 1}{(z - p_1)(z - p_2)}$$

where

$$p_1, p_2 = \frac{1}{2} \pm j\frac{1}{2}$$

Since

$$|p_1|, |p_2| < 1$$

the filter is stable.

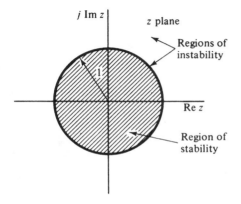

FIGURE 3.3
Permissible z-plane region for the location of the poles of $H(z)$.

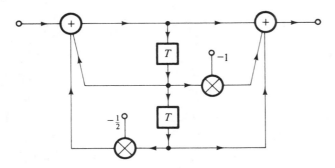

FIGURE 3.4
Second-order recursive filter (Example 3.2).

3.3.2 Constraint on Eigenvalues

The poles of $H(z)$ are the values of z for which $D(z)$, the denominator polynomial of $H(z)$, becomes zero. From Eq. (3.7), $D(z)$ is zero if and only if

$$\det(\mathbf{A} - z\mathbf{I}) = 0$$

Now the determinant of $(\mathbf{A} - z\mathbf{I})$ is the characteristic polynomial of matrix $\mathbf{A}$ [1] and, consequently, the poles of $H(z)$ are numerically equal to the *eigenvalues* $\lambda_1, \lambda_2, \ldots, \lambda_N$ of matrix $\mathbf{A}$. Therefore, a filter characterized by Eq. (3.7) is stable if and only if

$$|\lambda_i| < 1 \quad \text{for } i = 1, 2, \ldots, N$$

3.3.3 Stability Criteria

The stability of a digital filter can be checked by finding the roots of polynomial $D(z)$ or the eigenvalues of matrix $\mathbf{A}$. For a second- or third-order filter, this is easily accomplished. For higher-order filters, however, the use of a computer is necessary. Very often the designer needs to know whether a filter is stable or unstable and the exact values of the poles of $H(z)$ are unimportant. In such a case, one of several available *stability tests* or *criteria* like the Schur-Cohn or Jury-Marden criterion [2] can be used. These criteria are analogous to the Routh-Hurwitz criterion [3, 4] for continuous-time systems and usually involve an insignificant amount of computation relative to that required to find the roots of $D(z)$. The Schur-Cohn and other discrete-time criteria are based on the Routh-Hurwitz criterion which is, in turn, based on Sturm's theorem. The usefulness of Sturm's theorem in this context arises because of the fact that it provides the number of real roots of a given polynomial in some specified finite interval [3].

Some of the more important discrete-time stability criteria will now be described, and their principal properties discussed. Derivations and proofs are omitted for the sake of brevity, but the interested reader may consult the references at the end of the chapter.

Consider a filter characterized by the transfer function

$$H(z) = \frac{N(z)}{D(z)} \tag{3.10a}$$

where

$$N(z) = \sum_{i=0}^{M} a_i z^{M-i} \tag{3.10b}$$

$$D(z) = \sum_{i=0}^{N} b_i z^{N-i} \tag{3.10c}$$

and assume that $b_0 > 0$. If b_0 happens to be negative, a positive b_0 can be obtained by replacing all the coefficients of $D(z)$ by their negatives. This modification does not affect the stability of the filter since $D(z)$ and $-D(z)$ have the same roots. Assume also that $N(z)$ and $D(z)$ have no common factors which are not constants, i.e., $N(z)$ and $D(z)$ are *relatively prime*. If there are common factors in these polynomials, they should be canceled out before the application of any one of the aforementioned stability criteria since they do not affect the stability of the filter.

3.3.4 Test for Common Factors

The presence of common factors in $N(z)$ and $D(z)$ can be checked by applying the following test. The coefficients of $N(z)$ are used to construct the $N \times (N+M)$ matrix

$$\mathbf{R}_N = \begin{bmatrix} a_0 & a_1 & a_2 & \cdots & & a_M & 0 & \cdots & 0 & 0 \\ 0 & a_0 & a_1 & \cdots & & a_{M-1} & a_M & \cdots & 0 & 0 \\ \vdots & \vdots & \vdots & & & \vdots & \vdots & & \vdots & \vdots \\ 0 & 0 & 0 & \cdots & a_0 & a_1 & \cdots & & a_{M-1} & a_M \end{bmatrix}$$

and the coefficients of $D(z)$ are used to construct the $M \times (N+M)$ matrix

$$\mathbf{R}_M = \begin{bmatrix} 0 & 0 & 0 & \cdots & 0 & 0 & 0 & b_0 & \cdots & b_{N-1} & b_N \\ \vdots & \vdots & \vdots & & \vdots & \vdots & \vdots & \vdots & & \vdots & \vdots \\ 0 & b_0 & b_1 & \cdots & & & b_N & & \cdots & 0 & 0 & 0 \\ b_0 & b_1 & b_2 & \cdots & & b_N & 0 & 0 & \cdots & 0 & 0 & 0 \end{bmatrix}$$

Then the $(N+M) \times (N+M)$ matrix

$$\mathbf{R} = \begin{bmatrix} \mathbf{R}_N \\ \mathbf{R}_M \end{bmatrix}$$

is formed and its determinant is computed. If

$$\det \mathbf{R} = 0$$

then $N(z)$ and $D(z)$ have a common factor which is not a constant. Otherwise, the two polynomials are relatively prime [5, 6].

Example 3.3. Check the numerator and denominator polynomials of the transfer function

$$H(z) = \frac{N(z)}{D(z)} = \frac{z^2 + 3z + 2}{3z^3 + 5z^2 + 3z + 1}$$

for common factors.

Solution. Matrix **R** can be formed as

$$\mathbf{R} = \begin{bmatrix} 1 & 3 & 2 & 0 & 0 \\ 0 & 1 & 3 & 2 & 0 \\ 0 & 0 & 1 & 3 & 2 \\ 0 & 3 & 5 & 3 & 1 \\ 3 & 5 & 3 & 1 & 0 \end{bmatrix}$$

and since $\det \mathbf{R} = 0$, $N(z)$ and $D(z)$ have a common factor. In actual fact

$$H(z) = \frac{(z+1)(z+2)}{(z+1)(3z^2 + 2z + 1)}$$

3.3.5 Schur-Cohn Stability Criterion

The *Schur-Cohn* stability criterion was established during the early twenties [2], long before the era of digital filters, and its main application at that time was as a mathematical tool for the purpose of establishing whether or not a general polynomial of z has zeros inside the unit circle of the z plane. This criterion has been superseded in recent years by other more efficient criteria and is rarely used nowadays. Nevertheless, it is of interest since it is the basis of some of the modern criteria.

The Schur-Cohn criterion states that a polynomial $D(z)$ of the type given in Eq. (3.10c), whose coefficients may be complex, has roots inside the unit circle of the z plane if and only if

$$\det \mathbf{S}_k < 0 \quad \text{if } k \text{ is odd} \quad \text{and} \quad \det \mathbf{S}_k > 0 \quad \text{if } k \text{ is even}$$

for $k = 1, 2, \ldots, N$ where $\mathbf{S}_k$ is a $2k \times 2k$ matrix given by

$$\mathbf{S}_k = \begin{bmatrix} \mathbf{A}_k & \mathbf{B}_k \\ \mathbf{B}_k^T & \mathbf{A}_k^T \end{bmatrix}$$

with

$$\mathbf{A}_k = \begin{bmatrix} b_N & 0 & 0 & \cdots & 0 \\ b_{N-1} & b_N & 0 & \cdots & 0 \\ \vdots & \vdots & \vdots & & \vdots \\ b_{N-k+1} & b_{N-k+2} & b_{N-k+3} & \cdots & b_N \end{bmatrix}$$

and

$$\mathbf{B}_k = \begin{bmatrix} b_0 & b_1 & b_2 & \cdots & b_{k-1} \\ 0 & b_0 & b_1 & \cdots & b_{k-2} \\ \vdots & \vdots & \vdots & & \vdots \\ 0 & 0 & 0 & \cdots & b_0 \end{bmatrix}$$

that gives the Schur-Cohn determinants in terms of second-order determinants. This criterion is often referred to as the *Jury-Marden* criterion and, as is demonstrated below, it is both efficient and easy to apply. In this criterion, the coefficients of $D(z)$, which are assumed to be real, are used to construct an array of numbers known as the Jury-Marden array, as in Table 3.1. The first two rows of the array are formed by entering the coefficients of $D(z)$ directly in ascending order for the first row and in descending order for the second. The elements of the third and fourth rows are computed as

$$c_i = \begin{vmatrix} b_0 & b_{N-i} \\ b_N & b_i \end{vmatrix} \quad \text{for } i = 0, 1, \ldots, N-1$$

those of the fifth and sixth rows as

$$d_i = \begin{vmatrix} c_0 & c_{N-1-i} \\ c_{N-1} & c_i \end{vmatrix} \quad \text{for } i = 0, 1, \ldots, N-2$$

and so on until $2N - 3$ rows are obtained. The last row comprises three elements, say r_0, r_1, and r_2.

The Jury-Marden criterion states that polynomial $D(z)$ has roots inside the unit circle of the z plane if and only if the following conditions are satisfied:

(i) $D(1) > 0$
(ii) $(-1)^N D(-1) > 0$
(iii) $b_0 > |b_N|$
  $|c_0| > |c_{N-1}|$
  $|d_0| > |d_{N-2}|$
  . . . .
  $|r_0| > |r_2|$

TABLE 3.1
The Jury-Marden array

Row	Coefficients			
1	b_0	b_1	$\cdots$	b_N
2	b_N	b_{N-1}	$\cdots$	b_0
3	c_0	c_1	$\cdots$	c_{N-1}
4	c_{N-1}	c_{N-2}	$\cdots$	c_0
5	d_0	d_1	$\cdots$	d_{N-2}
6	d_{N-2}	d_{N-3}	$\cdots$	d_0
$\cdots$	. . . .			
$2N - 3$	r_0	r_1	r_2	

As can be seen, the Jury-Marden criterion involves determinants of 2×2 matrices and is easy to apply even without the use of a computer. Note that all three of the preceding three conditions must be satisfied for the filter to be stable and the Jury-Marden array need not be constructed if either of conditions (i) or (ii) is violated.

Example 3.5. Check the filters of Example 3.4 (*a*) and (*b*) for stability using the Jury-Marden criterion.

Solution. (*a*) The Jury-Marden array can be obtained as

1	4	3	2	1	1
2	1	1	2	3	4
3	15	11	6	1	
4	1	6	11	15	
5	224	159	79		

Since

$$D(1) = 11 \qquad (-1)^4 D(-1) = 3$$

and $b_0 > |b_4|$, $|c_0| > |c_3|$, $|d_0| > |d_2|$, the filter is stable.

(*b*) In this case

$$(-1)^4 D(-1) = -1$$

i.e., condition (ii) is violated and the filter is unstable.

3.3.8 Lyapunov Stability Criterion

Another stability criterion, which was originally developed for the analysis of discrete-time systems, states that a digital filter represented by the state-space equations is stable if and only if for any positive definite matrix **Q**, there exists a unique positive definite matrix **P** that satisfies the *Lyapunov equation* [10]

$$\mathbf{A}^T \mathbf{PA} - \mathbf{P} = -\mathbf{Q}$$

In this criterion, a positive definite matrix **Q** is assumed, say **Q** = **I**, and the Lyapunov equation is solved for **P** [11]. If **P** is found to be positive definite, the filter is classified as stable. This criterion is less practical to apply than the Jury-Marden criterion. Nevertheless, it finds some important applications in the study of parasitic oscillations in digital filters (see Chap. 11).

3.4 TIME-DOMAIN ANALYSIS

The *time-domain* response of a digital filter to any excitation $x(nT)$ can be readily obtained from Eq. (3.1) as

$$y(nT) = \mathcal{Z}^{-1}[H(z)X(z)]$$

Any one of the inversion techniques described in Sec. 2.5 can be used.

Example 3.6. Find the unit-step response of the filter shown in Fig. 3.4.

Solution. From Example 3.2

$$H(z) = \frac{z^2 - z + 1}{(z - p_1)(z - p_2)}$$

where
$$p_1 = \frac{1}{2} - j\frac{1}{2} = \frac{e^{-j\pi/4}}{\sqrt{2}} \qquad p_2 = \frac{1}{2} + j\frac{1}{2} = \frac{e^{j\pi/4}}{\sqrt{2}}$$

and from Table 2.1

$$X(z) = \frac{z}{z - 1}$$

On expanding $H(z)X(z)/z$ into partial fractions, we have

$$H(z)X(z) = \frac{Az}{z - 1} + \frac{B_1 z}{z - p_1} + \frac{B_2 z}{z - p_2}$$

where
$$A = 2 \qquad B_1 = \frac{e^{j5\pi/4}}{\sqrt{2}} \qquad B_2 = B_1^* = \frac{e^{-j5\pi/4}}{\sqrt{2}}$$

Hence

$$y(nT) = \mathcal{Z}^{-1}[H(z)X(z)]$$

$$= 2u(nT) + \frac{1}{(\sqrt{2})^{n+1}} u(nT)(e^{j(n-5)\pi/4} + e^{-j(n-5)\pi/4})$$

$$= 2u(nT) + \frac{1}{(\sqrt{2})^{n-1}} u(nT) \cos\left[(n - 5)\frac{\pi}{4}\right]$$

The response of the filter is plotted in Fig. 3.5.

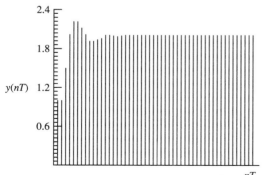

FIGURE 3.5
Unit-step response (Example 3.6).

3.5 FREQUENCY-DOMAIN ANALYSIS

An analog filter characterized by a continuous-time transfer function $H(s)$ has a steady-state sinusoidal response of the form

$$\tilde{y}(t) = \lim_{t \to \infty} y(t) = \lim_{t \to \infty} \mathcal{R}[u(t) \sin \omega t] = M(\omega) \sin[\omega t + \theta(\omega)]$$

where $\qquad M(\omega) = |H(j\omega)| \qquad$ and $\qquad \theta(\omega) = \arg H(j\omega)$

$M(\omega)$ is the *gain* and $\theta(\omega)$ is the *phase shift* of the filter at frequency ω. The functions $M(\omega)$ and $\theta(\omega)$, which are the *amplitude response* and *phase response*, respectively, constitute the basic frequency-domain characterization of the filter. If these plots are available, one can determine the steady-state response not only to any sinusoidal waveform but also to any other waveform that can be expressed as a linear combination of sinusoids.

3.5.1 Steady-State Sinusoidal Response

Let us consider a general Nth-order digital filter characterized by the transfer function of Eq. (3.3). The sinusoidal response of such a filter is

$$y(nT) = \mathcal{Z}^{-1}[H(z)X(z)]$$

where $\qquad X(z) = \mathcal{Z}[u(nT) \sin \omega nT] = \dfrac{z \sin \omega T}{(z - e^{j\omega T})(z - e^{-j\omega T})} \qquad (3.12)$

or
$$y(nT) = \frac{1}{2\pi j} \oint_\Gamma H(z)X(z)z^{n-1} dz$$
$$= \sum \text{res}\,[H(z)X(z)z^{n-1}] \qquad (3.13)$$

For $n > 0$, Eqs. (3.12) and (3.13) yield

$$y(nT) = \frac{1}{2j}[H(e^{j\omega T})e^{j\omega nT} - H(e^{-j\omega T})e^{-j\omega nT}] + \sum_{i=1}^{N} X(p_i) p_i^{n-1} \text{res}_{z=p_i} H(z)$$

Assuming that the filter is stable, we have $|p_i| < 1$ for $i = 1, 2, \ldots, N$ and hence as $n \to \infty$, the summation part in the above equation tends to zero since $p_i^{n-1} \to 0$. Therefore, the steady-state sinusoidal response can be expressed as

$$\tilde{y}(nT) = \lim_{n \to \infty} y(nT) = \frac{1}{2j}[H(e^{j\omega T})e^{j\omega nT} - H(e^{-j\omega T})e^{-j\omega nT}]$$

Now

$$H(e^{-j\omega T}) = \sum_{n=-\infty}^{\infty} h(nT)e^{j\omega nT} = \left[\sum_{n=-\infty}^{\infty} h(nT)e^{-j\omega nT}\right]^* = H^*(e^{j\omega T})$$

and if we let

$$H(e^{j\omega T}) = M(\omega)e^{j\theta(\omega)}$$

where
$$M(\omega) = |H(e^{j\omega T})| \quad \text{and} \quad \theta(\omega) = \arg H(e^{j\omega T})$$
then the steady-state response of the filter can be expressed as
$$\tilde{y}(nT) = M(\omega)\sin[\omega nT + \theta(\omega)]$$
Clearly, the effect of a digital filter on a sinusoidal excitation, like that of an analog filter, is to introduce a *gain* $M(\omega)$ and a *phase shift* $\theta(\omega)$. Therefore, a digital filter, like an analog filter, can be represented in the frequency domain by an *amplitude response* and a *phase response*. The main difference between analog and digital filters is that in the first case, the continuous-time transfer function $H(s)$ is evaluated on the imaginary axis of the s plane, whereas in the second case, the discrete-time transfer function $H(z)$ is evaluated on the unit circle $|z| = 1$ of the z plane.

3.5.2 Graphical Method for Amplitude and Phase Responses

For a transfer function expressed in terms of its zeros and poles as in Eq. (3.3)
$$H(e^{j\omega T}) = M(\omega)e^{j\theta(\omega)} = \frac{H_0 \prod_{i=1}^{N}(e^{j\omega T} - z_i)}{\prod_{i=1}^{N}(e^{j\omega T} - p_i)^{m_i}} \tag{3.14}$$
and by letting
$$e^{j\omega T} - z_i = M_{z_i} e^{j\psi_{z_i}} \tag{3.15}$$
$$e^{j\omega T} - p_i = M_{p_i} e^{j\psi_{p_i}} \tag{3.16}$$
we obtain
$$M(\omega) = \frac{|H_0| \prod_{i=1}^{N} M_{z_i}}{\prod_{i=1}^{N} M_{p_i}^{m_i}} \tag{3.17}$$
$$\theta(\omega) = \arg H_0 + \sum_{i=1}^{N} \psi_{z_i} - \sum_{i=1}^{N} m_i \psi_{p_i} \tag{3.18}$$
where $\arg H_0 = \pi$ if H_0 is negative. Thus $M(\omega)$ and $\theta(\omega)$ can be determined graphically by using the following procedure:

1. Mark the zeros and poles of the filter in the z plane.
2. Draw the unit circle.
3. Draw the phasor $e^{j\omega T}$ where ω is the frequency of interest.
4. Draw a phasor of the type in Eq. (3.15) for each simple zero of $H(z)$.
5. Draw m_i phasors of the type in Eq. (3.16) for each pole of order m_i.
6. Measure the magnitudes and angles of the phasors in steps 4 and 5 and use Eqs. (3.17) and (3.18) to calculate the gain $M(\omega)$ and phase shift $\theta(\omega)$, respectively.

The amplitude and phase responses of a filter can be determined by repeating the above procedure for frequencies $\omega = \omega_1, \omega_2, \ldots$ in the range 0 to π/T. This method of analysis is illustrated in Fig. 3.6 for a second-order filter with simple zeros and poles.

It should be mentioned that the modern approach for the analysis of digital filters is through the use of digital computers. Nevertheless, the above graphical method is of interest and merits consideration for two reasons. First, it illustrates some of the fundamental properties of digital filters. Second, it provides a certain degree of intuition about the expected amplitude or phase response of a filter. For example, if a filter has a pole close to the unit circle, then as ωT approaches the angle of the pole, the magnitude of the phasor from the pole to the unit circle decreases rapidly to a very small value and then increases as ωT increases above the angle of the pole. As a result, the amplitude response will exhibit a large peak in that frequency range. On the other hand, a zero close to the unit circle will lead to a notch in the amplitude response when ωT is equal to the angle of the zero.

3.5.3 Periodicity of Frequency Response

Points A and C in Fig. 3.6 correspond to frequencies 0 and π/T, and one complete revolution of $e^{j\omega T}$ about the origin corresponds to a frequency increment of $\omega_s =$

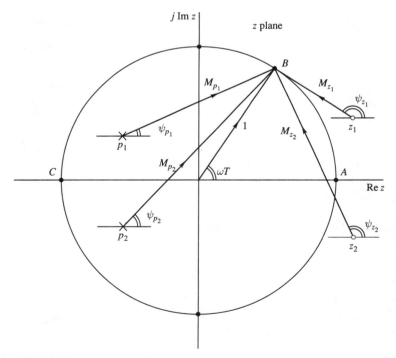

FIGURE 3.6
Phasor diagram for a second-order recursive filter.

$2\pi/T$ rad/s. Since T is the period between samples, ω_s is said to be the *sampling frequency*.

If phasor $e^{j\omega T}$ in Fig 3.6 is rotated k complete revolutions, the values of $M(\omega)$ and $\theta(\omega)$ will obviously remain unchanged, and as a result

$$H(e^{j(\omega+k\omega_s)T}) = H(e^{j\omega T})$$

In effect, the *frequency response*, namely $H(e^{j\omega T})$, is a *periodic function* of frequency with a period ω_s. The physical reason for this is easy to identify. Discrete-time signals $\sin \omega nT$ and $\sin[(\omega+k\omega_s)nT]$ are numerically identical, as demonstrated in Fig. 3.7, and therefore the filter must respond to both of them in exactly the same way.

A typical set of amplitude and phase responses is illustrated in Fig. 3.8. In future we shall be concerned almost exclusively with the fundamental period of $H(e^{j\omega T})$, which extends from $-\omega_s/2$ to $\omega_s/2$. This frequency range is called the *baseband*. The frequency $\omega_s/2$, which corresponds to point C in Fig. 3.6, is often called the *Nyquist frequency*. In later chapters we shall be dealing with lowpass, highpass, bandpass, and bandstop filters. In digital filters these terms are defined in relation to the baseband.

Example 3.7. The constants in Fig. 3.9 are

$$A_0 = 0.4 \qquad A_1 = 0.303 \qquad A_2 = 0.0935$$

Find the amplitude and phase responses of the filter if $\omega_s = 2$ rad/s.

Solution. The transfer function of the filter is

$$H(z) = \frac{A_2 z^2 + A_1 z + A_0 + A_1 z^{-1} + A_2 z^{-2}}{z^2}$$

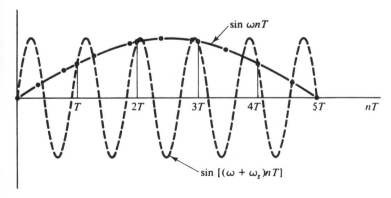

FIGURE 3.7
Plots of $\sin(\omega nT)$ and $\sin[(\omega+\omega_s)nT]$ versus nT.

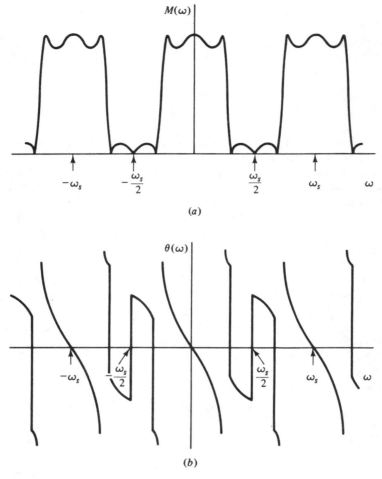

FIGURE 3.8
Typical frequency response for a digital filter: (*a*) amplitude response, (*b*) phase response.

Hence
$$H(e^{j\omega T}) = \frac{A_2(e^{j2\omega T} + e^{-j2\omega T}) + A_1(e^{j\omega T} + e^{-j\omega T}) + A_0}{e^{j2\omega T}}$$
$$= \frac{2A_2 \cos 2\omega T + 2A_1 \cos \omega T + A_0}{e^{j2\omega T}}$$

and so
$$M(\omega) = |2A_2 \cos 2\omega T + 2A_1 \cos \omega T + A_0|$$
$$\theta(\omega) = \theta_N - 2\omega T$$

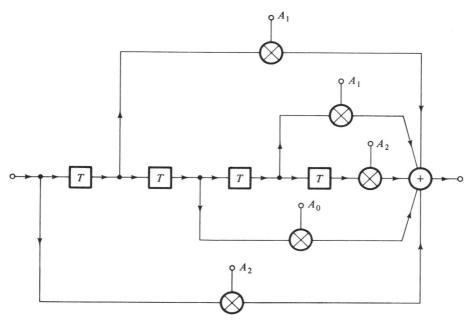

FIGURE 3.9
Fourth-order, nonrecursive filter (Example 3.7).

where $\qquad \theta_N = \begin{cases} 0 & \text{if } 2A_2 \cos 2\omega T + 2A_1 \cos \omega T + A_0 \geq 0 \\ \pi & \text{otherwise} \end{cases}$

The amplitude and phase responses are plotted in Fig. 3.10.

3.6 AMPLITUDE AND DELAY DISTORTION

In practice, a digital filter can distort the information content of the signal to be processed as is demonstrated below.

Consider a digital filter characterized by a transfer function $H(z)$ and assume that its input and output signals are $x(nT)$ and $y(nT)$, respectively. The frequency response of the filter is given by Eq. (3.14) where $M(\omega)$ and $\theta(\omega)$ are the gain and phase shift at frequency ω. The group delay of the filter at frequency ω is defined as

$$\tau(\omega) = -\frac{d\theta(\omega)}{d\omega}$$

and $\tau(\omega)$ as a function of ω is usually referred to as the *delay characteristic*.

The *frequency spectrum* of a discrete-time signal is its z transform evaluated on the unit circle of the z plane. Like the frequency response of a digital filter, it is a periodic function of ω with period ω_s. Assume that the information content of $x(nT)$

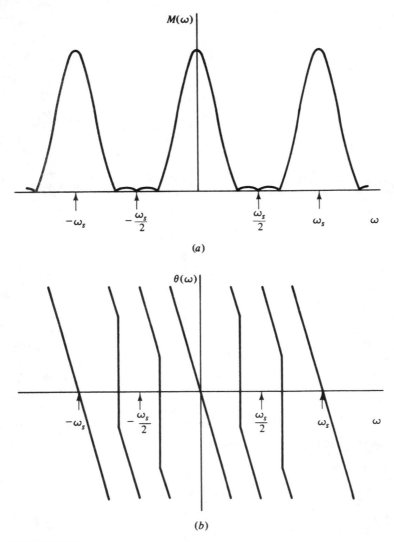

FIGURE 3.10
Frequency response (Example 3.7): (a) amplitude response, (b) phase response.

is concentrated in frequency band B given by the set

$$B = \{\omega : \omega_L \leq \omega \leq \omega_H\}$$

and that its frequency spectrum is zero elsewhere. If

$$M(\omega) = G_0 \quad \text{for } \omega \in B \tag{3.19}$$

and

$$\theta(\omega) = -\tau_g \omega + \theta_0 \quad \text{for } \omega \in B \tag{3.20}$$

or

$$\tau(\omega) = \tau_g \quad \text{for } \omega \in B$$

where τ_g is a constant, then the frequency spectrum of the output signal $y(nT)$ can be obtained from Eqs. (3.1), (3.14), (3.19), and (3.20) as

$$Y(e^{j\omega T}) = H(e^{j\omega T})X(e^{j\omega T})$$

$$= G_0 e^{-j\omega \tau_g + j\theta_0} X(e^{j\omega T})$$

Now if $\tau_g = mT$ where m is a constant, we have

$$Y(z) = G_0 e^{j\theta_0} z^{-m} X(z)$$

and from Theorems 2.5 and 2.6, we deduce

$$y(nT) = G_0 e^{j\theta_0} x(nT - mT)$$

In effect, if the amplitude response of the filter is flat and its phase response is a linear function of ω (i.e., the delay characteristic is flat) in frequency band B, then the output signal is a delayed replica of the input signal except that a gain G_0 and a phase shift θ_0 are introduced.

If the amplitude response of the filter is not flat in band B, then amplitude distortion will be introduced since different frequency components of the signal will be amplified by different amounts. On the other hand, if the delay characteristic is not flat, different frequency components will be delayed by different amounts, and delay (or phase) distortion will be introduced. Amplitude distortion can be quite objectionable in practice. Consequently, in each frequency band that carries information, the amplitude response is required to be constant to within a prescribed tolerance. If the ultimate receiver of the signal is the human ear, e.g., when a speech or music signal is to be processed, delay distortion is quite tolerable. However, in other applications it can be as objectionable as amplitude distortion, and the delay characteristic is required to be fairly flat. Applications of this type include data transmission where the signal is to be interpreted by digital hardware and image processing where the signal is used to reconstruct an image which is to be interpreted by the human eye.

3.7 INTRODUCTION TO THE DESIGN PROCESS

The design of digital filters comprises four general steps, as follows:

1. Approximation
2. Realization
3. Study of arithmetic errors
4. Implementation

The *approximation step* is the process of generating a transfer function that satisfies the desired specifications, which may concern the amplitude, phase, and possibly the time-domain response of the filter. The available methods for the solution of the approximation problem can be classified as *direct* or *indirect*. In direct methods, the problem is solved directly in the z domain. In indirect methods, a continuous-time transfer function is first obtained and then converted into a corresponding discrete-time transfer function. Nonrecursive filters are always designed through direct methods whereas recursive filters can be designed either through direct or indirect methods. Approximation methods can also be classified as *closed-form* or *iterative*. In closed-form methods, the problem is solved through a small number of design steps using a set of closed-form formulas. In iterative methods, an initial solution is assumed and, through the application of optimization methods, a series of progressively improved solutions are obtained until some design criterion is satisfied. In general, the designer is interested in simple and reliable approximation methods that yield precise designs with the minimum amount of computation.

The *synthesis* of a digital filter is the process of converting the transfer function or some other characterization of the filter into a network. This process is also referred to as the *realization step*. The network obtained is said to be the *realization* of the transfer function. As for approximation methods, realization methods can be classified as *direct* or *indirect*. In direct methods, the transfer function is put in some form that allows the identification of an interconnection of elemental digital-filter subnetworks. In indirect methods, an analog-filter network is converted into a topologically related digital-filter network. Many realization methods have been proposed in the past which lead to structures of varying complexity and properties. The designer is usually interested in realizations which require the minimum number of unit delays, adders, and multipliers, and which are not seriously affected by the use of finite-precision arithmetic in the implementation.

During the approximation step the coefficients of the transfer function are determined to a high degree of precision. In practice, however, digital hardware have a finite precision which depends on: the length of registers used to store numbers; the type of number system used (e.g., signed-magnitude, two's complement); the type of arithmetic used (e.g., fixed-point or floating-point), etc. Consequently, filter coefficients must be quantized (e.g., rounded or truncated) before they can be stored in registers. When the transfer function coefficients are quantized, errors are introduced in the amplitude and phase responses of the filter. In extreme cases, the required specifications can actually be violated. Similarly, signals to be processed, as well as the internal signals of a digital filter (e.g., the products generated by multipliers), must be quantized. Since errors introduced by the quantization of signals are actually sources of noise (see Chap. 11), they can have a dramatic effect on the performance of the filter. Under these circumstances, the design process cannot be deemed to be complete until the *effects of arithmetic errors* on the performance of the filter are investigated.

The *implementation* of a digital filter can assume two forms: *software* or *hardware*. In the first case, implementation involves the simulation of the filter network on a general-purpose digital computer, workstation, or DSP chip. In the second case, it involves the conversion of the filter network into a dedicated piece of hardware. The choice of implementation is usually critically dependent on the application at hand. In *nonreal-time* applications where a record of the data to be processed is available, a software implementation may be entirely satisfactory. In *real-time* applications, however, where data must be processed at a very high rate (e.g., in communication systems), a hardware implementation is mandatory. Often the best engineering solution might be partially in terms of software and partially in terms of hardware, since software and hardware are highly exchangeable nowadays.

The design of digital filters may often involve steps that do not appear explicitly in the above list. For example, if a digital filter is required to process continuous-time signals, the *effect of interfacing devices* (e.g., analog-to-digital and digital-to-analog converters), on the accuracy of processing must be investigated.

These design aspects occupy a considerable proportion of the remaining part of the book. Direct and indirect realization methods are considered in Chaps. 4 and 12, respectively. Approximation methods for analog filters are treated in Chap. 5 and applied to the design of recursive filters in Chaps. 7 and 8 after the relation between analog and digital filters in established in Chap. 6. Closed-form approximation methods for the design of nonrecursive filters are introduced in Chap. 9. Iterative approximations for recursive and nonrecursive filters are examined in Chaps. 14 and 15, respectively. Methods for the study of quantization effects are surveyed in Chap. 10 and applied to digital filters in Chap. 11. The implementation aspects of digital filters are considered in Chaps. 4, 12, 13, and 16. The application of digital filters for the processing of analog signals and the effects of interfacing are discussed in Chap. 6.

REFERENCES

1. F. E. Hohn, *Elementary Matrix Algebra*, MacMillan, London, 1964.
2. E. I. Jury, *Theory and Application of the z-Transform Method*, Wiley, New York, 1964.
3. L. de Pian, *Linear Active Network Theory*, Prentice-Hall, Englewood Cliffs, N.J., 1962.
4. R. J. Schwarz and B. Friedland, *Linear Systems*, McGraw-Hill, New York, 1965.
5. E. I. Jury, *Inners and Stability of Dynamical Systems*, Wiley-Interscience, New York, 1974.
6. N. K. Bose, *Digital Filters*, North-Holland, New York, 1985.
7. E. I. Jury, "A Simplified Stability Criterion for Linear Discrete Systems," *Proc. IRE*, vol. 50, pp. 1493–1500, June 1962.
8. B. D. O. Anderson and E. I. Jury, "A Simplified Schur-Cohn Test," *IEEE Trans. Automatic Control*, vol. AC-18, pp. 157–163, April 1973.
9. M. Marden, *The Geometry of the Zeros of a Polynomial in a Complex Variable*, Amer. Math. Soc., New York, pp. 152–157, 1949.
10. H. Freeman, *Discrete-Time Systems*, Wiley, New York, 1965.
11. S. J. Hammarling, "Numerical Solution of the Stable, Non-Negative Definite Lyapunov Equation," *IMA J. Numer. Anal.* vol. 2, pp. 303–323, 1982.

PROBLEMS

3.1. Derive the transfer functions of the filters in Fig. P1.4*a* and *b*.
3.2. Derive the transfer functions of the filters in Fig. P1.5*a* and *b*.
3.3. Derive the transfer functions of the filters in Figs. P1.6 and P1.7.

3.4. A recursive filter is characterized by the equations
$$y(nT) = y_1(nT) + \frac{7}{4}y(nT - T) - \frac{49}{32}y(nT - 2T)$$
$$y_1(nT) = x(nT) + \frac{1}{2}y_1(nT - T)$$
Obtain its transfer function.

3.5. A filter is represented by the state-space equations
$$\mathbf{q}(nT + T) = \mathbf{A}\mathbf{q}(nT) + \mathbf{b}x(nT)$$
$$y(nT) = \mathbf{c}^T\mathbf{q}(nT) + dx(nT)$$
where
$$\mathbf{A} = \begin{bmatrix} 0 & 1 & 0 \\ 0 & 0 & 1 \\ -\frac{1}{2} & -\frac{1}{2} & \frac{1}{2} \end{bmatrix} \quad \mathbf{b} = \begin{bmatrix} 0 \\ 0 \\ 1 \end{bmatrix} \quad \mathbf{c}^T = \begin{bmatrix} \frac{7}{2} & \frac{5}{2} & \frac{5}{2} \end{bmatrix} \quad d = 1$$
Deduce its transfer function.

3.6. Show that
$$H(z) = \frac{\sum_{i=0}^{M} a_i z^{M-i}}{z^N + \sum_{i=1}^{N} b_i z^{N-i}}$$
represents a causal filter only if $M \leq N$.

3.7. (a) Find the impulse response of the filter shown in Fig. 3.2.
(b) Repeat part (a) for the filter of Fig. 3.4.

3.8. Obtain the impulse response of the filter in Prob. 3.4. Sketch the response.

3.9. Starting from first principles, show that
$$H(z) = \frac{z}{\left(z - \frac{1}{4}\right)^4}$$
represents a stable filter.

3.10. (a) A recursive filter is represented by
$$H(z) = \frac{z^6}{6z^6 + 5z^5 + 4z^4 + 3z^3 + 2z^2 + z + 1}$$
Check the filter for stability.
(b) Repeat part (a) if
$$H(z) = \frac{(z+2)^2}{6z^6 + 5z^5 - 4z^4 + 3z^3 + 2z^2 + z + 1}$$

3.11. (a) Check the filter of Fig. P3.11a for stability.
(b) Check the filter of Fig. P3.11b for stability.

3.12. (a) A digital filter is characterized by the difference equation
$$y(nT) = x(nT) - \frac{1}{2}x(nT - T) - \frac{1}{3}x(nT - 2T)$$
$$- \frac{1}{4}x(nT - 3T) - \frac{1}{5}x(nT - 4T)$$
By using appropriate tests, check the stability of the filter.

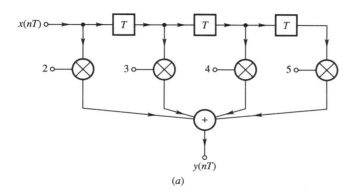

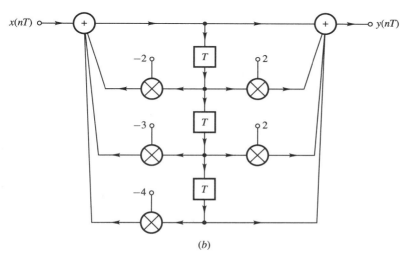

FIGURE P3.11 (*a*) and (*b*)

(b) Repeat part (*a*) for the filter represented by the equation

$$y(nT) = x(nT) - \frac{1}{2}y(nT - T) - \frac{1}{3}y(nT - 2T)$$
$$- \frac{1}{4}y(nT - 3T) - \frac{1}{5}y(nT - 4T)$$

3.13. Obtain (*a*) the transfer function, (*b*) the impulse response, and (*c*) the necessary condition for stability for the filter of Fig. P3.13. The constants m_1 and m_2 are given by

$$m_1 = 2r\cos\theta \quad \text{and} \quad m_2 = -r^2$$

3.14. A digital filter is characterized by the transfer function

$$H(z) = \frac{z^4}{4z^4 + 3z^3 + mz^2 + z + 1}$$

Find the range of *m* that will result in a stable filter.

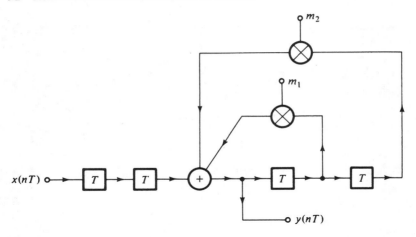

FIGURE P3.13

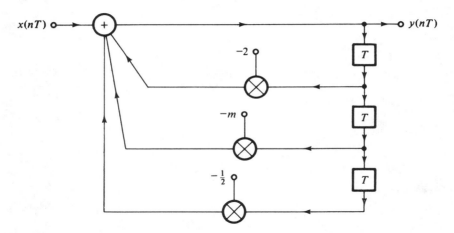

FIGURE P3.15

3.15. Find the permissible range for m in Fig. P3.15 if the filter is to be stable.

3.16. A digital filter is characterized by the transfer function

$$H(z) = \frac{1}{z^2 + 0.25}$$

Derive an expression for its unit-step response.

3.17. A digital filter is characterized by the transfer function

$$H(z) = \frac{32z}{z - 0.5}$$

(a) Find the response of the filter at $t = 4T$ using the convolution summation if the excitation is

$$x(nT) = (5+n)u(nT)$$

(b) Give a graphical construction for the convolution in part (a) indicating relevant quantities.

3.18. Repeat part (a) of Prob. 3.17 using the inversion formula in Eq. (2.8).

3.19. A digital filter is characterized by the transfer function

$$H(z) = \frac{z^2 - z + 1}{z^2 - z + 0.5}$$

Obtain its unit-step response.

3.20. Find the unit-step response of the filter shown in Fig. 3.2.

3.21. Find the unit-ramp response of the filter shown in Fig. 3.4 if $T = 1$.

3.22. The input excitation in Fig. 3.2 is

$$x(nT) = \begin{cases} n & \text{for } 0 \leq n \leq 2 \\ 4-n & \text{for } 2 < n \leq 4 \\ 0 & \text{for } n > 4 \end{cases}$$

Determine the response for $0 \leq n \leq 5$ by using the z transform.

3.23. Repeat Prob. 1.18 by using the z transform. For each of the three cases deduce the exact frequency of ringing if $T = 1$ and also the steady-state value of the response.

3.24. A filter has a transfer function

$$H(z) = \frac{1}{z^2 + \frac{1}{4}}$$

(a) Find the response if

$$x(nT) = u(nT)\sin \omega nT$$

(b) Deduce the steady-state sinusoidal response.

3.25. A filter is characterized by

$$H(z) = \frac{1}{(z-r)^2}$$

where $|r| < 1$. Show that the steady-state sinusoidal response is given by

$$y(nT) = M(\omega)\sin[\omega nT + \theta(\omega)] \quad \text{where} \quad \begin{aligned} M(\omega) &= |H(e^{j\omega T})| \\ \theta(\omega) &= \arg H(e^{j\omega T}) \end{aligned}$$

3.26. Fig. P3.26 depicts a nonrecursive filter.
(a) Derive an expression for its amplitude response.
(b) Derive an expression for its phase response.
(c) Calculate the gain in dB at $\omega = 0$, $\omega_s/4$, and $\omega_s/2$ (ω_s is the sampling frequency in rad/s).
(d) Calculate the phase-shift in degrees at $\omega = 0$, $\omega_s/4$, and $\omega_s/2$.

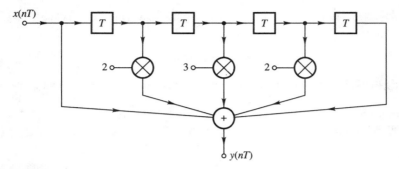

FIGURE P3.26

3.27. The discrete-time signal

$$x(nT) = u(nT) \sin \omega nT$$

is applied to the input of the filter in Fig. P3.27.
(a) Give the steady-state time-domain response of the filter.
(b) Derive an expression for the amplitude response.
(c) Derive an expression for the phase response.
(d) Calculate the gain and phase shift for $\omega = \pi/4T$ rad/s.

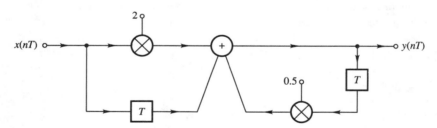

FIGURE P3.27

3.28. (a) Show that

$$M^2(\omega) = H(z)H(z^{-1})\big|_{z=e^{j\omega T}}$$

(b) By using this relation show that

$$H(z) = \frac{1 - az + bz^2}{b - az + z^2}$$

represents an allpass filter, i.e., one with constant gain at all frequencies.

3.29. Figure P3.29 shows a nonrecursive filter.
(a) Derive expressions for the gain and phase shift.
(b) Determine the transmission zeros of the filter, i.e., zero-gain frequencies.
(c) Sketch the amplitude and phase responses.

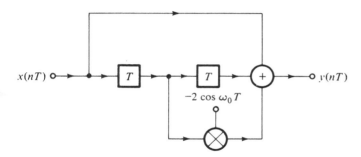

FIGURE P3.29

3.30. Show that the equation

$$y(nT) = x(nT) + 2x(nT - T) + 3x(nT - 2T) + 4x(nT - 3T) + 3x(nT - 4T)$$
$$+ 2x(nT - 5T) + x(nT - 6T)$$

represents a constant-delay filter.

3.31. Derive expressions for the gain and phase shift of the filter shown in Fig. 3.4.

3.32. Table P3.32 gives the transfer function coefficients of four filters labeled A to D. By plotting $20 \log M(\omega)$ versus ω in the range 0 to 5.0 rad/s, identify a lowpass, a highpass,

TABLE P3.32

Filter	i	a_{0i}	a_{1i}	b_{0i}	b_{1i}
A	1	2.222545×10^{-1}	-4.445091×10^{-1}	4.520149×10^{-2}	1.561833×10^{-1}
	2	3.085386×10^{-1}	-6.170772×10^{-1}	4.509715×10^{-1}	2.168171×10^{-1}
		$H_0 = 1.0$			
B	1	5.490566	9.752955	7.226400×10^{-1}	4.944635×10^{-1}
	2	5.871082×10^{-1}	-1.042887	7.226400×10^{-1}	-4.944634×10^{-1}
		$H_0 = 2.816456 \times 10^{-2}$			
C	1	1.747744×10^{-1}	1.517270×10^{-8}	5.741567×10^{-1}	1.224608
	2	1.399382	1.214846×10^{-7}	5.741567×10^{-1}	-1.224608
		$H_0 = 8.912509 \times 10^{-1}$			
D	1	9.208915	1.561801×10	5.087094×10^{-1}	-1.291110
	2	2.300089	1.721670	8.092186×10^{-1}	-1.069291
		$H_0 = 6.669086 \times 10^{-4}$			

a bandpass, and a bandstop filter. Each filter has a transfer function of the form

$$H(z) = H_0 \prod_{i=1}^{2} \frac{a_{0i} + a_{1i}z + a_{0i}z^2}{b_{0i} + b_{1i}z + z^2}$$

and the sampling frequency is 10 rad/s in each case.

3.33. Show that the gain and phase shift in a digital filter satisfy the relations

$$M(\omega_s - \omega) = M(\omega) \quad \text{and} \quad \theta(\omega_s - \omega) = -\theta(\omega)$$

CHAPTER 4

REALIZATION

4.1 INTRODUCTION

Once the required specifications for the application at hand are formulated, a suitable transfer function is obtained using one of the many approximation methods. The *realization* of the transfer function is then undertaken. It turns out that the realization step is much easier than the approximation step. For this reason, we consider these design tasks in the reverse order.

Two types of realization methods have been proposed over the past twenty years, namely, *direct* and *indirect*. In direct methods, the transfer function is put in some form that allows the identification of an interconnection of elemental digital-filter subnetworks of low order. The most frequently used realization methods of this class are [1–5]:

1. Direct
2. Direct canonic
3. State-space
4. Ladder
5. Lattice
6. Parallel
7. Cascade

In indirect methods, on the other hand, an analog-filter network is converted into a topologically related digital-filter network through the application of network-theoretic concepts in conjunction with some simple transformations [6–9].

Digital-filter structures obtained by different methods can differ quite significantly with respect to complexity, number of elements, and their properties. One structure might require a large number of multipliers and yet be relatively insensitive to coefficient quantization errors, and a second structure might be economical in terms of elements but generate parasitic oscillations when signals are quantized, and so on.

In this chapter, the various direct realization methods are described in detail. Then the *implementation* of some of the structures in terms of systolic arrays is considered. The chapter concludes with a discussion of some important topological properties of digital-filter networks in general, which are often quite useful. It is shown, for example, that given a digital-filter structure, another equivalent structure, called the *transpose structure*, can be readily derived.

Indirect realization methods, which are known to yield some interesting structures, and the merits and demerits of the various types of structures will be considered in Chap. 12.

4.2 DIRECT REALIZATION

A filter characterized by an Nth-order transfer function can be represented by the equation

$$\frac{Y(z)}{X(z)} = H(z) = \frac{N(z)}{D(z)} = \frac{N(z)}{1 + D'(z)} \tag{4.1}$$

where

$$N(z) = \sum_{i=0}^{N} a_i z^{-i} \tag{4.2}$$

and

$$D'(z) = \sum_{i=1}^{N} b_i z^{-i}$$

From Eq. (4.1), we can write

$$Y(z) = N(z)X(z) - D'(z)Y(z)$$

or

$$Y(z) = U_1(z) + U_2(z)$$

where
$$U_1(z) = N(z)X(z) \tag{4.3}$$

and
$$U_2(z) = -D'(z)Y(z)$$

and hence the realization of $H(z)$ can be broken down into the realization of two simpler transfer functions, $N(z)$ and $-D'(z)$, as illustrated in Fig. 4.1.

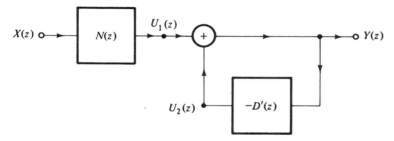

FIGURE 4.1
Decomposition of $H(z)$ into two simpler transfer functions.

Consider the realization of $N(z)$. From Eqs. (4.2) and (4.3)

$$U_1(z) = [a_0 + z^{-1}N_1(z)]X(z)$$

where

$$N_1(z) = \sum_{i=1}^{N} a_i z^{-i+1}$$

and thus $N(z)$ can be realized by using a multiplier with parameter a_0 in parallel with a network characterized by $z^{-1}N_1(z)$. In turn, $z^{-1}N_1(z)$ can be realized by using a unit delay in cascade with a network characterized by $N_1(z)$. Since the unit delay may precede or follow the realization of $N_1(z)$, two possibilities exist for $N(z)$, as depicted in Fig. 4.2.

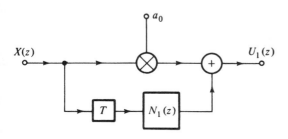

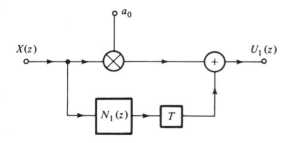

FIGURE 4.2
Two realizations of $N(z)$.

The above procedure can now be applied to $N_1(z)$. That is, $N_1(z)$ can be expressed as

$$N_1(z) = a_1 + z^{-1}N_2(z) \qquad \text{where} \quad N_2(z) = \sum_{i=2}^{N} a_i z^{-i+2}$$

and as before two networks can be obtained for $N_1(z)$. Clearly, there are four networks for $N(z)$. Two of them are shown in Fig. 4.3.

This cycle of events can be repeated N times, whereupon $N_N(z)$ will reduce to a single multiplier. In each cycle of the procedure there are two possibilities, and since there are N cycles, a total of 2^N distinct networks can be deduced for $N(z)$.

Three of the possibilities are depicted in Fig. 4.4a to c. These structures are obtained by placing the unit delays consistently at the left in the first case, consistently at the right in the second case, and alternately at the left and right in the third case.

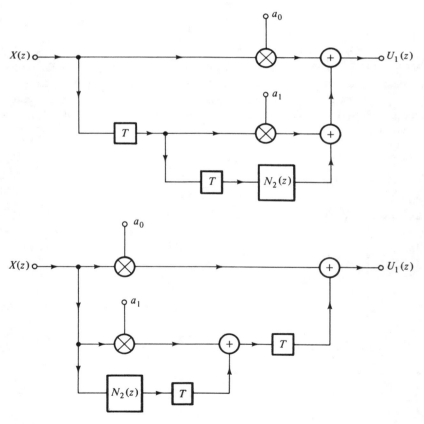

FIGURE 4.3
Two of four possible realizations of $N(z)$.

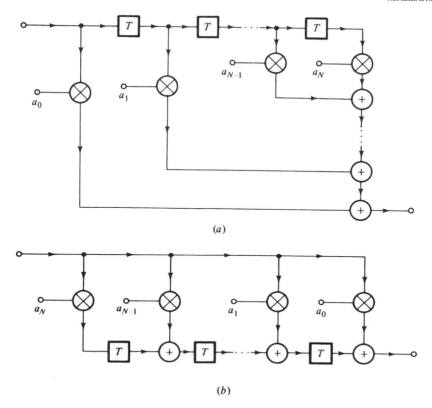

FIGURE 4.4a and 4.4b
Four possible realizations of $N(z)$.

Note that in the structure of Fig. 4.4a, the adders accumulate the products generated by the multipliers starting at the right and ending at the left of the structure. Since addition is associative, the products can be added in any other order. For example, if they are added starting at the left and ending at the right, the structure of Fig. 4.4d is obtained, which is quite useful (see Sec. 4.9).

$-D'(z)$ can be realized in exactly the same way. Networks for $-D'(z)$ can be obtained by replacing $a_0, a_1, a_2, \ldots$ in Fig. 4.4 by $0, -b_1, -b_2, \ldots$.

Finally, the realization of $H(z)$ can be accomplished by interconnecting realizations of $N(z)$ and $-D'(z)$ according to Fig. 4.1.

Example 4.1. Realize the transfer function

$$H(z) = \frac{a_0 + a_1 z^{-1} + a_2 z^{-2}}{1 + b_1 z^{-1} + b_2 z^{-2}}$$

Solution. Two realizations of $H(z)$ can be readily obtained from Fig. 4.4a and b, as shown in Fig. 4.5a and b.

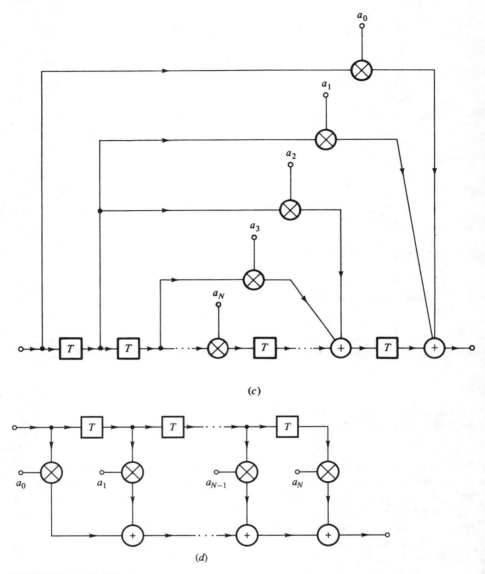

FIGURE 4.4c and 4.4d
Four possible realizations of $N(z)$.

4.3 DIRECT CANONIC REALIZATION

A digital network is said to be *canonic* if the number of unit delays employed is equal to the order of the transfer function.

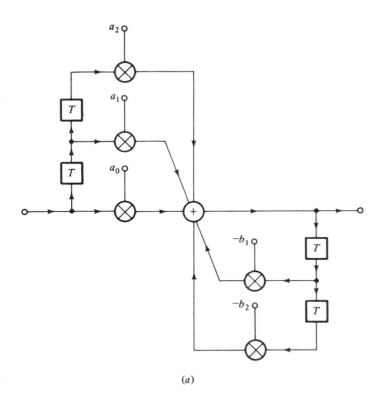

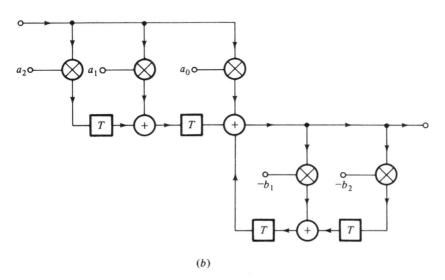

FIGURE 4.5
Two realizations of $H(z)$ (Example 4.1).

Equation (4.1) can be expressed as

$$Y(z) = N(z)Y'(z) \qquad \text{where } Y'(z) = X(z) - D'(z)Y'(z)$$

in which case $H(z)$ can be realized as in Fig. 4.6 by using the network of Fig. 4.4a for $N(z)$ as well as $-D'(z)$. Evidently, the signals at nodes $A', B', \ldots$ are equal to the corresponding signals at nodes $A, B, \ldots$. Therefore, nodes $A', B', \ldots$ can be merged with nodes $A, B, \ldots$, respectively, and one set of unit delays can be eliminated to yield a canonic realization.

4.4 STATE-SPACE REALIZATION

Another approach to the realization of digital filters is to start with a *state-space* characterization. For an Nth-order filter, Eqs. (1.17) and (1.18) give

$$q_i(nT + T) = \sum_{j=1}^{N} a_{ij} q_j(nT) + b_i x(nT) \qquad \text{for } i = 1, 2, \ldots, N \qquad (4.4)$$

and

$$y(nT) = \sum_{j=1}^{N} c_j q_j(nT) + d_0 x(nT) \qquad (4.5)$$

respectively. By assigning flow-graph nodes to $x(nT)$, $y(nT)$, $q_i(nT)$, and $q_i(nT+T)$ for $i = 1, 2, \ldots, N$, the state-space flow graph of Fig. 4.7 can be obtained, which can be readily converted into a network.

4.5 LADDER REALIZATION

Another realization method is one proposed by Mitra and Sherwood [4]. This is based on the configuration of Fig. 4.8a and leads to a class of *ladder* structures.

With $N = 4$ in Fig. 4.8a, straightforward analysis yields

$$Y(z) = H_1(z) Y'(z) + H_2(z) X(z) \qquad (4.6)$$

where
$$H_1(z) = \frac{N_1(z)}{D_1(z)} = \frac{m_4 m_3 m_2 z^3 + (m_4 + m_2) z}{m_4 m_3 m_2 m_1 z^4 + (m_4 m_3 + m_4 m_1 + m_2 m_1) z^2 + 1}$$

$$= \cfrac{1}{m_1 z + \cfrac{1}{m_2 z + \cfrac{1}{m_3 z + \cfrac{1}{m_4 z}}}}$$

$$H_2(z) = \frac{1}{D_1(z)}$$

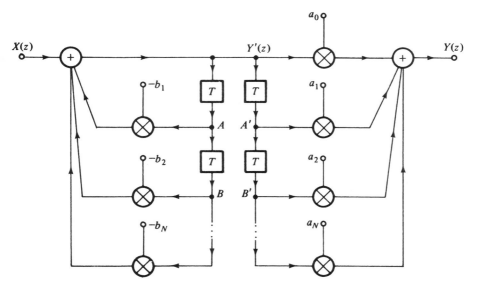

FIGURE 4.6
Derivation of the canonic realization of $H(z)$.

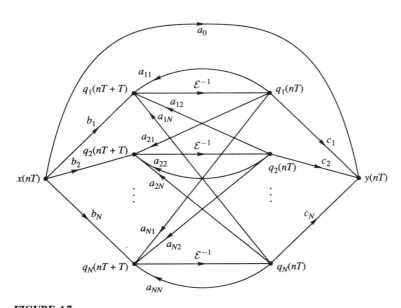

FIGURE 4.7
State-space flow graph.

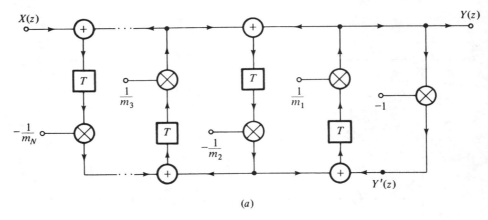

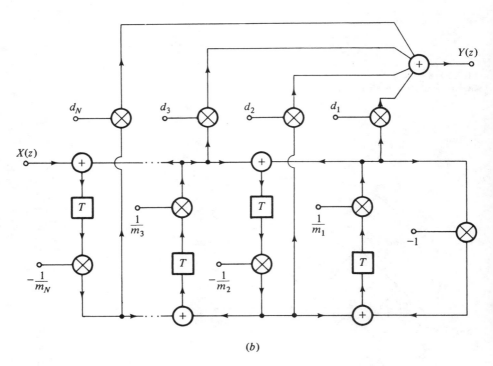

FIGURE 4.8
Ladder realization of (a) $H(z) = (-1)^k/D(z)$ and (b) $H(z) = N(z)/D(z)$.

and since
$$Y'(z) = -Y(z)$$

Eq. (4.6) gives
$$\frac{Y(z)}{X(z)} = H(z) = \frac{H_2(z)}{1 + H_1(z)} \tag{4.7}$$

Similarly, for any value of N it can be shown that

$$H_1(z) = \frac{N_1(z)}{D_1(z)} = \cfrac{1}{m_1 z + \cfrac{1}{m_2 z + \cfrac{1}{\ddots + \cfrac{1}{m_N z}}}}$$

$$H_2(z) = \frac{(-1)^k}{D_1(z)}$$

where k is the largest integer equal to or less than $N/2$; that is,

$$(-1)^k = \begin{cases} +1 & \text{for } N = 1, 4, 5, 8, 9, \ldots \\ -1 & \text{for } N = 2, 3, 6, 7, \ldots \end{cases}$$

Let us consider the transfer function

$$H(z) = \frac{(-1)^k}{D(z)} = \frac{(-1)^k}{1 + D'(z)}$$

where

$$D'(z) = \sum_{i=1}^{N} b_i z^i$$

We can write

$$H(z) = \frac{(-1)^k}{\text{Od } D(z) + \text{Ev } D(z)} \tag{4.8}$$

where Od $D(z)$ and Ev $D(z)$ denote the odd part and even part of $D(z)$, respectively. By comparing Eqs. (4.7) and (4.8) the following identifications can be made:

$$H_1(z) = \begin{cases} \dfrac{\text{Od } D(z)}{\text{Ev } D(z)} & \text{for even } N \\[2mm] \dfrac{\text{Ev } D(z)}{\text{Od } D(z)} & \text{for odd } N \end{cases}$$

$$H_2(z) = \begin{cases} \dfrac{(-1)^k}{\text{Ev } D(z)} & \text{for even } N \\[2mm] \dfrac{(-1)^k}{\text{Od } D(z)} & \text{for odd } N \end{cases}$$

Now by expressing $H_1(z)$ as a continued-fraction expansion about infinity we have

$$H_1(z) = \cfrac{1}{c_1 z + \cfrac{1}{c_2 z + \cfrac{1}{\ddots + \cfrac{1}{c_N z}}}} \tag{4.9}$$

and therefore if

$$m_i = c_i$$

the configuration of Fig. 4.8a becomes a realization of $H(z)$.

The synthesis can be extended to any transfer functions of the form

$$H(z) = \frac{N(z)}{D(z)} \quad \text{where } N(z) = \sum_{i=0}^{N} a_i z^i$$

by modifying the basic configuration as illustrated in Fig. 4.8b. For $N = 4$ and $m_i = c_i$, analysis yields

$$\frac{Y(z)}{X(z)} = H(z) = \frac{\sum_{i=1}^{4} d_i n_i(z)}{D(z)}$$

where
$$n_1(z) = 1$$
$$n_2(z) = c_1 z + 1$$
$$n_3(z) = -(c_1 c_2 z^2 + c_2 z + 1)$$
$$n_4(z) = -[c_1 c_2 c_3 z^3 + c_2 c_3 z^2 + (c_1 + c_3)z + 1]$$

By assigning

$$N(z) = \sum_{i=1}^{4} d_i n_i(z)$$

and then equating coefficients of like powers of z, the matrix equation

$$\begin{bmatrix} -c_1 c_2 c_3 & 0 & 0 & 0 \\ -c_2 c_3 & -c_1 c_2 & 0 & 0 \\ -(c_1 + c_3) & -c_2 & c_1 & 0 \\ -1 & -1 & 1 & 1 \end{bmatrix} \begin{bmatrix} d_4 \\ d_3 \\ d_2 \\ d_1 \end{bmatrix} = \begin{bmatrix} a_3 \\ a_2 \\ a_1 \\ a_0 \end{bmatrix}$$

can be formed. The solution of this equation yields the necessary values of $d_1, d_2, \ldots$.

The realization method relies on the existence of the continued-fraction expansion in Eq. (4.9), and as a consequence it sometimes breaks down, e.g., if $H(z)$ has poles on the imaginary axis of the z plane [4].

Example 4.2. Realize the transfer function

$$H(z) = \frac{10^{-2}(-3.517 + 0.665z + 0.665z^2 - 3.517z^3)}{1 - 3.266z + 3.739z^2 - 1.53z^3}$$

using the ladder method.

Solution. $H(z)$ can be expressed as

$$H(z) = 0.02299 + \frac{-0.0582 + 0.0817z - 0.0793z^2}{1 - 3.266z + 3.739z^2 - 1.53z^3}$$

$$= 0.02299 + H'(z)$$

and hence $H(z)$ can be realized by using a multiplier in parallel with a network characterized by $H'(z)$. Since the order of $H'(z)$ is odd, we can write

$$H_1(z) = -\frac{3.739z^2 + 1}{1.53z^3 + 3.266z}$$

$$= \cfrac{1}{-0.4092z + \cfrac{1}{-1.309z + \cfrac{1}{-2.856z}}}$$

and thus

$$m_1 = c_1 = -0.4092 \qquad m_2 = c_2 = -1.309 \qquad m_3 = c_3 = -2.856$$

For $N = 3$, analysis yields

$$n_1(z) = -1$$

$$n_2(z) = -(c_1 z + 1)$$

$$n_3(z) = c_1 c_2 z^2 + c_2 z + 1$$

and so d_1, d_2, and d_3 are given by

$$\begin{bmatrix} c_1 c_2 & 0 & 0 \\ c_2 & -c_1 & 0 \\ 1 & -1 & -1 \end{bmatrix} \begin{bmatrix} d_3 \\ d_2 \\ d_1 \end{bmatrix} = \begin{bmatrix} a_2 \\ a_1 \\ a_0 \end{bmatrix}$$

or

$$d_3 = \frac{a_2}{c_1 c_2} = -0.148 \qquad d_2 = \frac{d_3 c_2 - a_1}{c_1} = -0.274$$

$$d_1 = d_3 - d_2 - a_0 = 0.184$$

4.6 LATTICE REALIZATION

Yet another realization method is the so-called *lattice realization* method of Gray and Markel [5]. This is similar to the ladder realization method of the preceding section and is based on the configuration depicted in Fig. 4.9a. The filter sections represented by the blocks in Fig. 4.9a can assume a number of distinct forms. The most basic section is the 2-multiplier first-order lattice section depicted in Fig. 4.9b.

A transfer function of the type given by Eq. (4.1) can be realized by obtaining expressions for the multiplier constants $v_0, v_1, \ldots, v_N$ and $\mu_0, \mu_1, \ldots, \mu_{N-1}$ in terms of the transfer function coefficients $a_0, a_1, \ldots, a_N$ and $1, b_1, \ldots, b_N$. This can be accomplished by defining a series of polynomials of the form

$$N_j(z) = \sum_{i=0}^{j} \alpha_{ji} z^{-i} \qquad (4.10)$$

$$D_j(z) = \sum_{i=0}^{j} \beta_{ji} z^{-i} \qquad (4.11)$$

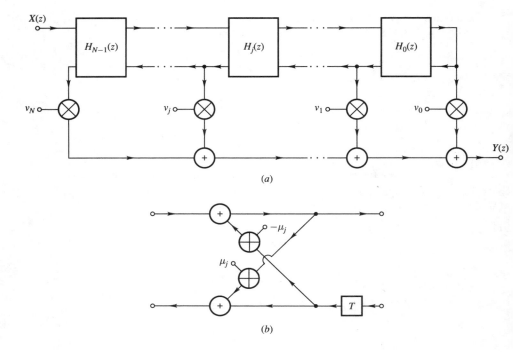

FIGURE 4.9
(a) General lattice configuration, (b) 2-multiplier filter section.

where

$$N_N(z) \equiv N(z) = \sum_{i=0}^{N} a_i z^{-i}$$

$$D_N(z) \equiv D(z) = \sum_{i=0}^{N} b_i z^{-i} \quad \text{with } b_0 = 1$$

and then using the recursive relations

$$v_j = \alpha_{jj} \tag{4.12a}$$

$$\mu_{j-1} = \beta_{jj} \tag{4.12b}$$

$$P_j(z) = D_j\left(\frac{1}{z}\right) z^{-j} \tag{4.12c}$$

$$N_{j-1}(z) = N_j(z) - v_j P_j(z) \tag{4.12d}$$

$$D_{j-1}(z) = \frac{D_j(z) - \mu_{j-1} P_j(z)}{1 - \mu_{j-1}^2} \tag{4.12e}$$

for $j = N, N-1, \ldots, 2$ and

$$v_1 = \alpha_{11} \tag{4.13a}$$

$$\mu_0 = \beta_{11} \tag{4.13b}$$

$$P_1(z) = D_1\left(\frac{1}{z}\right) z^{-1} \tag{4.13c}$$

$$N_0(z) = N_1(z) - v_1 P_1(z) \tag{4.13d}$$

$$v_0 = \alpha_{00} \tag{4.13e}$$

Example 4.3. Realize the transfer function of Example 4.1 using the lattice method.

Solution. From Eqs. (4.10) and (4.11), we can write

$$N_2(z) = \alpha_{20} + \alpha_{21} z^{-1} + \alpha_{22} z^{-2} \quad \text{with } \alpha_{2i} = a_i$$

$$D_2(z) = \beta_{20} + \beta_{21} z^{-1} + \beta_{22} z^{-2} \quad \text{with } \beta_{2i} = b_i$$

For $j = 2$, Eqs. (4.12a) to (4.12e) yield

$$v_2 = \alpha_{22}, \qquad \mu_1 = \beta_{22}$$

$$P_2(z) = \beta_{20} z^{-2} + \beta_{21} z^{-1} + \beta_{22}$$

$$N_1(z) = \alpha_{10} + \alpha_{11} z^{-1}$$

$$D_1(z) = \beta_{10} + \beta_{11} z^{-1}$$

where

$$\alpha_{10} = a_0 - a_2 b_2, \qquad \alpha_{11} = a_1 - a_2 b_1$$

$$\beta_{10} = 1, \qquad \beta_{11} = \frac{b_1}{1 + b_2}$$

Similarly from Eqs. (4.13a) to (4.13e), we have

$$v_1 = \alpha_{11}, \qquad \mu_0 = \beta_{11}$$

$$P_1(z) = \beta_{10} z^{-1} + \beta_{11}$$

$$N_0(z) = v_0 = \alpha_{00}$$

Therefore

$$v_0 = (a_0 - a_2 b_2) - \frac{(a_1 - a_2 b_1) b_1}{1 + b_2}$$

$$v_1 = a_1 - a_2 b_1, \qquad v_2 = a_2$$

$$\mu_0 = \frac{b_1}{1 + b_2}, \qquad \mu_1 = b_2$$

The 2-multiplier section of Fig. 4.9b yields structures that are canonic with respect to the number of unit delays. However, the number of multipliers can be quite large, as can be seen in Example 4.3. More economical realizations can be obtained by using *1-multiplier* first-order sections of the type shown in Fig. 4.10. Such realizations can be obtained by first realizing the transfer function in terms of 2-multiplier sections as described above and then replacing each of the 2-multiplier sections by either of the 1-multiplier sections of Fig. 4.10. The denominator multiplier constants $\mu_0, \mu_1, \ldots, \mu_{N-1}$ remain the same as before. However, the numerator multiplier constants $\nu_0, \nu_1, \ldots, \nu_N$ must be modified as

$$\tilde{\nu}_j = \frac{\nu_j}{\xi_j}$$

where

$$\xi_j = \begin{cases} 1 & \text{for } j = N \\ \prod_{i=j}^{N-1}(1 + \varepsilon_i \mu_i) & \text{for } j = 0, 1, \ldots, N-1 \end{cases}$$

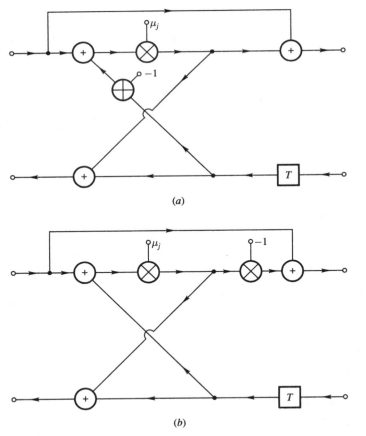

FIGURE 4.10
1-multiplier section: (*a*) for case where $\varepsilon_j = +1$, (*b*) for case where $\varepsilon_j = -1$.

Each parameter ε_i is a constant which is equal to $+1$ or -1 depending on whether the ith 2-multiplier section is replaced by the 1-multiplier section of Fig. 4.10a or that of Fig. 4.10b. The choice between the two types of sections is, in theory, arbitrary; however, in practice, it can be used to improve the performance of the structure in some respect. For example, by choosing the types of sections such that the signal levels at the internal nodes of the filter are maximized, improved signal scaling is achieved which results in increased signal-to-noise ratio (see [5] and Chap. 11).

4.7 CASCADE REALIZATION

When the transfer function coefficients are quantized, errors are introduced in the amplitude and phase responses of the filter. It turns out that when a transfer function is realized directly in terms of a single Nth-order network using any one of the methods described so far, the sensitivity of the structure to coefficient quantization increases rapidly with N. Consequently, small errors introduced by coefficient quantization give rise to large errors in the amplitude and phase responses. This problem can to some extent be overcome by realizing high-order filters as interconnections of first- and second-order networks. In this and the next section, it is shown that an arbitrary transfer function can be realized by connecting a number of first- and second-order structures in *cascade* or in *parallel*. Another approach to the reduction of coefficient quantization effects is to use the *wave* realization method, which is known to yield low-sensitivity structures. This possibility will be examined in Chap. 12.

An arbitrary transfer function can be factored into a product of first- and second-order transfer functions as

$$H(z) = \prod_{i=1}^{M} H_i(z)$$

where
$$H_i(z) = \frac{a_{0i} + a_{1i} z^{-1} + a_{2i} z^{-2}}{1 + b_{1i} z^{-1} + b_{2i} z^{-2}}$$

with $a_{2i} = b_{2i} = 0$ for a first-order transfer function.

Hence
$$Y(z) = [H_1(z) X(z)] H_2(z) \cdots H_M(z)$$
$$= [H_2(z) Y_1(z)] H_3(z) \cdots H_M(z)$$
$$\cdots\cdots\cdots\cdots\cdots\cdots\cdots\cdots\cdots$$
$$= H_M(z) Y_{M-1}(z)$$

where
$$Y_1(z) = H_1(z) X(z)$$
$$Y_i(z) = H_i(z) Y_{i-1}(z) \quad \text{for } i = 2, 3, \ldots, M-1$$

In this way, $H(z)$ can be realized by using the cascade configuration of Fig. 4.11a.

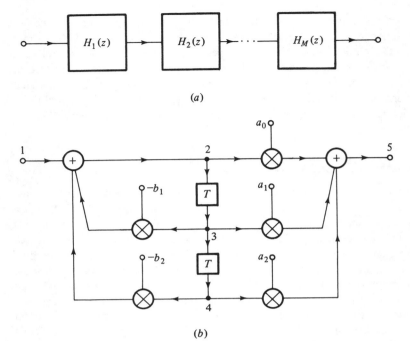

FIGURE 4.11
(a) Cascade realization of $H(z)$, (b) canonic second-order section.

The individual sections can be realized by employing any one of the methods described in this chapter. For example, using the direct canonic method yields the second-order section of Fig. 4.11b.

4.8 PARALLEL REALIZATION

Another realization comprising first- and second-order filter sections can be obtained by expanding the transfer function into partial fractions as

$$H(z) = \sum_{i=1}^{M} H_i(z)$$

where

$$H_i(z) = \frac{a_{0i} + a_{1i}z^{-1}}{1 + b_{1i}z^{-1} + b_{2i}z^{-2}}$$

In this case

$$Y(z) = \sum_{i=1}^{M} H_i(z) X(z)$$

and so the parallel configuration of Fig. 4.12 is obtained.

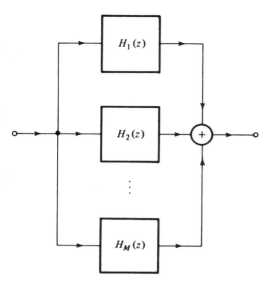

FIGURE 4.12
Parallel realization of $H(z)$.

An alternative parallel realization can be readily obtained by expanding $H(z)/z$ into partial fractions.

4.9 IMPLEMENTATION

As was stated in Sec. 3.7, the *implementation* of digital filters can assume two forms, namely, software and hardware. This classification is somewhat artificial, however, since software and hardware are nowadays highly interchangeable. In nonreal-time applications, speed is not of considerable importance and the implementation might assume the form of a computer program on a general-purpose computer, which will emulate the operation of the digital filter. On the other hand, if a digital filter is to be used in some communications system, speed is of the essence and the implementation might assume the form of a dedicated, highly specialized piece of hardware.

A hardware implementation can assume the form of a number of interconnected very-large-scale integrated (VLSI) circuit chips. Rapid progress is being made in this technology and the number of chips needed to implement a digital filter is decreasing as more and more functions are accommodated in each chip. It will soon be possible to implement entire digital filters on single chips. The design of highly complex chips of this type necessitates special considerations and requirements [10, 11].

4.9.1 Design Considerations

In practice, fabrication costs may be classified as recurring, e.g., the cost of parts, and nonrecurring, e.g., the design costs. For special-purpose systems like digital filters, demand is usually relatively small. Consequently, the design costs predominate over other costs and should be kept as low as possible. If the realization of the digital filter can be decomposed into a few types of basic building blocks that can be simply interconnected repetitively in a highly regular fashion, considerable savings in the design costs can be achieved. The reason is that the few types of building blocks need to be designed only once. A modular design of this type offers another advantage which can lead to cost reductions. By simply varying the number of modules used in a chip, a large selection of different digital filters can be easily designed that meet a variety of performance criteria or specifications. In this way, the nonrecurring design costs can be spread over a larger number of units fabricated and, therefore, the cost per unit can be reduced.

In certain real-time applications, high-order filters are required to operate at very high sampling rates. In such applications, a very large amount of computation needs to be carried out during each sampling period and the implementation must be very fast. While progress has been made in increasing the speed of gates and reducing the propagation delays by reducing the lengths of interconnection wires, progress is slowing down in these areas and the returns are diminishing rapidly. Therefore, any major improvement in the speed of computation must of necessity be achieved through the concurrent use of many *processing elements*. It turns out that the degree of concurrency is an underlying property of the digital-filter realization. For example, realizations which comprise parallel substructures allow a high degree of concurrency and, therefore, lead to fast implementations. When a large number of processing elements must operate simultaneously, communication among processing elements becomes critical. Since the cost, performance, and speed of the chip depend heavily on the delay and area of the interconnection network, a high degree of concurrency should be achieved in conjunction with simple, short, and regular communication paths among processing elements.

4.9.2 Systolic Implementations

Digital filters can be realized in terms of *systolic arrays* which are highly regular networks of simply connected processing elements that rhythmically process and pass data from one element to the next [10, 11]. The operation of these arrays is analogous to the rhythmical systolic operation of the heart and arteries by which blood is pumped forward from one artery to the next. Evidently, systolic realizations satisfy the design requirements alluded to earlier and are, as a consequence, highly amenable to VLSI implementation.

Close examination of the realizations considered so far reveals that most of them are not suitable for systolic implementation. However, some of them can be made suitable by simple modifications, as will be demonstrated below. A useful technique in this process is known as *pipelining*. In this technique, the computation is partitioned into smaller parcels which can be assigned to a series of different concurrent processing elements in such a way as to achieve a speed advantage. A pipeline in the present context is, in a way, analogous to a modern assembly line of automobiles in which the task of building an automobile is partitioned into a set of small subtasks carried out by concurrent workers working at different stations along the assembly line. Pipelining will introduce some delay into the system, but once the pipeline is filled an automobile will roll off the assembly line every few minutes. This sort of efficiency cannot be achieved by having all the workers working concurrently on one automobile.

Consider the realization of

$$y(nT) = \sum_{i=0}^{N} a_i x(nT - iT)$$

shown in Fig. 4.13a, and assume that each addition and multiplication can be performed in τ_a and τ_m seconds, respectively. This structure can be readily obtained from Fig. 4.4d. Processing elements may be readily identified, as illustrated by the dashed lines. The additional unit delay at the right and the adder at the left with zero input are included in order to improve the regularity of the structure; they serve no other purpose. There are two basic disadvantages associated with this implementation. First, there are two distinct types of elements: one type transfers its input signal to the next element and performs a multiplication, and the other performs an addition. Second, the *processing rate*, which is the maximum sampling rate allowed by the hardware, is limited. The processing rate of an implementation is the reciprocal of the time taken to perform all the required arithmetic operations between two successive samples. While the multiplications in Fig. 4.13a can be carried out concurrently, the N additions must be carried out sequentially from left to right. Therefore, a processing time of $\tau_m + N\tau_a$ seconds is required, which can be large in practice since N can be large. The first disadvantage can be alleviated to some extent by combining into one each multiplier element with its neighboring adder element. While the resulting processing element will be more complex, the interconnection network is simplified.

The processing rate in the structure of Fig. 4.13a can be increased by using faster adders. A more efficient approach, however, is to increase the degree of concurrency through the application of pipelining. Consider the possibility of adding unit delays between processing elements, as depicted in Fig. 4.13b. Since the top and bottom outputs of each processing element are delayed by the same amount by the additional unit delays, the two signals are not shifted relative to each other, and the operation of the structure is not destroyed. The only effect is that the overall output will be delayed by NT seconds, since there are N additional delays between processing elements. Indeed, straightforward analysis gives the output of the modified structure as

$$y_p(nT) = \sum_{i=0}^{N} a_i x(nT - iT - NT)$$

i.e.,

$$y_p(nT) = y(nT - NT)$$

where $y(nT)$ is the output of the original structure. The delay NT is said to be the *latency* of the structure. In the modified structure, only one multiplication and one addition is required per digital-filter cycle, and, therefore, the processing rate is $1/(\tau_m + \tau_a)$. In effect, the processing rate does not, in this case, decrease as the value of N is increased. The additional unit delays in Fig. 4.13b may be absorbed into the processing elements, as depicted in Fig. 4.13c.

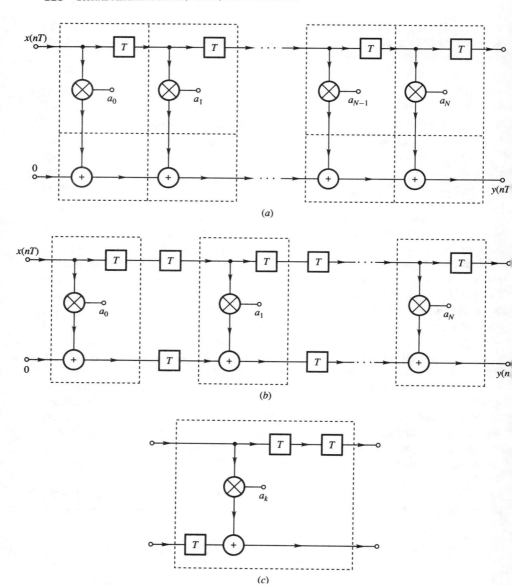

FIGURE 4.13
(a) Realization of Nth-order nonrecursive filter, (b) corresponding systolic realization, (c) typical processing element.

An alternative structure which is amenable to a systolic implementation is depicted in Fig. 4.14a. This is obtained from the structure of Fig. 4.4b. As can be seen, only one multiplication and one addition are required per digital-filter cycle, and so the processing rate is $1/(\tau_m + \tau_a)$. The basic disadvantage of this structure is that the input signal has to be communicated directly to all the processing elements simultaneously. Consequently, for large values of N, wires become long and the associated propagation delays are large, thereby imposing an upper limit on the sampling rate. The problem can be easily overcome by using padding delays, as in Fig. 4.14b.

4.10 TOPOLOGICAL PROPERTIES

Digital-filter networks can be represented in terms of flow graphs which have certain topological properties of theoretical as well as of practical interest. We discuss some of these properties here.

4.10.1 Signal Flow-Graph Representation

Consider the canonic network of Fig. 4.11b, and let $y_1(n), y_2(n), \ldots$ be the signals flowing out of nodes $1, 2, \ldots$. We can write

$$y_1(n) = i(n)$$
$$y_2(n) = i(n) - b_1 y_3(n) - b_2 y_4(n)$$
$$y_3(n) = \mathcal{E}^{-1} y_2(n)$$
$$y_4(n) = \mathcal{E}^{-1} y_3(n)$$
$$y_5(n) = a_0 y_2(n) + a_1 y_3(n) + a_2 y_4(n)$$
$$o(n) = y_5(n)$$

where $i(n)$ and $o(n)$ are the input and output, respectively. Thus a conventional flow graph can be constructed for the network comprising a number of nodes, a number of directed branches, a source $i(n)$, and a sink $o(n)$, as illustrated in Fig. 4.15a. The source and sink can be treated as branches by using the self-loop of Fig. 4.15b, where $q_i(n)$ is an independent quantity for a source and $q_i(n) = 0$ for a sink, and the transmittance of the branch between nodes i' and i'', represented by the dashed line, is zero. By designating the signal entering node j of the flow graph via path ij as x_{ij} and the signal flowing out of node j as y_j, the preceding equations can be put in the concise form

$$y_j = \sum_{i=1}^{5} x_{ij} \quad \text{for } j = 1, 2, \ldots, 5$$

where $x_{uv} = 0$ if no direct path exists between nodes u and v, and $x_{ww} = q_w(n)$ if there is a source $q_w(n)$ at node w.

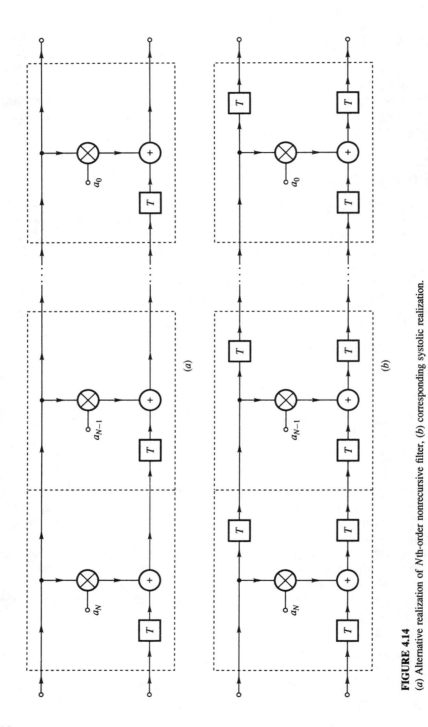

FIGURE 4.14
(a) Alternative realization of Nth-order nonrecursive filter, (b) corresponding systolic realization.

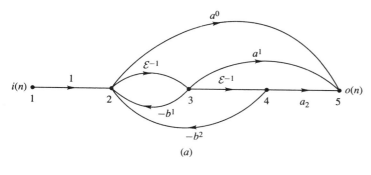

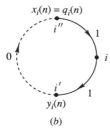

FIGURE 4.15
(a) Flow-graph representation of the filter depicted in Fig. 4.11b, (b) representation of a source or a sink by means of a self-loop.

Similarly, for a digital network with N nodes

$$y_j = \sum_{i=1}^{N} x_{ij} \quad \text{for } j = 1, 2, \ldots, N$$

Tellegen's theorem. Consider two distinct sets of signals $\mathbf{S}$ and $\mathbf{S}'$ defined by

$$\mathbf{S} = \{x_{11}, \ldots, x_{NN}, y_1, \ldots, y_N\} \qquad \mathbf{S}' = \{x'_{11}, \ldots, x'_{NN}, y'_1, \ldots, y'_N\}$$

These may pertain to one and the same network, in which case their differences may be due to differences in the input signals or possibly to variations in one or more multiplier constants. Alternatively, $\mathbf{S}$ and $\mathbf{S}'$ may pertain to distinct but topologically compatible networks (networks which have the same topology).

Tellegen's theorem as applied to digital networks [12, 13] states that the elements of $\mathbf{S}$ and $\mathbf{S}'$ satisfy the general relation

$$\sum_{i=1}^{N}\sum_{j=1}^{N}(y_j x'_{ij} - y'_i x_{ji}) = 0$$

Proof. The validity of this theorem can be established by writing the left-hand side of the relation as

$$\sum_{j=1}^{N} y_j \sum_{i=1}^{N} x'_{ij} - \sum_{i=1}^{N} y'_i \sum_{j=1}^{N} x_{ji}$$

and then replacing

$$\sum_{i=1}^{N} x'_{ij} \quad \text{and} \quad \sum_{j=1}^{N} x_{ji}$$

by y'_j and y_i, respectively.

Similarly, if

$$\mathcal{Z} x_{ij} = X_{ij} \quad \mathcal{Z} y_j = Y_j$$

the preceding steps yield

$$\sum_{i=1}^{N} \sum_{j=1}^{N} (Y_j X'_{ij} - Y'_i X_{ji}) = 0 \tag{4.14}$$

i.e., Tellegen's theorem holds in both the time and z domains.

4.10.2 Reciprocity

A flow graph with M accessible nodes is said to be a *multipole* (or M-pole). Such a flow graph can be represented as depicted in Fig. 4.16, where F is a source-free flow graph.

Consider an M-pole, and let

$$\mathbf{S} = \{X_1, \ldots, X_M, Y_1, \ldots, Y_M\} \quad \mathbf{S}' = \{X'_1, \ldots, X'_M, Y'_1, \ldots, Y'_M\}$$

be possible sets of signals. The M-pole is said to be *reciprocal* [12] if

$$\sum_{i=1}^{M} (X_i Y'_i - X'_i Y_i) = 0 \tag{4.15}$$

for all possible pairs of $\mathbf{S}$ and $\mathbf{S}'$.

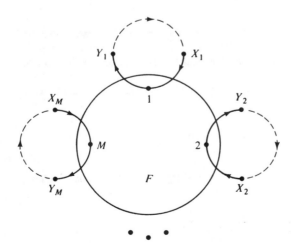

FIGURE 4.16
Multipole flow graph.

The M-pole can be represented by the set of equations

$$Y_i = \sum_{j=1}^{M} H_{ji} X_j \quad \text{for } i = 1, 2, \ldots, M \qquad (4.16)$$

where H_{ji} is the transfer function from node j to node i. From Eqs. (4.15) and (4.16)

$$\sum_{i=1}^{M} \left(X_i \sum_{j=1}^{M} H_{ji} X'_j - X'_i \sum_{j=1}^{M} H_{ji} X_j \right) = 0$$

or

$$\sum_{i=1}^{M} \sum_{j=1}^{M} H_{ji} X_i X'_j - \sum_{i=1}^{M} \sum_{j=1}^{M} H_{ji} X_j X'_i = 0$$

and on interchanging subscripts in the second summation we have

$$\sum_{i=1}^{M} \sum_{j=1}^{M} (H_{ji} - H_{ij}) X_i X'_j = 0$$

Therefore, an M-pole is a reciprocal flow graph if and only if

$$H_{ij} = H_{ji}$$

for all values of i and j.

4.10.3 Interreciprocity

Now consider two distinct M-poles G and G' having the same number of accessible nodes, and let **S** and **S'** as defined above be sets of signals pertaining to G and G', respectively. The two M-poles are said to be *interreciprocal* [12] if

$$\sum_{i=1}^{M} (X_i Y'_i - X'_i Y_i) = 0 \qquad (4.17)$$

for all possible pairs of **S** and **S'**.

G and G' can be described by

$$Y_i = \sum_{j=1}^{M} H_{ji} X_j \quad \text{for } i = 1, 2, \ldots, M \qquad (4.18)$$

and

$$Y'_i = \sum_{j=1}^{M} H'_{ji} X'_j \quad \text{for } i = 1, 2, \ldots, M \qquad (4.19)$$

respectively. From Eqs. (4.17) to (4.19)

$$\sum_{i=1}^{M} \sum_{j=1}^{M} (H'_{ji} - H_{ij}) X_i X'_j = 0$$

and therefore G and G' are interreciprocal flow graphs if and only if
$$H_{ij} = H'_{ji} \tag{4.20}$$
for all values of i and j.

4.10.4 Transposition

Given a flow graph G, a corresponding flow graph G' can be derived by reversing the direction in each and every branch in G. The flow graph so derived is said to be the *transpose* (or *adjoint*) of G [12–14]. If the transmittance of branch ij in G is designated as B_{ij} and that of branch ji in G' as B'_{ji}, then by definition
$$B_{ij} = B'_{ji} \tag{4.21}$$

Let
$$\mathbf{S} = \{X_{11}, \ldots, X_{NN}, Y_1, \ldots, Y_N\} \qquad \mathbf{S}' = \{X'_{11}, \ldots, X'_{NN}, Y'_1, \ldots, Y'_N\}$$
be possible sets of signals pertaining to G and G', respectively, and assume that nodes 1 to M are accessible. From Eq. (4.14)
$$\sum_{i=1}^{M}(Y_i X'_{ii} - Y'_i X_{ii}) + \sum_{i=M+1}^{N}(Y_i X'_{ii} - Y'_i X_{ii}) + \sum_{i=1}^{N}\sum_{\substack{j=1 \\ j \neq i}}^{N}(Y_j X'_{ij} - Y'_i X_{ji}) = 0 \tag{4.22}$$

The first summation represents contributions due to external self-loops, i.e., sources and sinks, whereas the second and third summations represent contributions due to internal self-loops and internal branches, respectively. For both internal self-loops and branches we can write
$$X_{ji} = B_{ji} Y_j \qquad \text{and} \qquad X'_{ij} = B'_{ij} Y'_i$$
and hence
$$\sum_{i=1}^{N}\sum_{\substack{j=1 \\ j \neq i}}^{N}(Y_j X'_{ij} - Y'_i X_{ji}) = \sum_{i=1}^{N}\sum_{\substack{j=1 \\ j \neq i}}^{N}(B'_{ij} - B_{ji})Y_j Y'_i = 0$$

$$\sum_{i=M+1}^{N}(Y_i X'_{ii} - Y'_i X_{ii}) = \sum_{i=M+1}^{N}(B'_{ii} - B_{ii})Y_i Y'_i = 0$$
according to Eq. (4.21). For external self-loops
$$X'_{ii} = X'_i \qquad X_{ii} = X_i$$
according to Fig. 4.16, and as a result Eq. (4.22) simplifies to Eq. (4.17). Therefore G and its *transpose* G' are *interreciprocal* flow graphs. Consequently, if H_{ij} is the transfer function from node i to node j in G and H'_{ji} is the transfer function from node j to node i in G', then
$$H_{ij} = H'_{ji}$$

Because of this property, transposition can serve as a means of deriving alternative digital-filter networks.

Example 4.4. Form the transpose of the canonic network of Fig. 4.11b.

Solution. The transpose flow graph is readily obtained, as shown in Fig. 4.17a, by using Fig. 4.15a. The corresponding network is shown in Fig. 4.17b.

4.10.5 Sensitivity Analysis

In the study of arithmetic errors (see Chap. 11) we shall be concerned with the *sensitivities* of digital-filter networks to variations in the multiplier constants. In a network characterized by

$$H(z) = f(z, m_1, m_2, \ldots)$$

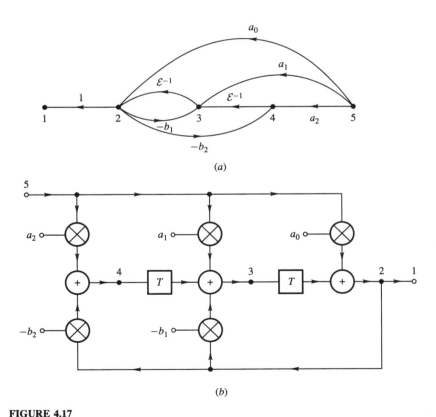

FIGURE 4.17
(a) Transpose flow graph of the network depicted in Fig. 4.11b (Example 4.4), (b) corresponding transpose network.

where $m_1, m_2, \ldots$ are multiplier constants, the sensitivities are given by

$$S_{m_i}^H(z) = \frac{\partial H(z)}{\partial m_i} \quad \text{for } i = 1, 2, \ldots$$

Although differentiation of the transfer function will ultimately yield the sensitivities, much manipulation is often necessary, especially in complicated networks. Fortunately, however, differentiation can be avoided through the concept of transposition [12, 13], as we shall now show.

The network of Fig. 4.18a, the transpose of this network, and the network with m changed to $m + \Delta m$ can be represented by the 2-poles of Fig. 4.18b, c, d, respectively. By virtue of Tellegen's theorem

$$\sum_{i=1}^{N} W_i = 0 \tag{4.23}$$

where

$$W_i = \sum_{j=1}^{N} (Y_j'' X_{ij}' - Y_i' X_{ji}'')$$

For $i = 1$

$$W_1 = Y_1'' X_{11}' - Y_1' X_{11}'' + \sum_{j=2}^{N} (B_{1j}' - B_{j1}) Y_j'' Y_1' = -Q Y_1' \tag{4.24}$$

since

$$X_{11}' = X_1' = 0 \qquad X_{11}'' = X_1'' = Q \qquad B_{1j}' = B_{j1}$$

For $i = 2$

$$W_2 = Y_2'' X_{22}' - Y_2' X_{22}'' + \sum_{\substack{j=1 \\ j \neq 2}}^{N} (B_{2j}' - B_{j2}) Y_j'' Y_2' = Q Y_2'' \tag{4.25}$$

since

$$X_{22}' = X_2' = Q \qquad X_{22}'' = X_2'' = 0 \qquad B_{2j}' = B_{j2}$$

For $i = 3$

$$W_3 = Y_4'' X_{34}' - Y_3' X_{43}'' + \sum_{\substack{j=1 \\ j \neq 4}}^{N} (B_{3j}' - B_{j3}) Y_j'' Y_3' = -\Delta m Y_3' Y_4'' \tag{4.26}$$

since

$$X_{34}' = m Y_3' \qquad X_{43}'' = (m + \Delta m) Y_4'' \qquad B_{3j}' = B_{j3}$$

Finally, for $i \geq 4$

$$W_i = \sum_{j=1}^{N} (B_{ij}' - B_{ji}) Y_j'' Y_i' = 0 \tag{4.27}$$

since

$$B_{ij}' = B_{ji} \quad \text{for } i \geq 4$$

REALIZATION 127

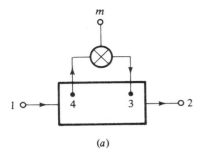

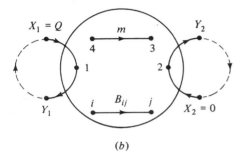

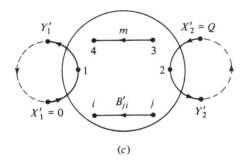

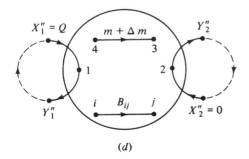

FIGURE 4.18
Derivation of sensitivities: (a) digital-filter network, (b) two-pole representation, (c) two-pole representation of the transpose network, (d) two-pole representation of the network with m changed to $m + \Delta m$.

Consequently, from Eqs. (4.23) to (4.27)

$$W_1 + W_2 + W_3 = 0$$

or

$$Y_2'' - Y_1' = \frac{\Delta m Y_3' Y_4''}{Q} \tag{4.28}$$

Now

$$\Delta H_{12} = H_{12}'' - H_{12} = H_{12}'' - H_{21}' = \frac{Y_2'' - Y_1'}{Q} \tag{4.29}$$

and from Eqs. (4.28) and (4.29)

$$\frac{\Delta H_{12}}{\Delta m} = \frac{Y_3' Y_4''}{Q^2} = H_{23}' H_{14}'' = H_{32} H_{14}''$$

As $\Delta m \to 0$, $H_{14}'' \to H_{14}$, and therefore

$$S_m^H(z) = \lim_{\Delta m \to 0} \frac{\Delta H_{12}}{\Delta m} = H_{14}(z) H_{32}(z) \tag{4.30}$$

i.e., the sensitivity to variations in multiplier constant m can be formed by multiplying the transfer function from the input of the network to the input of the multiplier by the transfer function from the output of the multiplier to the output of the network. This technique can eliminate considerable manipulation in complicated networks.

Example 4.5. Find the sensitivities for the network of Fig. 4.19.

Solution. Straightforward analysis yields

$$H_{14}(z) = \frac{z}{z^2 - b_1 z - b_2} \qquad H_{15}(z) = \frac{1}{z^2 - b_1 z - b_2}$$

$$H_{32}(z) = H_{62}(z) = \frac{(z+1)^2}{z^2 - b_1 z - b_2}$$

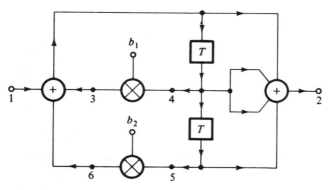

FIGURE 4.19
Second-order filter (Example 4.5).

Hence from Eq. (4.30)

$$S_{b_1}^H(z) = \frac{z(z+1)^2}{(z^2 - b_1 z - b_2)^2} \quad \text{and} \quad S_{b_2}^H(z) = \frac{(z+1)^2}{(z^2 - b_1 z - b_2)^2}$$

REFERENCES

1. B. Gold and C. M. Rader, *Digital Processing of Signals*, McGraw-Hill, New York, 1969.
2. A. Antoniou, "Realization of Digital Filters," *IEEE Trans. Audio Electroacoust.*, vol. AU-20, pp. 95–97, March 1972.
3. L. B. Jackson, A. G. Lindgren, and Y. Kim, "Synthesis of State-Space Digital Filters with Low Roundoff Noise and Coefficient Sensitivity," *Proc. 1977 Int. Symp. Circuits Syst.*, pp. 41–44.
4. S. K. Mitra and R. J. Sherwood, "Digital Ladder Networks," *IEEE Trans. Audio Electroacoust.*, vol. AU-21, pp. 30–36, February 1973.
5. A. H. Gray, Jr. and J. D. Markel, "Digital Lattice and Ladder Filter Synthesis," *IEEE Trans. Audio Electroacoust.*, vol. AU-21, pp. 491–500, December 1973.
6. A. Fettweis, "Digital Filter Structures Related to Classical Filter Networks," *Arch. Elektron. Übertrag.*, vol. 25, pp. 79–89, 1971.
7. A. Sedlmeyer and A. Fettweis, "Digital Filters with True Ladder Configuration," *Int. J. Circuit Theory Appl.*, vol. 1, pp. 5–10, March 1973.
8. L. T. Bruton, "Low-Sensitivity Digital Ladder Filters," *IEEE Trans. Circuits Syst.*, vol. CAS-22, pp. 168–176, March 1975.
9. A. Antoniou and M. G. Rezk, "Digital-Filter Synthesis Using Concept of Generalized-Immittance Convertor," *IEE J. Electron. Circuits Syst.*, vol. 1, pp. 207–216, November 1977.
10. H. T. Kung, "Why Systolic Architectures," *IEEE Computer*, vol. 15, pp. 37–46, January 1982.
11. S. Y. Kung, "VLSI Array Processors," *IEEE ASSP Magazine*, vol. 2, pp. 4–22, July 1985.
12. A. Fettweis, "A General Theorem for Signal-Flow Networks with Applications," *Arch. Elektron. Übertrag.*, vol. 25, pp. 557–561, 1971.
13. R. E. Seviora and M. Sablatash, "A Tellegen's Theorem for Digital Filters," *IEEE Trans. Circuit Theory*, vol. CT-18, pp. 201–203, January 1971.
14. L. B. Jackson, "On the Interaction of Roundoff Noise and Dynamic Range in Digital Filters," *Bell Syst. Tech. J.*, vol. 49, pp. 159–184, February 1970.

ADDITIONAL REFERENCES

Bomar, B. W.: "On the Design of Second-Order State-Space Digital Filter Sections," *IEEE Trans. Circuits Syst.*, vol. CAS-36, pp. 542–552, April 1989.

Director, S. W., and R. A. Rohrer: "The Generalized Adjoint Network and Network Sensitivities," *IEEE Trans. Circuit Theory*, vol. CT-16, pp. 318–323, August 1969.

Gray, Jr., A. H. and J. D. Markel: "A Normalized Digital Filter Structure," *IEEE Trans. Acoust., Speech, Signal Process.*, vol. ASSP-23, pp. 268–277, June 1975.

Ismail, M., and H. K. Kim: "Synthesis of New Canonical Digital Ladder Filters by Continued Fractions," *IEE Proc.*, vol. 132, pt. G, pp. 1–6, February 1985.

Mills, W. L., C. T. Mullis, and R. A. Roberts: "Low Roundoff Noise and Normal Realizations of Fixed Point IIR Digital Filters," *IEEE Trans. Acoust., Speech, Signal Process.*, vol. ASSP-29, pp. 893–903, August 1981.

Mitra, S. K., and K. Hirano: "Digital All-Pass Networks," *IEEE Trans. Circuits Syst.*, vol. CAS-21, pp. 688–700, September 1974.

Szczupak, J., and S. K. Mitra: "Digital Filter Realization Using Successive Multiplier-Extraction Approach," *IEEE Trans. Acoust., Speech, Signal Process.*, vol. ASSP-23, pp. 235–239, April 1975.

―――― and ――――: "Detection, Location, and Removal of Delay-Free Loops in Digital Filter Configurations," *IEEE Trans. Acoust., Speech, Signal Process.*, vol. ASSP-23, pp. 558–562, December 1975.

PROBLEMS

4.1. By using first the direct and then the canonic method, realize the following transfer functions:

(a) $H(z) = \dfrac{4(z-1)^4}{4z^4 + 3z^3 + 2z^2 + z + 1}$ (b) $H(z) = \dfrac{(z+1)^2}{4z^3 - 2z^2 + 1}$

4.2. A digital filter is characterized by the state-space equations

$$\mathbf{q}(nT+T) = \mathbf{A}\mathbf{q}(nT) + \mathbf{b}x(nT)$$

$$y(nT) = \mathbf{c}^T\mathbf{q}(nT) + dx(nT)$$

where

$$\mathbf{A} = \begin{bmatrix} 0 & 1 \\ -\frac{5}{16} & -1 \end{bmatrix} \quad \mathbf{b} = \begin{bmatrix} 0 \\ 1 \end{bmatrix} \quad \mathbf{c}^T = [\,-\tfrac{11}{8} \quad 2\,] \quad d = 2$$

(a) Obtain a state-space realization for the filter.
(b) Obtain a corresponding canonic realization.
(c) Compare the realizations in parts (a) and (b).

4.3. Repeat Prob. 4.2 if

$$\mathbf{A} = \begin{bmatrix} 0 & 1 & 0 \\ 0 & 0 & 1 \\ \frac{25}{64} & -\frac{29}{32} & \frac{3}{4} \end{bmatrix} \quad \mathbf{b} = \begin{bmatrix} 0 \\ 0 \\ 1 \end{bmatrix} \quad \mathbf{c}^T = [\,\tfrac{25}{64} \quad \tfrac{3}{32} \quad \tfrac{11}{4}\,] \quad d = 1$$

4.4. Realize the transfer function

$$H(z) = \dfrac{0.0154z^3 + 0.0462z^2 + 0.0462z + 0.0154}{z^3 - 1.990z^2 + 1.572z - 0.4582}$$

by using the ladder method.

4.5. Realize the transfer function

$$H(z) = \dfrac{z(z+1)}{\left(z^2 - \tfrac{1}{2}z + \tfrac{1}{4}\right)}$$

using a lattice structure.

4.6. A recursive digital filter is characterized by the state-space equations in Prob. 4.2 with

$$\mathbf{A} = \begin{bmatrix} 0 & 1 & 0 \\ 0 & 0 & 1 \\ -\tfrac{1}{2} & -m & -2 \end{bmatrix} \quad \mathbf{b} = \begin{bmatrix} 0 \\ 0 \\ 1 \end{bmatrix} \quad \mathbf{c}^T = [1 \quad 2 \quad -1] \quad d = 1$$

(a) Determine the range of m for which the filter is stable.
(b) Obtain a state-space realization for the filter.
(c) Obtain a lattice realization.
(d) Compare the realizations in parts (b) and (c).

4.7. Realize the transfer function

$$H(z) = \frac{4(z-1)(z+1)^2}{(2z+1)(2z^2-2z+1)}$$

using cascade canonic sections.

4.8. First-order filter sections of the type depicted in Fig. P4.8 are available. Using sections of this type, obtain a parallel realization of the transfer function

$$H(z) = \frac{216z^2 + 162z + 29}{(2z+1)(3z+1)(4z+1)}$$

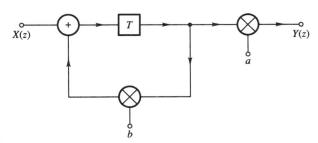

FIGURE P4.8

4.9. (a) Realize the transfer function

$$H(z) = \frac{16(z+1)z^2}{(4z+3)(4z^2-2z+1)}$$

using cascade canonic sections.
(b) Repeat with parallel canonic sections.

4.10. Obtain cascade and parallel realizations as in Prob. 4.9 for the transfer function

$$H(z) = \frac{6z}{6z^3 + 4z^2 + z - 1}$$

4.11. (a) Construct a flow chart for the software implementation of an N-section parallel filter.
(b) Repeat part (a) for an N-section cascade filter.

4.12. Given a continuous-time transfer function $H_A(s)$, a corresponding discrete-time transfer function $H_D(z)$ can be formed as

$$H_D(z) = H_A(s)\Big|_{s=\frac{2}{T}\left(\frac{z-1}{z+1}\right)}$$

(see Sec. 7.6).
(a) Obtain a digital-filter network by using the transfer fuction

$$H_A(s) = \frac{s^2}{s^2 + \sqrt{2}s + 1}$$

(b) Evaluate the gain of the filter for $\omega = 0$ and $\omega = \pi/T$.

4.13. (a) The flow graph of Fig. P4.13a represents a recursive filter. Deduce the transfer function of the filter.
(b) Repeat part (a) for the flow graph of Fig. P4.13b.

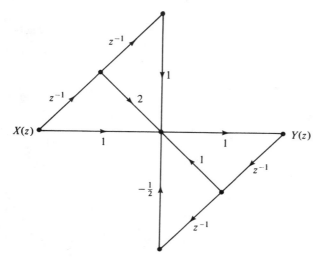

FIGURE P4.13a

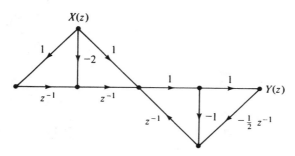

FIGURE P4.13b

4.14. (a) Convert the flow graph of Fig. P4.14 into a topologically equivalent network.
(b) Obtain an alternative realization by using the canonic method.

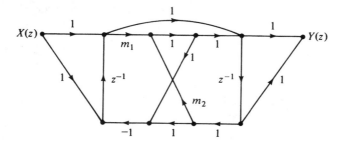

FIGURE P4.14

4.15. Derive flow-graph representations for the filters of Figs. 4.5b and 4.8a.

4.16. A flow graph is said to be *computable* if there are no closed delay-free loops. Check the flow graphs of Figs. P4.16a and P4.16b for computability.

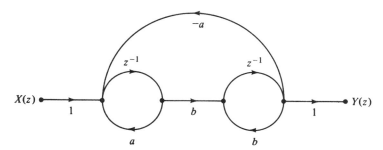

FIGURE P4.16a

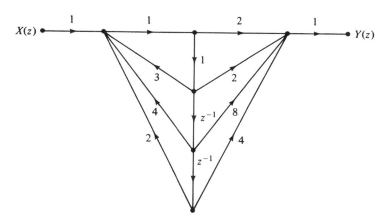

FIGURE P4.16b

4.17. The network in Fig. P4.17 is excited first with a unit step and then with a unit impulse. Show that the signal distributions in the two cases satisfy Tellegen's theorem.

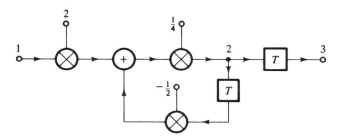

FIGURE P4.17

4.18. Figure P4.18 depicts two topologically compatible networks. Show that the signal distributions in the two networks satisfy Tellegen's theorem.

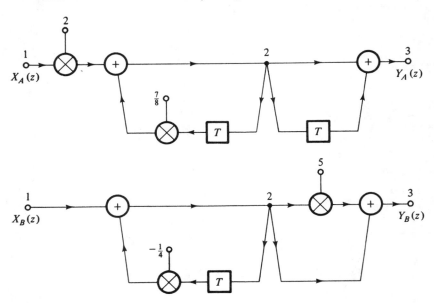

FIGURE P4.18

4.19. The 2-pole of Fig. P4.19a can be represented by the equations

$$Y_1(z) = H_{11}(z)X_1(z) + H_{21}(z)X_2(z)$$

$$Y_2(z) = H_{12}(z)X_1(z) + H_{22}(z)X_2(z)$$

(a) Find $H_{11}(z)$, $H_{12}(z)$, ... Hence check the flow graph for reciprocity.
(b) Repeat part (a) for the 2-pole of Fig. P4.19b.

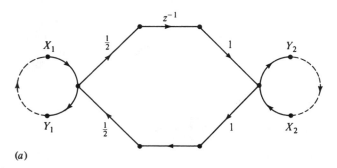

(a)

FIGURE P4.19a

4.20. (a) Figure P4.20a shows a pair of 2-poles. Check the pair for interreciprocity.
(b) Repeat part (a) for the 2-poles of Fig. P4.20b.

REALIZATION **135**

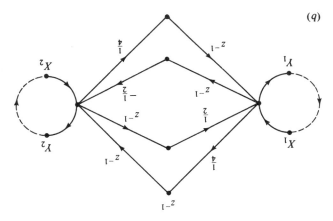

FIGURE P4.19b

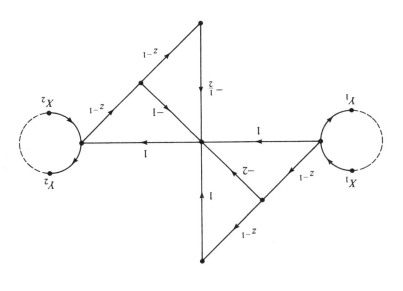

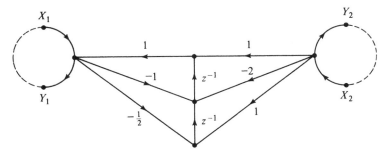

FIGURE P4.20a

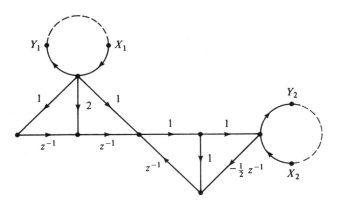

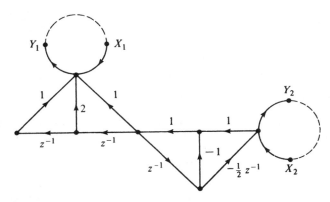

FIGURE P4.20b

4.21. (a) Obtain the transpose of the network of Fig. 3.9.
  (b) Repeat part (a) for the network of Fig. 3.2.
4.22. (a) Form the transpose of the network shown in Fig. 1.5b.
  (b) Obtain an alternative ladder structure by applying the transpose approach to the network of Fig. 4.8b.
4.23. Figure P4.23 depicts an allpass network. Obtain an alternative allpass network by using the transpose method.
4.24. (a) Find the sensitivities of the network shown in Fig. P4.24a.
  (b) Repeat part (a) for the network of Fig. P4.24b.
4.25. The gain and phase-shift sensitivities of a digital filter are defined as

$$S_{m_i}^M = \frac{\partial M(\omega)}{\partial m_i} \quad \text{and} \quad S_{m_i}^\theta = \frac{\partial \theta(\omega)}{\partial m_i}$$

respectively. Assuming that

$$\left.\frac{\partial H(z)}{\partial m_i}\right|_{z=e^{j\omega T}} = \text{Re } S_{m_i}^H(e^{j\omega T}) + j\text{Im } S_{m_i}^H(e^{j\omega T})$$

derive expressions for $S_{m_i}^M$ and $S_{m_i}^\theta$.

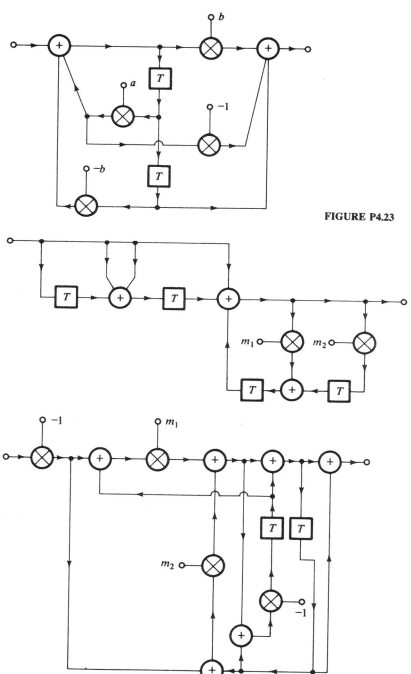

FIGURE P4.23

FIGURE P4.24

CHAPTER 5

ANALOG-FILTER APPROXIMATIONS

5.1 INTRODUCTION

The *approximation problem* in analog filters has been a subject of research throughout the past fifty years, and some powerful methods have been developed for its solution [1–5]. These methods yield a complete description of the continuous-time transfer function in *closed form*, either in terms of its zeros and poles or its coefficients. Through the application of some simple techniques or transformations, a continuous-time transfer function can be converted into a corresponding discrete-time transfer function. Consequently, the available analog-filter approximation methods can be used for the solution of the approximation problem in digital filters.

This chapter considers several analog-filter approximations that are suitable for the design of filters with piecewise-constant amplitude responses, i.e., filters whose passband and stopband gains are constant and zero, respectively, to within prescribed tolerances. The most frequently used approximations of this type are as follows:

1. Butterworth
2. Chebyshev
3. Inverse-Chebyshev
4. Elliptic
5. Bessel

No particular attention is paid to the phase response in the first four approximations. Hence, it turns out to be nonlinear, which may present a problem in applications where phase distortion is undesirable (see Sec. 3.6). In the last approximation, however, a constraint is imposed on the group delay, which results in a fairly *linear* phase response.

The chapter begins with an introductory section dealing with the terminology and characterization of analog filters and continues with the derivations and properties of the various approximations. The derivations provided deal with *lowpass* approximations since other types of approximations can be readily obtained through the application of transformations. Suitable transformations for the design of *highpass*, *bandpass*, and *bandstop* filters are described at the end of the chapter.

It should be mentioned that the derivation of the elliptic approximation is somewhat complicated as it entails a basic understanding of *elliptic* functions. Fortunately, the required results can be put in a simple form which is easy to apply, even for the uninitiated. The elliptic approximation is treated in detail here because this is the most efficient of the above five if piecewise-constant prescribed loss specifications are to be met; that is, it leads to the lowest-order transfer function for a given set of specifications. The reader who is interested in the application of the method may skip the derivations and proceed to Sec. 5.5.6 for a step-by-step procedure for the design. The reader who is also interested in the derivation of this very important method may start with Appendix A, which provides a brief review of the fundamentals of elliptic functions.

The application of analog-filter approximations in the design of recursive digital filters will be considered in Chaps. 7 and 8 after a link is established between analog and digital filters in Chap. 6. Chapter 8 considers, in addition, a *delay-equalization technique* that can be used in conjunction with the above methods for the design of digital filters with approximately linear phase response.

5.2 BASIC CONCEPTS

An analog filter like that in Fig. 5.1 can be represented by the equation

$$\frac{V_o(s)}{V_i(s)} = H(s) = \frac{N(s)}{D(s)}$$

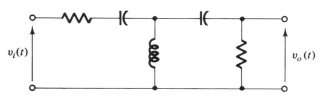

FIGURE 5.1
Analog filter.

where $V_i(s)$ and $V_o(s)$ are the Laplace transforms of the input and output voltages $v_i(t)$ and $v_o(t)$, respectively, $H(s)$ is the transfer function, and $N(s)$ and $D(s)$ are polynomials in $s \ (= \sigma + j\omega)$. The loss (or attenuation) of the filter in decibels is defined by

$$A(\omega) = 20 \log \left| \frac{V_i(j\omega)}{V_o(j\omega)} \right| = 20 \log \frac{1}{|H(j\omega)|} = 10 \log L(\omega^2)$$

where
$$L(\omega^2) = \frac{1}{H(j\omega)H(-j\omega)} \tag{5.1}$$

$A(\omega)$ as a function of ω is the *loss characteristic*.

The phase shift and group delay of the filter are given by

$$\theta(\omega) = \arg H(j\omega) \quad \text{and} \quad \tau(\omega) = -\frac{d\theta(\omega)}{d\omega}$$

respectively. As functions of ω, $\theta(\omega)$ and $\tau(\omega)$ are the *phase* and *delay characteristics*. With $\omega = s/j$ in Eq. (5.1) the function

$$L(-s^2) = \frac{D(s)D(-s)}{N(s)N(-s)}$$

can be formed. This is called the *loss function* of the filter, and, as is evident, its zeros are the poles of $H(s)$ and their negatives, whereas its poles are the zeros of $H(s)$ and their negatives. Typical zero-pole plots for $H(s)$ and $L(-s^2)$ are shown in Fig. 5.2.

5.2.1 Ideal and Practical Filters

An ideal lowpass filter is one that will pass only low-frequency components. Its loss characteristic is of the form depicted in Fig. 5.3a. The frequency ranges 0 to ω_c and

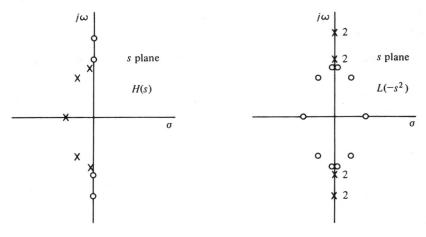

FIGURE 5.2
Typical zero-pole plots for $H(s)$ and $L(-s^2)$.

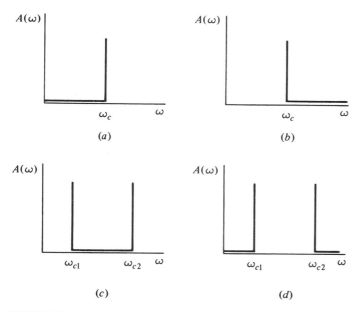

FIGURE 5.3
Ideal loss characteristics: (*a*) lowpass, (*b*) highpass, (*c*) bandpass, (*d*) bandstop.

ω_c to ∞ are the *passband* and *stopband*, respectively. The boundary between the passband and stopband, namely, ω_c, is the *cutoff frequency*.

Similarly, ideal highpass, bandpass, and bandstop filters can be identified having loss characteristics like those depicted in Fig. 5.3*b* to *d*.

A practical lowpass filter differs from an ideal one in that the passband loss is not zero, the stopband loss is not infinite, and the transition between passband and stopband is gradual. The loss characteristic might assume the form shown in Fig. 5.4*a*, where ω_p is the passband edge, ω_a is the stopband edge, A_p is the maximum passband loss, and A_a is the minimum stopband loss. The cutoff frequency ω_c is a hypothetical boundary between passband and stopband, which may be the 3-dB frequency or possibly the square root of $\omega_p \omega_a$ (in elliptic filters). Typical characteristics for practical highpass, bandpass, and bandstop filters are shown in Fig. 5.4*b* to *d*.

5.2.2 Realizability Constraints

A *filter approximation* is a realizable transfer function such that the loss characteristic approaches one of the ideal characteristics in Fig. 5.3. A transfer function is *realizable* if it characterizes a stable and causal network. Such a transfer function must satisfy the following constraints:

1. It must be a rational function of s with real coefficients.
2. Its poles must lie in the left-half s plane.

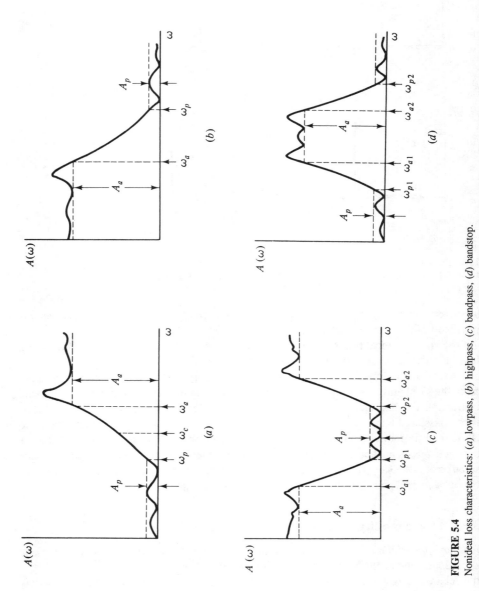

FIGURE 5.4
Nonideal loss characteristics: (a) lowpass, (b) highpass, (c) bandpass, (d) bandstop.

3. The degree of the numerator polynomial must be equal to or less than that of the denominator polynomial.

In the following four sections we focus our attention on normalized lowpass approximations; namely, Butterworth approximations in which the 3-dB cutoff frequency ω_c is equal to 1 rad/s, Chebyshev approximations in which the passband edge ω_p is equal to 1 rad/s, inverse-Chebyshev approximations in which the stopband edge ω_a is equal to 1 rad/s, elliptic approximations in which the cutoff frequency $\omega_c = \sqrt{(\omega_p \omega_a)}$ is equal to 1 rad/s, and Bessel approximations in which the group delay as $\omega \to 0$ is equal to 1 s.

The derivation of denormalized lowpass, highpass, bandpass, and bandstop approximations is almost invariably accomplished through transformations of normalized lowpass approximations. The appropriate transformations are described in Sec. 5.7.

5.3 BUTTERWORTH APPROXIMATION

The simplest lowpass approximation, the *Butterworth* approximation, is derived by assuming that $L(\omega^2)$ is a polynomial of the form

$$L(\omega^2) = b_0 + b_1 \omega^2 + \cdots + b_n \omega^{2n} \tag{5.2}$$

such that

$$\lim_{\omega^2 \to 0} L(\omega^2) = 1$$

in a maximally flat sense.

5.3.1 Derivation

The Taylor series of $L(x + h)$, where $x = \omega^2$, is

$$L(x + h) = L(x) + h \frac{dL(x)}{dx} + \cdots + \frac{h^k}{k!} \frac{d^k L(x)}{dx^k}$$

The polynomial $L(x)$ approaches unity in a *maximally flat* sense as $x \to 0$ if its first n derivatives are zero for $x = 0$. We may, therefore, assign

$$L(0) = 1$$

$$\left. \frac{d^k L(x)}{dx^k} \right|_{x=0} = 0 \qquad \text{for } k \leq n$$

Thus from Eq. (5.2)

$$b_0 = 1 \qquad \text{and} \qquad b_1 = b_2 = \cdots = b_{n-1} = 0$$

or

$$L(\omega^2) = 1 + b_n \omega^{2n}$$

Now for a normalized approximation in which

$$L(1) = 2$$

that is, $A(\omega) \approx 3\text{dB}$ at $\omega = 1$ rad/s, $b_n = 1$ and

$$L(\omega^2) = 1 + \omega^{2n} \qquad (5.3)$$

Hence the loss in a normalized lowpass Butterworth approximation is

$$A(\omega) = 10 \log(1 + \omega^{2n})$$

This is plotted in Fig. 5.5 for $n = 3, 6, 9$.

5.3.2 Normalized Transfer Function

With $\omega = s/j$ in Eq. (5.3) we have

$$L(-s^2) = 1 + (-s^2)^n = \prod_{k=1}^{2n}(s - s_k)$$

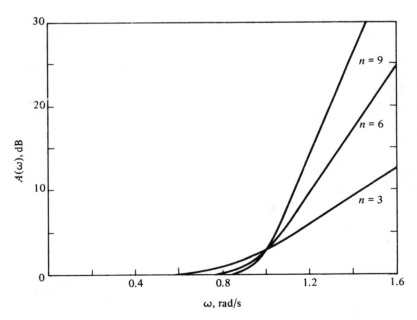

FIGURE 5.5
Typical Butterworth loss characteristics ($n = 3, 6, 9$).

where
$$s_k = \begin{cases} e^{j(2k-1)\pi/2n} & \text{for even } n \\ e^{j(k-1)\pi/n} & \text{for odd } n \end{cases} \quad (5.4)$$

and since $|s_k| = 1$, the zeros of $L(-s^2)$ are located on the unit circle $|s| = 1$. The normalized transfer function can be formed as

$$H_N(s) = \frac{1}{\prod_{i=1}^{n}(s - p_i)}$$

where p_i for $i = 1, 2, \ldots, n$ are the left-half s-plane zeros of $L(-s^2)$.

Example 5.1. Find $H_N(s)$ for (a) $n = 2$ and (b) $n = 3$.

Solution. (a) From Eq. (5.4)

$$s_k = \cos\frac{(2k-1)\pi}{4} + j\sin\frac{(2k-1)\pi}{4}$$

Hence
$$s_2, s_3 = -\frac{1}{\sqrt{2}} \pm \frac{j}{\sqrt{2}}$$

and
$$H_N(s) = \frac{1}{s^2 + \sqrt{2}s + 1}$$

(b) For $n = 3$

$$s_k = \cos\frac{(k-1)\pi}{3} + j\sin\frac{(k-1)\pi}{3}$$

Hence
$$s_4 = -1 \qquad s_3, s_5 = -\frac{1}{2} \pm \frac{j\sqrt{3}}{2}$$

and
$$H_N(s) = \frac{1}{(s+1)(s^2 + s + 1)}$$

The zero-pole plots for the two examples are shown in Fig. 5.6.

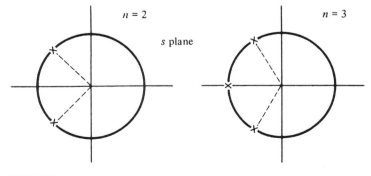

FIGURE 5.6
Zero-pole plots of $H(s)$ (Example 5.1).

5.4 CHEBYSHEV APPROXIMATION

In the Butterworth approximation, the loss is an increasing monotonic function of ω, and as a result the passband characteristic is lopsided, as can be seen in Fig. 5.5. A more balanced characteristic can be achieved by employing the *Chebyshev* approximation in which the passband loss oscillates between zero and a prescribed maximum A_p.

5.4.1 Derivation

The loss characteristic in a fourth-order normalized Chebyshev approximation is of the form illustrated in Fig. 5.7, where $\omega_p = 1$. The loss is given by

$$A(\omega) = 10 \log L(\omega^2)$$

where
$$L(\omega^2) = 1 + \varepsilon^2 F^2(\omega) \qquad (5.5)$$

and
$$\varepsilon^2 = 10^{0.1 A_p} - 1 \qquad (5.6)$$

$F(\omega)$, $L(\omega^2)$, and in turn $L(-s^2)$ are polynomials, and hence the normalized transfer function is of the form

$$H_N(s) = \frac{H_0}{D(s)}$$

where H_0 is a constant.

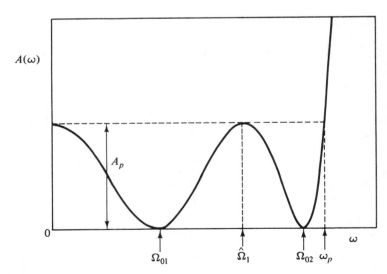

FIGURE 5.7
Loss characteristic of a fourth-order normalized Chebyshev filter.

The derivation of $H_N(s)$ involves three general steps:

1. The exact form of $F(\omega)$ is deduced such that the desired loss characteristic is achieved.
2. The exact form of $L(\omega^2)$ is obtained.
3. The zeros of $L(-s^2)$ and in turn the poles of $H_N(s)$ are found.

Close examination of the Chebyshev loss characteristic depicted in Fig. 5.7 reveals that $F(\omega)$ and $L(\omega^2)$ must have the following properties:

Property 1: $F(\omega) = 0$ if $\omega = \pm\Omega_{01}, \pm\Omega_{02}$
Property 2: $F^2(\omega) = 1$ if $\omega = 0, \pm\hat{\Omega}_1, \pm 1$
Property 3: $\dfrac{dL(\omega^2)}{d\omega} = 0$ if $\omega = 0, \pm\Omega_{01}, \pm\hat{\Omega}_1, \pm\Omega_{02}$

From property 1, $F(\omega)$ must be a polynomial of the form

$$F(\omega) = M_1(\omega^2 - \Omega_{01}^2)(\omega^2 - \Omega_{02}^2)$$

($M_1, M_2, \ldots, M_7$ represent constants in this analysis.) From property 2, $1 - F^2(\omega)$ has zeros at $\omega = 0, \pm\hat{\Omega}_1, \pm 1$. Furthermore, the derivative of $1 - F^2(\omega)$ with respect to ω, namely

$$\frac{d}{d\omega}[1 - F^2(\omega)] = -2F(\omega)\frac{dF(\omega)}{d\omega} = -\frac{1}{\varepsilon^2}\frac{dL(\omega^2)}{d\omega} \qquad (5.7)$$

has zeros at $\omega = 0, \pm\Omega_{01}, \pm\hat{\Omega}_1, \pm\Omega_{02}$, according to property 3. Consequently, $1 - F^2(\omega)$ must have at least double zeros at $\omega = 0, \pm\hat{\Omega}_1$. Therefore, we can write

$$1 - F^2(\omega) = M_2\omega^2(\omega^2 - \hat{\Omega}_1^2)^2(\omega^2 - 1)$$

Now from Eq. (5.7) and properties 1 and 3

$$\frac{dF(\omega)}{d\omega} = \frac{1}{2\varepsilon^2 F(\omega)}\frac{dL(\omega^2)}{d\omega} = M_3\omega(\omega^2 - \hat{\Omega}_1^2)$$

By combining the above results we can form the differential equation

$$\left[\frac{dF(\omega)}{d\omega}\right]^2 = \frac{M_4[1 - F^2(\omega)]}{1 - \omega^2}$$

which can be expressed in terms of definite integrals as

$$M_5\int_0^F \frac{dx}{\sqrt{1-x^2}} + M_6 = \int_0^\omega \frac{dy}{\sqrt{1-y^2}}$$

Hence F and ω are interrelated by the equation

$$M_5 \cos^{-1} F + M_7 = \cos^{-1}\omega = \theta \qquad (5.8)$$

i.e., for a given value of θ

$$\omega = \cos\theta \qquad \text{and} \qquad F = \cos\left(\frac{\theta}{M_5} - \frac{M_7}{M_5}\right)$$

What remains to be done is to determine constants M_5 and M_7. If $\omega = 0$, then $\theta = \pi/2$; and if $\omega = 1$, then $\theta = 0$, as depicted in Fig. 5.8. Now F will correspond to $F(\omega)$ only if it has two zeros in the range $0 \leq \theta \leq \pi/2$ (property 1), and its magnitude is unity if $\theta = 0, \pi/2$ (property 2). F must thus be of the form illustrated in Fig. 5.8. As can be seen, for $\theta = 0$

$$F = \cos\left(-\frac{M_7}{M_5}\right) = 1$$

or $M_7 = 0$. In addition, one period of F must be equal to one-quarter period of ω, i.e.,

$$2\pi M_5 = \frac{\pi}{2} \quad \text{or} \quad M_5 = \frac{1}{4}$$

Therefore, the exact form of $F(\omega)$ can be obtained from Eq. (5.8) as

$$F(\omega) = \cos(4\cos^{-1}\omega)$$

Alternatively, by expressing $\cos 4\theta$ in terms of $\cos \theta$, $F(\omega)$ can be put in the form

$$F(\omega) = 1 - 8\omega^2 + 8\omega^4$$

This polynomial is the fourth-order Chebyshev polynomial and is often designated as $T_4(\omega)$.

Similarly, for an nth-order Chebyshev approximation, one can show that

$$F(\omega) = T_n(\omega) = \cos(n\cos^{-1}\omega)$$

and hence from Eq. (5.5)

$$L(\omega^2) = 1 + \varepsilon^2[\cos(n\cos^{-1}\omega)]^2 \tag{5.9}$$

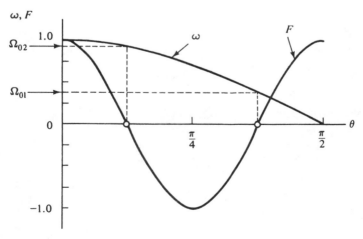

FIGURE 5.8
Plots of ω and F versus θ.

This relation gives the loss characteristic for $|\omega| \leq 1$. For $|\omega| > 1$, the quantity $\cos^{-1}\omega$ becomes complex, i.e.,

$$\cos^{-1}\omega = j\theta \tag{5.10}$$

and since

$$\omega = \cos j\theta = \frac{1}{2}(e^{j(j\theta)} + e^{-j(j\theta)}) = \cosh\theta$$

we have

$$\theta = \cosh^{-1}\omega$$

Now from Eq. (5.10)

$$\cos^{-1}\omega = j\cosh^{-1}\omega$$

and

$$\cos(n\cos^{-1}\omega) = \cos(jn\cosh^{-1}\omega) = \cosh(n\cosh^{-1}\omega)$$

Thus for $|\omega| > 1$, Eq. (5.9) becomes

$$L(\omega^2) = 1 + \varepsilon^2[\cosh(n\cosh^{-1}\omega)]^2 \tag{5.11}$$

In summary, the loss in a normalized lowpass Chebyshev approximation is given by

$$A(\omega) = 10\log[1 + \varepsilon^2 T_n^2(\omega)]$$

where

$$T_n(\omega) = \begin{cases} \cos(n\cos^{-1}\omega) & \text{for } |\omega| \leq 1 \\ \cosh(n\cosh^{-1}\omega) & \text{for } |\omega| > 1 \end{cases}$$

The loss characteristics for $n = 4$, $A_p = 1$ dB and $n = 7$, $A_p = 0.5$ dB are plotted in Fig. 5.9a. As can be seen

$$A(0) = \begin{cases} A_p & \text{for even } n \\ 0 & \text{for odd } n \end{cases}$$

as is generally the case in the Chebyshev approximation.

5.4.2 Normalized Transfer Function

With $\omega = s/j$, Eq. (5.11) becomes

$$L(-s^2) = 1 + \varepsilon^2\left[\cosh\left(n\cosh^{-1}\frac{s}{j}\right)\right]^2$$

and if $s_k = \sigma_k + j\omega_k$ is a zero of $L(-s^2)$, we can write

$$u_k + jv_k = \cosh^{-1}(-j\sigma_k + \omega_k) \tag{5.12a}$$

$$\cosh[n(u_k + jv_k)] = \pm\frac{j}{\varepsilon} \tag{5.12b}$$

From Eq. (5.12a)

$$-j\sigma_k + \omega_k = \cosh(u_k + jv_k) = \cosh u_k \cos v_k + j\sinh u_k \sin v_k$$

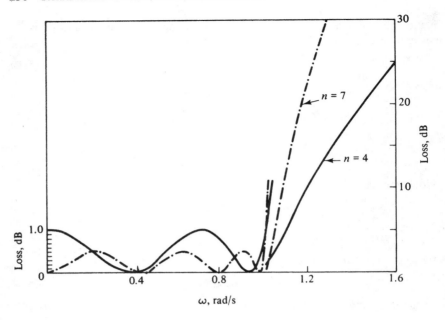

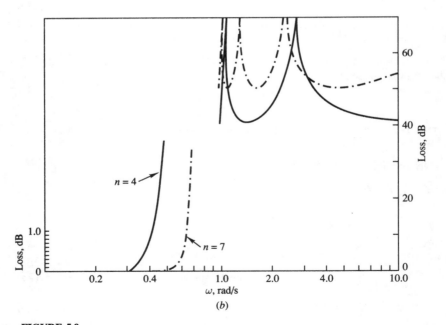

FIGURE 5.9
(a) Typical loss characteristics for Chebyshev filters ($n = 4$, $A_p = 1.0$ dB and $n = 7$, $A_p = 0.5$ dB), (b) typical loss characteristics for inverse-Chebyshev filters ($n = 4$, $A_a = 40$ dB and $n = 7$, $A_a = 50$ dB).

or
$$\sigma_k = -\sinh u_k \sin v_k \qquad (5.13)$$
and
$$\omega_k = \cosh u_k \cos v_k \qquad (5.14)$$

Similarly, from Eq. (5.12b)
$$\cosh nu_k \cos nv_k + j \sinh nu_k \sin nv_k = \pm \frac{j}{\varepsilon}$$
or
$$\cosh nu_k \cos nv_k = 0 \qquad (5.15a)$$
and
$$\sinh nu_k \sin nv_k = \pm \frac{1}{\varepsilon} \qquad (5.15b)$$

The solution of Eq. (5.15a) is
$$v_k = \frac{(2k-1)\pi}{2n} \quad \text{for } k = 1, 2, \ldots, n \qquad (5.16a)$$
and since $\sin(nv_k) = \pm 1$, Eq. (5.15b) yields
$$u_k = \pm \frac{1}{n} \sinh^{-1} \frac{1}{\varepsilon} \qquad (5.16b)$$

Therefore, from Eqs. (5.13), (5.14), and (5.16)
$$\sigma_k = \pm \sinh\left(\frac{1}{n} \sinh^{-1} \frac{1}{\varepsilon}\right) \sin \frac{(2k-1)\pi}{2n}$$
$$\omega_k = \cosh\left(\frac{1}{n} \sinh^{-1} \frac{1}{\varepsilon}\right) \cos \frac{(2k-1)\pi}{2n}$$

for $k = 1, 2, \ldots, n$. Evidently,
$$\frac{\sigma_k^2}{\sinh^2 u} + \frac{\omega_k^2}{\cosh^2 u} = 1$$

i.e., the zeros of $L(-s^2)$ are located on an ellipse, as depicted in Fig. 5.10.

The normalized transfer function can now be formed as
$$H_N(s) = \frac{H_0}{\prod_{i=1}^{n}(s - p_i)}$$

where p_i for $i = 1, 2, \ldots, n$ are the left-half s-plane zeros of $L(-s^2)$. Constant H_0 can be chosen as
$$H_0 = \begin{cases} 10^{-0.05 A_p} \prod_{i=1}^{n}(-p_i) & \text{for even } n \\ \prod_{i=1}^{n}(-p_i) & \text{for odd } n \end{cases}$$

so as to achieve zero minimum passband loss.

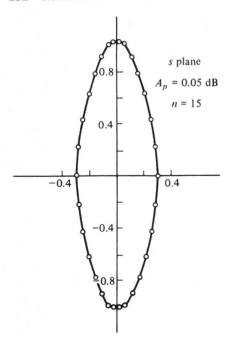

FIGURE 5.10
Zero-pole plot of $L(-s^2)$ for a fifteenth-order Chebyshev filter.

Example 5.2. Form $H_N(s)$ if $n = 4$ and $A_p = 1.0$ dB.

Solution. From Eq. (5.6)

$$x = \frac{1}{\varepsilon} = \frac{1}{\sqrt{10^{0.1} - 1}} = 1.965227$$

and

$$\sinh^{-1} \frac{1}{\varepsilon} = \ln(x + \sqrt{x^2 + 1}) = 1.427975$$

Hence

$$\sigma_k = \pm 0.364625 \sin \frac{(2k-1)\pi}{8}$$

$$\omega_k = 1.064402 \cos \frac{(2k-1)\pi}{8}$$

Thus

$$H_N(s) = \frac{H_0}{\prod_{i=1}^{2}(s - p_i)(s - p_i^*)}$$

where

$$p_1, p_1^* = -0.139536 \pm j0.983379$$

$$p_2, p_2^* = -0.336870 \pm j0.407329$$

$$H_0 = 0.245653$$

5.4.3 Inverse-Chebyshev Approximation

Another approximation which is closely related to the above is the *inverse-Chebyshev* approximation. The passband loss in this approximation is an increasing monotonic function of ω, as in the Butterworth approximation, while the stopband loss oscillates between infinity and a prescribed minimum A_a, as depicted in Fig. 5.9b. The loss is given by

$$A(\omega) = 10 \log \left[1 + \frac{1}{\delta^2 T_n^2(1/\omega)}\right]$$

where

$$\delta^2 = \frac{1}{10^{0.1A_a} - 1}$$

and the stopband extends from $\omega = 1$ to ∞. The normalized transfer function has a number of zeros on the $j\omega$ axis in this case and is given by

$$H_N(s) = H_0 \frac{\prod_{i=1}^n (s - 1/z_k)}{\prod_{i=1}^n (s - 1/p_k)}$$

where

$$H_0 = \prod_{i=1}^n \frac{z_k}{p_k}$$

$$z_k = j\Omega_{\infty k}$$

$$p_k = \sigma_k + j\omega_k$$

$$\Omega_{\infty k} = \cos \frac{(2k-1)\pi}{2n}$$

$$\sigma_k = -\sinh\left(\frac{1}{n}\sinh^{-1}\frac{1}{\delta}\right) \sin \frac{(2k-1)\pi}{2n}$$

$$\omega_k = \cosh\left(\frac{1}{n}\sinh^{-1}\frac{1}{\delta}\right) \cos \frac{(2k-1)\pi}{2n}$$

for $k = 1, 2, \ldots, n$. The derivation of $H_N(s)$ is left as an exercise for the reader (see Prob. 5.9).

5.5 ELLIPTIC APPROXIMATION

The Chebyshev approximation yields a much better passband characteristic and the inverse-Chebyshev approximation yields a much better stopband characteristic than the Butterworth approximation. An improved passband and an improved stopband characteristic can be achieved by using the *elliptic* approximation in which the passband loss oscillates between zero and a prescribed maximum A_p and the stopband loss oscillates between infinity and a prescribed minimum A_a.

The elliptic approximation is more efficient than the preceding two in that the transition between passband and stopband is steeper for a given approximation order.

Our approach to this approximation follows the formulation of Grossman [6], which, although involved, is probably the simplest.

5.5.1 Fifth-Order Approximation

The loss characteristic in a fifth-order normalized elliptic approximation is of the form depicted in Fig. 5.11, where

$$\omega_p = \sqrt{k} \qquad \omega_a = \frac{1}{\sqrt{k}} \qquad \omega_c = \sqrt{\omega_a \omega_p} = 1$$

The constants k and k_1 given by

$$k = \frac{\omega_p}{\omega_a}$$

and

$$k_1 = \left(\frac{10^{0.1 A_p} - 1}{10^{0.1 A_a} - 1} \right)^{1/2} \tag{5.17}$$

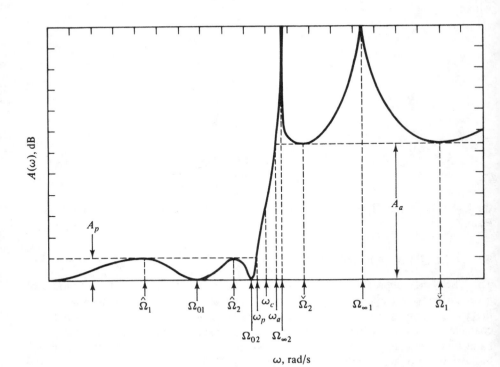

FIGURE 5.11
Loss characteristic of a fifth-order elliptic filter.

ANALOG-FILTER APPROXIMATIONS

are the *selectivity factor* and *discrimination factor*, respectively. The loss is given by

$$A(\omega) = 10 \log L(\omega^2)$$

where
$$L(\omega^2) = 1 + \varepsilon^2 F^2(\omega) \tag{5.18}$$

and
$$\varepsilon^2 = 10^{0.1 A_p} - 1 \tag{5.19}$$

Function $F(\omega)$ and in turn $L(\omega^2)$, $L(-s^2)$, and $H(s)$ are ratios of polynomials in this case.

According to the elliptic loss characteristic of Fig. 5.11, the prerequisite properties of $F(\omega)$ and $L(\omega^2)$ are as follows:

Property 1: $F(\omega) = 0$ if $\omega = 0, \pm\Omega_{01}, \pm\Omega_{02}$
Property 2: $F(\omega) = \infty$ if $\omega = \infty, \pm\Omega_{\infty 1}, \pm\Omega_{\infty 2}$
Property 3: $F^2(\omega) = 1$ if $\omega = \pm\hat{\Omega}_1, \pm\hat{\Omega}_2, \pm\sqrt{k}$
Property 4: $F^2(\omega) = \dfrac{1}{k_1^2}$ if $\omega = \pm\check{\Omega}_1, \pm\check{\Omega}_2, \pm\dfrac{1}{\sqrt{k}}$
Property 5: $\dfrac{dL(\omega^2)}{d\omega} = 0$ if $\omega = \pm\hat{\Omega}_1, \pm\hat{\Omega}_2, \pm\check{\Omega}_1, \pm\check{\Omega}_2$

By using these properties we shall attempt to derive the exact form of $F(\omega)$. The approach is analogous to that used earlier in the Chebyshev approximation.

From properties 1 and 2

$$F(\omega) = \frac{M_1 \omega (\omega^2 - \Omega_{01}^2)(\omega^2 - \Omega_{02}^2)}{(\omega^2 - \Omega_{\infty 1}^2)(\omega^2 - \Omega_{\infty 2}^2)} \tag{5.20}$$

(M_1 to M_7 represent constants), and from properties 2 and 3

$$1 - F^2(\omega) = \frac{M_2(\omega^2 - \hat{\Omega}_1^2)^2(\omega^2 - \hat{\Omega}_2^2)^2(\omega^2 - k)}{(\omega^2 - \Omega_{\infty 1}^2)^2(\omega^2 - \Omega_{\infty 2}^2)^2}$$

where the double zeros at $\omega = \pm\hat{\Omega}_1, \pm\hat{\Omega}_2$ are due to property 5 (see Sec. 5.4.1). Similarly, from properties 2, 4, and 5

$$1 - k_1^2 F^2(\omega) = \frac{M_3(\omega^2 - \check{\Omega}_1^2)^2(\omega^2 - \check{\Omega}_2^2)^2(\omega^2 - 1/k)}{(\omega^2 - \Omega_{\infty 1}^2)^2(\omega^2 - \Omega_{\infty 2}^2)^2}$$

and from property 5

$$\frac{dF(\omega)}{d\omega} = \frac{M_4(\omega^2 - \hat{\Omega}_1^2)(\omega^2 - \hat{\Omega}_2^2)(\omega^2 - \check{\Omega}_1^2)(\omega^2 - \check{\Omega}_2^2)}{(\omega^2 - \Omega_{\infty 1}^2)^2(\omega^2 - \Omega_{\infty 2}^2)^2}$$

By combining the above results we can form the important relation

$$\left[\frac{dF(\omega)}{d\omega}\right]^2 = \frac{M_5[1 - F^2(\omega)][1 - k_1^2 F^2(\omega)]}{(1 - \omega^2/k)(1 - k\omega^2)} \tag{5.21}$$

Alternatively, we can write

$$\int_0^F \frac{dx}{\sqrt{(1-x^2)(1-k_1^2 x^2)}} = \sqrt{M_5} \int_0^\omega \frac{dy}{\sqrt{(1-y^2/k)(1-ky^2)}} + M_7$$

and if $y = \sqrt{k}\, y'$, $y' = y$

$$\int_0^F \frac{dx}{\sqrt{(1-x^2)(1-k_1^2 x^2)}} = M_6 \int_0^{\omega/\sqrt{k}} \frac{dy}{\sqrt{(1-y^2)(1-k^2 y^2)}} + M_7$$

These are *elliptic integrals* of the first kind, and they can be put in the more convenient form

$$\int_0^{\phi_1} \frac{d\theta_1}{\sqrt{1-k_1^2 \sin^2 \theta_1}} = M_6 \int_0^{\phi} \frac{d\theta}{\sqrt{1-k^2 \sin^2 \theta}} + M_7$$

by using the transformations

$$x = \sin\theta_1 \qquad F = \sin\phi_1 \qquad y = \sin\theta \qquad \frac{\omega}{\sqrt{k}} = \sin\phi$$

The above two integrals can assume complex values if complex values are allowed for ϕ_1 and ϕ. By letting

$$\int_0^\phi \frac{d\theta}{\sqrt{1-k^2 \sin^2 \theta}} = z \qquad \text{where } z = u + jv$$

the solution of Eq. (5.21) can be expressed in terms of a pair of simultaneous equations as

$$\frac{\omega}{\sqrt{k}} = \sin\phi = \text{sn}\,(z, k) \tag{5.22}$$

$$F = \sin\phi_1 = \text{sn}\,(M_6 z + M_7, k_1) \tag{5.23}$$

The entities at the right-hand side are *elliptic functions*.

Further progress in this analysis can be made by using the properties of elliptic functions (see Appendix A).

As demonstrated in Sec. A.7, Eq. (5.22) is a transformation that maps trajectory $ABCD$ in Fig. 5.12a onto the positive real axis of the ω plane, as depicted in Fig. 5.12b. Since the behavior of $F(\omega)$ is known for all real values of ω, constants M_6 and M_7 can be determined. In turn, the exact form of $F(\omega)$ can be derived.

If $z = u$ and $0 \le u \le K$ (domain 1 in Sec. A.7), Eqs. (5.22) and (5.23) become

$$\omega = \sqrt{k}\, \text{sn}\,(u, k) \tag{5.24}$$

$$F = \text{sn}\,(M_6 u + M_7, k_1) \tag{5.25}$$

where ω and F have real periods of $4K$ and $4K_1/M_6$, respectively (see Sec. A.6). If $\omega = 0$, then $u = 0$; and if $\omega = \sqrt{k}$, then $u = K$, as illustrated in Fig. 5.13. Now F will correspond to $F(\omega)$ if it has zeros at $u = 0$ and at two other points in the range $0 < u \le K$ (property 1), and its magnitude is unity at $u = K$ (property 3).

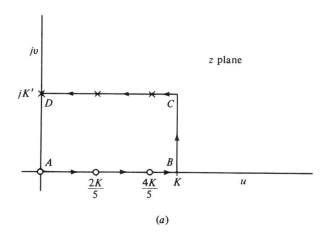

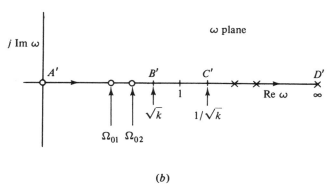

FIGURE 5.12
Mapping properties of Eq. (5.22).

Consequently, F must be of the form illustrated in Fig. 5.13. Clearly, for $u = 0$
$$F = \text{sn}\,(M_7, k_1) = 0$$
or $M_7 = 0$. Furthermore, five quarter periods of F must be equal to one quarter period of ω, i.e.,
$$M_6 = \frac{5K_1}{K}$$
and so from Eq. (5.25)
$$F = \text{sn}\left(\frac{5K_1 u}{K}, k_1\right)$$
Now F has z-plane zeros at
$$u = \frac{2Ki}{5} \quad \text{for } i = 0, 1, 2$$

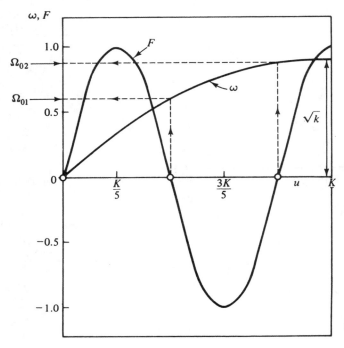

FIGURE 5.13
Plots of ω and F versus u.

and, therefore, $F(\omega)$ must have ω-plane zeros (zero-loss frequencies) at

$$\Omega_{0i} = \sqrt{k}\, \text{sn}\left(\frac{2Ki}{5}, k\right) \qquad \text{for } i = 0, 1, 2$$

according to Eq. (5.24) (see Fig. 5.12).

If $z = u + jK'$ and $0 \leq u \leq K$ (domain 3 in Sec. A.7), Eqs. (5.22) and (5.23) assume the form

$$\omega = \frac{1}{\sqrt{k}\, \text{sn}\,(u, k)} \tag{5.26}$$

$$F = \text{sn}\left[\frac{5K_1(u + jK')}{K}, k_1\right] \tag{5.27}$$

If $\omega = \infty$, $u = 0$ and F must be infinite (property 2), i.e.,

$$F = \text{sn}\left(\frac{j5K_1K'}{K}, k_1\right) = \infty$$

and from Eq. (A.19)

$$F = \frac{j\,\text{sn}\,(5K_1K'/K, k_1')}{\text{cn}\,(5K_1K'/K, k_1')} = \infty \qquad \text{where } k_1' = \sqrt{1 - k_1^2}$$

Hence it is necessary that

$$\operatorname{cn}\left(\frac{5K_1K'}{K}, k_1'\right) = 0$$

and therefore the relation

$$\frac{5K'}{K} = \frac{K_1'}{K_1} \tag{5.28}$$

must hold. The quantities K, K' are functions of k, and similarly K_1, K_1' are functions of k_1; in turn, k_1 is a function of A_p and A_a by definition. In effect, Eq. (5.28) constitutes an implicit constraint among filter specifications. We shall assume here that Eq. (5.28) holds. The implications of this assumption will be examined at a later point.

With Eq. (5.28) satisfied, Eq. (5.27) becomes

$$F = \operatorname{sn}\left(\frac{5K_1}{K}u + jK_1', k_1\right)$$

and after some manipulation

$$F = \frac{1}{k_1 \operatorname{sn}(5K_1 u/K, k_1)}$$

Evidently, $F = \infty$ if

$$u = 2Ki/5 \quad \text{for } i = 0, 1, 2 \tag{5.29}$$

that is, F has poles at

$$z = \frac{2Ki}{5} + jK' \quad \text{for } i = 0, 1, 2$$

as depicted in Fig. 5.12, and since line CD maps onto line $C'D'$, F corresponds to $F(\omega)$. That is, $F(\omega)$ has two poles in the range $1/\sqrt{k} \leq \omega < \infty$ and one at $\omega = \infty$ (property 2). The poles of $F(\omega)$ (infinite-loss frequencies) can be obtained from Eqs. (5.26) and (5.29) as

$$\Omega_{\infty i} = \frac{1}{\sqrt{k}\operatorname{sn}(2Ki/5, k)} \quad \text{for } i = 0, 1, 2$$

Therefore, the infinite-loss frequencies are the reciprocals of the zero-loss frequencies, i.e.,

$$\Omega_{\infty i} = \frac{1}{\Omega_{0i}}$$

and by eliminating $\Omega_{\infty i}$ in Eq. (5.20), we have

$$F(\omega) = \frac{M_1'\omega(\omega^2 - \Omega_{01}^2)(\omega^2 - \Omega_{02}^2)}{(1 - \omega^2\Omega_{01}^2)(1 - \omega^2\Omega_{02}^2)} \tag{5.30}$$

The only unknown at this point is constant M_1'. With $z = K + jv$ and $0 \leq v \leq K'$ (domain 2 in Sec. A.7), Eqs. (5.22) and (5.23) can be put in the form

$$\omega = \frac{\sqrt{k}}{\mathrm{dn}\,(v, k')} \qquad F = \mathrm{sn}\left[\frac{5K_1(K + jv)}{K}, k_1\right]$$

If $\omega = 1$, then $v = K'/2$ and $F(1) = M_1'$, according to Eq. (5.30). Hence

$$M_1' = \mathrm{sn}\left(5K_1 + j\frac{5K'K_1}{2K}, k_1\right)$$

or

$$M_1' = \mathrm{sn}\left(K_1 + \frac{jK_1'}{2}, k_1\right)$$

according to Eqs. (5.28) and (A.8) and after some manipulation

$$M_1' = \frac{1}{\mathrm{dn}\,(K_1'/2, k_1')} = \frac{1}{\sqrt{k_1}}$$

5.5.2 Nth-Order Approximation (n Odd)

For an nth-order approximation with n odd, constant M_7 in Eq. (5.23) is zero, and n quarter periods of F must correspond to one quarter period of ω, i.e.,

$$M_6 = \frac{nK_1}{K}$$

Therefore, Eq. (5.23) assumes the form

$$F = \mathrm{sn}\left(\frac{nK_1 z}{K}, k_1\right) \qquad (5.31)$$

where the relation

$$\frac{nK'}{K} = \frac{K_1'}{K_1}$$

must hold. The expression for $F(\omega)$ can be shown to be

$$F(\omega) = \frac{(-1)^r \omega}{\sqrt{k_1}} \prod_{i=1}^{r} \frac{\omega^2 - \Omega_i^2}{1 - \omega^2 \Omega_i^2}$$

where

$$r = \frac{n-1}{2}$$

and

$$\Omega_i = \sqrt{k}\,\mathrm{sn}\left(\frac{2Ki}{n}, k\right) \qquad \text{for } i = 1, 2, \ldots, r$$

5.5.3 Zeros and Poles of $L(-s^2)$

The next task is to determine the zeros and poles of $L(-s^2)$. From Eqs. (5.18) and (5.31), the z-domain representation of the loss function can be expressed as

$$L(z) = 1 + \varepsilon^2 \, \text{sn}^2\left(\frac{nK_1 z}{K}, k_1\right)$$

and by factoring

$$L(z) = \left[1 + j\varepsilon \, \text{sn}\left(\frac{nK_1 z}{K}, k_1\right)\right]\left[1 - j\varepsilon \, \text{sn}\left(\frac{nK_1 z}{K}, k_1\right)\right]$$

If z_1 is a root of the first factor, $-z_1$ must be a root of the second factor since the elliptic sine is an odd function of z. Consequently, the zeros of $L(z)$ can be determined by solving the equation

$$\text{sn}\left(\frac{nK_1 z}{K}, k_1\right) = \frac{j}{\varepsilon}$$

In practice, the value of k_1 is very small. For example, $k_1 \leq 0.0161$ if $A_p \leq 1$ dB and $A_a \geq 30$ dB and decreases further if A_p is reduced or A_a is increased. We can thus assume that $k_1 = 0$, in which case

$$\text{sn}\left(\frac{nK_1 z}{K}, 0\right) = \sin\frac{nK_1 z}{K} = \frac{j}{\varepsilon}$$

where $K_1 = \pi/2$, according to Eq. (A.2). Alternatively,

$$-j\frac{n\pi z}{2K} = \sinh^{-1}\frac{1}{\varepsilon}$$

and on using the identity

$$\sinh^{-1} x = \ln\left(x + \sqrt{x^2 + 1}\right)$$

and Eq. (5.19), we obtain one zero of $L(z)$ as

$$z_0 = j v_0$$

where

$$v_0 = \frac{K}{n\pi} \ln\frac{10^{0.05 A_p} + 1}{10^{0.05 A_p} - 1}$$

Now $\text{sn}\,(nK_1 z/K, k_1)$ has a real period of $4K/n$, and as a result all z_i given by

$$z_i = z_0 + \frac{4Ki}{n} \quad \text{for } i = 0, 1, 2, \ldots$$

must also be zeros of $L(z)$.

The zeros of $L(\omega^2)$ can be deduced by using the transformation between the z and ω planes, namely Eq. (5.22). In turn, the zeros of $L(-s^2)$ can be obtained by letting $\omega = s/j$. For $i = 0$, there is a real zero of $L(-s^2)$ at $s = \sigma_0$, where

$$\sigma_0 = j\sqrt{k} \, \text{sn}\,(j v_0, k) \qquad (5.32)$$

and for $i = 1, 2, \ldots, n - 1$ there are $n - 1$ distinct complex zeros at $s = \sigma_i + j\omega_i$, where

$$\sigma_i + j\omega_i = j\sqrt{k}\, \text{sn}\left(jv_0 + \frac{4Ki}{n}, k\right) \tag{5.33}$$

The remaining n zeros are negatives of zeros already determined.

For $n = 5$, the required values of the elliptic sine are

$$\text{sn}\left(jv_0 + \frac{4K}{5}\right)$$

$$\text{sn}\left(jv_0 + \frac{8K}{5}\right) = \text{sn}\left(jv_0 + 2K - \frac{2K}{5}\right) = -\text{sn}\left(jv_0 - \frac{2K}{5}\right)$$

$$\text{sn}\left(jv_0 + \frac{12K}{5}\right) = \text{sn}\left(jv_0 + 2K + \frac{2K}{5}\right) = -\text{sn}\left(jv_0 + \frac{2K}{5}\right)$$

$$\text{sn}\left(jv_0 + \frac{16K}{5}\right) = \text{sn}\left(jv_0 + 4K - \frac{4K}{5}\right) = \text{sn}\left(jv_0 - \frac{4K}{5}\right)$$

Hence Eq. (5.33) can be put in the form

$$\sigma_i + j\omega_i = j\sqrt{k}(-1)^i \text{sn}\left(jv_0 \pm \frac{2Ki}{5}, k\right) \qquad \text{for } i = 1, 2$$

Similarly, for any odd value of n

$$\sigma_i + j\omega_i = j\sqrt{k}(-1)^i \text{sn}\left(jv_0 \pm \frac{2Ki}{n}, k\right) \qquad \text{for } i = 1, 2, \ldots, \frac{n-1}{2}$$

Now with the aid of the addition formula we can show that

$$\sigma_i + j\omega_i = \frac{(-1)^i \sigma_0 V_i \pm j\Omega_i W}{1 + \sigma_0^2 \Omega_i^2} \qquad \text{for } i = 1, 2, \ldots, \frac{n-1}{2}$$

where

$$W = \sqrt{(1 + k\sigma_0^2)\left(1 + \frac{\sigma_0^2}{k}\right)} \tag{5.34}$$

$$V_i = \sqrt{(1 - k\Omega_i^2)\left(1 - \frac{\Omega_i^2}{k}\right)} \tag{5.35}$$

$$\Omega_i = \sqrt{k}\, \text{sn}\left(\frac{2Ki}{n}, k\right) \tag{5.36}$$

A complete description of $L(-s^2)$ is at this point available. It has zeros at $s = \pm\sigma_0, \pm(\sigma_i + j\omega_i)$ and double poles at $s = \pm j/\Omega_i$, which can be evaluated by using the series representation of elliptic functions given in Sec. A.8. From Eqs. (5.32) and (A.30).

$$\sigma_0 = \frac{-2q^{1/4} \sum_{m=0}^{\infty}(-1)^m q^{m(m+1)} \sinh\left[(2m+1)\Lambda\right]}{1 + 2\sum_{m=1}^{\infty}(-1)^m q^{m^2} \cosh 2m\Lambda} \tag{5.37}$$

where
$$\Lambda = \frac{1}{2n} \ln \frac{10^{0.05A_p} + 1}{10^{0.05A_p} - 1}$$

The parameter q, which is known as the *modular constant*, is given by

$$q = e^{-\pi K'/K} \quad (5.38)$$

Similarly, from Eqs. (5.36) and (A.30)

$$\Omega_i = \frac{2q^{1/4} \sum_{m=0}^{\infty} (-1)^m q^{m(m+1)} \sin \frac{(2m+1)\pi i}{n}}{1 + 2\sum_{m=1}^{\infty} (-1)^m q^{m^2} \cos \frac{2m\pi i}{n}} \quad \text{for } i = 1, 2, \ldots, \frac{n-1}{2} \quad (5.39)$$

The modular constant q can be determined by evaluating K and K' numerically. A quicker method, however, is to use the following procedure.

Since $\operatorname{dn}(0, k) = 1$, Eq. (A.32) gives

$$\sqrt{k'} = \frac{1 - 2q + 2q^4 - 2q^9 + \cdots}{1 + 2q + 2q^4 + 2q^9 + \cdots} \quad (5.40)$$

Now $q < 1$ since $K, K' > 0$, and hence a first approximation for q is

$$q_0 = \frac{1}{2}\left(\frac{1 - \sqrt{k'}}{1 + \sqrt{k'}}\right)$$

By eliminating $\sqrt{k'}$ using Eq. (5.40), rationalizing, and then performing long division we have

$$q \approx q_0 + 2q^5 - 5q^9 + 10q^{13}$$

Thus, if q_{m-1} is an approximation for q

$$q_m \approx q_0 + 2q_{m-1}^5 - 5q_{m-1}^9 + 10q_{m-1}^{13}$$

is a better approximation. By using this recursive relation repeatedly we can show that

$$q \approx q_0 + 2q_0^5 + 15q_0^9 + 150q_0^{13}$$

Since k is known, the quantities k', q_0, q, σ_0, Ω_i, σ_i, and ω_i can be evaluated. Subsequently, the normalized transfer function $H_N(s)$ can be formed.

5.5.4 Nth-Order Approximation (n Even)

So far we have been concerned with odd-order approximations. However, the results can be easily extended to the case of even n.

Function F is of the form

$$F = \operatorname{sn}\left(\frac{nK_1}{K}z + K_1, k_1\right)$$

where the relation

$$\frac{nK'}{K} = \frac{K_1'}{K_1}$$

must again hold. The expression for $F(\omega)$ in this case is given by

$$F(\omega) = \frac{(-1)^r}{\sqrt{k_1}} \prod_{i=1}^{r} \frac{\omega^2 - \Omega_i^2}{1 - \omega^2 \Omega_i^2}$$

where $\quad r = \dfrac{n}{2} \quad \Omega_i = \sqrt{k}\ \mathrm{sn}\left[\dfrac{(2i-1)K}{n}, k\right] \quad$ for $i = 1, 2, \ldots, r$

The zeros of $L(-s^2)$ are

$$s_i = \pm(\sigma_i + j\omega_i)$$

where $\quad \sigma_i + j\omega_i = \dfrac{\pm[\sigma_0 V_i + j(-1)^i \Omega_i W]}{1 + \sigma_0^2 \Omega_i^2}$

The parameters W, V_i, and σ_0 are given by Eqs. (5.34), (5.35), and (5.37), as in the case of odd n, and the values of Ω_i can be computed by replacing i by $i - \frac{1}{2}$ in the right-hand side of Eq. (5.39).

5.5.5 Specification Constraint

The results of the preceding sections are based on the assumption that the relation

$$\frac{nK'}{K} = \frac{K_1'}{K_1} \tag{5.41}$$

holds. As pointed out earlier, this equation constitutes a constraint among filter specifications of the form

$$f_1(n, k) = f_2(A_p, A_a)$$

Consequently, if three of the four parameters are specified, the fourth is automatically fixed. It is thus of interest to put Eq. (5.41) in a more useful form that can be used to evaluate the corresponding fourth parameter.

From the definition of the elliptic sine sn $(K_1, k_1) = 1$ and from Eq. (A.30)

$$k_1 = 4\sqrt{q_1}\left(\frac{1 + q_1^2 + q_1^6 + \cdots}{1 + 2q_1 + 2q_1^4 + \cdots}\right)^2 \qquad \text{where } q_1 = e^{-\pi K_1'/K_1}$$

In practice, k_1 is close to zero, k_1' is close to unity, K_1'/K_1 is large, and, as a result, $q_1 \ll 1$. Hence we can assume that

$$k_1 \approx 4\sqrt{q_1} \qquad \text{or} \qquad k_1^2 = 16 q_1 = 16 e^{-\pi K_1'/K_1}$$

By eliminating K_1'/K_1, using Eq. (5.41), we have

$$k_1^2 = 16 e^{-\pi n K'/K}$$

and from Eq. (5.38)

$$k_1^2 = 16 q^n$$

Therefore, from Eq. (5.17) the desired formula is

$$\frac{10^{0.1A_p} - 1}{10^{0.1A_a} - 1} = 16q^n \tag{5.42}$$

If n, k, and A_p are specified, the resulting minimum stopband loss is given by

$$A_a = 10 \log \left(\frac{10^{0.1A_p} - 1}{16q^n} + 1 \right) \tag{5.43}$$

This is plotted versus k in Fig. 5.14 for various values of n and A_p.

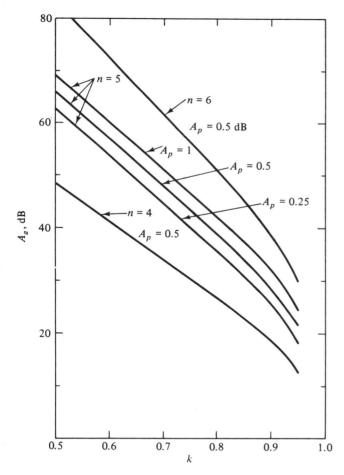

FIGURE 5.14
Plots of A_a versus k for various values of n and A_p.

Alternatively, if k, A_a, and A_p are specified, the required approximation order must satisfy the inequality

$$n \geq \frac{\log 16D}{\log (1/q)} \quad \text{where } D = \frac{10^{0.1A_a} - 1}{10^{0.1A_p} - 1}$$

5.5.6 Normalized Transfer Function

An elliptic normalized lowpass filter with a selectivity factor k, a maximum passband loss of A_p dB, and a minimum stopband loss equal to or in excess of A_a dB has a transfer function of the form

$$H_N(s) = \frac{H_0}{D_0(s)} \prod_{i=1}^{r} \frac{s^2 + a_{0i}}{s^2 + b_{1i}s + b_{0i}} \tag{5.44}$$

where

$$r = \begin{cases} \dfrac{n-1}{2} & \text{for odd } n \\ \dfrac{n}{2} & \text{for even } n \end{cases}$$

and

$$D_0(s) = \begin{cases} s + \sigma_0 & \text{for odd } n \\ 1 & \text{for even } n \end{cases}$$

The transfer-function coefficients and multiplier constant H_0 can be computed by using the following formulas in sequence:

$$k' = \sqrt{1 - k^2} \tag{5.45}$$

$$q_0 = \frac{1}{2}\left(\frac{1 - \sqrt{k'}}{1 + \sqrt{k'}}\right) \tag{5.46}$$

$$q = q_0 + 2q_0^5 + 15q_0^9 + 150q_0^{13} \tag{5.47}$$

$$D = \frac{10^{0.1A_a} - 1}{10^{0.1A_p} - 1} \tag{5.48}$$

$$n \geq \frac{\log 16D}{\log (1/q)} \tag{5.49}$$

$$\Lambda = \frac{1}{2n} \ln \frac{10^{0.05A_p} + 1}{10^{0.05A_p} - 1} \tag{5.50}$$

$$\sigma_0 = \left| \frac{2q^{1/4} \sum_{m=0}^{\infty} (-1)^m q^{m(m+1)} \sinh[(2m+1)\Lambda]}{1 + 2\sum_{m=1}^{\infty} (-1)^m q^{m^2} \cosh 2m\Lambda} \right| \tag{5.51}$$

$$W = \sqrt{(1 + k\sigma_0^2)\left(1 + \frac{\sigma_0^2}{k}\right)} \tag{5.52}$$

$$\Omega_i = \frac{2q^{1/4} \sum_{m=0}^{\infty} (-1)^m q^{m(m+1)} \sin \frac{(2m+1)\pi\mu}{n}}{1 + 2\sum_{m=1}^{\infty} (-1)^m q^{m^2} \cos \frac{2m\pi\mu}{n}} \tag{5.53}$$

where
$$\mu = \begin{cases} i & \text{for odd } n \\ i - \frac{1}{2} & \text{for even } n \end{cases} \quad i = 1, 2, \ldots, r$$

$$V_i = \sqrt{(1 - k\Omega_i^2)\left(1 - \frac{\Omega_i^2}{k}\right)} \tag{5.54}$$

$$a_{0i} = \frac{1}{\Omega_i^2} \tag{5.55}$$

$$b_{0i} = \frac{(\sigma_0 V_i)^2 + (\Omega_i W)^2}{(1 + \sigma_0^2 \Omega_i^2)^2} \tag{5.56}$$

$$b_{1i} = \frac{2\sigma_0 V_i}{1 + \sigma_0^2 \Omega_i^2} \tag{5.57}$$

$$H_0 = \begin{cases} \sigma_0 \prod_{i=1}^{r} \frac{b_{0i}}{a_{0i}} & \text{for odd } n \\ 10^{-0.05 A_p} \prod_{i=1}^{r} \frac{b_{0i}}{a_{0i}} & \text{for even } n \end{cases} \tag{5.58}$$

The actual minimum stopband loss is given by Eq. (5.43). The series in Eqs. (5.51) and (5.53) converge rapidly, and three or four terms are sufficient for most purposes.

Example 5.3. An elliptic filter is required satisfying the following specifications:

$$\omega_p = \sqrt{0.9} \text{ rad/s} \quad \omega_a = \frac{1}{\sqrt{0.9}} \text{ rad/s} \quad A_p = 0.1 \text{ dB} \quad A_a \geq 50.0 \text{ dB}$$

Form $H_N(s)$.

Solution. From Eqs. (5.45) to (5.49)

$$k = 0.9 \quad k' = 0.435890 \quad q_0 = 0.102330$$
$$q = 0.102352 \quad D = 4{,}293{,}090 \quad n \geq 7.92 \quad \text{or} \quad n = 8$$

From Eqs. (5.50) to (5.58) the transfer-function coefficients in Table 5.1 can be obtained. The corresponding loss characteristic is plotted in Fig. 5.15. The actual value of A_a is 50.82 dB according to Eq. (5.43).

TABLE 5.1
Coefficients of $H_N(s)$ (Example 5.3)

i	a_{0i}	b_{0i}	b_{1i}
1	1.434825×10	2.914919×10^{-1}	8.711574×10^{-1}
2	2.231643	6.123726×10^{-1}	4.729136×10^{-1}
3	1.320447	8.397386×10^{-1}	1.825141×10^{-1}
4	1.128832	9.264592×10^{-1}	4.471442×10^{-2}

$H_0 = 2.876332 \times 10^{-3}$

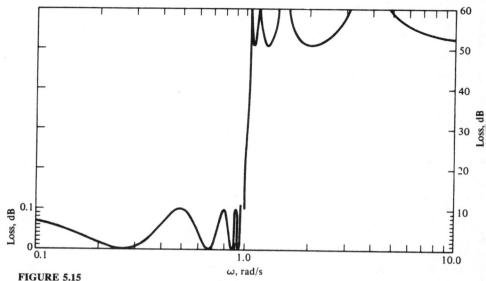

FIGURE 5.15
Loss characteristic of an eighth-order, elliptic filter (Example 5.3).

5.6 BESSEL APPROXIMATION

Ideally the group delay of a filter should be independent of frequency, or, equivalently, the phase shift should be a linear function of frequency. Since the only objective in the preceding three approximations is to achieve a specific loss characteristic, the phase characteristic turns out to be nonlinear. As a result the delay tends to vary with frequency, in particular in the elliptic approximation.

Consider the transfer function

$$H(s) = \frac{b_0}{\sum_{i=0}^{n} b_i s^i} = \frac{b_0}{s^n B(1/s)} \tag{5.59}$$

where

$$b_i = \frac{(2n-i)!}{2^{n-i} i! (n-i)!} \tag{5.60}$$

$B(s)$ is a Bessel polynomial, and $s^n B(1/s)$ can be shown to have zeros in the left-half s plane. $B(1/j\omega)$ can be expressed in terms of *Bessel* functions [2, 7] as

$$B\left(\frac{1}{j\omega}\right) = \frac{1}{j^n}\sqrt{\frac{\pi\omega}{2}}[(-1)^n J_{-v}(\omega) - j J_v(\omega)]e^{j\omega}$$

where $v = n + \frac{1}{2}$ and

$$J_v(\omega) = \omega^v \sum_{i=0}^{\infty} \frac{(-1)^i \omega^{2i}}{2^{2i+v} i! \Gamma(v+i+1)} \tag{5.61}$$

[$\Gamma(\cdot)$ is the *gamma* function]. Hence from Eq. (5.59)

$$|H(j\omega)|^2 = \frac{2b_0^2}{\pi\omega^{2n+1}[J_{-v}^2(\omega) + J_v^2(\omega)]}$$

$$\theta(\omega) = -\omega + \tan^{-1}\frac{(-1)^n J_v(\omega)}{J_{-v}(\omega)}$$

$$\tau(\omega) = -\frac{d\theta(\omega)}{d\omega} = 1 - \frac{(-1)^n(J_{-v}J_v' - J_v J_{-v}')}{J_{-v}^2(\omega) + J_v^2(\omega)}$$

Alternatively, from the properties of Bessel functions and Eq. (5.61) [2]

$$|H(j\omega)|^2 = 1 - \frac{\omega^2}{2n-1} + \frac{2(n-1)\omega^4}{(2n-1)^2(2n-3)} + \cdots$$

$$\tau(\omega) = 1 - \frac{\omega^{2n}}{b_0^2}|H(j\omega)|^2$$

Clearly, as $\omega \to 0$, $|H(j\omega)| \to 1$ and $\tau(\omega) \to 1$. Furthermore, the first $n-1$ derivatives of $\tau(\omega)$ with respect to ω^2 are zero if $\omega = 0$. This means that there is some frequency range $0 \le \omega < \omega_p$ for which the delay is approximately constant. On the other hand, if $\omega \to \infty$, $|H(j\omega)| \to 0$ and therefore $H(s)$ constitutes a lowpass constant-delay approximation. The delay is normalized to 1 s. However, any other delay can be achieved by replacing s by $\tau_0 s$ in Eq. (5.59).

Example 5.4. Form the Bessel transfer function for $n = 5$.

Solution. From Eq. (5.60)

$$H(s) = \frac{945}{945 + 945s + 420s^2 + 105s^3 + 15s^4 + s^5}$$

The corresponding loss and delay characteristics are depicted in Fig. 5.16.

5.7 TRANSFORMATIONS

In the preceding sections only normalized lowpass approximations have been considered. The reason is that denormalized lowpass, highpass, bandpass, and bandstop approximations can be easily derived by using *transformations* of the form

$$s = f(\bar{s})$$

5.7.1 Lowpass-to-Lowpass Transformation

Consider a normalized lowpass transfer function $H_N(s)$ with passband and stopband edges ω_p and ω_a, and let

$$s = \lambda \bar{s} \quad (5.62)$$

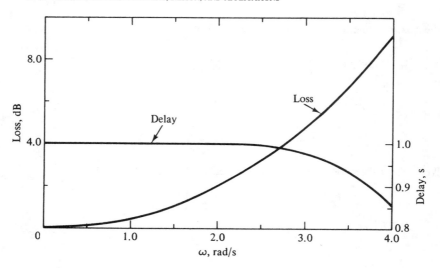

FIGURE 5.16
Loss and delay characteristics of fifth-order normalized Bessel filter (Example 5.4).

in $H_N(s)$. If $s = j\omega, \bar{s} = j\omega/\lambda$ and hence Eq. (5.62) maps the j axis of the s plane onto the j axis of the $\bar{s}$ plane. In particular, ranges 0 to $j\omega_p$ and $j\omega_a$ to $j\infty$ map onto ranges 0 to $j\omega_p/\lambda$ and $j\omega_a/\lambda$ to $j\infty$, respectively, as depicted in Fig. 5.17a. Therefore

$$H_{\text{LP}}(\bar{s}) = H_N(s)\Big|_{s=\lambda\bar{s}}$$

constitutes a denormalized lowpass approximation with passband and stopband edges ω_p/λ and ω_a/λ.

5.7.2 Lowpass-to-Bandpass Transformation

Now let

$$s = \frac{1}{B}\left(\bar{s} + \frac{\omega_0^2}{\bar{s}}\right)$$

in $H_N(s)$, where B and ω_0 are constants. If $s = j\omega$ and $\bar{s} = j\bar{\omega}$, we have

$$j\omega = \frac{j}{B}\left(\bar{\omega} - \frac{\omega_0^2}{\bar{\omega}}\right) \quad \text{or} \quad j\bar{\omega} = j\left[\frac{\omega B}{2} \pm \sqrt{\omega_0^2 + \left(\frac{\omega B}{2}\right)^2}\right]$$

Hence

$$\bar{\omega} = \begin{cases} \omega_0 & \text{if } \omega = 0 \\ \pm\bar{\omega}_{p1}, \pm\bar{\omega}_{p2} & \text{if } \omega = \pm\omega_p \\ \pm\bar{\omega}_{a1}, \pm\bar{\omega}_{a2} & \text{if } \omega = \pm\omega_a \end{cases}$$

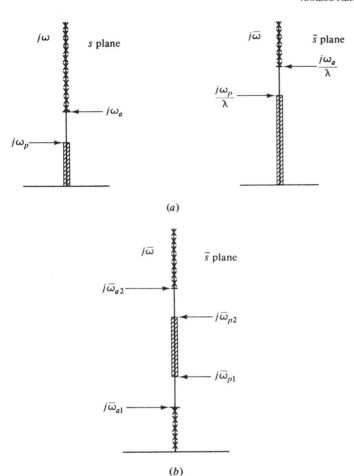

FIGURE 5.17
Analog-filter transformations: (a) lowpass to lowpass, (b) lowpass to bandpass.

where
$$\bar{\omega}_{p1}, \bar{\omega}_{p2} = \mp \frac{\omega_p B}{2} + \sqrt{\omega_0^2 + \left(\frac{\omega_p B}{2}\right)^2}$$

$$\bar{\omega}_{a1}, \bar{\omega}_{a2} = \mp \frac{\omega_a B}{2} + \sqrt{\omega_0^2 + \left(\frac{\omega_a B}{2}\right)^2}$$

The mapping for $s = j\omega$ is thus of the form illustrated in Fig. 5.17b, and consequently

$$H_{BP}(\bar{s}) = H_N(s)\bigg|_{s=\frac{1}{B}\left(\bar{s}+\frac{\omega_0^2}{\bar{s}}\right)}$$

is a bandpass approximation with passband edges ω_{p1}, ω_{p2} and stopband edges ω_{a1}, ω_{a2}.

TABLE 5.2
Analog-filter transformations

Type	Transformation
LP to LP	$s = \lambda \bar{s}$
LP to HP	$s = \dfrac{\lambda}{\bar{s}}$
LP to BP	$s = \dfrac{1}{B}\left(\bar{s} + \dfrac{\omega_0^2}{\bar{s}}\right)$
LP to BS	$s = \dfrac{B\bar{s}}{\bar{s}^2 + \omega_0^2}$

Similarly, the transformations in the second and fourth rows of Table 5.2 yield highpass and bandstop approximations.

REFERENCES

1. E. A. Guillemin, *Synthesis of Passive Networks*, Wiley, New York, 1957.
2. N. Balabanian, *Network Synthesis*, Prentice-Hall, Englewood Cliffs, N.J., 1958.
3. L. Weinberg, *Network Analysis and Synthesis*, McGraw-Hill, New York, 1962.
4. J. K. Skwirzynski, *Design Theory and Data for Electrical Filters*, Van Nostrand, London, 1965.
5. R. W. Daniels, *Approximation Methods for Electronic Filter Design*, McGraw-Hill, New York, 1974.
6. A. J. Grossman, "Synthesis of Tchebysheff Parameter Symmetrical Filters," *Proc. IRE*, vol. 45, pp. 454–473, April 1957.
7. G. N. Watson, *A Treatise on the Theory of Bessel Functions*, Cambridge University Press, London, 1948.

PROBLEMS

5.1. A fourth-order lowpass Butterworth analog filter is required.
   (a) Obtain the normalized transfer function $H_N(s)$.
   (b) Derive expressions for the loss and phase shift of the filter.
   (c) Calculate the loss and phase shift at $\omega = 0.5$ rad/s.
   (d) Obtain a corresponding denormalized transfer function $H_D(s)$ with a 3-dB cutoff frequency at 1000 rad/s.

5.2. A fifth-order Butterworth analog filter is required.
   (a) Form $H(s)$.
   (b) Plot the loss characteristic.

5.3. Filter specifications are often described pictorially as in Fig. P5.3, where $\tilde{\Omega}_p$ and $\tilde{\Omega}_a$ are desired passband and stopband edges, respectively, A_p is the maximum passband loss, and A_a is the minimum stopband loss. Find n, and in turn form $H(s)$, if $\tilde{\Omega}_p = 1$, $\tilde{\Omega}_a = 3$ rad/s, $A_p = 3.0$, $A_a \geq 45$ dB. Use the Butterworth approximation.

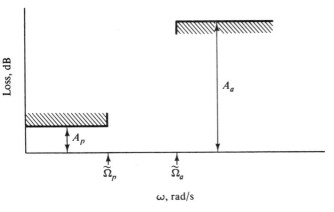

FIGURE P5.3

5.4. A third-order lowpass analog filter with passband edge $\omega_p = 1$ rad/s and passband ripple $A_p = 1.0$ dB is required. Obtain the poles and multiplier constant of the transfer function assuming a Chebyshev approximation.

5.5. A fifth-order normalized lowpass Chebyshev filter is required.
(a) Form $H(s)$ if $A_p = 0.1$ dB.
(b) Plot the loss characteristic of the filter.

5.6. A Chebyshev filter satisfying the specifications of Fig. P5.6 is required. Find n and in turn form $H(s)$.

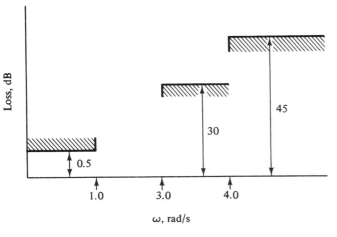

FIGURE P5.6

5.7. (a) Show that

$$T_{n+1}(\omega) = 2\omega T_n(\omega) - T_{n-1}(\omega)$$

(b) Hence demonstrate that the following relation [5] holds:

$$T_n(\omega) = \frac{n}{2} \sum_{r=0}^{K} \frac{(-1)^r (n-r-1)!}{r!(n-2r)!} (2\omega)^{n-2r} \qquad \text{where } K = \text{Int } \frac{n}{2}$$

(c) Obtain $T_{10}(\omega)$.

5.8. (a) Find $A(\omega)$ for the normalized lowpass Butterworth and Chebyshev approximations if $\omega \gg 1$.

(b) Show that $A(\omega)$ increases at the rate of $20n$ dB/decade in both cases.

5.9. The inverse-Chebyshev approximation can be derived by considering the loss function

$$A(\omega) = 10 \log \left[1 + \frac{1}{\delta^2 T_n^2(\omega)} \right]$$

where

$$\delta^2 = \frac{1}{10^{0.1 A_a} - 1}$$

(a) Show that $A(\omega)$ represents a highpass filter with an equiripple stopband loss, a monotonic decreasing passband loss, and a stopband edge $\omega_a = 1$ rad/s.

(b) Show that the filter represented by $A(\omega)$ has a transfer function of the form

$$H_{HP}(s) = \frac{\prod_{i=1}^{n}(s - z_k)}{\prod_{i=1}^{n}(s - p_k)}$$

where z_k and p_k for $k = 1, 2, \ldots, n$ are given at the end of Sec. 5.4.3.

(c) Show that $H_N(s) = H_{HP}(1/s)$ is the normalized lowpass transfer function for the inverse-Chebyshev approximation.

5.10. A fourth-order inverse-Chebyshev analog filter with a minimum stopband loss of 40 dB is required.

(a) Obtain the required transfer function.

(b) Find the 3-dB cutoff frequency.

5.11. Prove that

(a) $\sqrt{k} \, \text{sn}\,(u + jK', k) = \dfrac{1}{\sqrt{k} \, \text{sn}\,(u, k)}$

(b) $\text{sn}\left(\dfrac{5K_1 u}{K} + jK'_1, k_1 \right) = \dfrac{1}{k_1 \, \text{sn}\,(5K_1 u/K, k_1)}$

(c) $\text{sn}\left(K_1 + \dfrac{jK'_1}{2}, k_1 \right) = \dfrac{1}{\text{dn}\,(K'_1/2, k'_1)} = \dfrac{1}{\sqrt{k_1}}$

Parameter u is real in the range $0 \le u \le K$.

5.12. Show that

(a) $\cosh^{-1} x = \pm \ln (x + \sqrt{x^2 - 1})$

(b) $\sinh^{-1} x = \ln (x + \sqrt{x^2 + 1})$

5.13. Prove that

$$j\sqrt{k}(-1)^i \, \text{sn}\left(jv_0 \pm \frac{2Ki}{n}, k \right) = \frac{(-1)^i \sigma_0 V_i \pm j\Omega_i W}{1 + \sigma_0^2 \Omega_i^2}$$

where W, V_i and Ω_i are given by Eqs. (5.34) to (5.36).

5.14. Check the derivation of Eqs. (5.37) and (5.39).

5.15. Show that

$$q \approx q_0 + 2q_0^5 + 15q_0^9 + 150q_0^{13} \quad \text{where } q_0 = \frac{1}{2}\left(\frac{1-\sqrt{k'}}{1+\sqrt{k'}}\right)$$

5.16. (a) Sketch the plots of ω and F versus u for values of u in the range 0 to K, if $n = 7$ (see Fig. 5.13).
(b) Repeat (a) for $n = 8$.

5.17. Check the derivation of Eqs. (5.56) and (5.57).

5.18. (a) A lowpass elliptic filter is required satisfying the specifications

$$n = 4 \quad A_p = 1.0 \text{ dB} \quad k = 0.7$$

Form $H(s)$.
(b) Determine the corresponding minimum stopband loss.
(c) Plot the loss characteristic.

5.19. In a particular application an elliptic lowpass filter is required. The specifications are

$$k = 0.6 \quad A_p = 0.5 \text{ dB} \quad A_a \geq 40 \text{ dB}$$

Find n and form $H(s)$.

5.20. An elliptic lowpass analog filter satisfying the specifications

$$\omega_p = \sqrt{0.95} \quad \omega_a = 1/\sqrt{0.95} \text{ rad/s} \quad A_p = 0.3 \quad A_a \geq 60.0 \text{ dB}$$

is required.
(a) Determine the order of the transfer function.
(b) Determine the actual loss.
(c) Obtain the transfer function.

5.21. (a) Obtain $H(s)$ for the sixth-order normalized Bessel approximation.
(b) Plot the corresponding phase characteristic.

5.22. Show that

$$H(s) = \frac{\sum_{i=0}^{n} b_i (-s)^i}{\sum_{i=0}^{n} b_i s^i}$$

where

$$b_i = \frac{(2n-i)!}{2^{n-i} i! (n-i)!}$$

is a constant-delay, allpass transfer function.

5.23. A normalized elliptic transfer function for which $k = 0.8$ and $A_p = 0.1$ dB is subjected to the lowpass-to-bandpass transformation. Find the resulting passband and stopband edges and also the passband width if $B = 200$, $\omega_0 = 1000$ rad/s.

5.24. A highpass elliptic filter is required satisfying the following specifications:

$$\tilde{\Omega}_p = 3000 \text{ rad/s} \quad \tilde{\Omega}_a = 1000 \text{ rad/s} \quad A_p = 0.5 \text{ dB} \quad A_a \geq 40 \text{ dB}$$

(a) Find the necessary order n and the parameter of transformation λ.
(b) Hence form $H(s)$.

5.25. Figure P5.25 depicts a required bandpass-filter specification. Assuming that the elliptic approximation is to be employed, find ω_0, k, B, and n.

5.26. (a) Derive a highpass LC filter from the lowpass filter shown in Fig. P5.26.
(b) Now derive a bandpass LC filter.

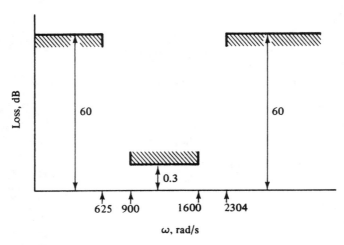

FIGURE P5.25

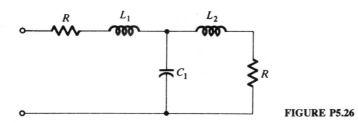

FIGURE P5.26

5.27. A constant-delay lowpass filter is required with a group delay of 1 ms. Form $H(s)$ using the sixth-order Bessel approximation.

CHAPTER 6

CONTINUOUS-TIME, SAMPLED, AND DISCRETE-TIME SIGNALS

6.1 INTRODUCTION

So far the digital filter has been treated as a distinct entity with its own methods of analysis. In this chapter, various *interrelations* are established between continuous-time, sampled, and discrete-time signals. Through these interrelations many analog techniques can be used in the analysis and design of digital filters. In addition, as will be demonstrated at the end of the chapter, digital filters can be used for the processing of continuous-time signals.

The chapter begins with a review of the Fourier transform and Fourier series and their properties and interrelations. *Poisson's summation formula* is then derived. This formula leads to a crucial interrelation between the spectra of the various types of signals, as will be shown in Sec. 6.6. Impulse functions and periodic signals are handled with some rigor in terms of *generalized functions* but every effort is made to keep abstraction to a minimum.

6.2 THE FOURIER TRANSFORM

The *Fourier transform* of a function $f(t)$ is defined by

$$F(j\omega) = \int_{-\infty}^{\infty} f(t) e^{-j\omega t} dt \tag{6.1}$$

In general $F(j\omega)$ is complex and can be written as
$$F(j\omega) = A(\omega)e^{j\phi(\omega)}$$
where $\quad A(\omega) = |F(j\omega)| \quad$ and $\quad \phi(\omega) = \arg F(j\omega)$

The functions $A(\omega)$ and $\phi(\omega)$ are called the *amplitude spectrum* and *phase spectrum* of $f(t)$, respectively.

The function $f(t)$ is said to be the *inverse Fourier transform* of $F(j\omega)$ and is given by
$$f(t) = \frac{1}{2\pi} \int_{-\infty}^{\infty} F(j\omega)e^{j\omega t} d\omega \tag{6.2}$$

Equations (6.1) and (6.2) can be represented by
$$F(j\omega) = \mathcal{F}f(t) \quad \text{and} \quad f(t) = \mathcal{F}^{-1}F(j\omega)$$
respectively. An alternative notation combining both equations is
$$f(t) \leftrightarrow F(j\omega)$$

The salient properties of the Fourier transform are summarized here through a list of theorems. A thorough treatment of the topic can be found in Papoulis [1] and Lighthill [2].

Theorem 6.1   Convergence. If
$$\lim_{T \to \infty} \int_{-T}^{T} |f(t)| dt < \infty$$
then the Fourier transform $F(j\omega)$ exists and satisfies Eq. (6.2).

This is a sufficient but not necessary condition for convergence; i.e., there are functions that are not absolutely integrable but have a Fourier transform satisfying Eq. (6.2) (see Sec. 6.3). It should be mentioned that if the inverse Fourier transform exists, then Eq. (6.2) holds at points where $f(t)$ is discontinuous if it is assumed that
$$f(t) = \frac{f(t+) + f(t-)}{2}$$
at these points, where
$$f(t+) = \lim_{\epsilon \to 0} f(t + \epsilon) \quad \text{and} \quad f(t-) = \lim_{\epsilon \to 0} f(t - \epsilon)$$
(see [1, pp. 29–31] for proof).

The following theorems hold if
$$f(t) \leftrightarrow F(j\omega) \quad \text{and} \quad g(t) \leftrightarrow G(j\omega)$$
The parameters a, b, t_0, and ω_0 are arbitrary real constants.

Theorem 6.2   Linearity
$$af(t) + bg(t) \leftrightarrow aF(j\omega) + bG(j\omega)$$

Theorem 6.3 Symmetry

$$F(jt) \leftrightarrow 2\pi f(-\omega)$$

Theorem 6.4 Time scaling

$$f(at) \leftrightarrow \frac{1}{|a|} F\left(\frac{j\omega}{a}\right)$$

Theorem 6.5 Time shifting

$$f(t - t_0) \leftrightarrow F(j\omega)e^{-j\omega t_0}$$

Theorem 6.6 Frequency shifting

$$e^{j\omega_0 t} f(t) \leftrightarrow F(j\omega - j\omega_0)$$

Theorem 6.7 Time convolution

$$f(t) \otimes g(t) \leftrightarrow F(j\omega)G(j\omega)$$

where

$$f(t) \otimes g(t) = \int_{-\infty}^{\infty} f(\tau)g(t - \tau)d\tau$$

The time convolution leads to the fundamental frequency-domain characterization of analog filters and systems in general. If the input, output, and impulse response of an analog filter are denoted by $x(t)$, $y(t)$, and $h(t)$, respectively, it can be shown that

$$y(t) = \int_{-\infty}^{\infty} x(\tau)h(t - \tau)d\tau$$

Hence the above theorem gives

$$Y(j\omega) = H(j\omega)X(j\omega)$$

where $H(j\omega)$ is the frequency response of the filter. Replacing $j\omega$ by s yields $H(s)$, the transfer function of the filter.

Theorem 6.8 Frequency convolution

$$f(t)g(t) \leftrightarrow F(j\omega) \otimes G(j\omega)$$

where

$$F(j\omega) \otimes G(j\omega) = \frac{1}{2\pi} \int_{-\infty}^{\infty} F(jv)G(j\omega - jv)dv$$

Theorem 6.9 Parseval's formula

$$\int_{-\infty}^{\infty} |f(t)|^2 dt = \frac{1}{2\pi} \int_{-\infty}^{\infty} |F(j\omega)|^2 d\omega$$

If $f(t)$ represents a voltage or current waveform, the left-hand integral represents the total energy that would be delivered to a 1-Ω resistor. Thus the quantity $|F(j\omega)|^2$ represents the energy density per unit bandwidth (in hertz), and is said to be the *energy spectral density*. As a function of ω, $|F(j\omega)|^2$ is called the *energy spectrum* of $f(t)$. Parseval's formula is the basic tool in obtaining a frequency-domain representation for random signals, as will be shown in Chap. 10.

Example 6.1. Find $F(j\omega)$ if

$$f(t) = p_{t_0}(t) = \begin{cases} 1 & \text{for } |t| \leq t_0 \\ 0 & \text{for } |t| > t_0 \end{cases}$$

Solution

$$F(j\omega) = \int_{-\infty}^{\infty} p_{t_0}(t) e^{-j\omega t}\, dt = \int_{-t_0}^{t_0} e^{-j\omega t}\, dt$$

$$= \frac{2 \sin \omega t_0}{\omega}$$

or
$$p_{t_0}(t) \leftrightarrow \frac{2 \sin \omega t_0}{\omega}$$

Function $p_{t_0}(t)$ and its amplitude spectrum are illustrated in Fig. 6.1.

Example 6.2. Find $F(j\omega)$ if

$$f(t) = \begin{cases} e^{-at} & \text{for } t \geq 0+ \\ 0 & \text{otherwise} \end{cases}$$

where a is a positive constant.

Solution. We can write

$$f(t) = u(t) e^{-at}$$

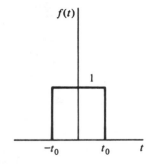
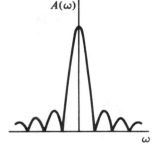

FIGURE 6.1
Function $f(t)$ and its amplitude spectrum (Example 6.1).

where $u(t)$ is the unit-step function defined by

$$u(t) = \begin{cases} 1 & \text{for } t \geq 0+ \\ 0 & \text{otherwise} \end{cases}$$

Hence

$$F(j\omega) = \int_{-\infty}^{\infty} [u(t)e^{-at}]e^{-j\omega t}\,dt = \int_{0}^{\infty} e^{-(a+j\omega)t}\,dt$$

$$= \frac{1}{a+j\omega}$$

or

$$u(t)e^{-at} \leftrightarrow \frac{1}{a+j\omega}$$

(see Fig. 6.2).

Example 6.3. Find $F(j\omega)$ if

$$f(t) = \begin{cases} e^{-at}\sin\omega_0 t & \text{for } t \geq 0+ \\ 0 & \text{otherwise} \end{cases}$$

and $a > 0$.

Solution

$$f(t) = \frac{u(t)}{2j}(e^{j\omega_0 t} - e^{-j\omega_0 t})e^{-at}$$

Hence from Theorem 6.6 and Example 6.2

$$F(j\omega) = \frac{1}{2j}\left[\frac{1}{a+j(\omega-\omega_0)} - \frac{1}{a+j(\omega+\omega_0)}\right] = \frac{\omega_0}{(a+j\omega)^2 + \omega_0^2}$$

or

$$u(t)e^{-at}\sin\omega_0 t \leftrightarrow \frac{\omega_0}{(a+j\omega)^2 + \omega_0^2}$$

(see Fig. 6.3).

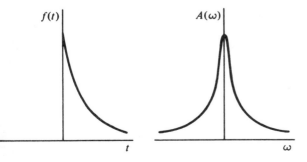

FIGURE 6.2
Function $f(t)$ and its amplitude spectrum (Example 6.2).

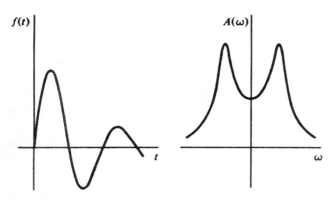

FIGURE 6.3
Function $f(t)$ and its amplitude spectrum (Example 6.3).

6.3 GENERALIZED FUNCTIONS

Some of the most important functions in signal analysis have no Fourier transform in ordinary function theory; i.e., either $F(j\omega)$ does not exist or it does not satisfy Eq. (6.2). An important function that leads to mathematical difficulties is the pulse function

$$p_0(t) = \begin{cases} \lim_{\epsilon \to 0} \dfrac{1}{2\epsilon} & \text{for } |t| \leq \epsilon \\ 0 & \text{otherwise} \end{cases} \quad (6.3)$$

depicted in Fig. 6.4a, which has often been used to define the continuous-time impulse function. From Eq. (6.1), we can readily obtain the Fourier transform of $p_0(t)$ as

$$\mathcal{F} p_0(t) = \int_{-\infty}^{\infty} p_0(t) e^{-j\omega t} dt = \lim_{\epsilon \to 0} \int_{-1/\epsilon}^{1/\epsilon} \frac{1}{2\epsilon} e^{-j\omega t} dt = 1$$

If we now attempt to obtain the inverse Fourier transform of unity, we have

$$f(t) = \frac{1}{2\pi} \int_{-\infty}^{\infty} e^{j\omega t} d\omega = \frac{1}{2\pi} \left[\int_{-\infty}^{\infty} \cos \omega t \, d\omega + j \int_{-\infty}^{\infty} \sin \omega t \, d\omega \right]$$

Evidently, these integrals do not converge and, therefore, we conclude that $F(j\omega)$ does not satisfy Eq. (6.2). A similar problem arises if we attempt to find the Fourier transform of any infinite-energy signal, such as a periodic signal. These difficulties can be overcome in a somewhat subtle mathematical way by defining $f(t)$ and $F(j\omega)$ in terms of sequences of well-behaved functions called *generalized functions*. A brief outline of this theory based on Lighthill's approach [2] follows.

Definition 1. (a) A function $\phi(t)$ is said to be a *good function* if it is everywhere differentiable any number of times such that the function itself and all its derivatives are

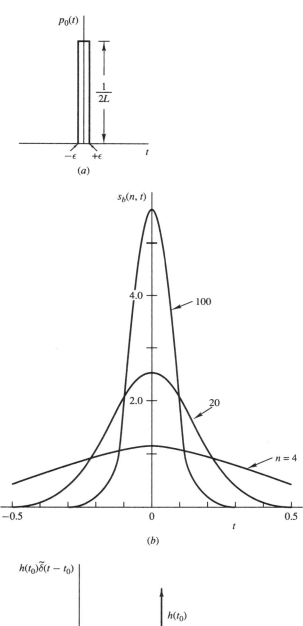
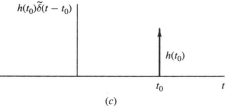

FIGURE 6.4
(a) Function $p_0(t)$, (b) a possible sequence of good functions defining $\tilde{\delta}(t)$, (c) simplified graph for an impulse.

of order $|t|^{-N}$ as $|t| \to \infty$, for all N, i.e.,

$$\lim_{t \to \infty} |t|^N |\phi(t)| < \infty \quad \text{and} \quad \lim_{t \to \infty} |t|^N \left| \frac{d^k \phi(t)}{dt^k} \right| < \infty$$

for all N and k.

Theorem 6.10. (a) The derivative of a good function is a good function.
 (b) The sum and product of two good functions are good functions.
 (c) The Fourier transform $F(j\omega)$ of a good function $f(t)$ exists, is a good function, and satisfies Eq. (6.2) (see [2, pp. 15–16] for proofs).

Definition 2. A sequence of good functions $s(n, t)$ is said to be *regular* if the limit

$$L = \lim_{n \to \infty} \int_{-\infty}^{\infty} s(n, t) \phi(t) dt \tag{6.4}$$

exists for any good function $\phi(t)$. For example, $s(n, t) = e^{-t^2/n^2}$ is regular since

$$\lim_{n \to \infty} \int_{-\infty}^{\infty} e^{-t^2/n^2} \phi(t) dt = \int_{-\infty}^{\infty} \phi(t) dt$$

Definition 3. Two or more sequences of good functions are called *equivalent* if for any good function $\phi(t)$ limit L in Eq. (6.4) exists and is the same for each sequence. For example, e^{-t^4/n^4} is equivalent to e^{-t^2/n^2}.

Definition 4. (a) A generalized function denoted by $\tilde{f}(t)$ is defined as a regular sequence $s_f(n, t)$ of good functions.
 (b) The integral of $\tilde{f}(t)\phi(t)$ between $-\infty$ and ∞, where $\phi(t)$ is any good function, is defined by

$$\int_{-\infty}^{\infty} \tilde{f}(t) \phi(t) dt = \lim_{n \to \infty} \int_{-\infty}^{\infty} s_f(n, t) \phi(t) dt \tag{6.5}$$

 (c) Two or more equivalent sequences define a unique generalized function.

Definition 5. If $\tilde{f}(t)$ and $\tilde{g}(t)$ are generalized functions defined by sequences $s_f(n, t)$ and $s_g(n, t)$, respectively, then $\tilde{f}(t) + \tilde{g}(t)$, $\tilde{f}'(t)$, $\tilde{f}(at + b)$, and $\tilde{f}(t)\phi(t)$ are defined by $s_f(n, t) + s_g(n, t)$, $s_f'(n, t)$, $s_f(n, at + b)$, and $s_f(n, t)\phi(t)$, respectively. Also, the Fourier transform of $\tilde{f}(t)$, namely,

$$\mathcal{F}\tilde{f}(t) = \tilde{F}(j\omega)$$

is defined by

$$\mathcal{F}s_f(n, t) = S_f(n, j\omega)$$

Theorem 6.11. A generalized function $\tilde{f}(t)$ has a Fourier transform $\tilde{F}(t)$ that satisfies Eq. (6.2), i.e.,

$$\tilde{f}(t) \leftrightarrow \tilde{F}(j\omega)$$

A generalized function $\tilde{f}(t)$ is defined by a regular sequence of good functions $f(n, t)$ which has a Fourier transform $S_f(n, j\omega)$ that satisfies Eq. (6.2) by virtue of Theorem 6.10c, i.e.,

$$s_f(n, t) \leftrightarrow S_f(n, j\omega)$$

Since the Fourier transform of $\tilde{f}(t)$ and inverse Fourier transform of $\tilde{F}(j\omega)$ are defined by the sequences $S_f(n, j\omega)$ and $s_f(n, t)$, respectively, according to Definition 5, Theorem 6.11 follows.

Definition 1 establishes a class of functions, referred to as good functions, that are everywhere differentiable any number of times and are well-behaved at infinity. Integrals of these functions from $-\infty$ to ∞ converge, as can be shown. Furthermore, functions of this class have Fourier transforms that satisfy Eq. (6.2). Generalized functions are defined in terms of sequences of good functions using Eq. (6.5) in Definition 4b. Unlike ordinary functions, generalized functions do not assume specific values. Their effect, characterization, and interaction with other functions are based on limit L defined in Eq. (6.4). Therefore, sequences used to define generalized functions must have a limit L; i.e., they must be regular according to Definition 2. It is often possible for two or more entirely different sequences to have the same limit L. Such sequences cannot define distinct generalized functions through Definition 4b. They must, therefore, define one and the same generalized function, according to Definition 4c. For this reason, they are called equivalent, according to Definition 3. Definition 5 provides a means of manipulating generalized functions, e.g., $\tilde{f}(t)$ and $\tilde{g}(t)$ are added by adding their defining sequences, their Fourier transforms are obtained by obtaining the Fourier transforms of their defining sequences, and so on.

6.3.1 Impulse and Unity Functions

An ordinary function $f(t)$ behaves as an *impulse function* if it assumes a large value as $|t| \to 0$ such that for any function $\phi(t)$ which is continuous at $t = 0$, the integral of $f(t)\phi(t)$ between $-\infty$ and ∞ exists and assumes the value $\phi(0)$. The pulse function in Eq. (6.3) has this property, but its inverse for the case where $\epsilon \to 0$ does not exist as was demonstrated earlier. This difficulty can be avoided by defining the impulse function as a generalized function $\tilde{\delta}(t)$ using Definition 4b, i.e.,

$$\int_{-\infty}^{\infty} \tilde{\delta}(t)\phi(t)dt = \lim_{n \to \infty} \int_{-\infty}^{\infty} s_\delta(n, t)\phi(t)dt = \phi(0) \quad (6.6)$$

A possible sequence of good functions that defines $\tilde{\delta}(t)$ is

$$s_\delta(n, t) = \sqrt{(n/\pi)} e^{-nt^2} \quad (6.7)$$

and is illustrated in Fig. 6.4b (see [2, p. 17] for proof).

Some fundamental properties of $\tilde{\delta}(t)$ follow readily from Eq. (6.6), for example

$$\int_{-\infty}^{\infty} \tilde{\delta}(t - t_0)\phi(t)dt = \int_{-\infty}^{\infty} \tilde{\delta}(t)\phi(t + t_0)dt = \phi(t_0) \quad (6.8)$$

Also for any good function $h(t)$

$$\int_{-\infty}^{\infty} \tilde{\delta}(t - t_0)h(t)\phi(t)dt = h(t_0)\phi(t_0)$$

and

$$\int_{-\infty}^{\infty} \tilde{\delta}(t - t_0)h(t_0)\phi(t)dt = h(t_0)\phi(t_0)$$

Therefore

$$\tilde{\delta}(t - t_0)h(t) = \tilde{\delta}(t - t_0)h(t_0) = h(t_0)\tilde{\delta}(t - t_0) \tag{6.9}$$

The quantity at the right-hand side is an impulse of strength $h(t_0)$ located at $t = t_0$. It can be represented by the simplified graph of Fig. 6.4c.

It should be mentioned that although the relation in Eq. (6.9) has been derived on the assumption that $h(t)$ is a good function, it holds for any function $h(t)$ which is continuous at $t = 0$, as can be shown by using Eqs. (6.6) and (6.7). This relation will be used extensively in the following sections.

Another important generalized function is the *unity function* $\tilde{i}(t)$ defined by

$$\int_{-\infty}^{\infty} \tilde{i}(t)\phi(t)dt = \lim_{n \to \infty} \int_{-\infty}^{\infty} s_i(n, t)\phi(t)dt = \int_{-\infty}^{\infty} \phi(t)dt \tag{6.10}$$

This has the effect of multiplying a good function $\phi(t)$ by unity. A possible sequence of good functions that defines $\tilde{i}(t)$ is

$$s_i(n, t) = e^{-t^2/4n} \tag{6.11}$$

and is illustrated in Fig. 6.5a. The unity function $\tilde{i}(t)$ can be represented by the simplified graph of Fig. 6.5b.

The impulse and unity functions have been defined in the time domain. Corresponding functions can be readily defined in the frequency domain by replacing t by ω in $s_\delta(n, t)$ and $s_i(n, t)$, respectively.

The Fourier transform of $s_\delta(n, t)$ can now be obtained. From Eq. (6.7)

$$\mathcal{F}s_\delta(n, t) = \int_{-\infty}^{\infty} s_\delta(n, t)e^{-j\omega t}dt = \sqrt{(n/\pi)} \int_{-\infty}^{\infty} e^{-nt^2 - j\omega t}dt$$

and from standard integral tables, we have

$$\mathcal{F}s_\delta(n, t) = e^{-\omega^2/4n}$$

Hence from Eq. (6.11)

$$s_\delta(n, t) \leftrightarrow s_i(n, \omega)$$

and therefore from Definition 5

$$\tilde{\delta}(t) \leftrightarrow \tilde{i}(\omega)$$

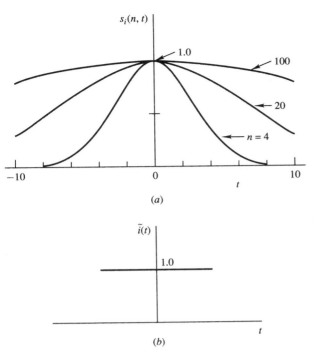

FIGURE 6.5
(a) A possible sequence of good functions defining $\tilde{i}(t)$, (b) simplified graph for the unity function.

6.3.2 Alternative Sequences for Impulse and Unity Functions

In the application of digital filters for the processing of continuous-time signals, a sampled signal is first obtained which consists of a train of impulses (see Secs. 6.6 and 6.9). The sampled signal is then converted into a discrete-time signal. At the output of the filter, the above sequence of events is reversed. The processed discrete-time signal is first converted into a sampled signal which is, in turn, converted into a continuous-time signal. It is, therefore, of significant practical importance to define the impulse function in terms of a sequence whose elements can be approximated by practical waveforms that can be easily generated by electronic devices.

Consider the sequence of functions

$$p_{\epsilon/2n}(n, t) = \begin{cases} \dfrac{n}{\epsilon} & \text{for } |t| \leq \dfrac{\epsilon}{2n} \\ 0 & \text{otherwise} \end{cases}$$

illustrated in Fig. 6.6a. The elements of this sequence are simple pulses like that in Fig. 6.4a and can be easily generated in practice. Since

$$\lim_{n \to \infty} \int_{-\infty}^{\infty} p_{\epsilon/2n}(n, t)\phi(t)dt = \lim_{n \to \infty} \int_{-\epsilon/2n}^{\epsilon/2n} \frac{n}{\epsilon}\phi(t)dt = \phi(0)$$

the sequence $p_{\epsilon/2n}(n, t)$ behaves just like $s_\delta(n, t)$ given by Eq. (6.7). Unfortunately,

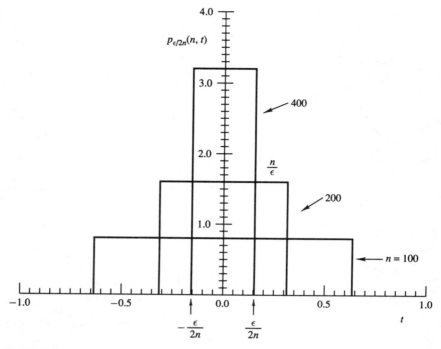

FIGURE 6.6a
Alternative sequences for impulse and unity functions: (*a*) sequence of pulse functions.

functions $p_{\epsilon/2n}(n, t)$ are not everywhere differentiable any number of times (i.e., they are not good functions according to Definition 1) and, consequently, cannot define a generalized function. This problem can be overcome by creating the sequence

$$s_\delta^i(n, t) = p_{\epsilon/2n}(t, n) \otimes \sqrt{(n/\pi)}e^{-nt^2} \qquad (6.12)$$

where the symbol $\otimes$ represents convolution. For any good function $\phi(t)$, we can write

$$\lim_{n\to\infty} \int_{-\infty}^{\infty} s_\delta^i(n, t)\phi(t)dt = \lim_{n\to\infty} \int_{-\infty}^{\infty} \left[\int_{-\infty}^{\infty} p_{\epsilon/2n}(\tau)\sqrt{(n/\pi)}e^{-n(t-\tau)^2}d\tau\right]\phi(t)dt$$

$$= \lim_{n\to\infty} \int_{-\infty}^{\infty} \sqrt{(n/\pi)}e^{-nt^2}\phi(t)dt = \phi(0)$$

and, hence, we conclude that the sequence $s_\delta^i(n, t)$ is equivalent to $s_\delta(n, t)$.
Now consider the sequence

$$s_i^i(n, \omega) = \frac{\sin(\omega\epsilon/2n)}{(\omega\epsilon/2n)} \times e^{-\omega^2/4n} \qquad (6.13)$$

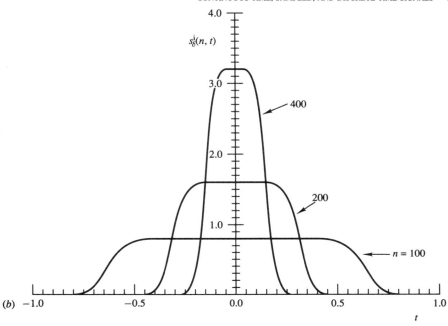
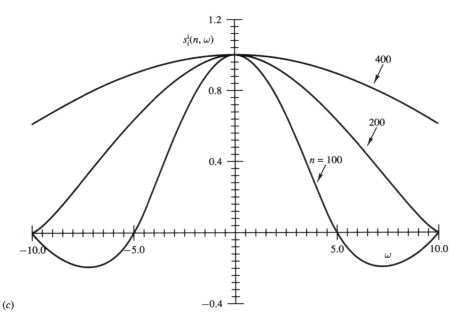

FIGURE 6.6b and 6.6c
Alternative sequences for impulse and unity functions: (b) sequence for impulse function; (c) sequence for unity function.

We can write

$$\lim_{n\to\infty}\int_{-\infty}^{\infty} s_i^i(n,\omega)\phi(\omega)d\omega = \lim_{n\to\infty}\int_{-\infty}^{-\Omega} + \lim_{n\to\infty}\int_{-\Omega}^{\Omega}$$
$$+ \lim_{n\to\infty}\int_{\Omega}^{\infty} \frac{\sin(\omega\epsilon/2n)}{(\omega\epsilon/2n)} \times e^{-\omega^2/2n}\phi(\omega)d\omega$$

and, as $\Omega \to \infty$, the first and last integrals at the right-hand side tend to zero, and the middle integral assumes the form

$$\lim_{n\to\infty}\int_{-\Omega}^{\Omega} \frac{\sin(\omega\epsilon/2n)}{(\omega\epsilon/2n)} \times e^{-\omega^2/2n}\phi(\omega)d\omega = \int_{-\Omega}^{\Omega} \phi(\omega)d\omega$$

Hence for any good function $\phi(\omega)$

$$\lim_{n\to\infty}\int_{-\infty}^{\infty} s_i^i(n,\omega)\phi(\omega)d\omega = \int_{-\infty}^{\infty} \phi(\omega)d\omega$$

and, from Eq. (6.10), we conclude that $s_i^i(n,\omega)$ is equivalent to $s_i(n,\omega)$. These sequences are illustrated in Fig. 6.6b and c.

From Eqs. (6.12) and (6.13), we note that

$$s_\delta^i(n,t) \leftrightarrow s_i^i(n,\omega)$$

and hence the transform pair $\tilde{\delta}(t) \leftrightarrow \tilde{i}(\omega)$ is again obtained.

Two other sequences that define the impulse and unity functions, which are often very useful, can be deduced by using the sequence

$$p_{n\Omega}(n,\omega) = \begin{cases} 1 & \text{for } |\omega| \leq n\Omega \\ 0 & \text{otherwise} \end{cases}$$

They are

$$s_\delta^{ii}(n,t) = \frac{\sin n\Omega t}{\pi t} \times e^{-t^2/4n} \tag{6.14}$$

and

$$s_i^{ii}(n,\omega) = p_{n\Omega}(n,\omega) \otimes 2\sqrt{(n\pi)}e^{-n\omega^2} \tag{6.15}$$

From Eq. (6.14), we can write

$$\lim_{n\to\infty}\int_{-\infty}^{\infty} s_\delta^{ii}(n,t)\phi(t)dt = \lim_{n\to\infty}\int_{-\infty}^{-\epsilon} + \lim_{n\to\infty}\int_{-\epsilon}^{\epsilon}$$
$$+ \lim_{n\to\infty}\int_{\epsilon}^{\infty} \frac{\sin n\Omega t}{\pi t} \times e^{-t^2/4n}\phi(t)dt$$

and, since $e^{-t^2/4n}\phi(t)/t$ is integrable, the first and last integrals at the right-hand side tend to zero by virtue of the Rieman-Lebesque Theorem (see [3, p. 565]). On the other hand, for a sufficiently small value of ϵ, the middle integral assumes the form

$$\lim_{n\to\infty}\int_{-\epsilon}^{\epsilon}\frac{\sin n\Omega t}{\pi t}\times e^{-t^2/4n}\phi(t)dt \approx \phi(0)\lim_{n\to\infty}\int_{-\epsilon}^{\epsilon}\frac{\sin n\Omega t}{\pi t}dt$$

and if we let $x = n\Omega t$, we have

$$\lim_{n\to\infty}\int_{-\infty}^{\infty}s_\delta^{ii}(n,t)\phi(t)dt = \phi(0)\lim_{n\to\infty}\int_{-n\Omega\epsilon}^{n\Omega\epsilon}\frac{\sin x}{\pi x}dx = \phi(0)$$

since

$$\int_{-\infty}^{\infty}\frac{\sin x}{\pi x}dx = 1$$

In effect, $s_\delta^{ii}(n,t)$ defines $\tilde{\delta}(t)$.

Now from Eq. (6.15)

$$\lim_{n\to\infty}\int_{-\infty}^{\infty}s_i^{ii}(n,\omega)\phi(\omega)d\omega = \lim_{n\to\infty}\int_{-\infty}^{\infty}\left[\int_{-\infty}^{\infty}p_{n\Omega}(n,v)\sqrt{(n/\pi)}e^{-n(\omega-v)^2}dv\right]\phi(\omega)d\omega$$

$$= \lim_{n\to\infty}\int_{-\infty}^{\infty}p_{n\Omega}(n,\omega)\phi(\omega)d\omega = \int_{-\infty}^{\infty}\phi(\omega)d\omega$$

i.e., $s_i^{ii}(n,\omega)$ defines $\tilde{i}(\omega)$. As in the previous case, we have

$$s_\delta^{ii}(n,t) \leftrightarrow s_i^{ii}(n,\omega)$$

and once again we obtain $\tilde{\delta}(t) \leftrightarrow \tilde{i}(\omega)$. These sequences are illustrated in Fig. 6.7a and b.

The above results are summarized in Table 6.1. In the following sections, $\tilde{\delta}(t)$ and $\tilde{i}(\omega)$ will be represented by the conventional symbols $\delta(t)$ and 1, respectively, unless otherwise stated.

6.3.3 Ordinary Functions as Generalized Functions

Definition 6. If $f(t)$ is an ordinary function such that

$$\lim_{T\to\infty}\int_{-T}^{T}\left|\frac{f(t)}{(1+t^2)^N}\right|dt < \infty$$

for some N, then a generalized function $\tilde{f}(t)$ can be defined for $f(t)$ in terms of a sequence of functions $s_f(n,t)$ such that for any good function $\phi(t)$

$$\int_{-\infty}^{\infty}\tilde{f}(t)\phi(t)dt = \lim_{n\to\infty}\int_{-\infty}^{\infty}s_f(n,t)\phi(t)dt = \int_{-\infty}^{\infty}f(t)\phi(t)dt$$

A possible sequence of good functions that defines $f(t)$ is given by

$$s_f(n,t) = f(t)ne^{-t^2/n^2} \otimes \xi[n(t)]$$

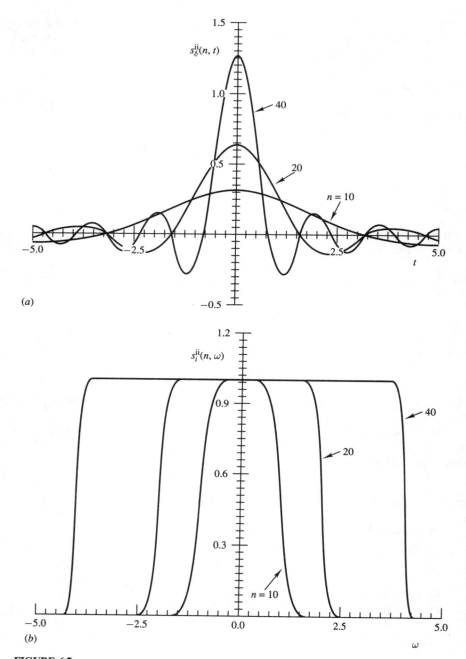

FIGURE 6.7
Another pair of sequences for impulse and unity functions: (*a*) sequence for impulse function, (*b*) sequence for unity function.

TABLE 6.1
Possible sequences for impulse and unity functions

$\tilde{\delta}(t)$	$\tilde{i}(\omega)$
$\sqrt{(n/\pi)}e^{-nt^2}$	$e^{-\omega^2/4n}$
$p_{\epsilon/2n}(t,n) \otimes \sqrt{(n/\pi)}e^{-nt^2}$	$\dfrac{\sin(\omega\epsilon/2n)}{(\omega\epsilon/2n)} \times e^{-\omega^2/4n}$
$\dfrac{\sin n\Omega t}{\pi t} \times e^{-t^2/4n}$	$p_{n\Omega}(n,\omega) \otimes 2\sqrt{(n\pi)}e^{-n\omega^2}$

where

$$\xi(t) = \begin{cases} \dfrac{e^{-1/(1-t^2)}}{\xi_0} & \text{for } |t| < 1 \\ 0 & \text{for } |t| \geq 1 \end{cases}$$

with

$$\xi_0 = \int_{-1}^{1} e^{-1/(1-x^2)} dx$$

(see [2, p. 22]).

Definition 7. If $h(t)$ is an ordinary function and $\tilde{f}(t)$ is a generalized function such that

$$\int_a^b h(t)\phi(t)dt = \int_{-\infty}^{\infty} \tilde{f}(t)\phi(t)dt$$

for every good function $\phi(t)$ which is zero outside the interval $a < t < b$, then for $a < t < b$

$$h(t) = \tilde{f}(t)$$

The left-hand integral in the above definition is assumed to exist as an ordinary integral for all $\phi(t)$ of the required type and, as a result, a restriction is imposed on function $h(t)$ with respect to interval $a < t < b$.

Definition 6 extends the range of application of generalized functions to many ordinary functions that present mathematical difficulties, whereas Definition 7 leads to relations between ordinary and generalized functions.

Example 6.4. Find the Fourier transforms of (a) $f(t) = \delta(t - t_0)$, (b) $f(t) = e^{j\omega_0 t}$, and (c) $f(t) = \cos \omega_0 t$.

Solution. (a) From Theorem 6.5

$$\delta(t - t_0) \leftrightarrow e^{-j\omega t_0}$$

(b) From Theorem 6.3 and part (a)

$$e^{-jt t_0} \leftrightarrow 2\pi \delta(-\omega - t_0)$$

and with $-t_0 = \omega_0$

$$e^{j\omega_0 t} \leftrightarrow 2\pi\delta(\omega - \omega_0)$$

(c)

$$\mathcal{F}\cos\omega_0 t = \mathcal{F}\frac{1}{2}(e^{j\omega_0 t} + e^{-j\omega_0 t}) = \pi[\delta(\omega - \omega_0) + \delta(\omega + \omega_0)]$$

or

$$\cos\omega_0 t \leftrightarrow \pi[\delta(\omega - \omega_0) + \delta(\omega + \omega_0)]$$

(see Fig. 6.8a to c).

A number of useful Fourier transform pairs can be found in Table 6.2. An important Fourier transform pair that will be used to establish a relation between the Fourier series and the Fourier transform and also to derive Poisson's summation formula is considered in the following example.

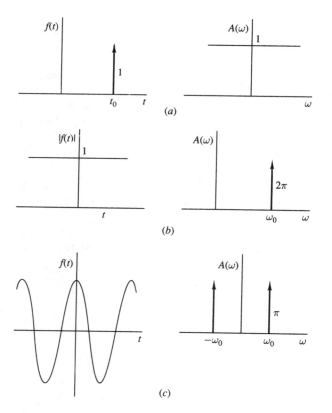

FIGURE 6.8
Function $f(t)$ and its amplitude spectrum: (a) to (c) parts (a) to (c) of Example 6.4, respectively.

TABLE 6.2
Fourier transforms

$f(t)$	$F(j\omega)$
$\delta(t)$	1
1	$2\pi\delta(\omega)$
$\delta(t-t_0)$	$e^{-j\omega t_0}$
$e^{j\omega_0 t}$	$2\pi\delta(\omega-\omega_0)$
$\cos\omega_0 t$	$\pi[\delta(\omega+\omega_0)+\delta(\omega-\omega_0)]$
$\sin\omega_0 t$	$j\pi[\delta(\omega+\omega_0)-\delta(\omega-\omega_0)]$
$u(t)e^{-at}$	$\dfrac{1}{a+j\omega}$
$u(t)e^{-at}\sin\omega_0 t$	$\dfrac{\omega_0}{(a+j\omega)^2+\omega_0^2}$
$p_{t_0}(t)=\begin{cases}1 & \|t\|\leq t_0 \\ 0 & \|t\|>t_0\end{cases}$	$\dfrac{2\sin\omega t_0}{\omega}$
$\dfrac{\sin\omega_0 t}{\pi t}$	$p_{\omega_0}(\omega)=\begin{cases}1 & \|\omega\|\leq\omega_0 \\ 0 & \|\omega\|>\omega_0\end{cases}$
$\sqrt{\dfrac{n}{\pi}}e^{-nt^2}$	$e^{-\omega^2/4n}$

Example 6.5. Show that

$$\sum_{n=-\infty}^{\infty}\delta(t-nT)\leftrightarrow\omega_s\sum_{n=-\infty}^{\infty}\delta(\omega-n\omega_s) \qquad (6.16)$$

where $\omega_s=2\pi/T$.

Solution. On applying the inverse Fourier transform to the right-hand side of Eq. (6.16), we get

$$\mathcal{F}^{-1}\omega_s\sum_{n=-\infty}^{\infty}\delta(\omega-n\omega_s)=\frac{1}{2\pi}\int_{-\infty}^{\infty}\left[\omega_s\sum_{n=-\infty}^{\infty}\delta(\omega-n\omega_s)\right]e^{j\omega t}d\omega$$

$$=\frac{1}{T}\sum_{n=-\infty}^{\infty}\int_{-\infty}^{\infty}\delta(\omega-n\omega_s)e^{j\omega t}d\omega$$

$$=\frac{1}{T}\sum_{n=-\infty}^{\infty}e^{jn\omega_s t}$$

Now let

$$k_N(t) = \frac{1}{T} \sum_{n=-N}^{N} e^{jn\omega_s t} = \frac{1}{T} \frac{e^{j(N+1)\omega_s t} - e^{-jN\omega_s t}}{e^{j\omega_s t} - 1}$$

$$= \frac{1}{T} \frac{\sin[(2N+1)\omega_s t/2]}{\sin(\omega_s t/2)} \tag{6.17}$$

This function is known as the *Fourier-series kernel* and is periodic with period T, as can be easily shown (see Fig. 6.9). To establish the required result, we need only demonstrate that in the interval $-T/2 < t < T/2$ the function $k_N(t)$ is equal to $\delta(t)$. From Eq. (6.17)

$$k_N(t) = \frac{1}{T} \frac{\pi t}{\sin(\omega_s t/2)} e^{t^2/4(2N+1)} \left\{ \frac{\sin[(2N+1)\omega_s t/2]}{\pi t} \times e^{-t^2/4(2N+1)} \right\}$$

and, with $2N + 1 = n$ and $\omega_s/2 = \Omega$, the last row of Table 6.1 gives

$$k_N(t) = h(t)\delta(t)$$

where function

$$h(t) = \frac{1}{T} \frac{\pi t}{\sin(\omega_s t/2)} e^{t^2/4(2N+1)}$$

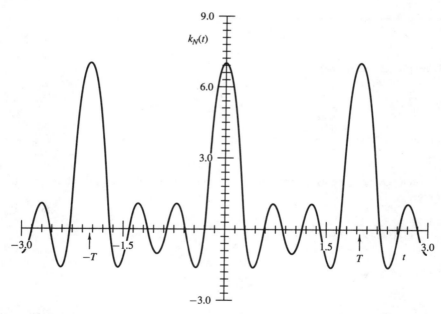

FIGURE 6.9
Fourier-series kernel.

is continuous and assumes the value of unity at $t = 0$. Now for any good function $\phi(t)$ which is zero outside the interval $-T/2 < t < T/2$, we can write

$$\int_{-T/2}^{T/2} k_N(t)\phi(t)dt = \int_{-T/2}^{T/2} h(t)\delta(t)\phi(t)dt$$

$$= \int_{-T/2}^{T/2} \delta(t)\phi(t)dt$$

$$= \int_{-\infty}^{\infty} \delta(t)\phi(t)dt$$

since $h(t)\delta(t) = h(0)\delta(t)$, according to Eq. (6.9). Therefore, from Definition 7, for $-T/2 < t < T/2$ we have

$$k_N(t) = \delta(t)$$

6.4 FOURIER SERIES

A periodic function $f(t)$ with period T such that

$$f(t + T) = f(t)$$

can be expressed in terms of an infinite sum of exponentials as

$$f(t) = \sum_{n=-\infty}^{\infty} a_n e^{jn\omega_s t} \tag{6.18}$$

where

$$a_n = \int_{-T/2}^{T/2} f(t)e^{-jn\omega_s t} dt \tag{6.19}$$

and $\omega_s = 2\pi/T$. Eq. (6.18) is known as the *Fourier series* of $f(t)$. The Fourier transform of $f(t)$ is obtained from Table 6.2 as

$$F(j\omega) = 2\pi \sum_{n=-\infty}^{\infty} a_n \delta(\omega - n\omega_s) \tag{6.20}$$

The Fourier series coefficients a_n are related to the Fourier transform of $f(t)$, as will now be demonstrated. Function $f(t)$ can be expressed as

$$f(t) = \sum_{n=-\infty}^{\infty} f_0(t + nT) \tag{6.21}$$

where

$$f_0(t) = \begin{cases} f(t) & \text{for } |t| < T/2 \\ 0 & \text{for } |t| \geq T/2 \end{cases}$$

and from Eq. (6.8)

$$f(t) = \sum_{n=-\infty}^{\infty} \int_{-\infty}^{\infty} f_0(\tau)\delta(t - \tau + nT)d\tau \qquad (6.22a)$$

$$= \int_{-\infty}^{\infty} f_0(\tau) \sum_{n=-\infty}^{\infty} \delta(t - \tau + nT)d\tau \qquad (6.22b)$$

$$= f_0(t) \otimes \sum_{n=-\infty}^{\infty} \delta(t - nT) \qquad (6.22c)$$

Now with

$$F_0(j\omega) = \int_{-\infty}^{\infty} f_0(t)e^{-j\omega t}dt = \int_{-T/2}^{T/2} f(t)e^{-j\omega t}dt \qquad (6.23)$$

Eqs. (6.22c), (6.16), and (6.9) yield

$$F(j\omega) = F_0(j\omega)\omega_s \sum_{n=-\infty}^{\infty} \delta(\omega - n\omega_s)$$

$$= 2\pi \sum_{n=-\infty}^{\infty} \frac{F_0(jn\omega_s)}{T} \delta(\omega - n\omega_s) \qquad (6.24)$$

and, on comparing Eqs. (6.20) and (6.24), we deduce the required result as

$$a_n = \frac{F_0(jn\omega_s)}{T} \qquad (6.25)$$

The Fourier series finds extensive applications in the analysis and design of digital filters. For example, it can be used for the solution of the approximation problem in nonrecursive digital filters, as will be shown in Chap. 9.

6.5 POISSON'S SUMMATION FORMULA

For an arbitrary periodic function $f(t)$, Eqs. (6.21), (6.18), and (6.25) give the relation

$$f(t) = \sum_{n=-\infty}^{\infty} f_0(t + nT) = \frac{1}{T} \sum_{n=-\infty}^{\infty} F_0(jn\omega_s)e^{jn\omega_s t} \qquad (6.26)$$

A similar result holds for an arbitrary nonperiodic function $f(t)$ with Fourier transform $F(j\omega)$, as will now be demonstrated. As in Eqs. (6.21) and (6.22a) to (6.22c), we can form the sum

$$\sum_{n=-\infty}^{\infty} f(t + nT) = f(t) \otimes \sum_{n=-\infty}^{\infty} \delta(t - nT)$$

and from Eqs. (6.16) and (6.9)

$$\mathcal{F}\sum_{n=-\infty}^{\infty} f(t+nT) = F(j\omega)\omega_s \sum_{n=-\infty}^{\infty} \delta(\omega - n\omega_s)$$

$$= \frac{2\pi}{T} \sum_{n=-\infty}^{\infty} F(jn\omega_s)\delta(\omega - n\omega_s)$$

Therefore

$$\sum_{n=-\infty}^{\infty} f(t+nT) = \frac{2\pi}{T} \sum_{n=-\infty}^{\infty} F(jn\omega_s)\mathcal{F}^{-1}\delta(\omega - n\omega_s)$$

$$= \frac{1}{T} \sum_{n=-\infty}^{\infty} F(jn\omega_s)e^{jn\omega_s t} \qquad (6.27)$$

This relation is known as *Poisson's summation formula*. It will be used in the next section to establish some fundamental interrelations between the various types of signals.

Two special cases of Poisson's formula are of interest. If $t = 0$, Eq. (6.27) assumes the form

$$\sum_{n=-\infty}^{\infty} f(nT) = \frac{1}{T} \sum_{n=-\infty}^{\infty} F(jn\omega_s) \qquad (6.28)$$

If $f(t)$ is discontinuous at $t = 0$ and $f(t) = 0$ for $t < 0$, then $f(t)$ is the inverse Fourier transform of $F(j\omega)$ if it is assumed that

$$\lim_{t \to 0} f(t) = \frac{f(0-) + f(0+)}{2} = \frac{f(0+)}{2}$$

In this case, Eq. (6.27) assumes the form

$$\sum_{n=0}^{\infty} f(nT) = \frac{f(0+)}{2} + \frac{1}{T} \sum_{n=-\infty}^{\infty} F(jn\omega_s) \qquad (6.29)$$

where $f(0) \equiv f(0+)$.

6.6 SAMPLED SIGNALS

A *sampled signal*, denoted as $\hat{x}(t)$, can be generated by sampling a continuous-time signal $x(t)$ using an ideal impulse sampler like that depicted in Fig. 6.10a. An *impulse sampler* is essentially a modulator characterized by the equation

$$\hat{x}(t) = c(t)x(t) \qquad (6.30)$$

where $c(t)$ is a carrier given by

$$c(t) = \sum_{n=-\infty}^{\infty} \delta(t - nT)$$

Hence from Eqs. (6.9) and (6.30)

$$\hat{x}(t) = \sum_{n=-\infty}^{\infty} x(nT)\delta(t - nT) \qquad (6.31)$$

In effect, a sampled signal is a sequence of continuous-time impulses, like that illustrated in Fig. 6.10b. Note that a sampled signal can be converted into a discrete-time signal by replacing each impulse of strength $x(nT)$ by a number $x(nT)$, as shown in Fig. 6.10c.

The Fourier transform of $\hat{x}(t)$ is

$$\hat{X}(j\omega) = \mathcal{F} \sum_{n=-\infty}^{\infty} x(nT)\delta(t - nT) = \sum_{n=-\infty}^{\infty} x(nT)\mathcal{F}\delta(t - nT)$$

$$= \sum_{n=-\infty}^{\infty} x(nT)e^{-j\omega nT} \qquad (6.32)$$

Clearly

$$\hat{X}(j\omega) = X_D(z)\Big|_{z=e^{j\omega T}} \qquad (6.33)$$

where

$$X_D(z) = \mathcal{Z}x(nT) \qquad (6.34)$$

i.e., *the Fourier transform of sampled signal $\hat{x}(t)$ is numerically equal to the z transform of the corresponding discrete-time signal $x(nT)$ evaluated on the unit circle $|z| = 1$.*

Now let $X(j\omega)$ be the Fourier transform of $x(t)$. From Theorem 6.6, the transform pair

$$x(t)e^{-j\Omega t} \leftrightarrow X(j\Omega + j\omega)$$

can be formed. On using Poisson's summation formula given by Eq. (6.28) and then letting $\Omega = \omega$, we obtain

$$\sum_{n=-\infty}^{\infty} x(nT)e^{-j\omega nT} = \frac{1}{T}\sum_{n=-\infty}^{\infty} X(j\omega + jn\omega_s) \qquad (6.35)$$

where $\omega_s = 2\pi/T$. Therefore, from Eqs. (6.32), (6.33), and (6.35)

$$\hat{X}(j\omega) = X_D(e^{j\omega T}) = \frac{1}{T}\sum_{n=-\infty}^{\infty} X(j\omega + jn\omega_s) \qquad (6.36)$$

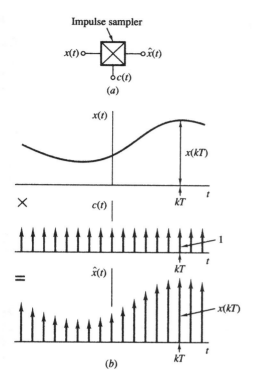

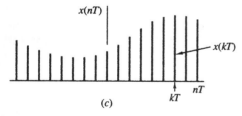

FIGURE 6.10
Generation of a sampled signal: (a) ideal impulse modulator, (b) sampled signal $\hat{x}(t)$, (c) discrete-time signal $x(nT)$.

i.e., *if the Fourier transform of $x(t)$ is known, that of $\hat{x}(t)$ is uniquely determined.* As is to be expected, $\hat{X}(j\omega)$ is a *periodic function* of ω with period ω_s. Indeed

$$\hat{X}(j\omega + jm\omega_s) = \frac{1}{T}\sum_{n=-\infty}^{\infty} X[j\omega + j(m+n)\omega_s] = \frac{1}{T}\sum_{n'=-\infty}^{\infty} X(j\omega + jn'\omega_s)$$
$$= \hat{X}(j\omega)$$

Often $x(t) = 0$ for $t \leq 0-$. If such is the case, Eq. (6.31) becomes

$$\hat{x}(t) = \sum_{n=0}^{\infty} x(nT)\delta(t - nT) \tag{6.37}$$

where $x(0) \equiv x(0+)$, and from Eqs. (6.32) and (6.33)

$$\hat{X}(j\omega) = \sum_{n=0}^{\infty} x(nT)e^{-j\omega nT} = X_D(e^{j\omega T})$$

Now from Eq. (6.29) we deduce

$$\hat{X}(j\omega) = X_D(e^{j\omega T}) = \frac{x(0+)}{2} + \frac{1}{T} \sum_{n=-\infty}^{\infty} X(j\omega + jn\omega_s) \qquad (6.38)$$

With $j\omega = s$ and $e^{sT} = z$, Eqs. (6.36) and (6.38) become

$$\hat{X}(s) = X_D(z) = \frac{1}{T} \sum_{n=-\infty}^{\infty} X(s + jn\omega_s) \qquad (6.39)$$

and

$$\hat{X}(s) = X_D(z) = \frac{x(0+)}{2} + \frac{1}{T} \sum_{n=-\infty}^{\infty} X(s + jn\omega_s) \qquad (6.40)$$

respectively, where $X(s)$ and $\hat{X}(s)$ are the *Laplace transforms* of $x(t)$ and $\hat{x}(t)$. If the value of $x(0+)$ is not available, it can be deduced from $X(s)$ as

$$x(0+) = \lim_{s \to \infty} [sX(s)]$$

by using the initial-value theorem of the one-sided Laplace transform.

The formula in Eq. (6.40) will be used in Sec. 6.9 to establish a relation between analog and digital filters. It will be applied in Chap. 7 in the so-called invariant-impulse-response method for the solution of the approximation problem in recursive filters.

Example 6.6. (a) Find $\hat{X}(j\omega)$ if $x(t) = \cos \omega_0 t$. (b) Repeat part (a) for $x(t) = u(t)e^{-t}$.

Solution. (a) From Eq. (6.36) and Table 6.2

$$\hat{X}(j\omega) = \frac{\pi}{T} \sum_{n=-\infty}^{\infty} [\delta(\omega + n\omega_s + \omega_0) + \delta(\omega + n\omega_s - \omega_0)]$$

(b) From Eq. 6.38 and Table 6.2

$$\hat{X}(j\omega) = \frac{1}{2} + \frac{1}{T} \sum_{n=-\infty}^{\infty} \frac{1}{1 + j(\omega + n\omega_s)}$$

(see Fig. 6.11a and b).

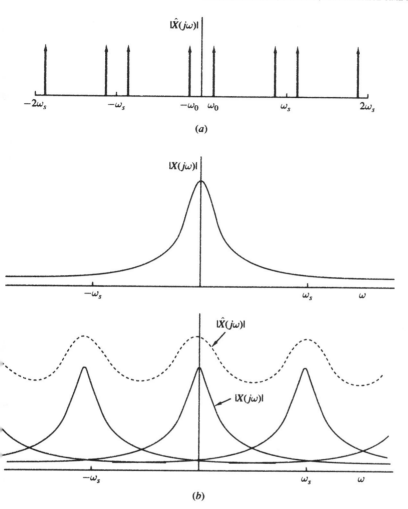

FIGURE 6.11
Amplitude spectrum of $\hat{x}(t)$: (a) Example 6.6(a), (b) Example 6.6(b).

6.7 THE SAMPLING THEOREM

The application of digital filters for the processing of continuous-time signals is made possible by the *sampling theorem*. This states that a bandlimited signal $x(t)$ for which

$$X(j\omega) = 0 \quad \text{for } |\omega| \geq \frac{\omega_s}{2} \tag{6.41}$$

where $\omega_s = 2\pi/T$, can be uniquely determined from its values $x(nT)$.

The validity of this theorem can be easily demonstrated. With Eq. (6.41) satisfied, $T\hat{X}(j\omega)$ given by Eq. (6.36) as

$$T\hat{X}(j\omega) = \sum_{n=-\infty}^{\infty} X(j\omega + jn\omega_s)$$

is a periodic continuation of $X(j\omega)$. Hence

$$X(j\omega) = H(j\omega)T\hat{X}(j\omega) \qquad (6.42)$$

where

$$H(j\omega) = \begin{cases} 1 & \text{for } |\omega| < \dfrac{\omega_s}{2} \\ 0 & \text{for } |\omega| \geq \dfrac{\omega_s}{2} \end{cases}$$

as depicted in Fig. 6.12. Thus from Eqs. (6.42) and (6.32) we can write

$$X(j\omega) = H(j\omega)T \sum_{n=-\infty}^{\infty} x(nT)e^{-j\omega nT}$$

and consequently

$$\begin{aligned} x(t) &= \mathcal{F}^{-1}\left[H(j\omega)T \sum_{n=-\infty}^{\infty} x(nT)e^{-j\omega nT}\right] \\ &= T \sum_{n=-\infty}^{\infty} x(nT)\mathcal{F}^{-1}[H(j\omega)e^{-j\omega nT}] \end{aligned} \qquad (6.43)$$

Now from Table 6.2

$$\frac{\sin(\omega_s t/2)}{\pi t} \leftrightarrow H(j\omega)$$

and by Theorem 6.5

$$\frac{\sin[\omega_s(t-nT)/2]}{\pi(t-nT)} \leftrightarrow H(j\omega)e^{-j\omega nT}$$

Therefore, from Eq. (6.43)

$$x(t) = \sum_{n=-\infty}^{\infty} x(nT) \frac{\sin[\omega_s(t-nT)/2]}{\omega_s(t-nT)/2} \qquad (6.44)$$

According to Eq. (6.42), signals $\hat{x}(t)$ and $x(t)$ can be regarded as the input and output of a lowpass filter characterized by $TH(j\omega)$. In effect, the sampling theorem provides a means by which a sampled signal can be converted back into the original continuous-time signal.

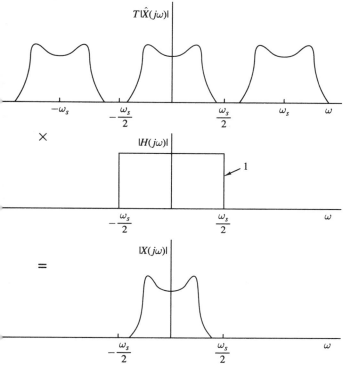

FIGURE 6.12
Derivation of $X(j\omega)$ from $\hat{X}(j\omega)$.

6.7.1 Aliasing

If
$$X(j\omega) \neq 0 \quad \text{for } |\omega| \geq \frac{\omega_s}{2}$$

as in the example of Fig. 6.11b, the tails of $X(j\omega)$ extend outside the baseband. As a result, $\hat{X}(j\omega)$ (dashed curve in Fig. 6.11b) is no longer a periodic continuation of $X(j\omega)/T$, and the use of an ideal lowpass filter will at best yield a distorted version of $x(t)$. This effect is known as *aliasing* or *frequency folding*.

6.8 INTERRELATIONS

Various important interrelations have been established in the preceding sections between continuous-time, sampled, and discrete-time signals. They are illustrated pictorially in Fig. 6.13. The two-directional paths between $\hat{x}(t)$ and $x(nT)$ and between $\hat{X}(j\omega)$ and $X_D(z)$ render the Fourier transform applicable to digital filters. On the

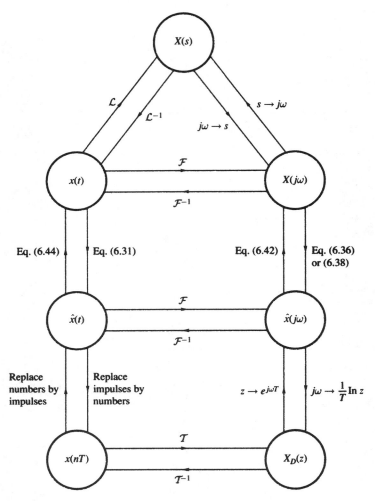

FIGURE 6.13
Interrelations between continuous-time, sampled, and discrete-time signals.

other hand, the two-directional paths between $x(t)$ and $x(nT)$ and between $X(j\omega)$ and $X_D(z)$ will allow us to use digital filters for the processing of continuous-time signals. Finally, the path between $X(s)$ and $X_D(z)$ will allow us to derive digital filters from analog filters.

---

†If $x(t) = p_{t_0}(t)$ (see Example 6.1) where $t_0 < T$, Eq. 6.9 yields $\hat{x}(t) = \hat{p}_{t_0}(t) = \delta(t)$ and hence $x(nT) = p_{t_0}(nT) = \delta(nT)$. In effect, $\delta(t)$ and $\delta(nT)$ are the sampled and discrete-time versions of $p_{t_0}(t)$, respectively.

6.9 THE PROCESSING OF CONTINUOUS-TIME SIGNALS

Consider the filtering scheme of Fig. 6.14a, where S_1 and S_2 are impulse samplers, F_A is an analog filter characterized by $H_A(s)$, and F_{LP} is a lowpass filter for which

$$H_{LP}(s) = \begin{cases} T^2 & \text{for } |\omega| < \frac{\omega_s}{2} \\ 0 & \text{otherwise} \end{cases} \quad (6.45)$$

F_A and S_2 constitute a *sampled-data filter* $\hat{F}_A$. By analogy with Eqs. (6.37) and (6.40) the impulse response and transfer function of $\hat{F}_A$ can be expressed as

$$\hat{h}_A(t) = \sum_{n=0}^{\infty} h_A(nT)\delta(t - nT)$$

and

$$\hat{H}_A(s) = H_D(z) = \frac{h_A(0+)}{2} + \frac{1}{T}\sum_{n=-\infty}^{\infty} H_A(s + jn\omega_s) \quad (6.46)$$

respectively, where

$$h_A(t) = \mathcal{L}^{-1} H_A(s) \qquad h_A(0+) = \lim_{s \to \infty}[sH_A(s)]$$

$$H_D(z) = \mathcal{Z} h_A(nT) \qquad z = e^{sT}$$

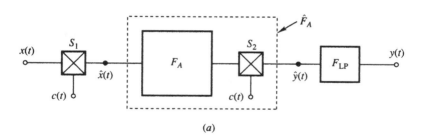

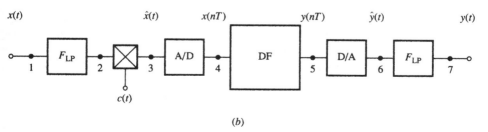

FIGURE 6.14
The processing of continuous-time signals: (*a*) using a sampled-data filter, (*b*) using a digital filter.

The Fourier transform of $y(t)$ in Fig. 6.14a is

$$Y(j\omega) = \hat{H}_A(j\omega) H_{LP}(j\omega) \hat{X}(j\omega) \qquad (6.47)$$

and if

$$x(0+) = h_A(0+) = 0$$

and

$$X(j\omega) = H_A(j\omega) = 0 \qquad \text{for } |\omega| \geq \frac{\omega_s}{2}$$

then $T\hat{X}(j\omega)$ and $T\hat{H}_A(j\omega)$ are *periodic continuations* of $X(j\omega)$ and $H_A(j\omega)$, respectively, in which case

$$\hat{X}(j\omega) = \frac{1}{T} X(j\omega) \quad \text{and} \quad \hat{H}_A(j\omega) = \frac{1}{T} H_A(j\omega) \quad \text{for} \quad |\omega| < \frac{\omega_s}{2}$$

Hence from Eqs. (6.45) and (6.47) we have

$$Y(j\omega) = \begin{cases} H_A(j\omega) X(j\omega) & \text{for } |\omega| < \frac{\omega_s}{2} \\ 0 & \text{for } |\omega| \geq \frac{\omega_s}{2} \end{cases}$$

and since

$$H_A(j\omega) X(j\omega) = 0 \qquad \text{for} \quad |\omega| \geq \frac{\omega_s}{2}$$

we conclude that for any frequency ω

$$Y(j\omega) = H_A(j\omega) X(j\omega)$$

In effect, the response of the configuration is identical with the response of filter F_A to excitation $x(t)$. We thus conclude that sampled-data filters can be used for the processing of continuous-time signals.

A sampled-data filter, like a digital filter, can be characterized by a *discrete-time transfer function*, according to Eq. (6.46). Hence, given a sampled-data filter, an equivalent digital filter can be derived and vice versa. Consequently, by implication, digital filters can be used for the processing of continuous-time signals. In addition, analog filters can be used to derive digital filters (see Chap. 7).

A digital-filter implementation of Fig. 6.14a can be obtained by replacing the sampled-data filter by a digital filter together with suitable interfacing devices, as shown in Fig. 6.14b. The *analog-to-digital* and *digital-to-analog* converters are required to convert impulses into numbers and numbers into impulses. The input lowpass filter is used to bandlimit $x(t)$ (if it is not already bandlimited) to prevent aliasing errors.

Example 6.7. The configuration of Fig. 6.14b is used to filter the periodic signal given by

$$x(t) = \begin{cases} \sin \omega_0 t & \text{for } 0 \leq \omega_0 t < \pi \\ 0 & \text{for } \pi \leq \omega_0 t < 2\pi \end{cases}$$

CONTINUOUS-TIME, SAMPLED, AND DISCRETE-TIME SIGNALS **209**

The lowpass filters are characterized by

$$H_{\text{LP}}(j\omega) = \begin{cases} 1 & \text{for } 0 \le |\omega| < 6\omega_0 \\ 0 & \text{otherwise} \end{cases}$$

and the digital filter has a baseband response

$$H_D(e^{j\omega T}) = \begin{cases} T & \text{for } 0.95\omega_0 < |\omega| < 1.05\omega_0 \\ 0 & \text{otherwise} \end{cases}$$

Assuming that $\omega_s = 12\omega_0$, find the time- and frequency-domain representations of the signals at nodes $1, 2, \ldots, 7$.

Solution. *Node 1* The Fourier series gives

$$x_1(t) = \sum_{k=-\infty}^{\infty} A_k e^{jk\omega_0 t}$$

where

$$A_k = \frac{1}{T_0} \int_0^{T_0} x_1(t) e^{-jk\omega_0 t} dt \qquad T_0 = \frac{2\pi}{\omega_0}$$

and hence $A_0 = \dfrac{1}{\pi}$ $\qquad A_1 = -A_{-1} = -\dfrac{j}{4}$

$$A_2 = A_{-2} = -\frac{1}{3\pi} \qquad A_3 = A_{-3} = 0$$

$$A_4 = A_{-4} = -\frac{1}{15\pi} \qquad A_5 = A_{-5} = 0 \qquad A_6 = A_{-6} = -\frac{1}{35\pi}$$

and so forth, or

$$x_1(t) = \frac{1}{\pi} + \frac{1}{2}\sin\omega_0 t - \frac{2}{3\pi}\cos 2\omega_0 t - \frac{2}{15\pi}\cos 4\omega_0 t$$
$$- \frac{2}{35\pi}\cos 6\omega_0 t - \cdots$$

and from Table 6.2

$$X_1(j\omega) = 2\pi \sum_{k=-\infty}^{\infty} A_k \delta(\omega - k\omega_0)$$

Node 2 The output of the bandlimiting filter is

$$X_2(j\omega) = 2\pi \sum_{k=-4}^{4} A_k \delta(\omega - k\omega_0)$$

in which case

$$x_2(t) = \frac{1}{\pi} + \frac{1}{2}\sin\omega_0 t - \frac{2}{3\pi}\cos 2\omega_0 t - \frac{2}{15\pi}\cos 4\omega_0 t$$

Nodes 3 and 4 The output of the sampler is obtained from Eq. (6.31) as

$$\hat{x}_3(t) = \sum_{n=-\infty}^{\infty} \left(\frac{1}{\pi} + \frac{1}{2}\sin\omega_0 nT - \frac{2}{3\pi}\cos 2\omega_0 nT - \frac{2}{15\pi}\cos 4\omega_0 nT \right) \delta(t - nT)$$

Thus
$$x_4(nT) = \frac{1}{\pi} + \frac{1}{2}\sin\omega_0 nT - \frac{2}{3\pi}\cos 2\omega_0 nT - \frac{2}{15\pi}\cos 4\omega_0 nT$$

From Eq. (6.36) and Fig. 6.13

$$\hat{X}_3(j\omega) = X_4(e^{j\omega T}) = \frac{2\pi}{T}\sum_{n=-\infty}^{\infty}\sum_{k=-4}^{4} A_k \delta(\omega + n\omega_s - k\omega_0)$$

Nodes 5 and 6 The digital filter will reject all components except those with frequencies $\pm\omega_0 \pm n\omega_s$, and so

$$X_5(e^{j\omega T}) = \hat{X}_6(j\omega) = 2\pi \sum_{n=-\infty}^{\infty}[A_1\delta(\omega + n\omega_s - \omega_0) + A_{-1}\delta(\omega + n\omega_s + \omega_0)]$$

$$= \frac{j\pi}{2}\sum_{n=-\infty}^{\infty}[\delta(\omega + n\omega_s + \omega_0) - \delta(\omega + n\omega_s - \omega_0)]$$

Thus from Table 6.2
$$x_5(nT) = \frac{1}{2}\sin\omega_0 nT$$

$$\hat{x}_6(nT) = \frac{1}{2}\sum_{n=-\infty}^{\infty}(\sin\omega_0 nT)\delta(t - nT)$$

Node 7 Finally, the lowpass filter at the output will reject all components with frequencies outside the baseband, and as a result

$$X_7(j\omega) = \frac{j\pi}{2}[\delta(\omega + \omega_0) - \delta(\omega - \omega_0)]$$

and
$$x_7(nT) = \frac{1}{2}\sin\omega_0 t$$

The various signal waveforms and amplitude spectra are illustrated in Fig. 6.15a and b, respectively.

6.9.1 Practical Considerations

A practical implementation of the *analog-to-digital* interface is shown in Fig. 6.16a. The function of the *sample-and-hold* device is to generate a signal

$$\tilde{x}(t) = \sum_{n=-\infty}^{\infty} x(nT) p_{T/2}\left(t - nT - \frac{T}{2}\right)$$

like that illustrated in Fig. 6.16b. The function of the *encoder*, on the other hand, is to convert each signal level $x(nT)$ into a corresponding binary number. Since the number of bits in the binary representation must be finite, the response of the encoder denoted by $x_q(nT)$ can assume only a finite number of discrete levels; that is, $x_q(nT)$ will be a quantized signal. Assuming that the encoder is designed so that each value

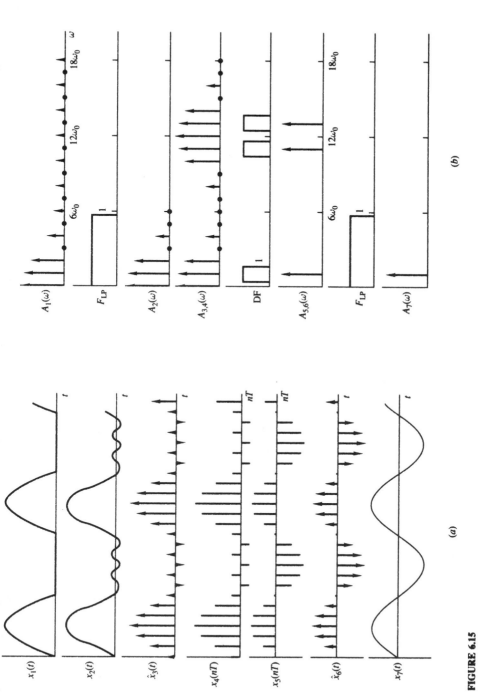

FIGURE 6.15
Example 6.7: (a) time-domain representations of signals at nodes 1, 2, ..., 7, (b) amplitude spectra of signals at nodes 1, 2, ..., 7.

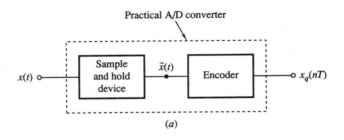

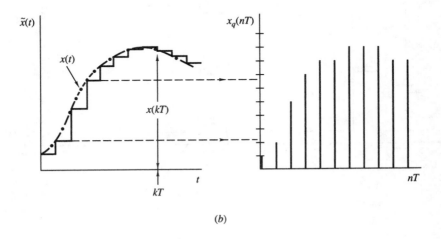

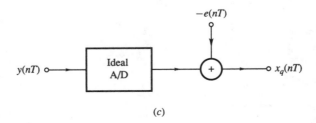

FIGURE 6.16
Analog-to-digital interface: (a) practical A/D converter, (b) response of a practical A/D converter, (c) model for a practical A/D converter.

of $x(nT)$ is rounded to the nearest discrete level, the response of the encoder will be of the form depicted in Fig. 6.16b. We can write

$$x_q(nT) = x(nT) - e(nT)$$

where $e(nT)$ is the quantization error. Hence a practical A/D converter can be represented by the model of Fig. 6.16c, where $-e(nT)$ can be regarded as a noise source. The effect of this noise source on the filter response will be considered in Chap. 11.

The function of the *D/A converter* in Fig. 6.14*b* is to generate a sampled signal $\tilde{y}(t)$ like that in Fig. 6.17*a*. In a practical D/A converter, however, the response is of the form illustrated in Fig. 6.17*b*, where

$$\tilde{y}(t) = \sum_{n=-\infty}^{\infty} y(nT) p_{\tau/2}\left(t - nT - \frac{\tau}{2}\right)$$

The Fourier transform of $\tilde{y}(t)$ is

$$\tilde{Y}(j\omega) = \sum_{n=-\infty}^{\infty} y(nT) \mathcal{F} p_{\tau/2}\left(t - nT - \frac{\tau}{2}\right)$$

$$= \frac{2\sin(\omega\tau/2)e^{-j\omega\tau/2}}{\omega} \sum_{n=-\infty}^{\infty} y(nT) e^{-j\omega nT}$$

and from Eq. (6.32)

$$\tilde{Y}(j\omega) = H_p(j\omega)\hat{Y}(j\omega)$$

where
$$H_p(j\omega) = \frac{\tau \sin(\omega\tau/2)e^{-j\omega\tau/2}}{\omega\tau/2} \quad \text{and} \quad \hat{Y}(j\omega) = \mathcal{F}\hat{y}(t)$$

Consequently, a practical D/A converter can be represented by an ideal D/A converter followed by a *fictitious filter* F_p characterized by $H_p(j\omega)$, as depicted in Fig. 6.17*c*. The amplitude response of this filter is given by

$$|H_p(j\omega)| = \tau \left|\frac{\sin(\omega\tau/2)}{\omega\tau/2}\right|$$

and is sketched in Fig. 6.17*d*. Clearly, a practical D/A converter will introduce distortion in the overall amplitude response. Fortunately, however, this effect can be eliminated to a large extent by designing the output lowpass filter so that

$$|H_p(j\omega)H_{\text{LP}}(j\omega)| \approx \begin{cases} 1 & \text{for } |\omega| \leq \dfrac{\omega_s}{2} \\ 0 & \text{otherwise} \end{cases}$$

The above effects and some others that may arise in the digital filtering of continuous-time signals are discussed by Stockham [4].

REFERENCES

1. A. Papoulis, *The Fourier Integral and Its Applications*, McGraw-Hill, New York, 1962.
2. M. J. Lighthill, *Introduction to Fourier Analysis and Generalised Functions*, Cambridge University Press, London, 1958.
3. J. M. Olmsted, *Real Variables* Appleton-Century-Crofts, New York, 1959.
4. T. G. Stockham Jr., "A-D and D-A Converters: Their Effect on Digital Audio Fidelity," *1971 Proc. 41st Conv. Audio Eng. Soc.*, New York, pp. 484–496.

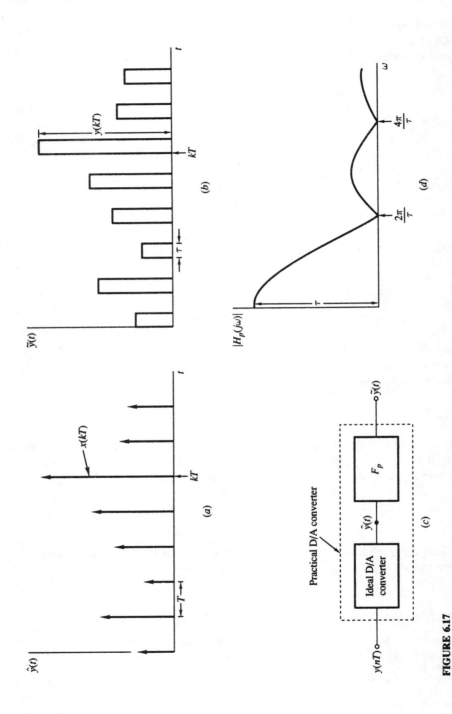

FIGURE 6.17
Digital-to-analog interface: (a) response of an ideal D/A converter, (b) response of a practical D/A converter, (c) model for a practical D/A converter, (d) amplitude response of fictitious filter F_p.

PROBLEMS

6.1. Prove Theorems 6.3, 6.4, and 6.7.

6.2. (a) Prove that
$$f^*(t) \leftrightarrow F^*(-j\omega)$$
(b) Derive Parseval's relation using the frequency convolution.

6.3. Show that for a real function $f(t)$
$$|F(j\omega)| = |F(-j\omega)| \quad \text{and} \quad \arg F(j\omega) = -\arg F(-j\omega)$$

6.4. (a) Show that for an even $f(t)$
$$\operatorname{Re} F(j\omega) = 2\int_0^\infty f(t)\cos\omega t\, dt \qquad \operatorname{Im} F(j\omega) = 0$$
(b) Show that for an odd $f(t)$
$$\operatorname{Re} F(j\omega) = 0 \qquad \operatorname{Im} F(j\omega) = 2\int_0^\infty f(t)\sin\omega t\, dt$$

6.5. Find the Fourier transforms of the following:
(a) $f(t) = u(t)e^{-at}\cos\omega_0 t$
(b) $f(t) = p_\tau(t - nT - \tau)$

6.6. (a) Assuming that the Fourier transform of $d^n f(t)/dt^n$ exists, prove that
$$\frac{d^n f(t)}{dt^n} \leftrightarrow (j\omega)^n F(j\omega)$$
(b) Hence show that
$$\mathcal{F}\int_{-\infty}^{t} f(\tau)d\tau = \frac{F(j\omega)}{j\omega}$$
(c) Use this relation to obtain the Fourier transform of
$$\phi(t) = \begin{cases} 1 - \dfrac{|t|}{T} & \text{for } |t| < T \\ 0 & \text{otherwise} \end{cases}$$

6.7. An analog filter is characterized by the transfer function
$$H(j\omega) = \sum_{i=1}^{N} \frac{A_i}{j\omega - p_i}$$
Find its impulse response.

6.8. Find the Fourier transforms of the following signals:
(a) $x_1(t) = 2[u(t + \tau) - u(t - \tau)]$
(b) $x_2(t) = \sum_{n=-\infty}^{\infty} x_1(t - nT)$

6.9. Find the Fourier transforms of the periodic signals shown in Fig. P6.9a and b. Sketch the amplitude spectra for the first two cases.

6.10. Repeat Prob. 6.9 for the signals shown in Fig. P6.10a and b.

6.11. Obtain the Fourier transforms of the following:
(a) $f(t) = \cos^2\omega_0 t$

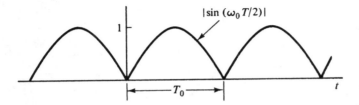

FIGURE P6.9a

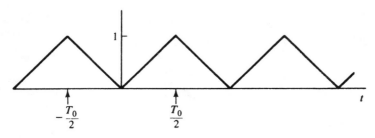

FIGURE P6.9b

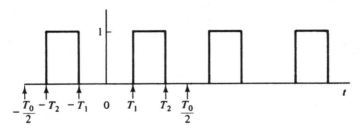

FIGURE P6.10a

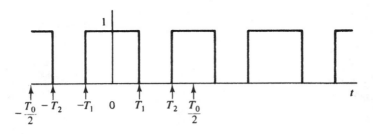

FIGURE P6.10b

(b) $f(t) = \cos at^2$

(c) $f(t) = \begin{cases} \frac{1}{2}(1 + \cos \omega_0 t) & \text{for } |t| \leq \dfrac{\pi}{\omega_0} \\ 0 & \text{otherwise} \end{cases}$

6.12. (a) Find the Fourier transform of $\hat{x}(t)$ in closed form if
$$x(t) = p_\tau(t)$$
where $\tau = (N-1)T/2$ and N is odd. The sampling frequency is $\omega_s = 2\pi/T$.

(b) Repeat part (a) if
$$x(t) = \begin{cases} \alpha + (1-\alpha)\cos\dfrac{\pi t}{\tau} & \text{for } |t| \leq \tau \\ 0 & \text{otherwise} \end{cases}$$

6.13. Signal $x(t)$ is given by
$$x(t) = \begin{cases} 1 - \dfrac{|t|}{\tau} & \text{for } |t| \leq \tau \\ 0 & \text{otherwise} \end{cases}$$
where $\tau = (N-1)T/2$ and $T = 2\pi/\omega_s$. Show that
$$\hat{X}(j\omega) \approx \frac{8}{\omega^2(N-1)T^2}\sin^2\frac{\omega(N-1)T}{4} \quad \text{for } |\omega| < \frac{\omega_s}{2}$$
if $\omega_s \gg 16/\pi N$.

6.14. The signal
$$x(t) = u(t)e^{-t}\cos 2t$$
is sampled at a rate of 2π rad/s.

(a) Show that
$$\hat{X}(j\omega) = X_D(e^{j\omega T}) = \frac{1}{2} + \sum_{k=-\infty}^{\infty}\frac{1 + j(\omega + 2\pi k)}{[1 + j(\omega + 2\pi k)]^2 + 4}$$

(b) Demonstrate the validity of this relation by computation [first evaluate the right-hand summation and then evaluate the z transform of $x(nT)$ on the unit circle $|z| = 1$].

6.15. The input and output of an ideal impulse modulator are $x(t)$ and $\hat{x}(t)$, respectively, where $\hat{x}(t)$ is given by Eq. (6.30). Find the Fourier transform of $\hat{x}(t)$ if
$$x(t) = u(t)e^{-t}$$

6.16. The signal
$$x(t) = u(t)e^{-t}\sin 2t$$
is sampled at a rate of 2π rad/s. Find the Fourier transform of sampled signal $\hat{x}(t)$.

6.17. (a) Find the Fourier transform of the signal depicted in Fig. P6.17.
(b) Obtain the Fourier transform of sampled signal $\hat{x}(t)$ in terms of an infinite summation.
(c) Obtain a closed-form expression for the Fourier transform of $\hat{x}(t)$.

6.18. The filter of Example 6.7 is used to process the signal of Fig. P6.9b.
(a) Assuming that $\omega_0 = 2\pi/T_0$, find the time- and frequency-domain representations of the signals at nodes $1, 2, \ldots, 7$.
(b) Sketch the various waveforms and amplitude spectra.

6.19. The configuration of Fig. 6.14b employs a digital filter with an amplitude response like that depicted in Fig. P6.19. The signal generated by the D/A converter is of the form shown in Fig. 6.17b, where $\tau = 3.0$ ms and the sampling frequency is 1000 rad/s.

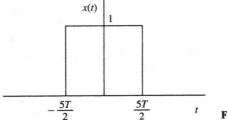

FIGURE P6.17

(a) Assuming ideal lowpass filters and A/D converter, sketch the overall amplitude response of the configuration. Indicate relevant quantities.

(b) The gain at $\omega = 300$ rad/s is required to be equal to or greater than 0.99 times the gain at $\omega = 200$ rad/s. Find the maximum permissible value of τ.

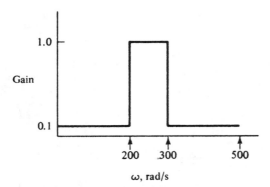

FIGURE P6.19

6.20. A bandpass digital filter is connected in cascade with a practical D/A converter as shown in Fig. 6.14b. The amplitude response of the filter is illustrated in Fig. P6.20a and the time-domain response of the D/A converter is a staircase waveform of the type shown in Fig. P6.20b. The sampling frequency is $\omega_s = 8$ rad/s.

(a) Calculate the overall gain of the cascade arrangement for $\omega = 0, 2 - \epsilon, 2 + \epsilon, 3 - \epsilon, 3 + \epsilon$, and 4 rad/s ($\epsilon$ is a small positive constant, say less than 0.0001).

(b) Sketch the amplitude response of the arrangement, indicating relevant quantities.

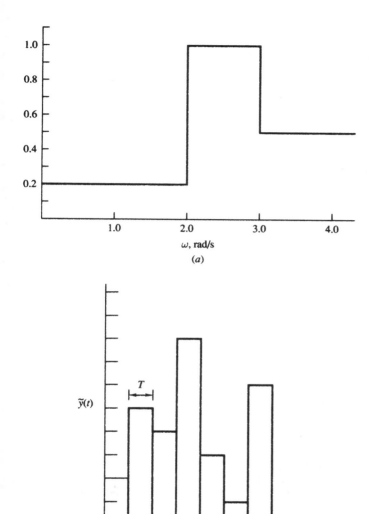

FIGURE P6.20a and b

CHAPTER 7

APPROXIMATIONS FOR RECURSIVE FILTERS

7.1 INTRODUCTION

Approximation methods for the design of recursive filters differ quite significantly from those used for the design of nonrecursive filters. The basic reason is that in the first case the transfer function is a ratio of polynomials of z whereas in the second case it is a polynomial of z^{-1}.

In *recursive filters*, the approximation problem is usually solved through indirect methods. First, a continuous-time transfer function that satisfies certain specifications is obtained using one of the standard analog-filter approximations. Then a corresponding discrete-time transfer function is obtained using one of the following methods [1–9]:

1. Invariant-impulse-response method
2. Modified version of method 1
3. Matched-z transformation
4. Bilinear transformation

In *nonrecursive filters*, on the other hand, the approximation problem is solved through direct methods which can involve the application of Fourier series, window functions, numerical analysis formulas, or the discrete-Fourier transform.

This chapter considers the approximation problem for recursive filters. The realizability constraints imposed on the transfer function are first outlined. Then the details of the above methods are discussed and design examples are supplied. The chapter concludes with a set of z-domain transformations that can be used to design filters with prescribed passband edges. Methods for the solution of the approximation problem in nonrecursive filters will be discussed in Chap 9.

Iterative methods that are suitable for the design of recursive and nonrecursive filters are considered in Chaps. 14 and 15, respectively.

7.2 REALIZABILITY CONSTRAINTS

In order to be *realizable* by a recursive filter, a transfer function must satisfy the following constraints:

1. It must be a rational function of z with real coefficients.
2. Its poles must lie within the unit circle of the z plane.
3. The degree of the numerator polynomial must be equal to or less than that of the denominator polynomial.

The first constraint is actually artificial and is imposed by our assumption in Chap. 1 that signals are real and that the constituent elements of a digital filter perform real arithmetic. If unit delays, adders, and multipliers are defined for complex signals in terms of complex arithmetic, then transfer functions with complex coefficients can be considered to be realizable [10, 11]. The second and third constraints will assure a stable and causal filter, respectively (see Sec. 3.3).

7.3 INVARIANT-IMPULSE-RESPONSE METHOD

Consider the sampled-data filter $\hat{F}_A$ of Fig. 7.1, where S is an ideal impulse sampler and F_A is an analog filter characterized by $H_A(s)$. $\hat{F}_A$ can be represented by a continuous-time transfer function $\hat{H}_A(s)$ or, equivalently, by a discrete-time transfer function $H_D(z)$, as shown in Sec. 6.9. From Eq. (6.46)

$$\hat{H}_A(j\omega) = H_D(e^{j\omega T}) = \frac{h_A(0+)}{2} + \frac{1}{T} \sum_{k=-\infty}^{\infty} H_A(j\omega + jk\omega_s) \qquad (7.1)$$

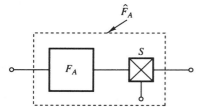

FIGURE 7.1
Sampled-data filter.

where $\omega_s = 2\pi/T$ is the sampling frequency and

$$h_A(t) = \mathcal{L}^{-1} H_A(s)$$
$$h_A(0+) = \lim_{s \to \infty} [s H_A(s)] \qquad (7.2)$$
$$H_D(z) = \mathcal{Z} h_A(nT)$$

Therefore, given an analog filter F_A, a corresponding digital filter, represented by $H_D(z)$, can be derived by using the following procedure:

1. Deduce $h_A(t)$, the impulse response of the analog filter.
2. Replace t by nT in $h_A(t)$.
3. Form the z transform of $h_A(nT)$.

If

$$H_A(j\omega) \approx 0 \qquad \text{for } |\omega| \geq \frac{\omega_s}{2} \qquad (7.3a)$$

then

$$\sum_{\substack{k=-\infty \\ k \neq 0}}^{\infty} H_A(j\omega + jk\omega_s) \approx 0 \qquad \text{for } |\omega| < \frac{\omega_s}{2} \qquad (7.3b)$$

If, in addition,

$$h_A(0+) = 0 \qquad (7.4)$$

Eqs. (7.1), (7.3b), and (7.4) yield

$$\hat{H}_A(j\omega) = H_D(e^{j\omega T}) \approx \frac{1}{T} H_A(j\omega) \qquad \text{for } |\omega| < \frac{\omega_2}{2} \qquad (7.5)$$

i.e., if $H_A(j\omega)$ is bandlimited, the baseband frequency response of the derived digital filter is approximately the same as that of the analog filter except that the gain of the digital filter is multiplied by the constant $1/T$. This constant can be eliminated by multiplying the numerator coefficients of $H_D(z)$ by T.

If the denominator degree in $H_A(s)$ exceeds the numerator degree by at least 2, the basic assumptions in Eqs. (7.3) and (7.4) hold for some value of ω_s. If, in addition, the poles of $H_A(s)$ are simple, we can write

$$H_A(s) = \sum_{i=1}^{N} \frac{A_i}{s - p_i} \qquad (7.6)$$

Hence from steps 1 and 2 above

$$h_A(t) = \mathcal{L}^{-1} H_A(s) = \sum_{i=1}^{N} A_i e^{p_i t} \qquad \text{and} \qquad h_A(nT) = \sum_{i=1}^{N} A_i e^{p_i nT}$$

Subsequently, from step 3

$$H_D(z) = \mathcal{Z} h_A(nT) = \sum_{i=1}^{N} \frac{A_i z}{z - e^{Tp_i}} \quad (7.7)$$

Since complex-conjugate pairs of poles in $H_A(s)$ yield complex-conjugate values of A_i and e^{Tp_i}, the coefficients in $H_D(z)$ are real. Pole $p_i = \sigma_i + j\omega_i$ gives rise to a pole z_i in $H_D(z)$, where

$$z_i = e^{Tp_i} = e^{T(\sigma_i + j\omega_i)}$$

and for $\sigma_i < 0$, $|z_i| < 1$. Hence a stable analog filter yields a stable digital filter. Also the numerator degree in $H_D(z)$ cannot exceed the denominator degree, and $H_D(z)$ is therefore realizable.

The above method, which is known as the *invariant-impulse-response* method, yields good results for Butterworth, Bessel, or Chebyshev lowpass and bandpass filters for which the basic assumptions of Eqs. (7.3a) and (7.4) hold. An advantage of the method is that it preserves the phase response as well as the loss characteristic of the analog filter.

Example 7.1. Design a digital filter using the Bessel transfer function

$$H_A(s) = \frac{105}{105 + 105s + 45s^2 + 10s^3 + s^4}$$

(see Sec. 5.6). Employ a sampling frequency $\omega_s = 8$ rad/s; repeat with $\omega_s = 16$ rad/s.

Solution. The poles of $H_A(s)$ and the residues in Eq. (7.6) are

$$p_1, p_1^* = -2.896211 \pm j0.8672341$$

$$p_2, p_2^* = -2.103789 \pm j2.657418$$

$$A_1, A_1^* = 1.663392 \mp j8.396299$$

$$A_2, A_2^* = -1.663392 \pm j2.244076$$

Hence from Eq. (7.7)

$$TH_D(z) = \sum_{j=1}^{2} \frac{a_{1j}z + a_{2j}z^2}{b_{0j} + b_{1j}z + z^2}$$

where coefficients a_{ij} and b_{ij} are given in Table 7.1. The transfer function is multiplied by T to eliminate constant $1/T$ in Eq. (7.5). The loss and delay characteristics obtained are plotted in Fig. 7.2a and b. The higher sampling frequency gives better results because aliasing errors are less pronounced.

TABLE 7.1
Coefficients of $TH_D(z)$ (Example 7.1)

ω_s	j	a_{1j}	a_{2j}	b_{0j}	b_{1j}
8	1	6.452333×10^{-1}	2.612851	1.057399×10^{-2}	-1.597700×10^{-1}
	2	-8.345233×10^{-1}	-2.612851	3.671301×10^{-2}	1.891907×10^{-1}
16	1	3.114550×10^{-1}	1.306425	1.028299×10^{-1}	-6.045080×10^{-1}
	2	-3.790011×10^{-1}	-1.306425	1.916064×10^{-1}	-4.404794×10^{-1}

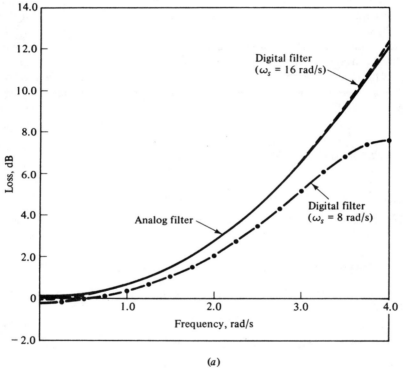

FIGURE 7.2a
Example 7.1: (a) loss characteristics.

7.4 MODIFIED INVARIANT-IMPULSE-RESPONSE METHOD

Aliasing errors tend to restrict the invariant-impulse-response method to the design of allpole filters. A *modified* version of the method, however, can be applied to filters with finite transmission zeros.

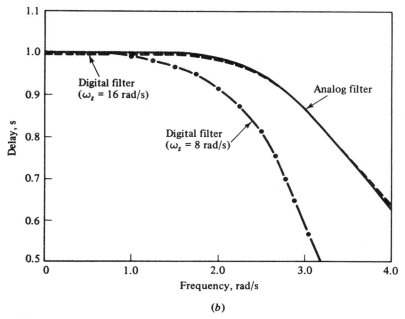

FIGURE 7.2b
Example 7.1: (b) delay characteristics.

Consider the transfer function

$$H_A(s) = \frac{H_0 N(s)}{D(s)} = \frac{H_0 \prod_{i=1}^{M}(s - s_i)}{\prod_{i=1}^{N}(s - p_i)} \quad (7.8)$$

where M can be as high as N. We can write

$$H_A(s) = \frac{H_0 H_{A1}(s)}{H_{A2}(s)}$$

where

$$H_{A1}(s) = \frac{1}{D(s)} \quad (7.9)$$

$$H_{A2}(s) = \frac{1}{N(s)} \quad (7.10)$$

Clearly, with $M, N \geq 2$ Eq. (7.2) yields

$$h_{A1}(0+) = 0 \quad h_{A2}(0+) = 0$$

and furthermore

$$\left. \begin{array}{l} H_{A1}(j\omega) \approx 0 \\ H_{A2}(j\omega) \approx 0 \end{array} \right\} \quad \text{for } |\omega| \geq \frac{\omega_s}{2}$$

for some value of ω_s. Consequently, from Eq. (7.1) we can write

$$\left.\begin{array}{l}\hat{H}_{A1}(j\omega) = H_{D1}(e^{j\omega T}) \approx \dfrac{1}{T} H_{A1}(j\omega) \\[6pt] \hat{H}_{A2}(j\omega) = H_{D2}(e^{j\omega T}) \approx \dfrac{1}{T} H_{A2}(j\omega)\end{array}\right\} \quad \text{for } |\omega| < \dfrac{\omega_s}{2}$$

Therefore, we can form

$$H_D(z) = \frac{H_0 H_{D1}(z)}{H_{D2}(z)} \tag{7.11}$$

such that

$$H_D(e^{j\omega T}) = \frac{H_0 H_{D1}(e^{j\omega T})}{H_{D2}(e^{j\omega T})} \approx H_A(j\omega) \quad \text{for } |\omega| < \frac{\omega_s}{2}$$

If the zeros and poles of $H_A(s)$ are simple, Eq. (7.7) gives

$$H_{D1}(z) = \sum_{i=1}^{N} \frac{A_i z}{z - e^{T p_i}} = \frac{N_1(z)}{D_1(z)} \tag{7.12}$$

$$H_{D2}(z) = \sum_{i=1}^{M} \frac{B_i z}{z - e^{T s_i}} = \frac{N_2(z)}{D_2(z)} \tag{7.13}$$

Thus from Eqs. (7.11) to (7.13)

$$H_D(z) = \frac{H_0 N_1(z) D_2(z)}{N_2(z) D_1(z)} \tag{7.14}$$

The derived filter can be unstable, as some of the zeros of $N_2(z)$ may be located on or outside the unit circle of the z plane, but the problem can be easily overcome. For an arbitrary pole of $H_D(z)$, say p_i, we can write

$$|(e^{j\omega T} - p_i)| = \left| -e^{j\omega T} p_i \left(e^{-j\omega T} - \frac{1}{p_i} \right) \right|$$

$$= |p_i| \left| \left(e^{j\omega T} - \frac{1}{p_i^*} \right)^* \right|$$

$$= |p_i| \left| \left(e^{j\omega T} - \frac{1}{p_i^*} \right) \right|$$

If p_i is real, we have

$$|(e^{j\omega T} - p_i)| = |p_i| \left| \left(e^{j\omega T} - \frac{1}{p_i} \right) \right|$$

and if p_i and p_i^* are a complex-conjugate pair of poles, then

$$|(e^{j\omega T} - p_i)(e^{j\omega T} - p_i^*)| = |p_i|^2 \left| \left(e^{j\omega T} - \frac{1}{p_i} \right)\left(e^{j\omega T} - \frac{1}{p_i^*} \right) \right|$$

Hence any poles of $H_D(z)$ located outside the unit circle can be replaced by their reciprocals without changing the shape of the loss characteristic, although a constant vertical shift will be introduced. On the other hand, poles on the unit circle, if any, can be moved inside the unit circle by decreasing their magnitudes slightly.

The method yields good results for lowpass and bandpass elliptic filters and, in addition, it provides some theoretical basis for the matched-z-transformation method of Sec. 7.5. Its main disadvantages are that polynomials $N_1(z)$ and $N_2(z)$ increase the order of $H_D(z)$ and that phase distortion is introduced if it becomes necessary to replace poles by their reciprocals. The vertical shift in the loss characteristic in the latter case can be eliminated by adjusting H_0, the multiplier constant of the transfer function.

Example 7.2. The transfer function

$$H_A(s) = H_0 \prod_{j=1}^{3} \frac{a_{0j} + s^2}{b_{01j} + b_{1j}s + s^2}$$

where H_0, a_{0j}, and b_{1j} are given in Table 7.2, represents a lowpass elliptic filter satisfying the following specifications:

Passband ripple: 0.1 dB
Minimum stopband loss: 43.46 dB
Passband edge: $\sqrt{0.8}$ rad/s
Stopband edge: $1/\sqrt{0.8}$ rad/s

Employing the modified invariant-impulse-response method, design a corresponding digital filter. Use $\omega_s = 7.5$ rad/s.

Solution. From Eqs. (7.9) and (7.10)

$$H_{A1}(s) = \prod_{j=1}^{3} \frac{1}{b_{0j} + b_{1j}s + s^2}$$

$$H_{A2}(s) = \prod_{j=1}^{3} \frac{1}{a_{0j} + s^2}$$

The design can be accomplished by using the following procedure:

1. Find the poles and residues of $H_{A1}(s)$ and $H_{A2}(s)$.

TABLE 7.2
Coefficients of $H_A(s)$ (Example 7.2)

j	a_{0j}	b_{0j}	b_{1j}
1	1.199341×10	3.581929×10^{-1}	9.508335×10^{-1}
2	2.000130	6.860742×10^{-1}	4.423164×10^{-1}
3	1.302358	8.633304×10^{-1}	1.088749×10^{-1}

$H_0 = 6.713267 \times 10^{-3}$

TABLE 7.3
Coefficients of $H_D(z)$ (Example 7.2)

j	a_{0j}	a_{1j}	b_{0j}	b_{1j}
1	1.0	1.942528	4.508735×10^{-1}	-1.281134
2	1.0	-7.529504×10^{-1}	6.903517×10^{-1}	-1.303834
3	1.0	-1.153491	9.128252×10^{-1}	-1.362371
4	6.603146×10^{-1}	2.193514	4.821261×10^{-1}	1.388706
5	6.552540×10^{-1}	1.775846×10	6.530851×10^{-3}	1.616274×10^{-1}

$H_0 = 3.846783 \times 10^{-4}$

2. Form $H_{D1}(z)$ and $H_{D2}(z)$ using Eqs. (7.12) and (7.13).
3. Replace zeros of $N_2(z)$ outside the unit circle by their reciprocals.
4. Adjust constant H_0 to achieve zero minimum passband loss.

With this procedure $H_D(z)$ can be deduced as

$$H_D(z) = H_0 \prod_{j=1}^{5} \frac{a_{0j} + a_{1j}z + z^2}{b_{0j} + b_{1j}z + z^2}$$

where H_0, a_{ij}, and b_{ij} are given in Table 7.3.

The loss characteristic achieved, plotted in Fig. 7.3a and b, is seen to be a faithful reproduction of the analog loss characteristic. For this filter, the conventional invariant-impulse-response method gives unsatisfactory results because the assumptions of Eqs. (7.3) and (7.4) are violated.

7.5 MATCHED-z-TRANSFORMATION METHOD

An alternative approximation method for the design of recursive filters is the so-called *matched-z-transformation method* [5, 8]. In this method, given a continuous-time transfer function like that in Eq. (7.8), a corresponding discrete-time transfer function can be formed as

$$H_D(z) = (z+1)^L \frac{H_0 \prod_{i=1}^{M}(z - e^{s_i T})}{\prod_{i=1}^{N}(z - e^{p_i T})} \quad (7.15)$$

where L is an integer. The method is closely related to the modified invariant-impulse-response method. The difference between the two is that $N_1(z)/N_2(z)$ in Eq. (7.14) is replaced by $(z+1)^L$. The value of L is equal to the number of zeros at $s = \infty$ in $H_A(s)$. Typical values for L are given in Table 7.4.

The method is very simple to apply and gives reasonable results for highpass and bandstop filters, but it tends to increase the amplitude of the passband error relative to that of the analog filter. For lowpass and bandpass filters, better passband characteristics can be achieved by using the modified invariant-impulse-response method, although the filter order is increased (see Example 7.4 below).

APPROXIMATIONS FOR RECURSIVE FILTERS **229**

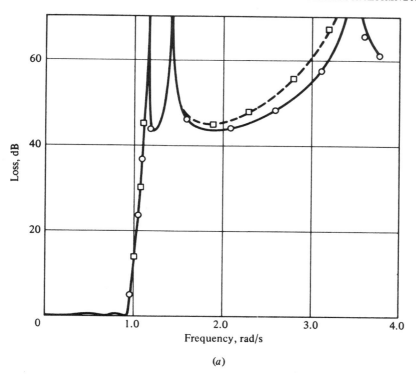

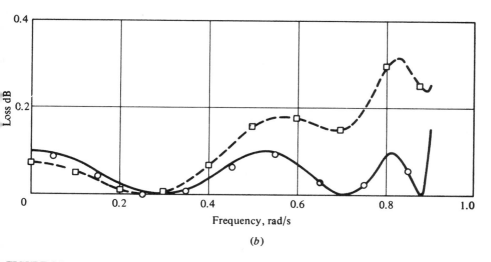

FIGURE 7.3
Examples 7.2 and 7.4: (a) loss characteristics, (b) passband characteristics. ——— analog filter. o o o modified impulse-invariant-response method, ---□---□---□--- matched-z-transformation method.

TABLE 7.4
Typical values of L in Eq. (7.15)

Type of filter	Lowpass	Highpass	Bandpass	Bandstop
Allpole	N	0	$N/2$	0
Elliptic,				
N odd	1	0		
N even	0	0	1 for $N/2$ odd	0
			0 for $N/2$ even	

Example 7.3. The transfer function

$$H_A(s) = \frac{H_0 s^4}{\prod_{j=1}^{2}(s-p_j)(s-p_j^*)}$$

where $\quad H_0 = 0.9885531 \quad p_1, p_1^* = -2.047535 \pm j1.492958$

$$p_2, p_2^* = -0.3972182 \pm j1.688095$$

represents a highpass Chebyshev filter with a passband edge of 2 rad/s and a passband ripple of 0.1 dB. Obtain a corresponding discrete-time transfer function employing the matched-z-transformation method. Use a sampling frequency of 10 rad/s.

Solution. The value of L in Eq. (7.15) is generally zero for highpass filters, according to Table 7.4. Hence $H_D(z)$ can be readily formed as

$$H_D(z) = H_0 \frac{(1-2z+z^2)^2}{\prod_{j=1}^{2}(b_{0j}+b_{1j}z+z^2)}$$

where $\quad b_{01} = 7.630567 \times 10^{-2} \quad b_{11} = -3.267079 \times 10^{-1}$

$\quad\quad\quad b_{02} = 6.070409 \times 10^{-1} \quad b_{12} = -7.608887 \times 10^{-1}$

$\quad\quad\quad H_0 = 2.076398 \times 10^{-1}$

The loss characteristic of the derived filter is compared with that of the analog filter in Fig. 7.4.

Example 7.4. Redesign the lowpass filter of Example 7.2 employing the matched-z-transformation method.

Solution. From Eq. (7.15) $H_D(z)$ can be formed as

$$H_D(z) = H_0 \prod_{j=1}^{3} \frac{a_{0j}+a_{1j}z+z^2}{b_{0j}+b_{1j}z+z^2}$$

where a_{1j} and b_{1j} are given by the first three rows in Table 7.3. For zero minimum passband loss H_0 is given by

$$H_0 = 8.604492 \times 10^{-3}$$

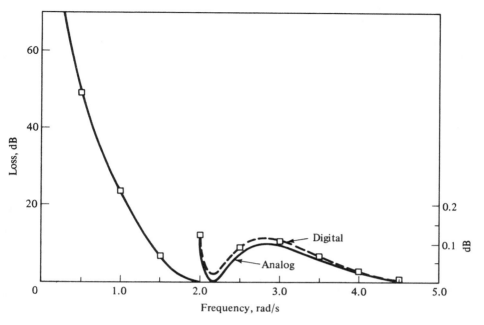

FIGURE 7.4
Loss characteristic (Example 7.3).

The loss characteristic achieved is shown in Fig. 7.3 (dashed curve). As can be seen, this is inferior to the loss characteristic obtained by using the modified invariant-impulse-response method.

7.6 BILINEAR-TRANSFORMATION METHOD

In the approximation method of Sec. 7.3 the derived digital filter has *exactly* the same impulse response as the original analog filter for $t = nT$. An approximation method will now be described whereby a digital filter is derived that has *approximately* the same time-domain response as the original analog filter *for any excitation*.

7.6.1 Derivation

Consider an *analog integrator* characterized by the transfer function

$$H_{AI}(s) = \frac{1}{s} \tag{7.16}$$

and assume that its response to an excitation $x(t)$ is $y(t)$, as depicted in Fig. 7.5. The impulse response of the integrator is given by

$$\mathcal{L}^{-1} H_I(s) = h_I(t) = \begin{cases} 1 & \text{for } t \geq 0+ \\ 0 & \text{for } t \leq 0- \end{cases}$$

FIGURE 7.5
Analog integrator.

and its response to an arbitrary excitation $x(t)$ is given by the convolution integral (see Theorem 6.7)

$$y(t) = \int_0^t x(\tau) h_I(t - \tau) d\tau$$

If $0+ < t_1 < t_2$, we can write

$$y(t_2) - y(t_1) = \int_0^{t_2} x(\tau) h_I(t_2 - \tau) d\tau - \int_0^{t_1} x(\tau) h_I(t_1 - \tau) d\tau \tag{7.17}$$

For $0+ < \tau \leq t_1, t_2$

$$h_I(t_2 - \tau) = h_I(t_1 - \tau) = 1$$

and thus Eq. (7.17) simplifies to

$$y(t_2) - y(t_1) = \int_{t_1}^{t_2} x(\tau) d\tau$$

As $t_1 \to t_2$, from Fig. 7.6

$$y(t_2) - y(t_1) \approx \frac{t_2 - t_1}{2}[x(t_1) + x(t_2)]$$

and on letting $t_1 = nT - T$ and $t_2 = nT$ the difference equation

$$y(nT) - y(nT - T) = \frac{T}{2}[x(nT - T) + x(nT)]$$

can be formed. This equation represents a *digital integrator* that has approximately the same time-domain response as the analog integrator for any excitation. By applying the z transform, we obtain

$$Y(z) - z^{-1} Y(z) = \frac{T}{2}[z^{-1} X(z) + X(z)]$$

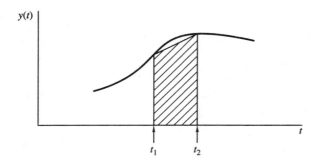

FIGURE 7.6
Response of analog integrator.

and hence the transfer function of the digital integrator can be derived as

$$H_{DI}(z) = \frac{Y(z)}{X(z)} = \frac{T}{2}\left(\frac{z+1}{z-1}\right) \tag{7.18}$$

Therefore, Eqs. (7.16) and (7.18) yield

$$H_{DI}(z) = H_{AI}(s)\Big|_{s=\frac{2}{T}\left(\frac{z-1}{z+1}\right)}$$

i.e., the transfer function of a digital integrator can be obtained by simply applying the bilinear transformation

$$s = \frac{2}{T}\left(\frac{z-1}{z+1}\right) \tag{7.19}$$

to the transfer function of the analog integrator.

Now consider an analog filter characterized by the transfer function

$$H_A(s) = \frac{\sum_{i=0}^{N} a_i s^{N-i}}{s^N + \sum_{i=0}^{N} b_i z^{N-i}}$$

From Eq. (7.16)

$$H_A(s) = \frac{\sum_{i=0}^{N} a_i \left[\frac{1}{H_{AI}(s)}\right]^{N-i}}{\left[\frac{1}{H_{AI}(s)}\right]^N + \sum_{i=0}^{N} b_i \left[\frac{1}{H_{AI}(s)}\right]^{N-i}}$$

and on replacing each $H_{AI}(s)$ by $H_{DI}(z)$, we obtain the discrete-time transfer function

$$H_D(z) = \frac{\sum_{i=0}^{N} a_i \left[\frac{1}{H_{DI}(z)}\right]^{N-i}}{\left[\frac{1}{H_{DI}(z)}\right]^N + \sum_{i=0}^{N} b_i \left[\frac{1}{H_{DI}(z)}\right]^{N-i}}$$

$$= H_A(s)\Big|_{s=\frac{2}{T}\left(\frac{z-1}{z+1}\right)} \tag{7.20}$$

Since the digital integrator has been assumed to have approximately the same time-domain response as the analog integrator for any excitation, it follows that the digital filter obtained will have approximately the same time-domain response as the analog filter for any excitation. The difference between the two responses tends to zero as T tends to zero, as may be expected.

7.6.2 Mapping Properties of Bilinear Transformation

The relation between the frequency response of the derived digital filter and that of the original analog filter can be established by examining the mapping properties of the bilinear transformation.

Equation (7.19) can be put in the form

$$z = \frac{2/T + s}{2/T - s}$$

and with $s = \sigma + j\omega$ we have

$$z = re^{j\theta}$$

where

$$r = \left[\frac{\left(\frac{2}{T}+\sigma\right)^2 + \omega^2}{\left(\frac{2}{T}-\sigma\right)^2 + \omega^2}\right]^{1/2}$$

and

$$\theta = \tan^{-1}\frac{\omega}{2/T+\sigma} + \tan^{-1}\frac{\omega}{2/T-\sigma} \qquad (7.21)$$

Clearly

$$\text{if } \sigma > 0, \quad \text{then } r > 1$$
$$\text{if } \sigma = 0, \quad \text{then } r = 1$$
$$\text{if } \sigma < 0, \quad \text{then } r < 1$$

i.e., the bilinear transformation maps

1. The open right-half s plane onto the region exterior to the unit circle $|z| = 1$ of the z plane
2. The j axis of the s plane onto the unit circle $|z| = 1$
3. The open left-half s plane onto the interior of the unit circle $|z| = 1$

For $\sigma = 0$, we have $r = 1$, and from Eq. (7.21) $\theta = 2\tan^{-1}(\omega T/2)$. Hence

$$\text{if } \omega = 0, \quad \text{then } \theta = 0$$
$$\text{if } \omega \to +\infty, \quad \text{then } \theta \to +\pi$$
$$\text{if } \omega \to -\infty, \quad \text{then } \theta \to -\pi$$

i.e., the origin of the s plane maps onto point $(1, 0)$ of the z plane and the positive and negative j axes of the s plane map onto the upper and lower semicircles $|z| = 1$, respectively. The transformation is illustrated in Fig. 7.7.

From property 2 above it follows that the maxima and minima of $|H_A(j\omega)|$ will be preserved in $|H_D(e^{j\Omega T})|$. Also if

$$M_1 \leq |H_A(j\omega)| \leq M_2$$

for some frequency range $\omega_1 \leq \omega \leq \omega_2$, then

$$M_1 \leq |H_D(e^{j\Omega T})| \leq M_2$$

for a corresponding frequency range $\Omega_1 \leq \Omega \leq \Omega_2$. Consequently, passbands or stopbands in the analog filter translate into passbands or stopbands in the digital filter.

From property 3 it follows that a stable analog filter will yield a stable digital filter, and since the transformation has real coefficients, $H_D(z)$ will have real coefficients. Finally, the numerator degree in $H_D(z)$ cannot exceed the denominator degree, and therefore $H_D(z)$ is a realizable transfer function.

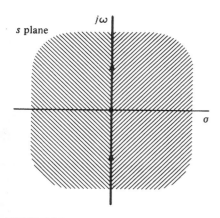
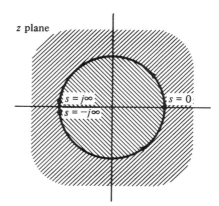

FIGURE 7.7
Bilinear transformation.

7.6.3 The Warping Effect

Let ω and Ω represent the frequency variable in the analog filter and the derived digital filter, respectively. From Eq. (7.20)

$$H_D(e^{j\Omega T}) = H_A(j\omega)$$

provided that

$$\omega = \frac{2}{T} \tan \frac{\Omega T}{2} \quad (7.22)$$

For $\Omega < 0.3/T$

$$\omega \approx \Omega$$

and, as a result, the digital filter has the same frequency response as the analog filter. For higher frequencies, however, the relation between ω and Ω becomes nonlinear, as illustrated in Fig. 7.8, and distortion is introduced in the frequency scale of the digital filter relative to that of the analog filter. This is known as the *warping effect* [2, 5].

The influence of the warping effect on the amplitude response can be demonstrated by considering an analog filter with a number of passbands centered at regular intervals, as in Fig. 7.8. The derived digital filter has the same number of passbands, but the center frequencies and bandwidths of higher-frequency passbands tend to be reduced disproportionately, as shown in Fig. 7.8.

If only the amplitude response is of concern, the warping effect can for all practical purposes be eliminated by *prewarping* the analog filter [2, 5]. Let $\omega_1, \omega_2, \ldots, \omega_i, \ldots$ be the passband and stopband edges in the analog filter. The corresponding passband and stopband edges in the digital filter are given by Eq. (7.22) as

$$\Omega_i = \frac{2}{T} \tan^{-1} \frac{\omega_i T}{2} \quad i = 1, 2, \ldots \quad (7.23)$$

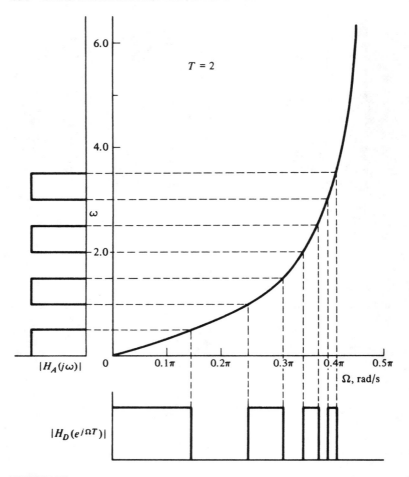

FIGURE 7.8
Influence of the warping effect on the amplitude response.

Consequently, if prescribed passband and stopband edges $\tilde{\Omega}_1, \tilde{\Omega}_2, \ldots, \tilde{\Omega}_i, \ldots$ are to be achieved in the digital filter, the analog filter must be prewarped before application of the bilinear transformation to ensure that

$$\omega_i = \frac{2}{T} \tan \frac{\tilde{\Omega}_i T}{2} \tag{7.24}$$

Under these circumstances

$$\Omega_i = \tilde{\Omega}_i$$

according to Eqs. (7.23) and (7.24), as required.

The bilinear transformation together with the above prewarping technique are used in Chap. 8 to develop a detailed procedure for the design of Butterworth, Chebyshev, inverse-Chebyshev, and elliptic filters satisfying prescribed loss specifications.

The influence of the warping effect on the phase response can be demonstrated by considering an analog filter with linear phase response. As illustrated in Fig. 7.9, the phase response of the derived digital filter is nonlinear. Furthermore, little can be done to linearize it except by employing delay equalization (see Sec. 8.5.1). Consequently, if it is mandatory to preserve a linear phase response, the alternative methods of Sec. 7.3 to 7.5 should be considered.

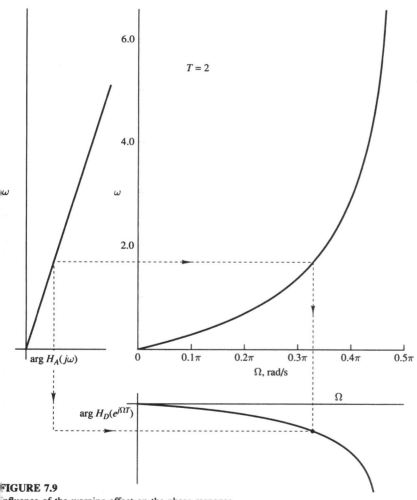

FIGURE 7.9
Influence of the warping effect on the phase response.

TABLE 7.5
Coefficients of $H_A(s)$ (Example 7.5)

j	a_{0j}	b_{0j}	b_{1j}
1	6.250	6.250	2.618910
2	8.013554	1.076433×10	3.843113×10^{-1}
3	4.874554	3.628885	2.231394×10^{-1}

TABLE 7.6
Coefficients of $H_D(z)$ (Example 7.5)

j	a'_{0j}	a'_{1j}	b'_{0j}	b'_{1j}
1	6.627508×10^{-1}	-3.141080×10^{-1}	3.255016×10^{-1}	-3.141080×10^{-1}
2	8.203382×10^{-1}	-1.915542×10^{-1}	8.893929×10^{-1}	5.716237×10^{-2}
3	1.036997	-7.266206×10^{-1}	9.018366×10^{-1}	-8.987781×10^{-1}

Example 7.5. The transfer function

$$H_A(s) = \prod_{j=1}^{3} \frac{a_{0j} + s^2}{b_{0j} + b_{1j}s + s^2}$$

where a_{0j} and b_{ij} are given in Table 7.5, represents an elliptic bandstop filter with a passband ripple of 1 dB and a minimum stopband loss of 34.45 dB. Use the bilinear transformation to obtain a corresponding digital filter. Assume a sampling frequency of 10 rad/s.

Solution. From Eq. (7.20)

$$H_D(z) = \prod_{j=1}^{3} \frac{a'_{0j} + a'_{1j}z + a'_{0j}z^2}{b'_{0j} + b'_{1j}z + z^2}$$

where
$$a'_{0j} = \frac{a_{0j} + 4/T^2}{c_j} \qquad a'_{1j} = \frac{2(a_{0j} - 4/T^2)}{c_j}$$

$$b'_{0j} = \frac{b_{0j} - 2b_{1j}/T + 4/T^2}{c_j} \qquad b'_{1j} = \frac{2(b_{0j} - 4/T^2)}{c_j}$$

$$c_j = b_{0j} + \frac{2b_{1j}}{T} + \frac{4}{T^2}$$

The numerical values of a'_{ij} and b'_{ij} are given in Table 7.6. The loss characteristic of the derived digital filter is compared with that of the analog filter in Fig. 7.10. The expected lateral displacement in the characteristic of the digital filter is evident.

7.7 DIGITAL-FILTER TRANSFORMATIONS

A normalized lowpass analog filter can be transformed into a denormalized lowpass, highpass, bandpass, or bandstop filter by employing the transformations described in

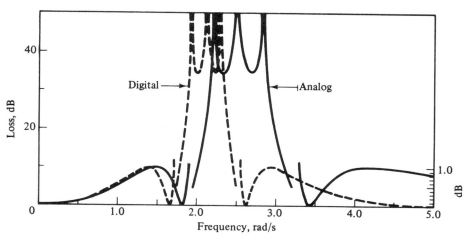

FIGURE 7.10
Loss characteristic (Example 7.5).

Sec. 5.7. Analogous transformations can be derived for digital filters as we shall now show. These are due to Constantinides [12].

7.7.1 General Transformation

Consider the transformation

$$z = f(\bar{z}) = e^{j\zeta\pi} \prod_{i=1}^{m} \frac{\bar{z} - a_i^*}{1 - a_i \bar{z}} \qquad (7.25)$$

where ζ and m are integers and a_i^* is the complex conjugate of a_i. With $z = Re^{j\Omega T}$, $\bar{z} = re^{j\omega T}$, and $a_i = c_i e^{j\psi_i}$, Eq. (7.25) becomes

$$Re^{j\Omega T} = e^{j\zeta\pi} \prod_{i=1}^{m} \frac{re^{j\omega T} - c_i e^{-j\psi_i}}{1 - rc_i e^{j(\omega T + \psi_i)}}$$

and hence

$$R^2 = \prod_{i=1}^{m} \frac{r^2 + c_i^2 - 2rc_i \cos(\omega T + \psi_i)}{1 + (rc_i)^2 - 2rc_i \cos(\omega T + \psi_i)} \qquad (7.26)$$

Evidently,

if $R > 1$, then $r^2 + c_i^2 > 1 + (rc_i)^2$ or $r > 1$
if $R = 1$, then $r^2 + c_i^2 = 1 + (rc_i)^2$ or $r = 1$
if $R < 1$, then $r^2 + c_i^2 < 1 + (rc_i)^2$ or $r < 1$

In effect, Eq. (7.25) maps

1. The unit circle $|z| = 1$ onto the unit circle $|\bar{z}| = 1$
2. The interior of $|z| = 1$ onto the interior of $|\bar{z}| = 1$
3. The exterior of $|z| = 1$ onto the exterior of $|\bar{z}| = 1$

as illustrated in Fig. 7.11.

Now consider a normalized lowpass filter characterized by $H_N(z)$ with a passband extending from 0 to Ω_p. On applying the above transformation we can form

$$H(\bar{z}) = H_N(z)\Big|_{z=f(\bar{z})} \tag{7.27}$$

With the poles of $H_N(z)$ located inside the unit circle $|z| = 1$ those of $H(\bar{z})$ will be located inside the unit circle $|\bar{z}| = 1$; that is, $H(\bar{z})$ will represent a stable filter. Furthermore, from item 1 above, if

$$M_1 \leq |H_N(e^{j\Omega T})| \leq M_2$$

for some frequency range $\Omega_1 \leq \Omega \leq \Omega_2$, then

$$M_1 \leq |H(e^{j\omega T})| \leq M_2$$

for one or more corresponding ranges of ω; that is, the passband (stopband) in $H_N(z)$ will translate into one or more passbands (stopbands) in $H(\bar{z})$. Therefore, the above transformation can form the basis of a set of transformations that can be used to derive denormalized lowpass, highpass, bandpass, and bandstop filters from a given normalized lowpass filter.

7.7.2 Lowpass-to-Lowpass Transformation

The appropriate values for ζ, m, and a_i in Eq. (7.25) can be determined by examining the details of the necessary mapping. If $H(\bar{z})$ is to represent a *lowpass* filter with a

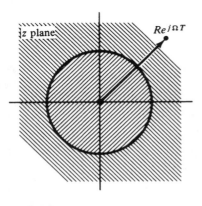
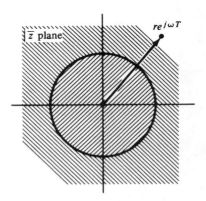

FIGURE 7.11
General z-domain transformation.

passband edge ω_p, the mapping must be of the form shown in Fig. 7.12a, where solid lines denote passbands. As phasor $e^{j\Omega T}$ traces the unit circle in the z plane once, $e^{j\omega T}$ must trace the unit circle in the $\bar{z}$ plane once in the same sense. The transformation must thus be bilinear ($m = 1$) of the form

$$z = e^{j\zeta\pi}\frac{\bar{z} - a^*}{1 - a\bar{z}} \tag{7.28}$$

At points A and A', $z = \bar{z} = 1$, and at C and C', $z = \bar{z} = -1$. Hence Eq. (7.28) gives

$$1 = e^{j\zeta\pi}\frac{1 - a^*}{1 - a} \quad \text{and} \quad 1 = e^{j\zeta\pi}\frac{1 + a^*}{1 + a}$$

By solving these equations we obtain

$$a = a^* \equiv \alpha \qquad \zeta = 0$$

where α is a real constant. Thus Eq. (7.28) becomes

$$z = \frac{\bar{z} - \alpha}{1 - \alpha\bar{z}}$$

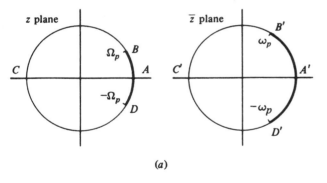

(a)

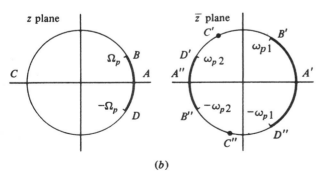

(b)

FIGURE 7.12
(a) Lowpass-to-lowpass transformation, (b) lowpass-to-bandstop transformation.

The necessary value for α can be determined by noting that at points B and B', we have $\Omega = \Omega_p$ and $\omega = \omega_p$, in which case

$$e^{j\Omega_p T} = \frac{e^{j\omega_p T} - \alpha}{1 - \alpha e^{j\omega_p T}}$$

or

$$\alpha = \frac{\sin[(\Omega_p - \omega_p)T/2]}{\sin[(\Omega_p + \omega_p)T/2]}$$

7.7.3 Lowpass-to-Bandstop Transformation

If a *bandstop* filter is required with passband edges ω_{p1} and ω_{p2}, the mapping must have the form shown in Fig. 7.12*b*. In order to introduce an upper passband in $H(\bar{z})$, $e^{j\Omega T}$ must trace the unit circle of the z plane twice for each revolution of $e^{j\omega T}$ in the $\bar{z}$ plane. Consequently, in this case, the transformation must be biquadratic ($m = 2$) of the form

$$z = e^{j\zeta\pi} \frac{\bar{z}^2 + \beta\bar{z} + \gamma}{1 + \beta\bar{z} + \gamma\bar{z}^2}$$

where β and γ are real constants. At points A and A', $z = \bar{z} = 1$ and

$$e^{j\zeta\pi} = 1$$

so that

$$z = \frac{\bar{z}^2 + \beta\bar{z} + \gamma}{1 + \beta\bar{z} + \gamma\bar{z}^2}$$

With $z = e^{j\Omega T}$ and $\bar{z} = e^{j\omega T}$

$$e^{j\Omega T} = \frac{e^{j2\omega T} + \beta e^{j\omega T} + \gamma}{e^{j2\omega T}(e^{-j2\omega T} + \beta e^{-j\omega T} + \gamma)}$$

Hence

$$\frac{\Omega T}{2} = \tan^{-1} \frac{\sin 2\omega T + \beta \sin \omega T}{\cos 2\omega T + \beta \cos \omega T + \gamma} - \omega T$$

and after some manipulation

$$\tan \frac{\Omega T}{2} = \frac{(1-\gamma)\sin \omega T}{(1+\gamma)\cos \omega T + \beta}$$

At points B and B', $\Omega = \Omega_p$ and $\omega = \omega_{p1}$, respectively, and as a result

$$\tan \frac{\Omega_p T}{2} = \frac{(1-\gamma)\sin \omega_{p1} T}{(1+\gamma)\cos \omega_{p1} T + \beta} \tag{7.29}$$

Likewise, at points D and D', $\Omega = -\Omega_p$ and $\omega = \omega_{p2}$ so that

$$\tan \frac{-\Omega_p T}{2} = \frac{(1-\gamma)\sin \omega_{p2} T}{(1+\gamma)\cos \omega_{p2} T + \beta} \tag{7.30}$$

Now by solving Eqs. (7.29) and (7.30), β and γ can be deduced as

$$\beta = -\frac{2\alpha}{1+k} \qquad \gamma = \frac{1-k}{1+k}$$

where $\alpha = \dfrac{\cos\left[(\omega_{p2}+\omega_{p1})T/2\right]}{\cos\left[(\omega_{p2}-\omega_{p1})T/2\right]}$ and $k = \tan\dfrac{\Omega_p T}{2}\tan\dfrac{(\omega_{p2}-\omega_{p1})T}{2}$

Lowpass-to-highpass and *lowpass-to-bandpass* transformations can similarly be derived. The complete set of transformations is summarized in Table 7.7.

7.7.4 Application

The Constantinides transformations can be readily applied to design filters with prescribed passband edges. The following procedure can be employed:

1. Obtain a lowpass transfer function $H_N(z)$ using any approximation method.
2. Determine the passband edge Ω_p in $H_N(z)$.
3. Form $H(\bar{z})$ according to Eq. (7.27) using the appropriate transformation.

An important feature of filters designed by using the above procedure is that the passband edge in lowpass or highpass filters can be varied by varying a single parameter, namely, α. Similarly, both the lower and upper passband edges in bandpass or bandstop filters can be varied by varying only a pair of parameters, namely, α and k [13, 14].

An alternative design procedure by which prescribed passband as well as stopband edges can be achieved is described in Chap. 8.

TABLE 7.7
Constantinides transformations

Type	Transformation	α, k
LP to LP	$z = \dfrac{\bar{z}-\alpha}{1-\alpha\bar{z}}$	$\alpha = \dfrac{\sin\left[(\Omega_p-\omega_p)T/2\right]}{\sin\left[(\Omega_p+\omega_p)T/2\right]}$
LP to HP	$z = -\dfrac{\bar{z}-\alpha}{1-\alpha\bar{z}}$	$\alpha = \dfrac{\cos\left[(\Omega_p-\omega_p)T/2\right]}{\cos\left[(\Omega_p+\omega_p)T/2\right]}$
LP to BP	$z = -\dfrac{\bar{z}^2 - \dfrac{2\alpha k}{k+1}\bar{z} + \dfrac{k-1}{k+1}}{1 - \dfrac{2\alpha k}{k+1}\bar{z} + \dfrac{k-1}{k+1}\bar{z}^2}$	$\alpha = \dfrac{\cos\left[(\omega_{p2}+\omega_{p1})T/2\right]}{\cos\left[(\omega_{p2}-\omega_{p1})T/2\right]}$ $k = \tan\dfrac{\Omega_p T}{2}\cot\dfrac{(\omega_{p2}-\omega_{p1})T}{2}$
LP to BS	$z = \dfrac{\bar{z}^2 - \dfrac{2\alpha}{1+k}\bar{z} + \dfrac{1-k}{1+k}}{1 - \dfrac{2\alpha}{1+k}\bar{z} + \dfrac{1-k}{1+k}\bar{z}^2}$	$\alpha = \dfrac{\cos\left[(\omega_{p2}+\omega_{p1})T/2\right]}{\cos\left[(\omega_{p2}-\omega_{p1})T/2\right]}$ $k = \tan\dfrac{\Omega_p T}{2}\tan\dfrac{(\omega_{p2}-\omega_{p1})T}{2}$

REFERENCES

1. J. F. Kaiser, "Design Methods for Sampled Data Filters," *Proc. 1st Allerton Conf. Circuit Syst. Theory*, pp. 221–236, November, 1963.
2. R. M. Golden and J. F. Kaiser, "Design of Wideband Sampled-Data Filters," *Bell Syst. Tech. J.*, vol. 43, pp. 1533–1546, July 1964.
3. C. M. Rader and B. Gold, "Digital Filter Design Techniques in the Frequency Domain," *Proc. IEEE*, vol. 55, pp. 149–171, February 1967.
4. D. J. Nowak and P. E. Schmid, "Introduction to Digital Filters," *IEEE Trans. Electromagn. Compat.*, vol. EMC-10, pp. 210–220, June 1968.
5. R. M. Golden, "Digital Filter Synthesis by Sampled-Data Transformation," *IEEE Trans. Audio Electroacoust.*, vol. AU-16, pp. 321–329, September 1968.
6. A. J. Gibbs, "An Introduction to Digital Filters," *Aust. Telecommun. Res.*, vol. 3, pp. 3–14, November 1969.
7. A. J. Gibbs, "The Design of Digital Filters," *Aust. Telecommun. Res.*, vol. 4, pp. 29–34, March 1970.
8. L. R. Rabiner and B. Gold, *Theory and Application of Digital Signal Processing*, Prentice-Hall, Englewood Cliffs, N.J., 1975.
9. A. V. Oppenheim and R. W. Schafer, *Discrete-Time Signal Processing*, Prentice-Hall, Englewood Cliffs, N.J., 1989.
10. T. H. Crystal and L. Ehrman, "The Design and Applications of Digital Filters with Complex Coefficients," *IEEE Trans. Audio Electroacoust.*, vol. AU-16, pp. 315–320, September 1968.
11. P. A. Regalia, S. K. Mitra, and J. Fadavi-Ardekani, "Implementation of Real Coefficient Digital Filters Using Complex Arithmetic," *IEEE Trans. Circuits Syst.*, vol. CAS-34, pp. 345–353, April 1987.
12. A. G. Constantinides, "Spectral Transformations for Digital Filters," *IEE Proc.*, vol. 117, pp. 1585–1590, August 1970.
13. R. E. Crochiere and P. Penfield, Jr., "On the Efficient Design of Bandpass Digital Filter Structures," *IEEE Trans. Acoust., Speech, Signal Process.*, vol. ASSP-23, pp. 380–381, August 1975.
14. M. N. S. Swamy and K. S. Thyagarajan, "Digital Bandpass and Bandstop Filters with Variable Center Frequency and Bandwidth," *Proc. IEEE*, vol. 64, pp. 1632–1634, November 1976.
15. S. S. Haykin and R. Carnegie, "New Method of Synthetising Linear Digital Filters Based on Convolution Integral," *IEE Proc.*, vol. 117, pp. 1063–1072, June 1970.
16. S. A. White, "New Method of Synthetising Linear Digital Filters Based on Convolution Integral," *IEE Proc. (Corr.)*, vol. 118, p. 348, February 1971.
17. A. Antoniou and C. Shekher, "Invariant-Sinusoid Approximation Method for Recursive Digital Filters," *Electron. Lett.*, vol. 9, pp. 498–500, October 1973.

PROBLEMS

7.1. By using the invariant-impulse-response method, derive a discrete-time transfer function from the continuous-time transfer function.

$$H_A(s) = \frac{1}{(s+1)(s^2+s+1)}$$

The sampling frequency is 10 rad/s.

7.2. An analog filter is characterized by the continuous-time transfer function

$$H_A(s) = \frac{1}{s^3 + 6s^2 + 11s + 6}$$

Obtain a discrete-time transfer function by using the invariant-impulse-response method. The sampling frequency is 6π.

7.3. The sixth-order normalized Bessel transfer function can be expressed as

$$H_A(s) = \sum_{i=1}^{3} \left(\frac{A_i}{s - p_i} + \frac{A_i^*}{s - p_i^*} \right)$$

where A_i and p_i are given in Table P7.3.
(a) Design a digital filter by using the invariant-impulse-response method, assuming a sampling frequency of 10 rad/s.
(b) Plot the phase response of the resulting filter.

TABLE P7.3

i	p_i	A_i
1	$-4.248359 + j8.675097 \times 10^{-1}$	$1.095923 \times 10 - j3.942517 \times 10$
2	$-3.735708 + j2.626272$	$-1.412677 \times 10 + j1.270117 \times 10$
3	$-2.515932 + j4.492673$	$3.167539 - j2.024596 \times 10^{-1}$

7.4. The transfer function

$$H_A(s) = \frac{s^2 - 3s + 3}{s^2 + 3s + 3}$$

is a constant-delay approximation. Can one design a corresponding digital filter by using the invariant-impulse-response method? If so, carry out the design employing a sampling frequency of 10 rad/s. Otherwise, explain the reasons for the failure of the method.

7.5. A bandpass filter is required with passband edges of 900 and 1600 rad/s and a maximum passband loss of 1.0 dB. Obtain a design by employing the invariant-impulse-response method. Start with a second-order normalized lowpass Chebyshev approximation and neglect the effects of aliasing. A suitable sampling frequency is 10,000 rad/s.

7.6. Given an analog filter characterized by

$$H_A(s) = \sum_{i=1}^{N} \frac{A_i}{s - p_i}$$

a corresponding digital filter characterized by $H_D(z)$ can be derived such that

$$\mathcal{R}_A u(t)\Big|_{t=nT} = \mathcal{R}_D u(nT)$$

This is called the *invariant-unit-step-response approximation method* [15, 16].
(a) Show that

$$H_D(e^{j\omega T}) \approx H_A(j\omega) \quad \text{for } |\omega| < \frac{\omega_s}{2}$$

if $\omega \ll 1/T$ and

$$\frac{H_A(j\omega)}{j\omega} \approx 0 \quad \text{for } |\omega| \geq \frac{\omega_s}{2}$$

(b) Show that

$$H_D(z) = \sum_{i=1}^{N} \frac{A_i'}{z - e^{p_i T}} \quad \text{where } A_i' = \frac{(e^{p_i T} - 1) A_i}{p_i}$$

7.7. Given an analog filter characterized by

$$H_A(s) = H_0 + \sum_{i=1}^{N} \frac{A_i}{s - p_i}$$

a corresponding digital filter characterized by $H_D(z)$ can be derived such that

$$\mathcal{R}_A u(t) \sin \omega_0 t \Big|_{t=nT} = \mathcal{R}_D u(nT) \sin \omega_0 nT$$

This is the so-called *invariant-sinusoid-response approximation method* [17].

(a) Show that

$$H_D(e^{j\omega T}) \approx \frac{2\omega_0 (\cos \omega T - \cos \omega_0 T)}{(\omega_0^2 - \omega^2) T \sin \omega_0 T} H_A(j\omega) \quad \text{for } |\omega| < \frac{\omega_s}{2}$$

if $\omega \ll 1/T$ and

$$\frac{\omega_0 H_A(j\omega)}{\omega_0^2 - \omega^2} \approx 0 \quad \text{for } |\omega| \geq \frac{\omega_s}{2}$$

(b) Show that

$$H_D(z) = H_0 + \sum_{i=1}^{N} \frac{U_i z + V_i}{z - e^{p_i T}}$$

where

$$U_i = (\omega_0 e^{p_i T} - p_i \sin \omega_0 T - \omega_0 \cos \omega_0 T) A'_i$$

$$V_i = [e^{p_i T}(p_i \sin \omega_0 T - \omega_0 \cos \omega_0 T) + \omega_0] A'_i$$

$$A'_i = \frac{A_i}{(p_i^2 + \omega_0^2) \sin \omega_0 T}$$

7.8. (a) Design a third-order digital filter by applying the invariant-unit-step-response method (see Prob. 7.6) to the transfer function in Prob. 7.1. Assume that $\omega_s = 10$ rad/s.
 (b) Compare the design with that obtained in Prob. 7.1.

7.9. (a) Redesign the filter of Prob. 7.3 by employing the invariant-sinusoid-response method (see Prob. 7.7). The value of ω_0 may be assumed to be 1 rad/s.
 (b) Plot the resulting phase response.

7.10. A lowpass filter is required satisfying the specifications of Fig. P7.10. Obtain a design by applying the modified invariant-impulse-response method to an elliptic approximation. The sampling frequency is 20,000 rad/s.

7.11. Redesign the filter of Prob. 7.10 by using the matched-z-transformation method.

7.12. Design a highpass filter satisfying the specifications of Fig. P7.12. Use the matched-z-transformation method along with an elliptic approximation. Assume that $\omega_s = 6000$ rad/s.

7.13. (a) Obtain a discrete-time transfer function by applying the bilinear transformation of the transfer function of Prob. 7.1. The sampling frequency is 4π rad/s.
 (b) Determine the gain and phase-shift of the filter at $\omega = 0$, and $\omega = \pi$ rad/s.

7.14. Design a digital filter by applying the bilinear transformation to the Chebyshev transfer function of Example 5.2. The sampling frequency is 10 rad/s. Adjust the multiplier constant to achieve a minimum passband loss of zero.

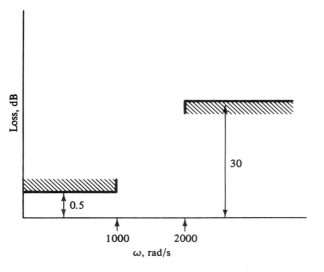

FIGURE P7.10

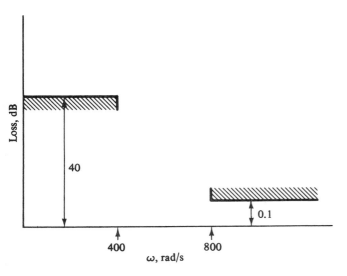

FIGURE P7.12

7.15. The lowpass transfer function of Example 5.3 is subjected to the bilinear transformation.
   (a) Assuming that $\omega_s = 10$ rad/s, find the resulting passband and stopband edges and also the infinite-loss frequencies.
   (b) Determine the effective selectivity factor for the digital filter.
   (c) Find the minimum value of ω_s if the passband and stopband edges in the digital filter are to be within ± 1 percent of the corresponding values in the analog filter.

7.16. (a) Obtain a continuous-time fourth-order highpass transfer function with a 3-dB cutoff frequency at 1 rad/s using the Butterworth approximation.

(b) Obtain a corresponding discrete-time transfer function using the bilinear transformation. Assume a sampling frequency $\omega_s = 10$ rad/s.

(c) Determine the exact 3-dB cutoff frequency of the digital filter.

7.17. $H_D(z)$ represents a lowpass filter with a passband edge Ω_p. Show that $H_D(-z)$ represents a highpass filter with a passband edge $\Omega_p - \omega_s/2$.

7.18. The transfer function

$$H(z) = H_0 \prod_{j=1}^{2} \frac{a_{0j} + a_{1j}z + a_{0j}z^2}{b_{0j} + b_{1j}z + z^2}$$

where a_{ij} and b_{ij} are given in Table P7.18, represents a lowpass filter with a passband edge of 1 rad/s if $\omega_s = 2\pi$ rad/s. By using the lowpass-to-highpass transformation, design a highpass filter with a passband edge of 2 rad/s if $\omega_s = 2\pi$ rad/s.

TABLE P7.18

j	a_{0j}	a_{1j}	b_{0j}	b_{1j}
1	1.722415×10^{-1}	3.444829×10^{-1}	4.928309×10^{-1}	-1.263032
2	1.727860×10^{-1}	3.455720×10^{-1}	7.892595×10^{-1}	-9.753164×10^{-1}

$H_0 = 3.500865 \times 10^{-1}$

7.19. By using the transfer function of Prob. 7.18 obtain a lowpass cascade canonic structure whose passband edge can be varied by varying a single parameter.

7.20. Derive the lowpass-to-highpass transformation of Table 7.7.

7.21. Derive the lowpass-to-bandpass transformation of Table 7.7.

CHAPTER 8

RECURSIVE FILTERS SATISFYING PRESCRIBED SPECIFICATIONS

8.1 INTRODUCTION

The previous chapter has shown that given an analog filter, a corresponding digital filter can be readily obtained by using the bilinear transformation. This design method preserves the maxima and minima of the amplitude response and, as a consequence, passbands and stopbands in the analog filter translate into corresponding passbands and stopbands in the digital filter; furthermore, the passband ripple and minimum stopband attenuation in the analog filter are preserved in the digital filter, and the latter filter is stable if the former is stable. Owing to these important advantages, the bilinear transformation method is one of the most important methods for the design of digital filters, if not the most important. As was demonstrated in Chap. 7, the main problem with the method is the warping effect which introduces frequency-scale distortion. If $\omega_1, \omega_2, \ldots, \omega_i, \ldots$ are the passband and stopband edges in the analog filter, the corresponding passband and stopband edges in the derived digital filter are given by

$$\Omega_i = \frac{2}{T} \tan^{-1} \frac{\omega_i T}{2} \qquad i = 1, 2, \ldots$$

according to Eq. (7.23). Consequently, if prescribed passband and stopband edges $\tilde{\Omega}_1, \tilde{\Omega}_2, \ldots, \tilde{\Omega}_i, \ldots$ are to be achieved, the analog filter must be prewarped before the application of the bilinear transformation to ensure that

$$\omega_i = \frac{2}{T} \tan \frac{\tilde{\Omega}_i T}{2}$$

in which case

$$\Omega_i = \tilde{\Omega}_i$$

The design of lowpass, highpass, etc., filters is usually accomplished in two steps. First a normalized lowpass transfer function is transformed into a denormalized lowpass, highpass, etc., transfer function employing the standard analog-filter transformations described in Sec. 5.7. Then the bilinear transformation is applied. Prewarping can be effected by choosing the parameters in the analog-filter tranformations appropriately.

This chapter considers the details of the above design procedure. Formulas are derived for the parameters of the analog-filter transformations for Butterworth, Chebyshev, inverse-Chebyshev, and elliptic filters, which simplify the design of filters satisfying prescribed specifications [1].

8.2 DESIGN PROCEDURE

Consider a normalized analog filter characterized by $H_N(s)$, with a loss

$$A_N(\omega) = 20 \log \frac{1}{|H_N(j\omega)|}$$

and assume that

$$0 \leq A_N(\omega) \leq A_p \quad \text{for } 0 \leq |\omega| \leq \omega_p$$
$$A_N(\omega) \geq A_a \quad \text{for } \omega_a \leq |\omega| \leq \infty$$

as illustrated in Fig. 8.1. A corresponding lowpass, highpass, bandpass, or bandstop filter with the same passband ripple and the same minimum stopband loss can be derived by using the following steps:

1. Form

$$H_X(\bar{s}) = H_N(s)\Big|_{s=f_X(\bar{s})} \tag{8.1}$$

where $f_X(\bar{s})$ is given in Table 8.1 (see Sec. 5.7).

2. Form

$$H_D(z) = H_X(\bar{s})\Big|_{\bar{s}=\frac{2}{T}\left(\frac{z-1}{z+1}\right)} \tag{8.2}$$

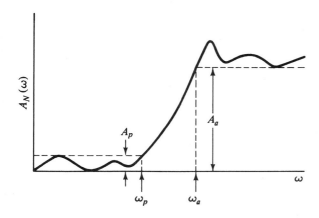

FIGURE 8.1
Loss characteristic of normalized lowpass filter.

TABLE 8.1
Standard forms of $f_X(\bar{s})$

X	$f_X(\bar{s})$
LP	$\lambda \bar{s}$
HP	$\dfrac{\lambda}{\bar{s}}$
BP	$\dfrac{1}{B}\left(\bar{s} + \dfrac{\omega_0^2}{\bar{s}}\right)$
BS	$\dfrac{B\bar{s}}{\bar{s}^2 + \omega_0^2}$

If the derived filter is to have prescribed passband and stopband edges, the parameters λ, ω_0, and B in Table 8.1 and the order of $H_N(s)$ must be chosen appropriately. Formulas for these parameters for the various types of analog-filter approximations are derived in the following section.

8.3 DESIGN FORMULAS

8.3.1 Lowpass and Highpass Filters

Consider the *lowpass-filter* specification of Fig. 8.2, where $\tilde{\Omega}_p$ and $\tilde{\Omega}_a$ are the desired passband and stopband edges, and assume that the above design procedure yields a transfer function $H_D(z)$ such that

$$A_D(\Omega) = 20 \log \frac{1}{|H_D(e^{j\Omega T})|}$$

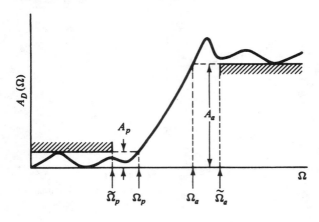

FIGURE 8.2
Loss characteristic of lowpass digital filter.

where
$$0 \leq A_D(\Omega) \leq A_p \quad \text{for } 0 \leq |\Omega| \leq \Omega_p$$
$$A_D(\Omega) \geq A_a \quad \text{for } \Omega_a \leq |\Omega| \leq \frac{\omega_s}{2}$$

(see Fig. 8.2).
From Eq. (8.1) and Table 8.1

$$|H_{\text{LP}}(j\bar{\omega})| = |H_N(j\omega)|$$

if

$$\omega = \lambda\bar{\omega}$$

and hence
$$\omega_p = \lambda\bar{\omega}_p \tag{8.3}$$
$$\omega_a = \lambda\bar{\omega}_a \tag{8.4}$$

where $\bar{\omega}_p$ and $\bar{\omega}_a$ denote the passband and stopband edges, respectively, in $H_{\text{LP}}(\bar{s})$. From Eq. (8.2)

$$|H_D(e^{j\Omega T})| = |H_{\text{LP}}(j\bar{\omega})|$$

if

$$\bar{\omega} = \frac{2}{T}\tan\frac{\Omega T}{2}$$

and thus
$$\bar{\omega}_p = \frac{2}{T}\tan\frac{\Omega_p T}{2} \tag{8.5}$$
$$\bar{\omega}_a = \frac{2}{T}\tan\frac{\Omega_a T}{2} \tag{8.6}$$

Hence Eqs. (8.3) to (8.6) yield

$$\omega_p = \frac{2}{T}\lambda \tan \frac{\Omega_p T}{2} \tag{8.7}$$

$$\omega_a = \frac{2}{T}\lambda \tan \frac{\Omega_a T}{2} \tag{8.8}$$

Now on assigning

$$\Omega_p = \tilde{\Omega}_p \qquad \Omega_a \leq \tilde{\Omega}_a$$

the desired specifications will be met. The appropriate value for λ is obtained from Eq. (8.7) as

$$\lambda = \frac{T\omega_p}{2\tan(\tilde{\Omega}_p T/2)}$$

Since $\tan(\tilde{\Omega}_a T/2) \geq \tan(\Omega_a T/2)$ for $0 < \Omega_a, \tilde{\Omega}_a < \omega_s/2$, Eqs. (8.7) and (8.8) give

$$\omega_a \leq \frac{\omega_p}{K_0}$$

where

$$K_0 = \frac{\tan(\tilde{\Omega}_p T/2)}{\tan(\tilde{\Omega}_a T/2)} \tag{8.9}$$

This inequality imposes a lower limit on the order of $H_N(s)$, as will be shown later.

The preceding approach can be readily extended to *highpass filters*. The resulting formulas for λ and ω_a are given in Table 8.2.

8.3.2 Bandpass and Bandstop Filters

Now consider the *bandpass-filter* specification of Fig. 8.3, where $\tilde{\Omega}_{p1}$, $\tilde{\Omega}_{p2}$ and $\tilde{\Omega}_{a1}$, $\tilde{\Omega}_{a2}$ represent the desired passband and stopband edges, respectively, and assume that the

TABLE 8.2
Lowpass and highpass filters

LP	$\omega_a \leq \dfrac{\omega_p}{K_0}$	
	$\lambda = \dfrac{\omega_p T}{2\tan(\tilde{\Omega}_p T/2)}$	
HP	$\omega_a \leq \omega_p K_0$	
	$\lambda = \dfrac{2\omega_p \tan(\tilde{\Omega}_p T/2)}{T}$	
where	$K_0 = \dfrac{\tan(\tilde{\Omega}_p T/2)}{\tan(\tilde{\Omega}_a T/2)}$	

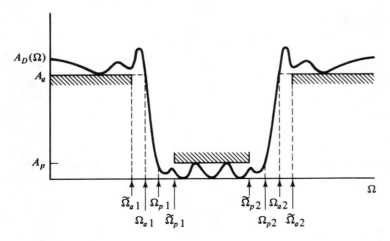

FIGURE 8.3
Loss characteristic of bandpass digital filter.

derived filter has

$$0 \leq A_D(\Omega) \leq A_p \quad \text{for } \Omega_{p1} \leq |\Omega| \leq \Omega_{p2}$$

$$A_D(\Omega) \geq A_a \begin{cases} \text{for } 0 \leq |\Omega| \leq \Omega_{a1} \\ \text{and } \Omega_{a2} \leq |\Omega| \leq \dfrac{\omega_s}{2} \end{cases}$$

as shown in Fig. 8.3.

From Eq. (8.1) and Table 8.1

$$|H_{\text{BP}}(j\bar{\omega})| = |H_N(j\omega)|$$

provided that

$$\omega = \frac{1}{B}\left(\bar{\omega} - \frac{\omega_0^2}{\bar{\omega}}\right)$$

or, by solving for $\bar{\omega}$, if

$$\bar{\omega} = \frac{\omega B}{2} \pm \sqrt{\omega_0^2 + \left(\frac{\omega B}{2}\right)^2} \tag{8.10}$$

With

$$\omega = \pm\omega_p \text{ or } \pm\omega_a$$

the positive passband and stopband edges in $H_{\text{BP}}(\bar{s})$ can be deduced from Eq. (8.10) as

$$\bar{\omega}_{p1}, \bar{\omega}_{p2} = \mp\frac{\omega_p B}{2} + \sqrt{\omega_0^2 + \omega_p^2\left(\frac{B}{2}\right)^2}$$

and
$$\bar{\omega}_{a1}, \bar{\omega}_{a2} = \mp \frac{\omega_a B}{2} + \sqrt{\omega_0^2 + \omega_a^2 \left(\frac{B}{2}\right)^2}$$

respectively. Evidently

$$\bar{\omega}_{p1}\bar{\omega}_{p2} = \omega_0^2 \tag{8.11}$$

$$\bar{\omega}_{a1}\bar{\omega}_{a2} = \omega_0^2 \tag{8.12}$$

$$\bar{\omega}_{p2} - \bar{\omega}_{p1} = \omega_p B \tag{8.13}$$

$$\bar{\omega}_{a2} - \bar{\omega}_{a1} = \omega_a B \tag{8.14}$$

From Eq. (8.2)
$$|H_D(e^{j\Omega T})| = |H_{\mathrm{BP}}(j\bar{\omega})|$$

if
$$\bar{\omega} = \frac{2}{T} \tan \frac{\Omega T}{2}$$

and hence
$$\bar{\omega}_{p1} = \frac{2}{T} \tan \frac{\Omega_{p1} T}{2} \tag{8.15}$$

$$\bar{\omega}_{p2} = \frac{2}{T} \tan \frac{\Omega_{p2} T}{2} \tag{8.16}$$

$$\bar{\omega}_{a1} = \frac{2}{T} \tan \frac{\Omega_{a1} T}{2} \tag{8.17}$$

$$\bar{\omega}_{a2} = \frac{2}{T} \tan \frac{\Omega_{a2} T}{2} \tag{8.18}$$

We can now assign
$$\Omega_{p1} = \tilde{\Omega}_{p1} \quad \text{and} \quad \Omega_{p2} = \tilde{\Omega}_{p2}$$

From Eqs. (8.13), (8.15), and (8.16)
$$B = \frac{2K_A}{T\omega_p} \tag{8.19}$$

where
$$K_A = \tan \frac{\tilde{\Omega}_{p2} T}{2} - \tan \frac{\tilde{\Omega}_{p1} T}{2} \tag{8.20}$$

Also from Eqs. (8.11), (8.15), and (8.16)
$$\omega_0 = \frac{2\sqrt{K_B}}{T} \tag{8.21}$$

where
$$K_B = \tan \frac{\tilde{\Omega}_{p1} T}{2} \tan \frac{\tilde{\Omega}_{p2} T}{2} \tag{8.22}$$

From Eqs. (8.11) and (8.12)
$$\bar{\omega}_{a1}\bar{\omega}_{a2} = \bar{\omega}_{p1}\bar{\omega}_{p2}$$

but since $\bar{\omega}_{p1}$ and $\bar{\omega}_{p2}$ have already been fixed by assigning values to Ω_{p1} and Ω_{p2}, it is not in general possible to achieve the prescribed stopband edges exactly. The alternative is to assign

$$\Omega_{a1} \geq \tilde{\Omega}_{a1} \quad \text{and} \quad \Omega_{a2} \leq \tilde{\Omega}_{a2}$$

without violating Eqs. (8.12) and (8.14). In this way the loss at $\Omega = \tilde{\Omega}_{a1}$ and $\tilde{\Omega}_{a2}$ will equal or exceed the minimum specified.

If

$$\Omega_{a1} \geq \tilde{\Omega}_{a1}$$

then Eq. (8.17) gives

$$\bar{\omega}_{a1} = \frac{2}{T} \tan \frac{\Omega_{a1} T}{2} \geq \frac{2}{T} \tan \frac{\tilde{\Omega}_{a1} T}{2} \tag{8.23}$$

since $\Omega_{a1}T/2$, $\tilde{\Omega}_{a1}T/2 < \pi/2$. If Ω_{a2} is chosen such that Eqs. (8.12), (8.14), and (8.17) are satisfied, we have

$$\begin{aligned}\omega_a &= \frac{\omega_0^2 - \bar{\omega}_{a1}^2}{B\bar{\omega}_{a1}} \\ &= \frac{\omega_0^2 - (2/T)^2 \tan^2(\Omega_{a1}T/2)}{(2B/T)\tan(\Omega_{a1}T/2)}\end{aligned} \tag{8.24}$$

Therefore, from Eqs. (8.23) and (8.24)

$$\omega_a \leq \frac{\omega_0^2 - (2/T)^2 \tan^2(\tilde{\Omega}_{a1}T/2)}{(2B/T)\tan(\tilde{\Omega}_{a1}T/2)}$$

and on eliminating ω_0 and B using Eqs. (8.21) and (8.19) we deduce

$$\omega_a \leq \frac{\omega_p}{K_1}$$

where

$$K_1 = \frac{K_A \tan(\tilde{\Omega}_{a1}T/2)}{K_B - \tan^2(\tilde{\Omega}_{a1}T/2)} \tag{8.25}$$

On the other hand, if

$$\Omega_{a2} \leq \tilde{\Omega}_{a2}$$

then from Eq. (8.18)

$$\bar{\omega}_{a2} = \frac{2}{T} \tan \frac{\Omega_{a2}T}{2} \leq \frac{2}{T} \tan \frac{\tilde{\Omega}_{a2}T}{2} \tag{8.26}$$

since $\Omega_{a2}T/2$, $\tilde{\Omega}_{a2}T/2 < \pi/2$. Now if Ω_{a1} is chosen such that Eqs. (8.12), (8.14), and (8.18) are satisfied, we have

$$\omega_a = \frac{\bar{\omega}_{a2}^2 - \omega_0^2}{B\bar{\omega}_{a2}}$$

$$= \frac{(2/T)^2 \tan^2(\Omega_{a2}T/2) - \omega_0^2}{(2B/T) \tan(\Omega_{a2}T/2)} \quad (8.27)$$

Therefore, from Eqs. (8.26) and (8.27)

$$\omega_a \leq \frac{(2/T)^2 \tan^2(\tilde{\Omega}_{a2}T/2) - \omega_0^2}{(2B/T) \tan(\tilde{\Omega}_{a2}T/2)}$$

and on eliminating ω_0 and B we have

$$\omega_a \leq \frac{\omega_p}{K_2}$$

where

$$K_2 = \frac{K_A \tan(\tilde{\Omega}_{a2}T/2)}{\tan^2(\tilde{\Omega}_{a2}T/2) - K_B} \quad (8.28)$$

Summarizing, if $\Omega_{a1} \geq \tilde{\Omega}_{a1}$ then $\omega_a \leq \omega_p/K_1$; and if $\Omega_{a2} \leq \tilde{\Omega}_{a2}$ then $\omega_a \leq \omega_p/K_2$. These relations also hold in the reverse order, that is, if $\omega_a \leq \omega_p/K_1$ then $\Omega_{a1} \geq \tilde{\Omega}_{a1}$, and if $\omega_a \leq \omega_p/K_2$ then $\Omega_{a2} \leq \tilde{\Omega}_{a2}$. Therefore, if we ensure that

$$\omega_a \leq \min\left(\frac{\omega_p}{K_1}, \frac{\omega_p}{K_2}\right) \quad \text{or} \quad \omega_a \leq \frac{\omega_p}{K} \quad \text{where } K = \max(K_1, K_2)$$

then $\Omega_{a1} \geq \tilde{\Omega}_{a1}$ and $\Omega_{a2} \leq \tilde{\Omega}_{a2}$

as required. The appropriate value for K is easily deduced from Eqs. (8.25) and (8.28) as

$$K = \begin{cases} K_1 & \text{if } K_C \geq K_B \\ K_2 & \text{if } K_C < K_B \end{cases}$$

where

$$K_C = \tan\frac{\tilde{\Omega}_{a1}T}{2} \tan\frac{\tilde{\Omega}_{a2}T}{2} \quad (8.29)$$

The same approach can also be applied to *bandstop filters*. The relevant design formulas are summarized in Table 8.3.

The formulas derived so far are very general and apply to any normalized lowpass filter that has a loss characteristic of a form illustrated in Fig. 8.1. Let us now consider the specific requirements for Butterworth, Chebyshev, inverse-Chebyshev, and elliptic filters.

8.3.3 Butterworth Filters

The loss in a normalized *Butterworth* filter is given by

$$A_N(\omega) = 10\log(1 + \omega^{2n})$$

TABLE 8.3
Bandpass and bandstop filters

BP	$\omega_0 = \dfrac{2\sqrt{K_B}}{T}$	
	$\omega_a \leq \begin{cases} \dfrac{\omega_p}{K_1} & \text{if } K_C \geq K_B \\ \dfrac{\omega_p}{K_2} & \text{if } K_C < K_B \end{cases}$	
	$B = \dfrac{2K_A}{T\omega_p}$	
BS	$\omega_0 = \dfrac{2\sqrt{K_B}}{T}$	
	$\omega_a \leq \begin{cases} \omega_p K_2 & \text{if } K_C \geq K_B \\ \omega_p K_1 & \text{if } K_C < K_B \end{cases}$	
	$B = \dfrac{2K_A \omega_p}{T}$	
where	$K_A = \tan \dfrac{\tilde{\Omega}_{p2} T}{2} - \tan \dfrac{\tilde{\Omega}_{p1} T}{2}$ $\quad K_B = \tan \dfrac{\tilde{\Omega}_{p1} T}{2} \tan \dfrac{\tilde{\Omega}_{p2} T}{2}$	
	$K_C = \tan \dfrac{\tilde{\Omega}_{a1} T}{2} \tan \dfrac{\tilde{\Omega}_{a2} T}{2}$ $\quad K_1 = \dfrac{K_A \tan(\tilde{\Omega}_{a1} T/2)}{K_B - \tan^2(\tilde{\Omega}_{a1} T/2)}$	
	$K_2 = \dfrac{K_A \tan(\tilde{\Omega}_{a2} T/2)}{\tan^2(\tilde{\Omega}_{a2} T/2) - K_B}$	

(see Sec. 5.3.1), where n is the order of the transfer function. For $\omega = \omega_p$ or ω_a

$$A_N(\omega_p) = A_p = 10 \log (1 + \omega_p^{2n}) \qquad A_N(\omega_a) = A_a = 10 \log (1 + \omega_a^{2n})$$

or

$$\omega_p = (10^{0.1A_p} - 1)^{1/2n} \qquad \omega_a = (10^{0.1A_a} - 1)^{1/2n}$$

Thus from Tables 8.2 and 8.3

$$\frac{\omega_a}{\omega_p} = \left(\frac{10^{0.1A_a} - 1}{10^{0.1A_p} - 1} \right)^{1/2n} \leq \frac{1}{K}$$

where K is given in Table 8.4. Therefore, the necessary value for n in order to meet the prescribed specifications is given by

$$n \geq \frac{\log D}{2 \log (1/K)}$$

where

$$D = \frac{10^{0.1A_a} - 1}{10^{0.1A_p} - 1} \tag{8.30}$$

TABLE 8.4
Butterworth filters

$$n \geq \frac{\log D}{2 \log (1/K)}$$

$$\omega_p = (10^{0.1 A_p} - 1)^{1/2n}$$

LP	$K = K_0$
HP	$K = \dfrac{1}{K_0}$
BP	$K = \begin{cases} K_1 & \text{if } K_C \geq K_B \\ K_2 & \text{if } K_C < K_B \end{cases}$
BS	$K = \begin{cases} \dfrac{1}{K_2} & \text{if } K_C \geq K_B \\ \dfrac{1}{K_1} & \text{if } K_C < K_B \end{cases}$

where $\quad D = \dfrac{10^{0.1 A_a} - 1}{10^{0.1 A_p} - 1}$

8.3.4 Chebyshev Filters

In normalized *Chebyshev* filters

$$A_N(\omega) = 10 \log [1 + \varepsilon^2 T_n^2(\omega)]$$

where
$$T_n(\omega) = \cosh (n \cosh^{-1} \omega) \quad \text{for } \omega_p \leq \omega < \infty$$

$$\varepsilon^2 = 10^{0.1 A_p} - 1 \qquad \omega_p = 1$$

(see Sec. 5.4.2). For $\omega = \omega_a$

$$A_N(\omega_a) = A_a = 10 \log \{1 + (10^{0.1 A_p} - 1)[\cosh (n \cosh^{-1} \omega_a)]^2\}$$

or
$$\omega_a = \cosh \left(\frac{1}{n} \cosh^{-1} \sqrt{D} \right)$$

Thus from Tables 8.2 and 8.3

$$\frac{\omega_a}{\omega_p} = \cosh \left(\frac{1}{n} \cosh^{-1} \sqrt{D} \right) \leq \frac{1}{K}$$

where K is given in Table 8.5. Therefore

$$n \geq \frac{\cosh^{-1} \sqrt{D}}{\cosh^{-1}(1/K)}$$

where $\cosh^{-1} x$ can be evaluated using the identity

$$\cosh^{-1} x = \ln (x + \sqrt{x^2 - 1})$$

TABLE 8.5
Chebyshev filters

$$n \geq \frac{\cosh^{-1} \sqrt{D}}{\cosh^{-1}(1/K)}$$

$$\omega_p = 1$$

LP	$K = K_0$
HP	$K = \dfrac{1}{K_0}$
BP	$K = \begin{cases} K_1 & \text{if } K_C \geq K_B \\ K_2 & \text{if } K_C < K_B \end{cases}$
BS	$K = \begin{cases} \dfrac{1}{K_2} & \text{if } K_C \geq K_B \\ \dfrac{1}{K_1} & \text{if } K_C < K_B \end{cases}$

8.3.5 Inverse-Chebyshev Filters

In normalized *inverse-Chebyshev* filters, the loss is given by

$$A_N(\omega) = 10 \log \left[1 + \frac{1}{\delta^2 T_n^2(1/\omega)} \right]$$

where

$$T_n(1/\omega) = \cosh [n \cosh^{-1}(1/\omega)] \qquad \text{for } 0 < \omega < \omega_a$$

$$\delta^2 = \frac{1}{10^{0.1 A_a} - 1}$$

and

$$\omega_a = 1$$

(see Sec. 5.4.3). For $\omega = \omega_p$, we can write

$$A_N(\omega_p) = 10 \log \left[1 + \frac{1}{\delta^2 T_n^2(1/\omega_p)} \right]$$

Now solving for ω_p, we obtain

$$\omega_p = \frac{1}{\cosh (\frac{1}{n} \cosh^{-1} \sqrt{D})} \tag{8.31}$$

Thus, as in the Chebyshev approximation, we have

$$\frac{\omega_a}{\omega_p} = \cosh \left(\frac{1}{n} \cosh^{-1} \sqrt{D} \right) \leq \frac{1}{K}$$

and, therefore, the order of the filter must satisfy the inequality

$$n \geq \frac{\cosh^{-1}\sqrt{D}}{\cosh^{-1}(1/K)}$$

where K is given in Table 8.5. Evidently, inverse-Chebyshev filters can be designed in the same way as Chebyshev filters except that the value of ω_p is given by Eq. (8.31) instead of being equal to unity.

8.3.6 Elliptic Filters

The selectivity factor in *elliptic* filters is defined as

$$k = \frac{\omega_p}{\omega_a}$$

(see Sec. 5.5.1). Thus from Tables 8.2 and 8.3

$$k \geq K$$

where $K = K_0, 1/K_0, \ldots$. Since any value in the range 0 to 1, except for unity, is a permissible value for k, we can assign

$$k = K$$

as in Table 8.6. With k chosen, the value of ω_p is fixed, i.e.,

$$\omega_p = \sqrt{k}$$

TABLE 8.6
Elliptic filters

	k	ω_p
LP	K_0	$\sqrt{K_0}$
HP	$\dfrac{1}{K_0}$	$\dfrac{1}{\sqrt{K_0}}$
BP	K_1 if $K_C \geq K_B$ K_2 if $K_C < K_B$	$\sqrt{K_1}$ $\sqrt{K_2}$
BS	$\dfrac{1}{K_2}$ if $K_C \geq K_B$ $\dfrac{1}{K_1}$ if $K_C < K_B$	$\dfrac{1}{\sqrt{K_2}}$ $\dfrac{1}{\sqrt{K_1}}$
	$n \geq \dfrac{\log 16D}{\log(1/q)}$	

Finally, with k, A_p, and A_a known the necessary value for n can be computed by using the formula in Table 8.6 (see Sec. 5.5.5).

8.4 DESIGN USING THE FORMULAS AND TABLES

The formulas and tables developed in the preceding section lead to the following simple design procedure:

1. Using the prescribed specifications, determine n, ω_p, and for elliptic filters k, from Tables 8.4 to 8.6 (use Eq. 8.31 to calculate ω_p for inverse-Chebyshev filters).
2. Determine λ for lowpass and highpass filters using Table 8.2 or B and ω_0 for bandpass and bandstop filters using Table 8.3.
3. Form the normalized transfer function (see Chap. 5).
4. Apply the transformation in Eq. (8.1).
5. Apply the transformation in Eq. (8.2).

The procedure yields lowpass and highpass filters that satisfy the specifications exactly at the passband edge and oversatisfy the specifications at the stopband edge. In the case of bandpass and bandstop filters, the specifications are satisfied exactly at the two passband edges and at the stopband edge at the narrower transition region, and oversatisfy the specifications at the other stopband edge.

Example 8.1. Design a highpass filter satisfying the following specifications:

$$A_p = 1 \text{ dB} \qquad A_a = 45 \text{ dB} \qquad \tilde{\Omega}_p = 3.5 \text{ rad/s}$$

$$\tilde{\Omega}_a = 1.5 \text{ rad/s} \qquad \omega_s = 10 \text{ rad/s}$$

Use a Butterworth, a Chebyshev, and then an elliptic approximation.

Solution. Butterworth filter From Eqs. (8.9) and (8.30)

$$K_0 = \frac{\tan(3.5\pi/10)}{\tan(1.5\pi/10)} = 3.85184 \qquad D = \frac{10^{4.5} - 1}{10^{0.1} - 1} = 1.22127 \times 10^5$$

Hence from Table 8.4

$$n \geq \frac{\log D}{2 \log K_0} \approx 4.34$$

$$= 5$$

$$\omega_p = (10^{0.1} - 1)^{0.1} = 0.8736097$$

Now from Table 8.2

$$\lambda = \frac{2}{T} \omega_p \tan \frac{\tilde{\Omega}_p T}{2} = 5.4576$$

Chebyshev filter From Table 8.5

$$n \geq \frac{\cosh^{-1}\sqrt{D}}{\cosh^{-1} K_0} = \frac{\ln\left(\sqrt{D} + \sqrt{D-1}\right)}{\ln\left(K_0 + \sqrt{K_0^2 - 1}\right)} \approx 3.24$$

$$= 4$$

$$\omega_p = 1$$

Hence from Table 8.2

$$\lambda = 6.247183$$

Elliptic filter From Table 8.6

$$k = \frac{1}{K_0} = 0.2596162$$

From Eqs. (5.45) to (5.47)

$$k' = \sqrt{1 - k^2} = 0.9657119$$

$$q_0 = \frac{1}{2}\left(\frac{1 - \sqrt{k'}}{1 + \sqrt{k'}}\right) = 4.361108 \times 10^{-3}$$

$$q = q_0 + 2q_0^5 + \cdots \approx q_0$$

Hence

$$n \geq \frac{\log 16D}{\log(1/q)} \approx 2.67$$

$$= 3$$

$$\omega_p = \sqrt{k} = 0.5095255$$

and from Table 8.2

$$\lambda = 3.183099$$

The transfer functions $H_N(s)$, $H_{HP}(\bar{s})$, and $H_D(z)$ can be put in the form

$$H_0 \prod_{j=1}^{J} \frac{a_{0j} + a_{1j}w + a_{2j}w^2}{b_{0j} + b_{1j}w + b_{2j}w^2}$$

where H_0 is a multiplier constant and $w = s$, $\bar{s}$, or z. The coefficients of $H_N(s)$ can be computed as in Table 8.7, those of $H_{HP}(\bar{s})$ as in Table 8.8, and those of $H_D(z)$ as in Table 8.9.

The loss characteristics of the three filters are plotted in Fig. 8.4.

Example 8.2. Design an elliptic bandpass filter satisfying the following specifications:

$A_p = 1$ dB $\quad A_a = 45$ dB $\quad \tilde{\Omega}_{p1} = 900$ rad/s $\quad \tilde{\Omega}_{p2} = 1100$ rad/s

$\tilde{\Omega}_{a1} = 800$ rad/s $\quad \tilde{\Omega}_{a2} = 1200$ rad/s $\quad \omega_s = 6000$ rad/s

TABLE 8.7
Coefficients of $H_N(s)$ (Example 8.1)

	j	a_{0j}	a_{1j}	a_{2j}	b_{0j}	b_{1j}	b_{2j}
Butterworth	1	1	0	0	1	1	0
	2	1	0	0	1	1.618034	1
	3	1	0	0	1	0.618034	1
	$H_0 = 1.0$						
Chebyshev	1	1	0	0	0.279398	0.673739	1
	2	1	0	0	0.986505	0.279072	1
	$H_0 = 0.245653$						
Elliptic	1	1	0	0	0.257305	1	0
	2	5.091668	0	1	0.259234	0.244205	1
	$H_0 = 0.0131003$						

TABLE 8.8
Coefficients of $H_{HP}(\bar{s})$ (Example 8.1)

	j	a_{0j}	a_{1j}	a_{2j}	b_{0j}	b_{1j}	b_{2j}
Butterworth	1	0	1	0	5.45760	1	0
	2	0	0	1	29.7854	8.83058	1
	3	0	0	1	29.7854	3.37298	1
	$H_0 = 1.0$						
Chebyshev	1	0	0	1	139.684	15.0644	1
	2	0	0	1	39.5612	1.76726	1
	$H_0 = 0.891251$						
Elliptic	1	0	1	0	12.3709	1	0
	2	1.98994	0	1	39.0848	2.99855	1
	$H_0 = 1.0$						

Solution. From Eqs. (8.20), (8.22), and (8.29)

$$K_A = \tan\frac{1100\pi}{6000} - \tan\frac{900\pi}{6000} = 0.1398821$$

$$K_B = \tan\frac{900\pi}{6000} \tan\frac{1100\pi}{6000} = 0.3308897$$

$$K_C = \tan\frac{800\pi}{6000} \tan\frac{1200\pi}{6000} = 0.3234776$$

TABLE 8.9
Coefficients of $H_D(z)$ (Example 8.1)

	j	a_{0j}	a_{1j}	a_{2j}	b_{0j}	b_{1j}	b_{2j}
Butterworth	1	−0.368384	$-a_{01}$	0	0.263231	1	0
	2	0.148945	−0.297889	a_{02}	0.173594	0.577816	1
	3	0.200026	−0.400052	a_{03}	0.576084	0.775981	1
	$H_0 = 1.0$						
Chebyshev	1	0.0512326	−0.102465	a_{01}	0.515070	1.31014	1
	2	0.183159	−0.366318	a_{02}	0.796619	1.06398	1
	$H_0 = 0.891251$						
Elliptic	1	−0.204648	$-a_{01}$	0	0.590704	1	0
	2	0.206292	−0.277126	a_{02}	0.675139	0.985428	1
	$H_0 = 1.0$						

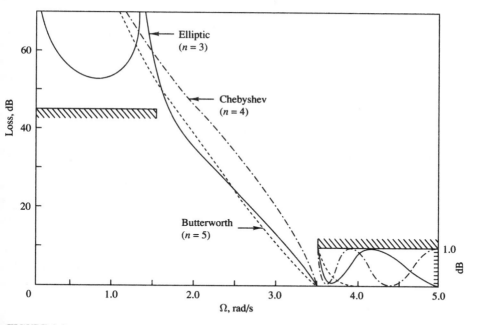

FIGURE 8.4
Loss characteristics of highpass filters (Example 8.1).

Hence $K_B > K_C$ and from Table 8.6

$$k = K_2 = \frac{K_A \tan(\tilde{\Omega}_{a2}T/2)}{\tan^2(\tilde{\Omega}_{a2}T/2) - K_B} = 0.5159572 \qquad \omega_p = \sqrt{K_2} = 0.7183016$$

D is the same as in Example 8.1, and

$$k' = \sqrt{1 - k^2} = 0.8566144$$

$$q_0 = \frac{1}{2}\left(\frac{1 - \sqrt{k'}}{1 + \sqrt{k'}}\right) = 0.01933628$$

$$q = q_0 + 2q_0^5 + \cdots \approx 0.01933629$$

Hence
$$n \geq \frac{\log 16D}{\log(1/q)} \approx 3.67$$

$$= 4$$

From Table 8.3

$$\omega_0 = \frac{2\sqrt{K_B}}{T} = 1098.609 \qquad B = \frac{2K_A}{T\omega_p} = 371.9263$$

The transfer function of the filter can be formed as

$$H_D(z) = H_0 \prod_{j=1}^{4} \frac{a_{0j} + a_{1j}z + a_{0j}z^2}{b_{0j} + b_{1j}z + z^2}$$

The numerical values of the coefficients are given in Table 8.10. The loss characteristic of the filter is plotted in Fig. 8.5.

Example 8.3. Design a Chebyshev bandstop filter satisfying the following specifications:

$A_p = 0.5$ dB $\qquad A_a = 40$ dB $\qquad \tilde{\Omega}_{p1} = 350$ rad/s $\qquad \tilde{\Omega}_{p2} = 700$ rad/s

$\tilde{\Omega}_{a1} = 430$ rad/s $\qquad \tilde{\Omega}_{a2} = 600$ rad/s $\qquad \omega_s = 3000$ rad/s

Solution. From Tables 8.3 and 8.5

$$n = 5 \qquad \omega_0 = 561.4083 \qquad B = 493.2594$$

The transfer function is of the form

$$H_D(z) = \prod_{j=1}^{5} \frac{a_{0j} + a_{1j}z + a_{0j}z^2}{b_{0j} + b_{1j}z + z^2}$$

TABLE 8.10
Coefficients of $H_D(z)$ (Example 8.2)

j	a_{0j}	a_{1j}	b_{0j}	b_{1j}
1	1.40266	−0.0102157	0.926867	−0.888660
2	0.826391	−1.32443	0.930606	−1.04661
3	1.07347	−0.631800	0.973854	−0.804891
4	0.941754	−1.25358	0.976782	−1.16031

$H_0 = 0.00293912$

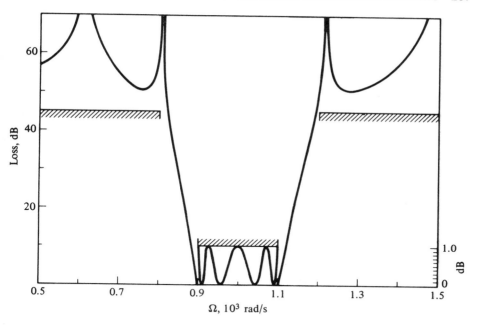

FIGURE 8.5
Loss characteristic of elliptic bandpass filter (Example 8.2).

TABLE 8.11
Coefficients of $H_D(z)$ (Example 8.3)

j	a_{0j}	a_{1j}	b_{0j}	b_{1j}
1	0.485564	−0.472249	−0.0288728	−0.472249
2	0.529076	−0.514569	0.623010	0.0502889
3	1.06121	−1.03211	0.754357	−1.40016
4	0.718033	−0.698344	0.916899	−0.217511
5	1.13666	−1.10549	0.942893	−1.43593

where a_{ij} and b_{ij} are given in Table 8.11. The loss characteristic of the filter is plotted in Fig. 8.6.

8.5 CONSTANT GROUP DELAY

The phase response in filters designed by using the preceding method is in general quite nonlinear because of two reasons. First, the Butterworth, Chebyshev, inverse-Chebyshev, and elliptic approximations are inherently nonlinear-phase approximations. Second, the warping effect tends to increase the nonlinearity of the phase re-

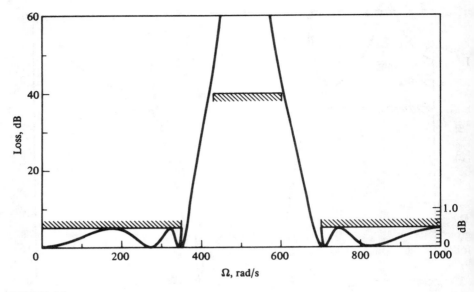

FIGURE 8.6
Loss characteristic of Chebyshev bandstop filter (Example 8.3).

sponse. As a consequence, the group delay tends to vary with frequency and the application of these filters tends to introduce delay distortion (see Sec. 3.6).

Constant group-delay filters can sometimes be designed by using constant-delay approximations such as the Bessel approximation with design methods that preserve the linearity in the phase response, e.g., the invariant-impulse-response method. However, a constant delay and prescribed loss specifications are usually difficult to achieve simultaneously, particularly if bandpass or bandstop high-selectivity filters are desired.

8.5.1 Delay Equalization

The design of constant-delay analog filters satisfying prescribed loss specifications is almost invariably accomplished in two steps. First a filter is designed satisfying the loss specifications ignoring the group delay. Then a delay equalizer is designed which can be used in cascade with the filter to compensate for variations in the group delay of the filter. The same technique can also be used in digital filters.

Let $H_F(z)$ and $H_E(z)$ be the transfer functions of the filter and equalizer, respectively. The group delays of the filter and equalizer are given by

$$\tau_F(\omega) = -\frac{d\theta_F(\omega)}{d\omega} \quad \text{and} \quad \tau_E(\omega) = -\frac{d\theta_E(\omega)}{d\omega}$$

respectively, where

$$\theta_F(\omega) = \arg H_F(e^{j\omega T}) \quad \text{and} \quad \theta_E(\omega) = \arg H_E(e^{j\omega T})$$

The overall transfer function of the filter-equalizer combination is

$$H_{FE}(z) = H_F(z)H_E(z)$$

Hence

$$|H_{FE}(e^{j\omega T})| = |H_F(e^{j\omega T})||H_E(e^{j\omega T})|$$

and

$$\theta_{FE}(\omega) = \theta_F(\omega) + \theta_E(\omega) \tag{8.32}$$

Now from Eq. (8.32) the overall group delay of the filter-equalizer combination can be obtained as

$$\tau_{FE}(\omega) = \tau_F(\omega) + \tau_E(\omega)$$

Therefore, a digital filter that satisfies prescribed loss specifications and has constant group delay with respect to some passband $\omega_{p1} \leq \omega \leq \omega_{p2}$ can be designed using the following steps:

1. Design a filter satisfying the loss specifications using the procedure in Sec. 8.4.
2. Design an equalizer with

$$|H_E(e^{j\omega T})| = 1 \quad \text{for } 0 \leq \omega \leq \frac{\omega_s}{2}$$

and

$$\tau_E(\omega) = \tau - \tau_F(\omega) \quad \text{for } \omega_{p1} \leq \omega \leq \omega_{p2} \tag{8.33}$$

where τ is a constant.

From step 2, $H_E(z)$ must be an *allpass* transfer function of the form

$$H_E(z) = \prod_{j=1}^{M} \frac{1 + c_{1j}z + c_{0j}z^2}{c_{0j} + c_{1j}z + z^2}$$

The equalizer can be designed by finding a set of values for c_{0j}, c_{1j}, τ, and M such that: (a) Eq. (8.33) is satisfied to within a prescribed error in order to achieve approximately constant group delay with respect to the passband, and (b) the poles of $H_E(z)$ are inside the unit circle of the z plane to ensure that the equalizer is stable. Equalizers can be designed by using optimization methods as will be demonstrated in Chap. 14 (see Sec. 14.8).

8.5.2 Zero-Phase Filters

In nonreal-time applications, the problem of delay distortion can be eliminated in a fairly simple manner by designing the filter as a cascade arrangement of two filters characterized by $H(z)$ and $H(z^{-1})$, as depicted in Fig. 8.7a. Since $H(e^{-j\omega T})$ is the complex conjugate of $H(e^{j\omega T})$, the frequency response of the cascade arrangement

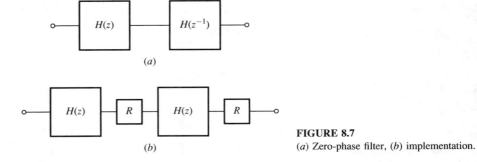

FIGURE 8.7
(*a*) Zero-phase filter, (*b*) implementation.

can be expressed as

$$H_0(e^{j\omega T}) = H(e^{j\omega T})H(e^{-j\omega T}) = |H(e^{j\omega T})|^2$$

In other words, the frequency response of the arrangement is real and, as a result, the filter has *zero phase response*. If a filter with passband ripple A_p and minimum stopband loss A_a is required, the design can be readily completed by obtaining a transfer function with passband ripple $A_p/2$ and minimum stopband loss $A_a/2$, since the two filters in Fig. 8.7*a* have identical amplitude responses.

If the impulse response of the first filter is $h(n)$, then that of the second filter is $h(-n)$, as can be readily demonstrated. Hence the cascade of Fig. 8.7*a* can be implemented, as depicted in Fig. 8.7*b*, where devices R are used to reverse the signals at the input and output of the second filter. In this arrangement, the first filter introduces certain delay which depends on the frequency, and thus a certain amount of delay distortion is introduced. The second filter introduces exactly the same delay as the first, but, since the signal is fed backwards, the delay is actually a time advance and, therefore, cancels the delay of the first filter.

An alternative approach for the design of constant-delay filters is to use nonrecursive approximations. This possibility is explored in Chaps. 9 and 15.

8.6 AMPLITUDE EQUALIZATION

In many applications, a filter is required to operate in cascade with a channel or system that does not have a constant amplitude response (e.g., a D/A converter, see Fig. 6.17*d*). If the transfer function of such a channel is $H_C(z)$ and the passband of the channel-filter combination extends from ω_{p1} to ω_{p2}, then the transfer function of the filter must be chosen such that

$$|H_C(e^{j\omega T})H_F(e^{j\omega T})| = 1 \qquad \text{for } \omega_{p1} \leq \omega \leq \omega_{p2}$$

to within a prescribed tolerance in order to keep the amplitude distortion to an acceptable level (see Sec. 3.6). If the variation in the amplitude response of the channel is small, it may be possible to solve the problem by taking the channel loss into account when the filter specifications are formulated. Alternatively, if the variation of the amplitude response of the channel is large, then the filter may have to be tuned or redesigned using one of the optimization methods described in Chap. 14 (e.g., see Example 14.3).

REFERENCE

1. A. Antoniou, "Design of Elliptic Digital Filters: Prescribed Specifications," *IEE Proc.*, vol. 124, pp. 341–344, April 1977 (see vol. 125, p. 504, June 1978 for errata).

PROBLEMS

8.1. Design a lowpass digital filter satisfying the specifications of Fig. P8.1. Use a Butterworth approximation.

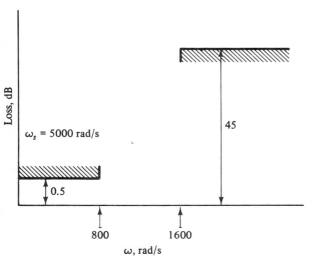

FIGURE P8.1

8.2. Redesign the filter of Prob. 8.1 using a Chebyshev approximation.
8.3. Redesign the filter of Prob. 8.1 using an inverse-Chebyshev approximation.
8.4. Redesign the filter of Prob. 8.1 using an elliptic approximation.
8.5. Design a highpass digital filter satisfying the specifications of Fig. P8.5. Use a Butterworth approximation.
8.6. Redesign the filter of Prob. 8.5 using a Chebyshev approximation.
8.7. Redesign the filter of Prob. 8.5 using an inverse-Chebyshev approximation.
8.8. Redesign the filter of Prob. 8.5 using an elliptic approximation.
8.9. Design a bandpass digital filter satisfying the specifications of Fig. P8.9. Use a Butterworth approximation.

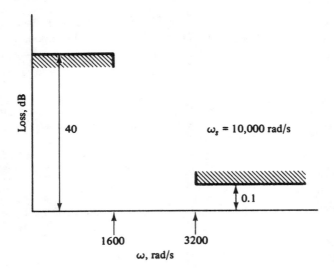

FIGURE P8.5

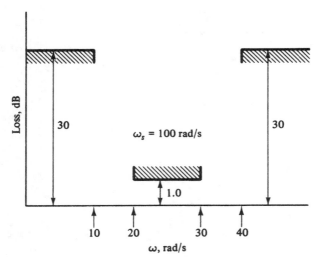

FIGURE P8.9

8.10. Redesign the filter of Prob. 8.9 using a Chebyshev approximation.
8.11. Redesign the filter of Prob. 8.9 using an inverse-Chebyshev approximation.
8.12. Redesign the filter of Prob. 8.9 using an elliptic approximation.
8.13. Design a bandstop digital filter satisfying the specifications of Fig. P8.13. Use a Butterworth approximation.
8.14. Redesign the filter of Prob. 8.13 using a Chebyshev approximation.
8.15. Redesign the filter of Prob. 8.13 using an inverse-Chebyshev approximation.
8.16. Redesign the filter of Prob. 8.13 using an elliptic approximation.
8.17. Derive the formulas of Table 8.2 for highpass filters.
8.18. Derive the formulas of Table 8.3 for bandstop filters.

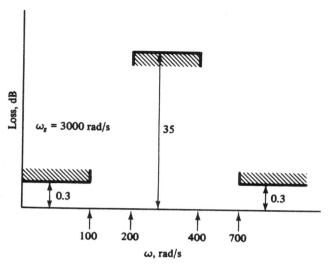

FIGURE P8.13

CHAPTER 9

DESIGN OF NONRECURSIVE FILTERS

9.1 INTRODUCTION

The preceding two chapters have shown that some highly selective and computationally efficient filters can be easily designed as recursive filters. The methods presented involve a minimal amount of computation and can be used to design filters with arbitrary piecewise-constant amplitude responses. Unfortunately, recursive filters with constant group delay or with amplitude responses that are not piecewise constant are difficult to design in practice. Constant group delay can be achieved by using the Bessel approximation in conjunction with the invariant-impulse-response method (see Secs. 5.6 and 7.3), but this approach is limited to low-selectivity lowpass and bandpass filters. High-selectivity filters with approximately constant group delay and filters with arbitrary amplitude responses can be designed, but advanced optimization methods are required, as is demonstrated in Chap. 14, which involve a considerable amount of computation. In contrast, nonrecursive filters can be easily designed to have *constant group delay*, and a variety of amplitude responses can be readily achieved. Filters of this class are, in addition, naturally suited for certain applications, e.g., for the design of differentiators and interpolators. Furthermore, since the impulse response of nonrecursive filters is of finite duration, they can be implemented in terms of fast-Fourier transforms.

The approximation problem in nonrecursive filters can be solved by using Fourier series or numerical analysis formulas. The details of these methods are examined in this chapter. An alternative approach is based on the use of the discrete Fourier transform [1, 2] and is described in Chap. 13. A third possibility is to use a powerful multivariable optimization algorithm known as the *Remez exchange algorithm*. This approach is examined in Chap. 15.

The methods of this chapter are in terms of *closed-form* solutions and, as a result, they are straightforward and involve a minimal amount of computation. Unfortunately, the designs obtained are suboptimal; that is, the required filter order to satisfy a set of prescribed specifications is not minimum. On the other hand, the use of the Remez exchange algorithm yields optimal designs, but a large amount of computation is required to complete a design, as may be expected.

9.2 PROPERTIES OF NONRECURSIVE FILTERS

9.2.1 Constant-Delay Filters

A nonrecursive causal filter can be characterized by the transfer function

$$H(z) = \sum_{n=0}^{N-1} h(nT)z^{-n} \tag{9.1}$$

Its frequency response is given by

$$H(e^{j\omega T}) = M(\omega)e^{j\theta(\omega)} = \sum_{n=0}^{N-1} h(nT)e^{-j\omega nT} \tag{9.2}$$

where
$$M(\omega) = |H(e^{j\omega T})|$$

and
$$\theta(\omega) = \arg H(e^{j\omega T}) \tag{9.3}$$

The *phase* and *group* delays of a filter are given by

$$\tau_p = -\frac{\theta(\omega)}{\omega} \quad \text{and} \quad \tau_g = -\frac{d\theta(\omega)}{d\omega}$$

respectively.

For constant phase delay as well as group delay the phase response must be linear, i.e.,

$$\theta(\omega) = -\tau\omega$$

and thus from Eqs. (9.2) and (9.3)

$$\theta(\omega) = -\tau\omega = \tan^{-1} \frac{-\sum_{n=0}^{N-1} h(nT) \sin \omega nT}{\sum_{n=0}^{N-1} h(nT) \cos \omega nT}$$

Consequently

$$\tan \omega\tau = \frac{\sum_{n=0}^{N-1} h(nT) \sin \omega nT}{\sum_{n=0}^{N-1} h(nT) \cos \omega nT}$$

and accordingly

$$\sum_{n=0}^{N-1} h(nT)(\cos \omega nT \sin \omega \tau - \sin \omega nT \cos \omega \tau) = 0$$

or

$$\sum_{n=0}^{N-1} h(nT) \sin (\omega \tau - \omega nT) = 0$$

The solution of this equation can be shown to be

$$\tau = \frac{(N-1)T}{2} \tag{9.4}$$

$$h(nT) = h[(N-1-n)T] \quad \text{for } 0 \le n \le N-1 \tag{9.5}$$

Therefore, a nonrecursive filter, unlike a recursive filter, can have constant phase and group delays *over the entire baseband*. It is only necessary for the impulse response to be *symmetrical* about the midpoint between samples $(N-2)/2$ and $N/2$ for even N or about sample $(N-1)/2$ for odd N. The required symmetry is illustrated in Fig. 9.1 for $N = 10$ and 11.

In many applications only the group delay need be constant, in which case the phase response can have the form

$$\theta(\omega) = \theta_0 - \tau \omega$$

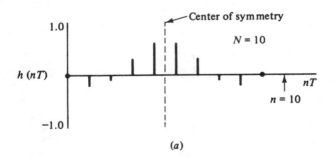

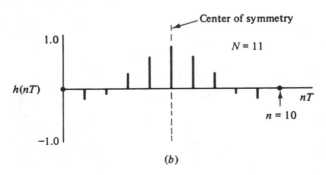

FIGURE 9.1
Impulse response for constant phase and group delays: (*a*) even N, (*b*) odd N.

where θ_0 is a constant. On using the above procedure a second class of constant-delay nonrecursive filters can be obtained. With $\theta_0 = \pm\pi/2$, the solution is

$$\tau = \frac{(N-1)T}{2} \tag{9.6}$$

$$h(nT) = -h[(N-1-n)T] \tag{9.7}$$

In this case the impulse response is *antisymmetrical* about the midpoint between samples $(N-2)/2$ and $N/2$ for even N or about sample $(N-1)/2$ for odd N, as illustrated in Fig. 9.2.

9.2.2 Frequency Response

Equations (9.5) and (9.7) lead to some simple expressions for the *frequency response*. For a *symmetrical* impulse response with N odd, Eq. (9.2) can be expressed as

$$H(e^{j\omega T}) = \sum_{n=0}^{(N-3)/2} h(nT)e^{-j\omega nT} + h\left[\frac{(N-1)T}{2}\right]e^{-j\omega(N-1)T/2}$$

$$+ \sum_{n=(N+1)/2}^{N-1} h(nT)e^{-j\omega nT} \tag{9.8}$$

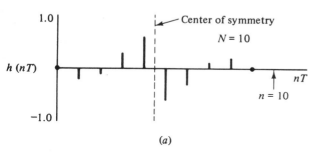

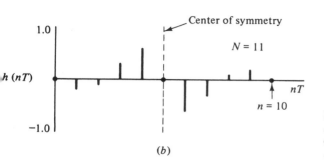

FIGURE 9.2
Alternative impulse response for constant group delay: (*a*) even N, (*b*) odd N.

By using Eq. (9.5) and then letting $N - 1 - n = m$, $m = n$ the last summation in the above equation can be expressed as

$$\sum_{n=(N+1)/2}^{N-1} h(nT)e^{-j\omega nT} = \sum_{n=(N+1)/2}^{N-1} h[(N-1-n)T]e^{-j\omega nT}$$

$$= \sum_{n=0}^{(N-3)/2} h(nT)e^{-j\omega(N-1-n)T} \quad (9.9)$$

Now from Eqs. (9.8) and (9.9)

$$H(e^{j\omega T}) = e^{-j\omega(N-1)T/2} \left\{ h\left[\frac{(N-1)T}{2}\right] + \sum_{n=0}^{(N-3)/2} 2h(nT) \cos\left[\omega\left(\frac{N-1}{2} - n\right)T\right] \right\}$$

and hence with $(N-1)/2 - n = k$ we have

$$H(e^{j\omega T}) = e^{-j\omega(N-1)T/2} \sum_{k=0}^{(N-1)/2} a_k \cos \omega kT$$

where

$$a_0 = h\left[\frac{(N-1)T}{2}\right] \quad (9.10)$$

$$a_k = 2h\left[\left(\frac{N-1}{2} - k\right)T\right] \quad (9.11)$$

Similarly, the frequency responses for the case of symmetrical impulse response with N even and for the two cases of *antisymmetrical* response simplify to the expressions summarized in Table 9.1.

TABLE 9.1
Frequency response of constant-delay nonrecursive filters

$h(nT)$	N	$H(e^{j\omega T})$
Symmetrical	Odd	$e^{-j\omega(N-1)T/2} \sum_{k=0}^{(N-1)/2} a_k \cos \omega kT$
	Even	$e^{-j\omega(N-1)T/2} \sum_{k=1}^{N/2} b_k \cos[\omega(k-\frac{1}{2})T]$
Antisymmetrical	Odd	$e^{-j[\omega(N-1)T/2 - \pi/2]} \sum_{k=1}^{(N-1)/2} a_k \sin \omega kT$
	Even	$e^{-j[\omega(N-1)T/2 - \pi/2]} \sum_{k=1}^{N/2} b_k \sin[\omega(k-\frac{1}{2})T]$

where $a_0 = h\left[\frac{(N-1)T}{2}\right] \quad a_k = 2h\left[\left(\frac{N-1}{2} - k\right)T\right] \quad b_k = 2h\left[\left(\frac{N}{2} - k\right)T\right]$

9.2.3 Location of Zeros

The impulse response constraints of Eqs. (9.5) and (9.7) impose certain restrictions on the zeros of $H(z)$. For odd N, Eqs. (9.1), (9.5), and (9.7) yield

$$H(z) = \frac{1}{z^{(N-1)/2}} \sum_{n=0}^{(N-3)/2} h(nT)(z^{(N-1)/2-n} \pm z^{-[(N-1)/2-n]})$$

$$+ \frac{1}{2} h\left[\frac{(N-1)T}{2}\right] (z^0 \pm z^0) \qquad (9.12)$$

where the negative sign applies to the case of antisymmetrical impulse response. With $(N-1)/2 - n = k$ Eq. (9.12) can be put in the form

$$H(z) = \frac{N(z)}{D(z)} = \frac{1}{z^{(N-1)/2}} \sum_{k=0}^{(N-1)/2} \frac{a_k}{2}(z^k \pm z^{-k})$$

where a_0 and a_k are given by Eqs. (9.10) and (9.11).

The zeros of $H(z)$ are the roots of $N(z)$ given by

$$N(z) = \sum_{k=0}^{(N-1)/2} a_k(z^k \pm z^{-k})$$

If z is replaced by z^{-1}, we have

$$N(z^{-1}) = \sum_{k=0}^{(N-1)/2} a_k(z^{-k} \pm z^k)$$

$$= \pm \sum_{k=0}^{(N-1)/2} a_k(z^k \pm z^{-k}) = \pm N(z)$$

The same relation holds for even N, as can be easily shown, and therefore if $z_i = r_i e^{j\psi_i}$ is a zero of $H(z)$, then $z_i^{-1} = e^{-j\psi_i}/r_i$ must also be a zero of $H(z)$. This has the following implications on the zero locations:

1. An arbitrary number of zeros can be located at $z_i = \pm 1$ since $z_i^{-1} = \pm 1$.
2. An arbitrary number of complex-conjugate pairs of zeros can be located on the unit circle since

$$(z - z_i)(z - z_i^*) = (z - e^{j\psi_i})(z - e^{-j\psi_i}) = \left(z - \frac{1}{z_i^*}\right)\left(z - \frac{1}{z_i}\right)$$

3. Real zeros off the unit circle must occur in reciprocal pairs.
4. Complex zeros off the unit circle must occur in groups of four, namely, z_i, z_i^*, and their reciprocals.

Polynomials with the above properties are often called *mirror-image polynomials*. A typical zero-pole plot for a constant-delay nonrecursive filter is shown in Fig. 9.3.

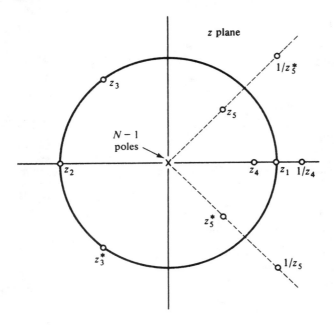

FIGURE 9.3
Typical zero-pole plot for a constant-delay nonrecursive filter.

9.3 DESIGN USING THE FOURIER SERIES

As the frequency response of a nonrecursive filter is a periodic function of ω with period ω_s, it can be expressed as a *Fourier series* (see Sec. 6.4). We can write

$$H(e^{j\omega T}) = \sum_{n=-\infty}^{\infty} h(nT) e^{-j\omega nT} \qquad (9.13)$$

where
$$h(nT) = \frac{1}{\omega_s} \int_{-\omega_s/2}^{\omega_s/2} H(e^{j\omega T}) e^{j\omega nT} \, d\omega \qquad (9.14)$$

and if $e^{j\omega T} = z$, Eq. (9.13) gives

$$H(z) = \sum_{n=-\infty}^{\infty} h(nT) z^{-n} \qquad (9.15)$$

Hence with an analytic representation for the frequency response available, a corresponding transfer function can be readily derived. Unfortunately, however, this is *noncausal* and of *infinite order*. For a finite-order transfer function, the series of Eq. (9.15) can be truncated by assigning

$$h(nT) = 0 \quad \text{for } |n| > \frac{N-1}{2}$$

in which case

$$H(z) = h(0) + \sum_{n=1}^{(N-1)/2} [h(-nT)z^n + h(nT)z^{-n}] \qquad (9.16)$$

Causality can be brought about by multiplying $H(z)$ by $z^{-(N-1)/2}$ so that

$$H'(z) = z^{-(N-1)/2} H(z) \qquad (9.17)$$

This modification is permissible since the amplitude response will remain unchanged and the group delay will be increased by a constant $(N-1)T/2$.

Note that if $H(e^{j\omega T})$ in Eq. (9.14) is an even function of ω, then the impulse response obtained is symmetrical about $n = 0$, and hence the filter has zero group delay. Consequently, the filter represented by the transfer function of Eq. (9.17) has constant group delay equal to $(N-1)T/2$.

Example 9.1. Design a lowpass filter with a frequency response

$$H(e^{j\omega T}) \approx \begin{cases} 1 & \text{for } |\omega| \leq \omega_c \\ 0 & \text{for } \omega_c < |\omega| \leq \dfrac{\omega_s}{2} \end{cases}$$

where ω_s is the sampling frequency.

Solution. From Eq. (9.14)

$$h(nT) = \frac{1}{\omega_s} \int_{-\omega_c}^{\omega_c} e^{j\omega nT} \, d\omega = \frac{1}{n\pi} \sin \omega_c nT$$

Hence Eqs. (9.16) and (9.17) yield

$$H(z) = z^{-(N-1)/2} \sum_{n=0}^{(N-1)/2} \frac{a_n}{2}(z^n + z^{-n})$$

where

$$a_0 = h(0) \qquad a_n = 2h(nT)$$

The amplitude response of the preceding filter with ω_c and ω_s assumed to be 2 and 10 rad/s, respectively, is plotted in Fig. 9.4 for $N = 11, 21$, and 31. The passband and stopband oscillations observed are due to *slow convergence* in the Fourier series, which in turn, is caused by the discontinuity at the passband edge. These are known as *Gibbs' oscillations*. As N is increased, the frequency of these oscillations is seen to increase, and at both low and high frequencies their amplitude is decreased. The amplitude of the last passband ripple, however, and that of the first stopband ripple tend to remain virtually unchanged. This type of performance is often objectionable in practice, and ways must be sought for the reduction of Gibbs' oscillations.

A rudimentary method is to avoid discontinuities in the frequency response by introducing transition bands between passbands and stopbands [3]. For example, the

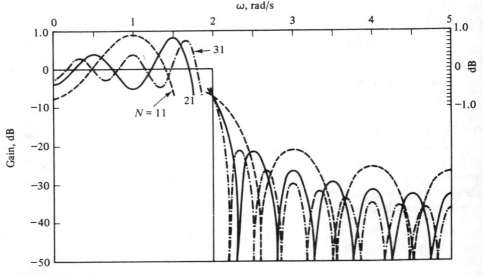

FIGURE 9.4
Amplitude response of lowpass filter (Example 9.1).

response of the above lowpass filter could be redefined as

$$H(e^{j\omega T}) \approx \begin{cases} 1 & \text{for} \quad |\omega| \leq \omega_p \\ -\dfrac{\omega - \omega_a}{\omega_a - \omega_p} & \text{for } \omega_p < |\omega| < \omega_a \\ 0 & \text{for } \omega_a \leq |\omega| \leq \dfrac{\omega_s}{2} \end{cases}$$

9.4 USE OF WINDOW FUNCTIONS

An alternative and easy-to-apply technique for the reduction of Gibbs' oscillations is to precondition $h(nT)$ as given by Eq. (9.14) using a class of time-domain functions known as *window functions*.

Let

$$H(z) = \mathcal{Z}h(nT) = \sum_{n=-\infty}^{\infty} h(nT)z^{-n} \tag{9.18}$$

$$W(z) = \mathcal{Z}w(nT) = \sum_{n=-\infty}^{\infty} w(nT)z^{-n} \tag{9.19}$$

$$H_w(z) = \mathcal{Z}[w(nT)h(nT)] \tag{9.20}$$

where $w(nT)$ represents a window function. The use of the complex convolution (Theorem 2.10) gives

$$H_w(z) = \frac{1}{2\pi j} \oint_\Gamma H(v) W\left(\frac{z}{v}\right) v^{-1} \, dv \tag{9.21}$$

where Γ represents a contour in the common region of convergence of $H(v)$ and $W(z/v)$. With

$$v = e^{j\Omega T} \quad \text{and} \quad z = e^{j\omega T}$$

and $H(v)$ as well as $W(z/v)$ convergent on the unit circle of the v plane, Eq. (9.21) can be expressed as

$$H_w(e^{j\omega T}) = \frac{T}{2\pi} \int_0^{2\pi/T} H(e^{j\Omega T}) W(e^{j(\omega-\Omega)T}) \, d\Omega \qquad (9.22)$$

For the sake of exposition let

$$H(e^{j\Omega T}) = \begin{cases} 1 & \text{for } 0 \leq |\Omega| \leq \omega_c \\ 0 & \text{for } \omega_c < |\Omega| \leq \dfrac{\omega_s}{2} \end{cases}$$

as illustrated in Fig. 5a. Also let $W(e^{j\Omega T})$ be real with the form illustrated in Fig. 9.5b and assume that

$$W(e^{j\Omega T}) = 0 \quad \text{for } \omega_m \leq |\Omega| \leq \frac{\omega_s}{2} \qquad (9.23)$$

According to Eq. (9.22), $H_w(e^{j\omega T})$ can be formed by using the following graphical procedure:

1. Shift $W(e^{j\Omega T})$ to the right by ω, as in Fig. 9.5c.
2. Multiply $H(e^{j\Omega T})$ by $W(e^{j(\omega-\Omega)T})$, as in Fig. 9.5d.
3. Find the area in Fig. 9.5d.

As ω is varied in the range ω_1 to ω_5 through the point of discontinuity of $H(e^{j\Omega T})$, the successive values of $H_w(e^{j\omega T})$ can be determined as in Fig. 9.6. Evidently, with Eq. (9.23) satisfied and the area under the curve in Fig. 9.5b equal to unity, the derived function $H_w(e^{j\omega T})$ is a close approximation for $H(e^{j\omega T})$, and furthermore it is free of Gibbs' oscillations.

If $H(z)$ as given by Eq. (9.15) represents a constant-delay filter and $H_w(z)$ is to represent a finite-order constant-delay filter, $w(nT)$ must have the following time-domain properties: (1) it must be zero for $|n| > (N-1)/2$; (2) for odd N, it must be symmetrical about sample $n = 0$. A typical window function is illustrated in Fig. 9.7.

In practice, the spectrum of $w(nT)$ is of the form shown in Fig. 9.8, where $k\omega_s/N$ is the width of the *main lobe* (k is a constant). This deviates from the ideal spectrum of Fig. 9.5b in that Eq. (9.23) is only approximately satisfied.

The effect of *side lobes* in $W(e^{j\Omega T})$ can be deduced from Fig. 9.5. As ω is varied in the passband of the filter, left- and right-hand side lobes are swept in and out of the passband, respectively, and as their amplitudes vary with frequency, the area in Fig. 9.5d oscillates about unity. Similarly, as ω is varied in the stopband, a number of left-hand-tail side lobes are swept through the passband and the area in

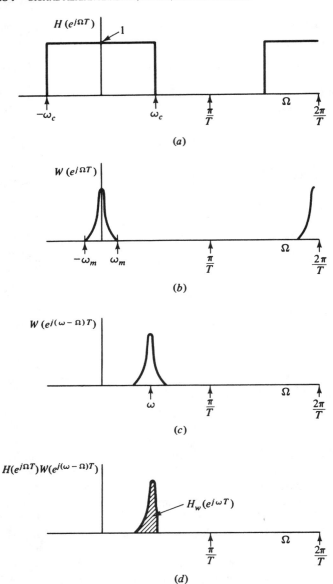

FIGURE 9.5
Complex convolution ($T = 2\pi$).

Fig. 9.5d oscillates about zero. Clearly, side lobes in the spectrum of $w(nT)$ give rise to Gibbs' oscillations in the amplitude response of the filter. Therefore, for low passband ripple and large stopband attenuation the area under the side lobes should be a small proportion of that under the main lobe.

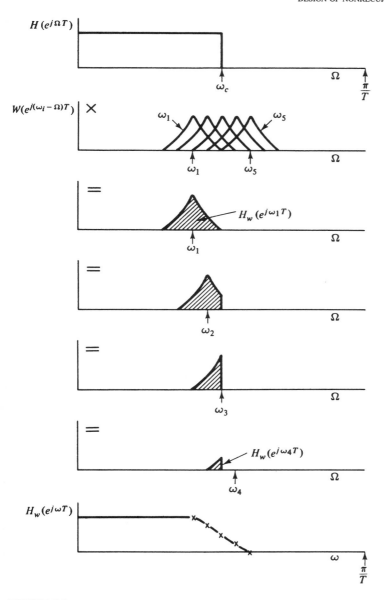

FIGURE 9.6
The effect of a window function.

As ω is varied about the passband edge, $H_w(e^{j\omega T})$ begins to decrease rapidly when

$$\omega = \omega_c - \frac{k\omega_s}{2N}$$

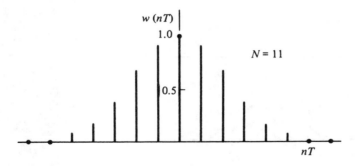

FIGURE 9.7
Typical window function.

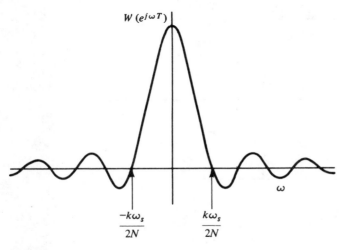

FIGURE 9.8
Typical spectrum of a window function.

and slows down when

$$\omega = \omega_c + \frac{k\omega_s}{2N}$$

i.e., the main-lobe width determines the transition width of the resulting filter. For high selectivity, N and hence the order of the filter should be large, as may be expected, and k should be small.

The most frequently used window functions are [3, 4]:

1. Rectangular
2. von Hann[†]

[†]Due to Julius von Hann and often referred to inaccurately as the Hanning window function.

3. Hamming
4. Blackman
5. Dolph-Chebyshev
6. Kaiser

9.4.1 Rectangular Window

The *rectangular* window is given by

$$w_R(nT) = \begin{cases} 1 & \text{for } |n| \leq \dfrac{N-1}{2} \\ 0 & \text{otherwise} \end{cases} \quad (9.24)$$

This corresponds to the direct truncation of the Fourier series, and its effect on $H(e^{j\omega T})$ has been noted earlier.

The *spectrum* of $w_R(nT)$ can be deduced from Eqs. (9.19) and (9.24) as

$$W_R(e^{j\omega T}) = \sum_{n=-(N-1)/2}^{(N-1)/2} e^{-j\omega nT} = \frac{e^{j\omega(N-1)T/2} - e^{-j\omega(N+1)T/2}}{1 - e^{-j\omega T}}$$

$$= \frac{e^{j\omega NT/2} - e^{-j\omega NT/2}}{e^{j\omega T/2} - e^{-j\omega T/2}} = \frac{\sin(\omega NT/2)}{\sin(\omega T/2)} \quad (9.25)$$

This is plotted in Fig. 9.9 for $N = 11$ and $\omega_s = 10$ rad/s.

Since $W(e^{j\omega T}) = 0$ at $\omega = m\omega_s/N$ for $m = \pm 1, \pm 2, \ldots$, the *main-lobe width* is $2\omega_s/N$. The *ripple ratio* defined as

$$r = \frac{100 \text{ (maximum side-lobe amplitude)}}{\text{main-lobe amplitude}} \%$$

is 22.34 percent for $N = 11$ and decreases to 21.70 as N is increased to 101 (see Table 9.2).

9.4.2 von Hann and Hamming Windows

The *von Hann and Hamming* windows are given by

$$w_H(nT) = \begin{cases} \alpha + (1-\alpha)\cos\dfrac{2\pi n}{N-1} & \text{for } |n| \leq \dfrac{N-1}{2} \\ 0 & \text{otherwise} \end{cases} \quad (9.26)$$

The two differ in the choice of α. In the von Hann window $\alpha = 0.5$, and in the Hamming window $\alpha = 0.54$.

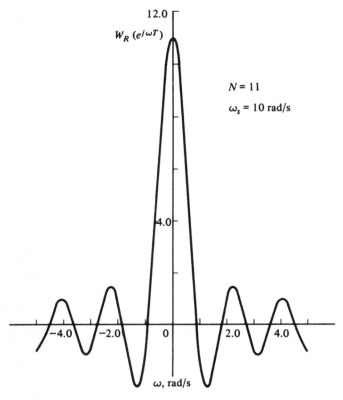

FIGURE 9.9
Spectrum of rectangular window.

TABLE 9.2
Summary of window parameters

Type of window	Main-lobe width	Ripple ratio, %		
		$N = 11$	$N = 21$	$N = 101$
Rectangular	$\dfrac{2\omega_s}{N}$	22.34	21.89	21.70
von Hann	$\dfrac{4\omega_s}{N}$	2.62	2.67	2.67
Hamming	$\dfrac{4\omega_s}{N}$	1.47	0.93	0.74
Blackman	$\dfrac{6\omega_s}{N}$	0.08	0.12	0.12

The spectra of these windows can be related to that of the rectangular window. Equation (9.26) can be expressed as

$$w_H(nT) = w_R(nT)\left[\alpha + (1-\alpha)\cos\frac{2\pi n}{N-1}\right]$$

$$= \alpha w_R(nT) + \frac{1-\alpha}{2} w_R(nT)(e^{j2\pi n/(N-1)} + e^{-j2\pi n/(N-1)})$$

and on using Theorem 2.7 we have

$$W_H(e^{j\omega T}) = \mathcal{Z}[w_H(nT)]\Big|_{z=e^{j\omega T}}$$

$$= \alpha W_R(e^{j\omega T}) + \frac{1-\alpha}{2} W_R(e^{j[\omega T - 2\pi/(N-1)]})$$

$$+ \frac{1-\alpha}{2} W_R(e^{j[\omega T + 2\pi/(N-1)]})$$

Now from Eq. (9.25)

$$W_H(e^{j\omega T}) = \frac{\alpha \sin(\omega NT/2)}{\sin(\omega T/2)} + \frac{1-\alpha}{2}\frac{\sin[\omega NT/2 - N\pi/(N-1)]}{\sin[\omega T/2 - \pi/(N-1)]}$$

$$+ \frac{1-\alpha}{2}\frac{\sin[\omega NT/2 + N\pi/(N-1)]}{\sin[\omega T/2 + \pi/(N-1)]} \qquad (9.27)$$

Consequently, the spectra for the von Hann and Hamming windows can be formed by shifting $W_R(e^{j\omega T})$ first to the right and then to the left by $2\pi/(N-1)T$ and subsequently forming the sum in Eq. (9.27), as in Fig. 9.10. As can be observed, the second and third terms tend to cancel the right and left side lobes in $\alpha W_R(e^{j\omega T})$, and as a result both the von Hann and Hamming windows have reduced side lobes compared with those of the rectangular window. For $N = 11$ and $\omega_s = 10$ rad/s, the ripple ratios for the two windows are 2.62 and 1.47 percent and change to 2.67 and 0.74 percent, respectively, for $N = 101$ (see Table 9.2).

The first term in Eq. (9.27) is zero at

$$\omega = \frac{m\omega_s}{N}$$

and, similarly, the second and third terms are zero at

$$\omega = \left(m + \frac{N}{N-1}\right)\frac{\omega_s}{N} \quad \text{and} \quad \omega = \left(m - \frac{N}{N-1}\right)\frac{\omega_s}{N}$$

respectively, for $m = \pm 1, \pm 2, \ldots$. If $N \gg 1$, all three terms in Eq. (9.27) have their first common zero at $|\omega| \approx 2\omega_s/N$, and hence the main-lobe width for the von Hann and Hamming windows is approximately $4\omega_s/N$.

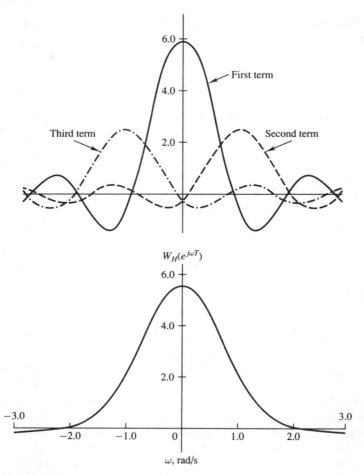

FIGURE 9.10
Spectrum of von Hann or Hamming window.

9.4.3 Blackman Window

The *Blackman* window is similar to the preceding two and is given by

$$w_B(nT) = \begin{cases} 0.42 + 0.5\cos\dfrac{2\pi n}{N-1} + 0.08\cos\dfrac{4\pi n}{N-1} & \text{for } |n| \leq \dfrac{N-1}{2} \\ 0 & \text{otherwise} \end{cases}$$

The additional cosine term leads to a further reduction in the amplitude of Gibbs' oscillations. The ripple ratio for $N = 11$ and $\omega_s = 10$ rad/s is 0.08 percent and changes to 0.12 percent for $N = 101$. The main-lobe width, however, is increased to about $6\omega_s/N$ (see Table 9.2).

Example 9.2. Redesign the lowpass filter of Example 9.1 using the von Hann, Hamming, and Blackman windows.

Solution. The impulse response is the same as in Example 9.1, i.e.,

$$h(nT) = \frac{1}{n\pi} \sin \omega_c nT$$

On multiplying $h(nT)$ by the appropriate window function and then using Eqs. (9.20) and (9.17) we obtain

$$H'_w(z) = z^{-(N-1)/2} \sum_{n=0}^{(N-1)/2} \frac{a'_n}{2}(z^n + z^{-n})$$

where
$$a'_0 = w(0)h(0) \qquad a'_n = 2w(nT)h(nT)$$

The amplitude responses for the three filters are given by

$$M(\omega) = \left| \sum_{n=0}^{(N-1)/2} a'_n \cos \omega nT \right|$$

These are plotted in Fig. 9.11 for $N = 21$ and $\omega_s = 10$. As expected, the passband ripple is reduced, and the minimum stopband attenuation as well as the transition width are increased progressively from the von Hann to the Hamming to the Blackman window.

9.4.4 Dolph-Chebyshev Window

The windows considered so far have a ripple ratio which is practically independent of N, as can be seen in Table 9.2, and as a result the usefulness of these windows is

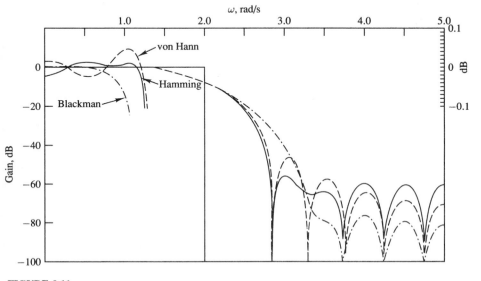

FIGURE 9.11
Amplitude response of lowpass filter (Example 9.2).

limited. A more versatile window is the *Dolph-Chebyshev* window. This window is given by

$$w_{DC}(nT) = \frac{1}{N}\left[\frac{1}{r} + 2\sum_{i=1}^{(N-1)/2} T_{N-1}\left(x_0 \cos\frac{i\pi}{N}\right)\cos\frac{2n\pi i}{N}\right]$$

for $n = 0, 1, 2, \ldots, (N-1)/2$ where r is the required ripple ratio as a fraction and

$$x_0 = \cosh\left(\frac{1}{N-1}\cosh^{-1}\frac{1}{r}\right)$$

Function $T_k(x)$ is the kth-order Chebyshev polynomial associated with the Chebyshev approximation for recursive filters (see Sec. 5.4.1) and is given by

$$T_k(x) = \begin{cases} \cos(k\cos^{-1}x) & \text{for } |x| \leq 1 \\ \cosh(\cosh^{-1}x) & \text{for } |x| > 1 \end{cases}$$

Evidently, an arbitrary ripple ratio can be achieved and, as in other windows, the main-lobe width can be controlled by choosing the value of N.

The Dolph-Chebyshev window has two additional properties of interest. First, with N fixed, the main-lobe width is the *smallest* that can be achieved for a given ripple ratio; second, all the side lobes have the same amplitude. A consequence of the first property is that filters designed by using this window have a narrow transition band. A consequence of the second property is that the approximation error tends to be more *uniformly distributed* with respect to frequency.

9.4.5 Kaiser Window

Another window that allows independent control of the ripple ratio and the main-lobe width is the *Kaiser* window [5]. This window is given by

$$w_K(nT) = \begin{cases} \dfrac{I_0(\beta)}{I_0(\alpha)} & \text{for } |n| \leq \dfrac{N-1}{2} \\ 0 & \text{otherwise} \end{cases} \qquad (9.28)$$

where α is an independent parameter and

$$\beta = \alpha\sqrt{1 - \left(\frac{2n}{N-1}\right)^2}$$

$I_0(x)$ is the *zeroth-order modified Bessel function of the first kind*. This can be evaluated to any desired degree of accuracy by using the rapidly converging series

$$I_0(x) = 1 + \sum_{k=1}^{\infty}\left[\frac{1}{k!}\left(\frac{x}{2}\right)^k\right]^2$$

The spectrum of $w_K(nT)$ can be readily obtained from Eq. (9.19) as

$$W_K(e^{j\omega T}) = w_K(0) + 2\sum_{n=1}^{(N-1)/2} w_K(nT)\cos\omega nT$$

The continuous-time counterpart of the Kaiser window is

$$w_K(t) = \begin{cases} \dfrac{I_0(\beta)}{I_0(\alpha)} & \text{for } |t| \leq \tau \\ 0 & \text{otherwise} \end{cases}$$

where
$$\beta = \alpha\sqrt{1 - \left(\dfrac{t}{\tau}\right)^2} \quad \text{and} \quad \tau = \dfrac{(N-1)T}{2}$$

The spectrum of $w_K(t)$ can be shown to be [5]

$$W_K(j\omega) = \dfrac{2}{I_0(\alpha)} \dfrac{\sin(\tau\sqrt{\omega^2 - \omega_a^2})}{\sqrt{\omega^2 - \omega_a^2}} \tag{9.29}$$

where
$$\omega_a = \dfrac{\alpha}{\tau}$$

If
$$W_K(j\omega) \approx 0 \quad \text{for } |\omega| \geq \dfrac{\omega_s}{2}$$

the spectrum of sampled signal $\hat{w}_K(t)$ or equivalently the spectrum of $w_K(nT)$ can be expressed as

$$\hat{W}_K(j\omega) = W_K(e^{j\omega T}) \approx \dfrac{1}{T} W_K(j\omega) \quad \text{for } 0 \leq |\omega| < \dfrac{\omega_s}{2} \tag{9.30}$$

according to Eq. (6.36). Hence from Eqs. (9.29) and (9.30) a *closed-form* but approximate expression for $W_K(e^{j\omega T})$ can be deduced as

$$W_K(e^{j\omega T}) \approx \dfrac{N-1}{\alpha I_0(\alpha)} \dfrac{\sin[\alpha\sqrt{(\omega/\omega_a)^2 - 1}]}{\sqrt{(\omega/\omega_a)^2 - 1}}$$

The ripple ratio can be varied continuously from the low value in the Blackman window to the high value in the rectangular window by simply varying the parameter α. Also, as in other windows, the main-lobe width, designated by B_m, can be adjusted by varying N. The influence of α on the ripple ratio and main-lobe width is illustrated in Fig. 9.12a and b for $N = 21$ and $\omega_s = 10$ rad/s. An important advantage of the Kaiser window is that some empirical formulas are available [5] which can be used to determine the required values of α and N that will yield prescribed filter specifications. A design method based on these formulas is described below.

9.4.6 Prescribed Filter Specifications

Consider the *lowpass* specification of Fig. 9.13a. The passband ripple and minimum stopband attenuation in decibels are given by

$$A_p = 20\log\dfrac{1+\delta}{1-\delta} \tag{9.31}$$

and
$$A_a = -20\log\delta \tag{9.32}$$

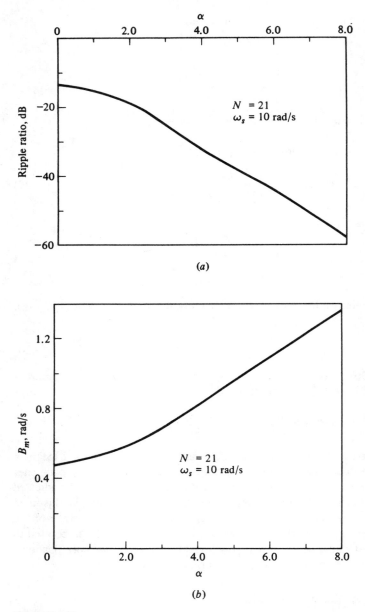

FIGURE 9.12
Kaiser window: (*a*) ripple ratio versus α, (*b*) main-lobe width versus α.

respectively, and the transition width in radians per second is

$$B_t = \omega_a - \omega_p$$

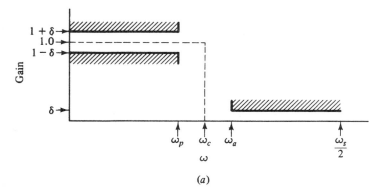

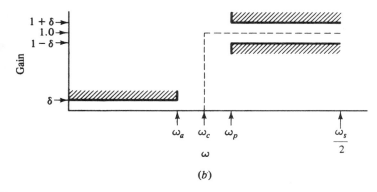

FIGURE 9.13
Idealized frequency responses: (*a*) lowpass filter, (*b*) highpass filter.

A filter with a passband ripple equal to or less than A'_p, a minimum stopband attenuation equal to or greater than A'_a, and a transition width B_t can be readily designed using the following procedure:

1. Determine $h(nT)$ using the Fourier-series approach of Sec. 9.3 assuming an idealized frequency response

$$H(e^{j\omega T}) = \begin{cases} 1 & \text{for} \quad |\omega| \leq \omega_c \\ 0 & \text{for} \quad \omega_c < |\omega| \leq \dfrac{\omega_s}{2} \end{cases}$$

(dashed line in Fig. 9.13*a*), where

$$\omega_c = \tfrac{1}{2}(\omega_p + \omega_a)$$

2. Choose δ in Eqs. (9.31) and (9.32) such that $A_p \leq A'_p$ and $A_a \geq A'_a$. A suitable value is

$$\delta = \min(\delta_1, \delta_2)$$

where
$$\delta_1 = 10^{-0.05A'_a} \qquad \delta_2 = \frac{10^{0.05A'_p} - 1}{10^{0.05A'_p} + 1}$$

3. Calculate A_a using Eq. (9.32).
4. Choose parameter α as follows:
$$\alpha = \begin{cases} 0 & \text{for} \quad A_a \leq 21 \\ 0.5842(A_a - 21)^{0.4} + 0.07886(A_a - 21) & \text{for } 21 < A_a \leq 50 \\ 0.1102(A_a - 8.7) & \text{for} \quad A_a > 50 \end{cases}$$

5. Choose parameter D as follows:
$$D = \begin{cases} 0.9222 & \text{for } A_a \leq 21 \\ \dfrac{A_a - 7.95}{14.36} & \text{for } A_a > 21 \end{cases}$$

Then select the lowest odd value of N satisfying the inequality

$$N \geq \frac{\omega_s D}{B_t} + 1$$

6. Form $w_K(nT)$ using Eq. (9.28).
7. Form

$$H'_w(z) = z^{-(N-1)/2} H_w(z) \qquad \text{where } H_w(z) = \mathcal{Z}[w_K(nT)h(nT)]$$

Example 9.3. Design a lowpass filter satisfying the following specifications:

Maximum passband ripple in frequency range 0 to 1.5 rad/s: 0.1 dB
Minimum stopband attenuation in frequency range 2.5 to 5.0 rad/s: 40 dB
Sampling frequency: 10 rad/s

Solution. From step 1 and Example 9.1

$$h(nT) = \frac{1}{n\pi} \sin \omega_c nT \qquad \text{where } \omega_c = \tfrac{1}{2}(1.5 + 2.5) = 2.0 \text{ rad/s}$$

Step 2 gives

$$\delta_1 = 10^{-0.05(40)} = 0.01$$

$$\delta_2 = \frac{10^{0.05(0.1)} - 1}{10^{0.05(0.1)} + 1} = 5.7564 \times 10^{-3}$$

Hence
$$\delta = 5.7564 \times 10^{-3}$$

and from step 3

$$A_a = 44.797 \text{ dB}$$

Steps 4 and 5 yield

$$\alpha = 3.9524 \qquad D = 2.5660$$

Hence
$$N \geq \frac{10(2.566)}{1} + 1 = 26.66$$

or
$$N = 27$$

Finally steps 6 and 7 give

$$H'_w(z) = z^{-(N-1)/2} \sum_{n=0}^{(N-1)/2} \frac{a'_n}{2}(z^n + z^{-n})$$

where $\qquad a'_0 = w_K(0)h(0) \qquad a'_n = 2w_K(nT)h(nT)$

The numerical values of $h(nT)$ and $w_K(nT)h(nT)$ are given in Table 9.3, and the amplitude response achieved is plotted in Fig. 9.14. This satisfies the prescribed specifications.

The above procedure can be readily applied to the design of highpass, bandpass, and bandstop filters. For the *highpass* specification of Fig. 9.13b, the transition width and idealized frequency response in step 1 can be taken as

$$B_t = \omega_p - \omega_a \quad \text{and} \quad H(e^{j\omega T}) = \begin{cases} 0 & \text{for} \quad |\omega| < \omega_c \\ 1 & \text{for} \quad \omega_c \leq |\omega| \leq \frac{\omega_s}{2} \end{cases}$$

TABLE 9.3
Numerical values of $h(nT)$ and $w_K(nT)h(nT)$ (Example 9.3)

n	$h(nT)$	$w_K(nT)h(nT)$
0	4.00000×10^{-1}	4.00000×10^{-1}
1	3.02731×10^{-1}	2.99692×10^{-1}
2	9.35489×10^{-2}	8.98359×10^{-2}
3	-6.23660×10^{-2}	-5.69018×10^{-2}
4	-7.56827×10^{-2}	-6.42052×10^{-2}
5	0.0	0.0
6	5.04551×10^{-2}	3.45003×10^{-2}
7	2.67283×10^{-2}	1.57769×10^{-2}
8	-2.33872×10^{-2}	-1.15598×10^{-2}
9	-3.36367×10^{-2}	-1.34373×10^{-2}
10	0.0	0.0
11	2.75210×10^{-2}	6.23505×10^{-3}
12	1.55915×10^{-2}	2.39574×10^{-3}
13	-1.43921×10^{-2}	-1.32685×10^{-3}

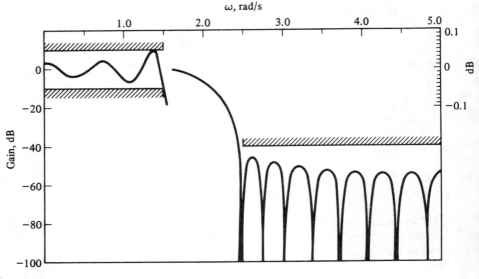

FIGURE 9.14
Amplitude response of lowpass filter (Example 9.3).

where
$$\omega_c = \frac{1}{2}(\omega_a + \omega_p)$$

The remaining steps apply without modification.

For the *bandpass* specification of Fig. 9.15a, the design must be based on the narrower of the two transition bands, i.e.,

$$B_t = \min\left[(\omega_{p1} - \omega_{a1}), (\omega_{a2} - \omega_{p2})\right] \qquad (9.33)$$

Hence
$$H(e^{j\omega T}) = \begin{cases} 0 & \text{for } 0 \le |\omega| < \omega_{c1} \\ 1 & \text{for } \omega_{c1} \le |\omega| \le \omega_{c2} \\ 0 & \text{for } \omega_{c2} < |\omega| \le \frac{\omega_s}{2} \end{cases} \qquad (9.34)$$

where
$$\omega_{c1} = \omega_{p1} - \frac{B_t}{2} \qquad \omega_{c2} = \omega_{p2} + \frac{B_t}{2} \qquad (9.35)$$

Similarly, for the *bandstop* specification of Fig. 9.15b

$$B_t = \min\left[(\omega_{a1} - \omega_{p1}), (\omega_{p2} - \omega_{a2})\right]$$

and
$$H(e^{j\omega T}) = \begin{cases} 1 & \text{for } 0 \le |\omega| \le \omega_{c1} \\ 0 & \text{for } \omega_{c1} < |\omega| < \omega_{c2} \\ 1 & \text{for } \omega_{c2} \le |\omega| \le \frac{\omega_s}{2} \end{cases}$$

where
$$\omega_{c1} = \omega_{p1} + \frac{B_t}{2} \qquad \omega_{c2} = \omega_{p2} - \frac{B_t}{2}$$

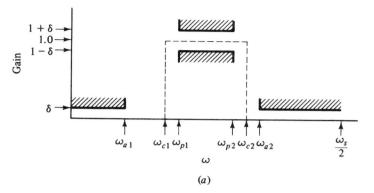

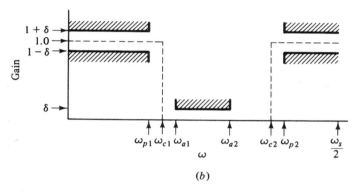

FIGURE 9.15
Idealized frequency responses: (a) bandpass filter, (b) bandstop filter.

Example 9.4. Design a bandpass filter satisfying the following specifications:

Minimum attenuation for $0 \leq \omega \leq 200$: 45 dB
Maximum passband ripple for $400 < \omega < 600$: 0.2 dB
Minimum attenuation for $700 \leq \omega \leq 1000$: 45 dB
Sampling frequency: 2000 rad/s

Solution. From Eq. (9.33)

$$B_t = \min\left[(400 - 200), \ (700 - 600)\right] = 100$$

Hence from Eq. (9.35)

$$\omega_{c1} = 400 - 50 = 350 \qquad \omega_{c2} = 600 + 50 = 650$$

Step 1 of the design procedure yields

$$h(nT) = \frac{1}{\omega_s} \int_{-\omega_s/2}^{\omega_s/2} H(e^{j\omega T}) e^{j\omega nT} d\omega$$

$$= \frac{1}{\omega_s} \int_0^{\omega_s/2} [H(e^{j\omega T}) e^{j\omega nT} + H(e^{-j\omega T}) e^{-j\omega nT}] d\omega$$

and from Eq. (9.34)

$$h(nT) = \frac{1}{\omega_s} \int_{\omega_{c1}}^{\omega_{c2}} 2\cos(\omega nT) d\omega$$

or

$$h(nT) = \frac{1}{\pi n}(\sin \omega_{c2} nT - \sin \omega_{c1} nT)$$

Now according to step 2,

$$\delta_1 = 10^{-0.05(45)} = 5.6234 \times 10^{-3}$$

$$\delta_2 = \frac{10^{0.05(0.2)} - 1}{10^{0.05(0.2)} + 1} = 1.1512 \times 10^{-2}$$

and
$$\delta = 5.6234 \times 10^{-3} \quad \text{or} \quad A_a = 45 \text{ dB}$$

The design can be completed as in Example 9.3. The resulting values for α, D, and N are

$$\alpha = 3.9754 \quad D = 2.580 \quad N = 53$$

The amplitude response achieved is plotted in Fig. 9.16.

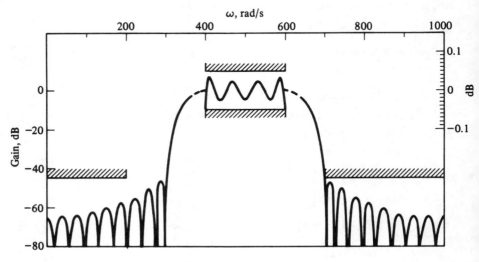

FIGURE 9.16
Amplitude response of bandpass filter (Example 9.4).

Note that the above expression for $h(nT)$ can also be used to design lowpass or highpass filters by letting $\omega_{c1} = 0$ and $\omega_{c2} = \omega_c$ or $\omega_{c1} = \omega_c$ and $\omega_{c2} = \omega_s/2$.

9.5 DESIGN BASED ON NUMERICAL-ANALYSIS FORMULAS

In signal processing, the need often arises for a signal to be interpolated, extrapolated, or differentiated at some instant $t = t_1$, or to be integrated between two distinct instants t_1 and t_2. Such mathematical operations can be performed by using the many available numerical-analysis formulas [6, 7]. Formulas of this type, which are derived from the Taylor series, can be readily used for the design of nonrecursive filters.

The most fundamental numerical formulas are the formulas for interpolation since they form the basis of many other formulas, including formulas for differentiation and integration. The most commonly used interpolation formulas are the *Gregory-Newton*, forward- and backward-difference formulas and the *Bessel*, *Everett*, and *Stirling* central-difference formulas. The value of $x(t)$ at $t = nT + pT$, where $0 \le p < 1$, is given by the Gregory-Newton formulas as

$$x(nT + pT) = (1 + \Delta)^p x(nT) = \left[1 + p\Delta + \frac{p(p-1)}{2!}\Delta^2 + \cdots\right] x(nT)$$

or $\quad x(nT + pT) = (1 - \nabla)^{-p} x(nT) = \left[1 + p\nabla + \frac{p(p+1)}{2!}\nabla^2 + \cdots\right] x(nT)$

where $\Delta x(nT) = x(nT + T) - x(nT) \quad$ and $\quad \nabla x(nT) = x(nT) - x(nT - T)$

On the other hand, the Stirling formula yields

$$\begin{aligned} x(nT + pT) = & \left[1 + \frac{p^2}{2!}\delta^2 + \frac{p^2(p^2-1)}{4!}\delta^4 + \cdots\right] x(nT) \\ & + \frac{p}{2}[\delta x(nT - \tfrac{1}{2}T) + \delta x(nT + \tfrac{1}{2}T)] \\ & + \frac{p(p^2-1)}{2(3!)}[\delta^3 x(nT - \tfrac{1}{2}T) + \delta^3 x(nT + \tfrac{1}{2}T)] \\ & + \frac{p(p^2-1)(p^2-2^2)}{2(5!)}[\delta^5 x(nT - \tfrac{1}{2}T) + \delta^5 x(nT + \tfrac{1}{2}T)] + \cdots \end{aligned}$$

(9.36)

where $\quad \delta x(nT + \tfrac{1}{2}T) = x(nT + T) - x(nT) \quad$ (9.37)

The differential of $x(t)$ at $t = nT + pT$ can be expressed as

$$\left.\frac{dx(t)}{dt}\right|_{t=nT+pT} = \frac{dx(nT + pT)}{dp} \times \frac{dp}{dt}$$

$$= \frac{1}{T}\frac{dx(nT + pT)}{dp} \quad (9.38)$$

and therefore the above interpolation formulas lead directly to corresponding differentiation formulas. Similarly, integration formulas can be derived by writing

$$\int_{nT}^{t_2} x(t)\,dt = T \int_{0}^{p_2} x(nT + pT)\,dp$$

where $\quad nT < t_2 \leq nT + T \quad$ and $\quad p_2 = \dfrac{t_2 - nT}{T}$

that is, $0 < p_2 \leq 1$.

Nonrecursive filters that perform *interpolation, differentiation,* or *integration* can now be obtained. Let $x(nT)$ and $y(nT)$ be the input and output in a nonrecursive filter and assume that $y(nT)$ is equal to the desired function of $x(t)$, i.e.,

$$y(nT) = f[x(t)] \tag{9.39}$$

For example, if $y(nT)$ is required to be the differential of $x(t)$ at $t = nT + pT$, where $0 \leq p \leq 1$, we can write

$$y(nT) = \left.\frac{dx(t)}{dt}\right|_{t=nT+pT} \tag{9.40}$$

By choosing an appropriate formula for $f[x(t)]$ and then eliminating operators using their definitions, Eq. (9.39) can be put in the form

$$y(nT) = \sum_{i=-K}^{M} a_i x(nT - iT)$$

Thus the desired transfer function is obtained as

$$H(z) = \sum_{n=-K}^{M} h(nT) z^{-n}$$

For the case of a forward- or central-difference formula, $H(z)$ is noncausal. Hence for real-time applications it will be necessary to multiply $H(z)$ by an appropriate negative power of z.

Example 9.5. A signal $x(t)$ is sampled at a rate of $1/T$ Hz. Design a sixth-order differentiator with a time-domain response

$$y(nT) = \left.\frac{dx(t)}{dt}\right|_{t=nT}$$

Use the Stirling formula.

Solution. From Eqs. (9.36) and (9.38)

$$y(nT) = \left.\frac{dx(t)}{dt}\right|_{t=nT} = \frac{1}{2T}[\delta x(nT - \tfrac{1}{2}T) + \delta x(nT + \tfrac{1}{2}T)]$$

$$- \frac{1}{12T}[\delta^3 x(nT - \tfrac{1}{2}T) + \delta^3 x(nT + \tfrac{1}{2}T)]$$

$$+ \frac{1}{60T}[\delta^5 x(nT - \tfrac{1}{2}T) + \delta^5 x(nT + \tfrac{1}{2}T)] + \cdots$$

Now on using Eq. (9.37)

$$\delta x(nT - \tfrac{1}{2}T) + \delta x(nT + \tfrac{1}{2}T) = x(nT + T) - x(nT - T)$$

$$\delta^3 x(nT - \tfrac{1}{2}T) + \delta^3 x(nT + \tfrac{1}{2}T) = x(nT + 2T) - 2x(nT + T)$$
$$+ 2x(nT - T) - x(nT - 2T)$$

$$\delta^5 x(nT - \tfrac{1}{2}T) + \delta^5 x(nT + \tfrac{1}{2}T) = x(nT + 3T) - 4x(nT + 2T)$$
$$+ 5x(nT + T) - 5x(nT - T)$$
$$+ 4x(nT - 2T) - x(nT - 3T)$$

Hence

$$y(nT) = \frac{1}{60T}[x(nT + 3T) - 9x(nT + 2T) + 45x(nT + T)$$
$$- 45x(nT - T) + 9x(nT - 2T) - x(nT - 3T)]$$

and therefore

$$H(z) = \frac{1}{60T}(z^3 - 9z^2 + 45z - 45z^{-1} + 9z^{-2} - z^{-3})$$

The filter has an antisymmetrical impulse response and is noncausal. A causal filter can be obtained by multiplying $H(z)$ by z^{-3}. The amplitude response of the filter is plotted in Fig. 9.17 for $\omega_s = 2\pi$.

Differentiators can also be designed by employing the Fourier series method of Sec. 9.3. An analog differentiator is characterized by

$$H(s) = s$$

Hence a corresponding digital differentiator can be designed by assigning

$$H(e^{j\omega T}) = j\omega \qquad \text{for} \quad 0 \le |\omega| < \frac{\omega_s}{2} \qquad (9.41)$$

Then on assuming a periodic frequency response, the appropriate impulse response can be determined from Eq. (9.14). Gibbs' oscillations due to the transition in $H(e^{j\omega T})$ at $\omega = \omega_s/2$ can be reduced as before, by using the window technique.

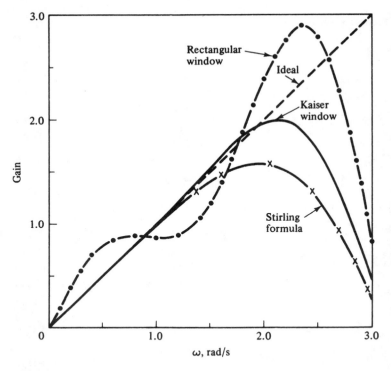

FIGURE 9.17
Amplitude response of differentiators (Examples 9.5 and 9.6).

Example 9.6. Redesign the differentiator of Example 9.5 employing the Fourier-series method. Use (*a*) a rectangular window and (*b*) the Kaiser window with $\alpha = 3.0$.

Solution. (*a*) From Eqs. (9.41) and (9.14)

$$h(nT) = \frac{1}{\omega_s} \int_{-\omega_s/2}^{\omega_s/2} j\omega e^{j\omega nT} \, d\omega = -\frac{1}{\omega_s} \int_0^{\omega_s/2} 2\omega \sin(\omega nT) d\omega$$

On integrating by parts

$$h(nT) = \frac{1}{nT} \cos \pi n - \frac{1}{n^2 \pi T} \sin \pi n$$

or
$$h(nT) = \begin{cases} 0 & \text{for } n = 0 \\ \dfrac{1}{nT} \cos \pi n & \text{otherwise} \end{cases}$$

Now by using the rectangular window with $N = 7$, we deduce

$$H_w(z) = \frac{1}{6T}(2z^3 - 3z^2 + 6z - 6z^{-1} + 3z^{-2} - 2z^{-3})$$

(b) Similarly, the Kaiser window yields

$$H_w(z) = \sum_{n=-3}^{3} w_K(nT)h(nT)z^{-n}$$

where $w_K(nT)$ can be computed from Eq. (9.28).

The amplitude responses for the two filters are compared in Fig. 9.17 with the corresponding response obtained in Example 9.5.

As before, the parameter α in the Kaiser window can be increased to increase the in-band accuracy or decreased to increase the bandwidth. Thus the Kaiser differentiator has the important advantage that it can be adjusted to suit the application. The design of digital differentiators satisfying prescribed specifications is considered in [8, 9] (see also Sec. 15.9.3).

9.6 COMPARISON BETWEEN RECURSIVE AND NONRECURSIVE DESIGNS

Before a solution is sought for the approximation problem, a choice must be made between a recursive and a nonrecursive design. In recursive filters the poles of the transfer function can be placed anywhere inside the unit circle. A consequence of this degree of freedom is that high selectivity (i.e., narrow transition bands) can easily be achieved with low-order transfer functions. In nonrecursive filters, on the other hand, with the poles fixed at the origin, high selectivity can be achieved only by using a relatively high order for the transfer function. For the same filter specification the required order in a nonrecursive design can be as high as 5 to 10 times that in a recursive design. For example, the bandpass-filter specification in Example 9.4 can be met using a nonrecursive filter of order 52 or a recursive elliptic filter of order 8. In practice, the cost of a digital filter tends to increase and its speed tends to decrease as the order of the transfer function is increased. Hence, for *high-selectivity* applications where the delay characteristic is of secondary importance, the choice is expected to be a recursive design.

Constant group delay is mandatory for certain applications, e.g., in data transmission and image processing (see Sec. 3.6). For such applications, the choice is between a nonrecursive design and an *equalized* recursive design. If computational efficiency is unimportant (e.g., if the amount of data to be processed is small), a nonrecursive design based on the methods considered in this chapter may be entirely acceptable. However, if computational efficiency is of prime importance (e.g., in real-time applications or in applications where massive amounts of data are to be processed), an optimal nonrecursive design based on the Remez exchange algorithm described in Chap. 15 or an equalized recursive design based on the method of Sec. 14.8 must be selected. Optimal nonrecursive designs are easier to obtain than equalized recursive designs. However, computational efficiency is significantly better in equalized recursive designs, particularly if a high selectivity is required.

Nonrecursive filters are naturally suited for certain applications, e.g., to perform numerical operations like interpolation, extrapolation, differentiation and integration.

Further, owing to the fact that their impulse response is of finite duration, nonrecursive filters can be implemented in terms of *fast-Fourier transforms*. This possibility is considered in Sec. 13.12.

In certain applications, the choice between a nonrecursive and a recursive design may be determined by other factors. For example, nonrecursive filters are less sensitive to quantization errors and, owing to the absence of feedback, they cannot become unstable. Further, their realizations are simple and regular and are highly attractive for VLSI implementation (see Sec. 4.9).

REFERENCES

1. L. R. Rabiner and B. Gold, *Theory and Application of Digital Signal Processing*, Prentice-Hall, Englewood Cliffs, N.J., 1975.
2. A. V. Oppenheim and R. W. Schafer, *Discrete-Time Signal Processing*, Prentice-Hall, Englewood Cliffs, N.J., 1989.
3. F. F. Kuo and J. F. Kaiser, *System Analysis by Digital Computer*, Chap. 7, Wiley, New York, 1966.
4. R. B. Blackman, *Data Smoothing and Prediction*, Addison-Wesley, Reading, Mass., 1965.
5. J. F. Kaiser, "Nonrecursive Digital Filter Design Using the I_0-sinh Window Function," *Proc. 1974 IEEE Int. Symp. Circuit Theory*, pp. 20–23.
6. R. Butler and E. Kerr, *An Introduction to Numerical Methods*, Pitman, London, 1962.
7. C. E. Fröberg, *Introduction to Numerical Analysis*, Addison-Wesley, Reading, Mass., 1965.
8. A. Antoniou, "Design of Digital Differentiators Satisfying Prescribed Specifications," *IEE Proc.*, vol. 127, pt. E, pp. 24–30, January 1980.
9. A. Antoniou and C. Charalambous, "Improved Design Method for Kaiser Differentiators and Comparison with Equiripple Method," *IEE Proc.*, vol. 128, pt. E, pp. 190–196, September 1981.

ADDITIONAL REFERENCES

Crochiere, R. E., and L. R. Rabiner: "Optimum FIR Digital Filter Implementations for Decimation, Interpolation, and Narrow-Band Filtering," *IEEE Trans. Acoust., Speech, Signal Process.*, vol. ASSP-23, pp. 444-456, October 1975.

Rabiner, L. R.: "Approximate Design Relationships for Low-Pass FIR Digital Filters," *IEEE Trans. Audio Electroacoust.*, vol. AU-21, pp. 456–460, October 1973.

———, J. F. Kaiser, O. Herrmann, and M. T. Dolan: "Some Comparisons between FIR and IIR Digital Filters," *Bell Syst. Tech. J.*, vol. 53, pp. 305–331, February 1974.

Shafer, R. W., and L. R. Rabiner: "A Digital Signal Processing Approach to Interpolation," *Proc. IEEE*, vol. 61, pp. 692–702, June 1973.

PROBLEMS

9.1. (*a*) A nonrecursive filter is characterized by the transfer function

$$H(z) = \frac{1 + 2z + 3z^2 + 4z^3 + 3z^4 + 2z^5 + z^6}{z^6}$$

Find the group delay of the filter.

(*b*) Repeat part (*a*) if

$$H(z) = \frac{1 - 2z + 3z^2 - 4z^3 + 3z^4 - 2z^5 + z^6}{z^6}$$

9.2. Figure P9.2 shows the zero-pole plots of two nonrecursive filters. Check each filter for phase-response linearity.

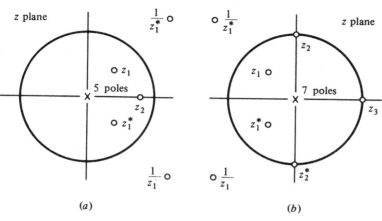

FIGURE P9.2

9.3. Design a nonrecursive bandstop filter assuming an idealized frequency response
$$H(e^{j\omega T}) = \begin{cases} 1 & \text{for } |\omega| \le \omega_{c1} \\ 0 & \text{for } \omega_{c1} < |\omega| < \omega_{c2} \\ 1 & \text{for } \omega_{c2} \le |\omega| \le \dfrac{\omega_s}{2} \end{cases}$$

9.4. A digital filter is required with a frequency response
$$H(e^{j\omega T}) \approx \begin{cases} 0 & \text{for } |\omega| < \omega_{c1} \\ 1 & \text{for } \omega_{c1} \le |\omega| \le \omega_{c2} \\ 0 & \text{for } \omega_{c2} < |\omega| < \omega_{c3} \\ 1 & \text{for } \omega_{c3} \le |\omega| \le \omega_{c4} \\ 0 & \text{for } \omega_{c4} < |\omega| \le \dfrac{\omega_s}{2} \end{cases}$$
Obtain a causal transfer function by using the Fourier-series method.

9.5. (a) Derive an exact expression for the spectrum of the Blackman window.
(b) By using the result in part (a) and assuming that $N \gg 1$, show that the main-lobe width for the Blackman window is approximately $6\omega_s/N$.

9.6. The Bartlett (or triangular) window is given by
$$w_{BA}(nT) = \begin{cases} 1 - \dfrac{2|n|}{N-1} & \text{for } |n| \le \dfrac{N-1}{2} \\ 0 & \text{otherwise} \end{cases}$$

(a) Assuming that $w_{BA}(t)$ is bandlimited, obtain an approximate expression for $W_{BA}(e^{j\omega T})$.
(b) Estimate the main-lobe width if $N \gg 1$.
(c) Estimate the ripple ratio if $N \gg 1$.
   Hint: See Prob. 6.13.

9.7. Show that the Kaiser window includes the rectangular window as a special case.

9.8. (*a*) Design a nonrecursive highpass filter in which

$$H(e^{j\omega T}) \approx \begin{cases} 0 & \text{for} \quad |\omega| < 2.5 \text{ rad/s} \\ 1 & \text{for } 2.5 \leq |\omega| \leq 5.0 \text{ rad/s} \end{cases}$$

Use the rectangular window and assume that $\omega_s = 10$ rad/s and $N = 11$.
(*b*) Repeat part (*a*) with $N = 21$ and $N = 31$. Compare the three designs.

9.9. Redesign the filter of Prob. 9.8 using the von Hann, Hamming, and Blackman windows in turn. Assume that $N = 21$. Compare the three designs.

9.10. Design a nonrecursive bandpass filter in which

$$H(e^{j\omega T}) \approx \begin{cases} 0 & \text{for} \quad |\omega| < 400 \text{ rad/s} \\ 1 & \text{for } 400 \leq |\omega| \leq 600 \text{ rad/s} \\ 0 & \text{for } 600 < |\omega| \leq 1000 \text{ rad/s} \end{cases}$$

Use the von Hann window and assume that $\omega_s = 2000$ rad/s and $N = 21$.

9.11. Design a nonrecursive bandstop filter with a frequency response

$$H(e^{j\omega T}) \approx \begin{cases} 1 & \text{for} \quad |\omega| \leq 300 \text{ rad/s} \\ 0 & \text{for } 300 < |\omega| < 700 \text{ rad/s} \\ 1 & \text{for } 700 \leq |\omega| \leq 1000 \text{ rad/s} \end{cases}$$

Use the Hamming window and assume that $\omega_s = 2000$ rad/s and $N = 21$.

9.12. A digital filter is required with a frequency response like that depicted in Fig. P9.12. Obtain a nonrecursive design using the rectangular window. Assume that $\omega_s = 10$ rad/s and $N = 21$.

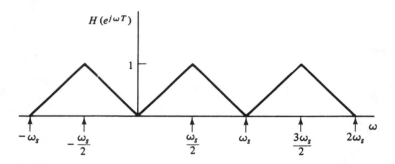

FIGURE P9.12

9.13. Design a nonrecursive filter with a frequency response like that depicted in Fig. P9.13. Use a sampling frequency $\omega_s = 10$ rad/s and assume that $N = 21$.

9.14. Design a nonrecursive lowpass filter satisfying the following specifications:

$$A_p \leq 0.1 \text{ dB} \qquad A_a \geq 44.0 \text{ dB}$$

$$\omega_p = 20 \text{ rad/s} \qquad \omega_a = 30 \text{ rad/s} \qquad \omega_s = 100 \text{ rad/s}$$

9.15. Design a nonrecursive highpass filter satisfying the following specifications:

$$A_p \leq 0.3 \text{ dB} \qquad A_a \geq 45.0 \text{ dB} \qquad \omega_p = 3 \text{ rad/s} \qquad \omega_a = 2 \text{ rad/s} \qquad \omega_s = 10 \text{ rad/s}$$

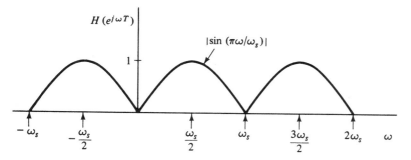

9.16. Design a nonrecursive bandpass filter satisfying the following specifications:

$A_p \leq 0.5$ dB $\quad A_a \geq 35.0$ dB $\quad \omega_{p1} = 40$ rad/s $\quad \omega_{p2} = 60$ rad/s

$\omega_{a1} = 20$ rad/s $\quad \omega_{a2} = 80$ rad/s $\quad \omega_s = 200$ rad/s

9.17. Design a nonrecursive bandstop filter satisfying the following specifications:

$A_p \leq 0.2$ dB $\quad A_a \geq 40$ dB $\quad \omega_{p1} = 1000$ rad/s $\quad \omega_{p2} = 4000$ rad/s

$\omega_{a1} = 2000$ rad/s $\quad \omega_{a2} = 3000$ rad/s $\quad \omega_s = 10{,}000$ rad/s

9.18. Redesign the filter of Prob. 9.16 as a recursive filter. Use an elliptic approximation. Compare this design with the nonrecursive one.

9.19. (a) Show that

$$\mathcal{Z}\nabla^k x(nT) = (1 - z^{-1})^k X(z)$$

(b) A signal $x(t)$ is sampled at a rate of 2π rad/s. Design a sixth-order differentiator in which

$$y(nT) \approx \left. \frac{dx(t)}{dt} \right|_{t=nT}$$

Use the Gregory-Newton backward-difference formula.

9.20. The phase response $\theta(\omega)$ of an analog filter is sampled at $\omega = n\Omega$ for $n = 0, 1, 2, \ldots$. Design a sixth-order digital filter which can be used to generate the group delay of the analog filter. Use the Stirling formula.

9.21. A signal $x(t)$ is sampled at a rate of 2π rad/s. Design a sixth-order integrator filter in which

$$y(nT) \approx \int_{nT}^{(n+1)T} x(t)\, dt$$

Use the Gregory-Newton backward-difference formula.

9.22. Two digital filters are to be cascaded. The sampling frequency in the first filter is 2π rad/s, and that in the second is 4π rad/s. Design a sixth-order interface using the Gregory-Newton backward-difference formula. *Hint*: Design an interpolating filter.

CHAPTER 10

RANDOM SIGNALS

10.1 INTRODUCTION

The methods of analysis considered so far assume deterministic signals. Frequently in digital filters and communication systems in general random signals are encountered, e.g., the noise generated by an A/D converter or the noise generated by an amplifier. Signals of this type can assume an infinite number of waveforms, and measurement will at best yield a set of typical waveforms. Despite the lack of a complete description many statistical attributes of a random signal can be determined from a statistical description of the signal.

The time- and frequency-domain statistical attributes of random signals as well as the effect of filtering on such signals can be studied by using the concept of a *random process*.

This chapter provides a brief description of random processes. The main results are presented in terms of continuous-time random signals and are then extended to discrete-time signals by using the interrelation between the Fourier and z transforms. The chapter begins with a brief summary of the essential features of random variables. Detailed discussions of random variables and processes can be found in [1–4].

10.2 RANDOM VARIABLES

Consider an experiment which may have a finite or infinite number of random outcomes, and let $\zeta_1, \zeta_2, \ldots$ be the possible outcomes. A set S comprising all ζ can be constructed, and a number $\mathbf{x}(\zeta)$ can be assigned to each ζ according to some rule. The function $\mathbf{x}(\zeta)$ or simply $\mathbf{x}$ whose domain is set S and whose range is a set of numbers is called a *random variable*. Typical random variables are the outcome of a throw of dice and the hit position in a game of darts. Specific random variables that will be studied in some detail in Chap. 11 are the errors introduced by the quantization of signals and filter coefficients.

10.2.1 Probability-Distribution Function

A random variable $\mathbf{x}$ may assume values in a certain range (x_1, x_2) where x_1 can be as low as $-\infty$ and x_2 as high as $+\infty$. The probability of observing random variable $\mathbf{x}$ below or at value x is referred to as the *probability-distribution function* of $\mathbf{x}$ and is denoted by

$$P_\mathbf{x}(x) = \Pr[\mathbf{x} \leq x]$$

10.2.2 Probability-Density Function

The derivative of $P_\mathbf{x}(x)$ with respect to x is called the *probability-density function* of $\mathbf{x}$ and is denoted by

$$p_\mathbf{x}(x) = \frac{dP_\mathbf{x}(x)}{dx}$$

A fundamental property of $p_\mathbf{x}(x)$ is

$$\int_{-\infty}^{\infty} p_\mathbf{x}(x)\,dx = 1$$

since the range $(-\infty, +\infty)$ must necessarily include the value of $\mathbf{x}$. Also

$$\Pr[x_1 \leq \mathbf{x} \leq x_2] = \int_{x_1}^{x_2} p_\mathbf{x}(x)\,dx$$

10.2.3 Uniform Probability Density

In many situations there is no preferred range for the random variable. If such is the case, the probability density is said to be *uniform* and is given by

$$p_\mathbf{x}(x) = \begin{cases} \frac{1}{x_2 - x_1} & \text{for } x_1 \leq x \leq x_2 \\ 0 & \text{otherwise} \end{cases}$$

10.2.4 Gaussian Probability Density

Very common in nature is the *gaussian* probability density given by

$$p_\mathbf{x}(x) = \frac{1}{\sigma\sqrt{2\pi}} e^{-(x-\eta)^2/2\sigma^2} \qquad -\infty \leq x \leq \infty \qquad (10.1)$$

The parameters σ and η are constants.

There are many other important probability-density functions, e.g., *binomial*, *Poisson*, and *Rayleigh* [1], but these are beyond the scope of this book.

10.2.5 Joint Distributions

An experiment may have two sets of random outcomes, say $\zeta_{x1}, \zeta_{x2}, \ldots$ and $\zeta_{y1}, \zeta_{y2}, \ldots$. For example, in an experiment of target practice, the hit position can be described in terms of two coordinates. Experiments of this type necessitate two random variables, say **x** and **y**. The probability of observing **x** and **y** below or at x and y, respectively, is said to be the *joint distribution function* of **x** and **y** and is denoted by

$$P_{xy}(x, y) = \Pr[\mathbf{x} \leq x, \mathbf{y} \leq y]$$

The joint probability-density function of **x** and **y** is denoted by

$$p_{xy}(x, y) = \frac{\partial^2 P_{xy}(x, y)}{\partial x \partial y}$$

The range $(-\infty, \infty)$ must include **x** and **y**, and hence

$$\int_{-\infty}^{\infty} \int_{-\infty}^{\infty} p_{xy}(x, y) dx dy = 1$$

The probability of observing **x** and **y** in the ranges $x_1 \leq x \leq x_2$ and $y_1 \leq y \leq y_2$, respectively, is given by

$$\Pr[x_1 \leq \mathbf{x} \leq x_2, y_1 \leq \mathbf{y} \leq y_2] = \int_{y_1}^{y_2} \int_{x_1}^{x_2} p_{xy}(x, y) dx dy$$

Two random variables **x** and **y** representing outcomes $\zeta_{x1}, \zeta_{x2}, \ldots$ and $\zeta_{y1}, \zeta_{y2}, \ldots$ of an experiment are said to be *statistically independent* if the occurrence of any outcome ζ_x does not influence the occurrence of any outcome ζ_y and vice versa. A necessary and sufficient condition for statistical independence is

$$p_{xy}(x, y) = p_x(x) p_y(y) \tag{10.2}$$

10.2.6 Mean Values and Moments

The *mean* or *expected value* of random variable **x** is defined as

$$E\{\mathbf{x}\} = \int_{-\infty}^{\infty} x p_x(x) dx$$

Similarly, if

$$\mathbf{z} = f(\mathbf{x}, \mathbf{y})$$

then

$$E\{\mathbf{z}\} = \int_{-\infty}^{\infty} z p_z(z) dz \tag{10.3}$$

If **z** is a single-valued function of **x** and **y** and $x \leq \mathbf{x} \leq x + dx$, $y \leq \mathbf{y} \leq y + dy$, then $z \leq \mathbf{z} \leq z + dz$. Hence

$$\Pr[z \leq \mathbf{z} \leq z + dz] = \Pr[x \leq \mathbf{x} \leq x + dx, y \leq \mathbf{y} \leq y + dy]$$

or
$$p_z(z)dz = p_{xy}(x,y)dxdy$$

and from Eq. (10.3)
$$E\{z\} = \int_{-\infty}^{\infty}\int_{-\infty}^{\infty} f(x,y)p_{xy}(x,y)dxdy$$

Actually this is a general relation that holds for multivalued functions as well [1]. For
$$z = xy$$
we have
$$E\{xy\} = \int_{-\infty}^{\infty}\int_{-\infty}^{\infty} xy p_{xy}(x,y)dxdy$$

and if x and y are statistically independent variables, the use of Eq. (10.2) yields
$$E\{xy\} = \int_{-\infty}^{\infty} x p_x(x)dx \int_{-\infty}^{\infty} y p_y(y)dy = E\{x\}E\{y\} \tag{10.4}$$

The nth moment of x is defined as
$$E\{x^n\} = \int_{-\infty}^{\infty} x^n p_x(x)dx$$

Similarly, the nth central moment of x is
$$E\{(x - E\{x\})^n\} = \int_{-\infty}^{\infty} (x - E\{x\})^n p_x(x)dx \tag{10.5}$$

The second central moment is also known as the *variance* and is given by
$$\sigma_x^2 = E\{(x - E\{x\})^2\}$$
$$= E\{x^2 - 2xE\{x\} + (E\{x\})^2\}$$
$$= E\{x^2\} - (E\{x\})^2 \tag{10.6}$$

If
$$z = a_1 x_1 + a_2 x_2$$

where a_1, a_2 are constants and x_1, x_2 are statistically independent random variables, then from Eqs. (10.4) and (10.5)
$$\sigma_z^2 = a_1^2 \sigma_{x_1}^2 + a_2^2 \sigma_{x_2}^2$$

In general, if
$$z = \sum_{i=1}^{n} a_i x_i$$

and $x_1, x_2, \ldots, x_n$ are statistically independent random variables, then

$$\sigma_z^2 = \sum_{i=1}^{n} a_i^2 \sigma_{x_i}^2 \tag{10.7}$$

Example 10.1. (*a*) Find the mean and variance for the case of uniform probability density

$$p_x(x) = \begin{cases} \frac{1}{x_2 - x_1} & \text{for } x_1 \leq x \leq x_2 \\ 0 & \text{otherwise} \end{cases}$$

(*b*) Repeat part (*a*) for the gaussian density

$$p_x(x) = \frac{1}{\sigma\sqrt{2\pi}} e^{-(x-\eta)^2/2\sigma^2} \qquad -\infty \leq x \leq \infty$$

Solution. (*a*)

$$E\{x\} = \int_{x_1}^{x_2} \frac{x}{x_2 - x_1} dx = \frac{1}{2}(x_1 + x_2) \tag{10.8}$$

$$E\{x^2\} = \int_{x_1}^{x_2} \frac{x^2}{x_2 - x_1} dx = \frac{x_2^3 - x_1^3}{3(x_2 - x_1)} \tag{10.9}$$

Hence from Eq. (10.6)

$$\sigma_x^2 = \frac{(x_2 - x_1)^2}{12} \tag{10.10}$$

(*b*)

$$E\{x\} = \frac{1}{\sigma\sqrt{2\pi}} \int_{-\infty}^{\infty} x e^{-(x-\eta)^2/2\sigma^2} dx$$

With $x = y + \eta$

$$E\{x\} = \frac{1}{\sigma\sqrt{2\pi}} \left(\int_{-\infty}^{\infty} y e^{-y^2/2\sigma^2} dy + \eta \int_{-\infty}^{\infty} e^{-y^2/2\sigma^2} dy \right)$$

The first integral is zero because the integrand is an odd function of y, whereas the second integral is $\sigma\sqrt{2\pi}$, according to standard tables. Hence

$$E\{x\} = \eta$$

Now

$$E\{x^2\} = \frac{1}{\sigma\sqrt{2\pi}} \int_{-\infty}^{\infty} x^2 e^{-(x-\eta)^2/2\sigma^2} dx$$

and, as before,

$$E\{x^2\} = \sigma^2 + \eta^2 \qquad \text{or} \qquad \sigma_x^2 = \sigma^2$$

10.3 RANDOM PROCESSES

A random process is an extension of the concept of a random variable. Consider an experiment with possible random outcomes $\zeta_1, \zeta_2, \ldots$. A set S comprising all ζ can be constructed and a waveform $\mathbf{x}(t, \zeta)$ can be assigned to each ζ according to some rule. The set of waveforms obtained is called an *ensemble*, and each individual waveform is said to be a *sample function*. Set S, the ensemble, and the probability description associated with S constitute a *random process*.

The concept of a random process can be illustrated by an example. Suppose that a large number of radio receivers of a particular model are receiving a carrier signal transmitted by a station. With the receivers located at different distances from the station, the amplitude and phase of the received carrier will be different at each receiver. As a result, the set of the received waveforms, illustrated in Fig. 10.1, can be described by

$$\mathbf{x}(t, \zeta) = \mathbf{z} \cos(\omega_c t + \mathbf{y})$$

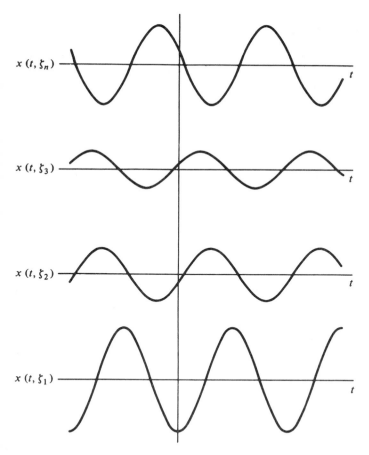

FIGURE 10.1
A random process.

where **z** and **y** are random variables and $\zeta = \zeta_1, \zeta_2, \ldots$. The set of all possible waveforms that might be received constitutes an ensemble and the ensemble together with the probability densities of **z** and **y** constitute a random process.

10.3.1 Notation

A random process can be represented by $\mathbf{x}(t, \zeta)$ or in a simplified notation by $\mathbf{x}(t)$. Depending on the circumstances, $\mathbf{x}(t, \zeta)$ can represent one of four things:

1. The *ensemble*, if t and ζ are variables.
2. A *sample function*, if t is variable and ζ is fixed.
3. A *random variable*, if t is fixed and ζ is variable.
4. A *single number*, if t and ζ are fixed.

10.4 FIRST- AND SECOND-ORDER STATISTICS

For a fixed value of t, $\mathbf{x}(t)$ is a random variable representing the instantaneous values of the various sample functions over the ensemble. The probability distribution and probability density of $\mathbf{x}(t)$ are denoted by

$$P(x; t) = \Pr[\mathbf{x}(t) \leq x] \quad \text{and} \quad p(x; t) = \frac{\partial P(x; t)}{\partial x}$$

respectively. These two equations constitute the *first-order statistics* of the random process.

At any two instants t_1 and t_2, $\mathbf{x}(t_1)$ and $\mathbf{x}(t_2)$ are distinct random variables. Their joint probability distribution and joint probability density depend on t_1 and t_2 in general, and are denoted by

$$P(x_1, x_2; t_1, t_2) = \Pr[\mathbf{x}(t_1) \leq x_1, \mathbf{x}(t_2) \leq x_2]$$

and

$$p(x_1, x_2; t_1, t_2) = \frac{\partial^2 P(x_1, x_2; t_1, t_2)}{\partial x_1 \partial x_2}$$

respectively. These two equations constitute the *second-order statistics* of the random process.

Similarly, at any k instants $t_1, t_2, \ldots, t_k$, the quantities $\mathbf{x}_1, \mathbf{x}_2, \ldots,$ and $\mathbf{x}_k$ are distinct random variables. Their joint probability distribution and joint probability density depend on $t_1, t_2, \ldots, t_k$ and can be defined as before. These quantities constitute the *kth-order statistics* of the random process.

Example 10.2. Find the first-order probability density $p(x; t)$ for random process

$$\mathbf{x}(t) = \mathbf{y}t - 2$$

where **y** is a random variable with a probability density

$$p_\mathbf{y}(y) = \frac{1}{\sqrt{2\pi}} e^{-y^2/2} \qquad -\infty \leq y \leq \infty.$$

Solution. If x and y are possible values of $\mathbf{x}(t)$ and $\mathbf{y}$, then
$$x = yt - 2 \quad \text{or} \quad y = \frac{1}{t}(x+2)$$

From Fig. 10.2
$$\Pr[x \leq \mathbf{x} \leq x + |dx|] = \Pr[y \leq \mathbf{y} \leq y + |dy|]$$
i.e.,
$$p_{\mathbf{x}}(x)|dx| = p_{\mathbf{y}}(y)|dy| \quad \text{or} \quad p_{\mathbf{x}}(x) = \frac{p_{\mathbf{y}}(y)}{|dx/dy|}$$

Since
$$\frac{dx}{dy} = t$$

we obtain
$$p(x;\ t) = p_{\mathbf{x}}(x) = \frac{1}{|t|\sqrt{2\pi}} e^{-(x+2)^2/2t^2} \qquad -\infty \leq x \leq \infty$$

Example 10.3. Find the first-order probability density $p(x;\ t)$ of the random process
$$\mathbf{x}(t) = \cos(\omega_c t + \mathbf{y})$$
where $\mathbf{y}$ is a random variable with probability density
$$p_{\mathbf{y}}(y) = \begin{cases} \dfrac{1}{2\pi} & \text{for } 0 \leq y \leq 2\pi \\ 0 & \text{otherwise} \end{cases}$$

Solution. If x and y are possible values of $\mathbf{x}(t)$ and $\mathbf{y}$, then
$$x = \cos(\omega_c t + y)$$

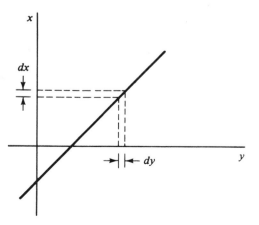

FIGURE 10.2
Function $x = yt - 2$ (Example 10.2).

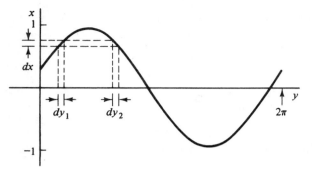

FIGURE 10.3
Function $x = \cos(\omega_c t + y)$ (Example 10.3).

and from Fig. 10.3

$$\Pr[x \leq \mathbf{x} \leq x + |dx|] = \Pr[y_1 \leq \mathbf{y} \leq y_1 + |dy_1|]$$
$$+ \Pr[y_2 \leq \mathbf{y} \leq y_2 + |dy_2|]$$

or
$$p_{\mathbf{x}}(x)|dx| = p_{\mathbf{y}}(y_1)|dy_1| + p_{\mathbf{y}}(y_2)|dy_2|$$

Hence
$$p_{\mathbf{x}}(x) = \frac{p_{\mathbf{y}}(y_1)}{|x'(y_1)|} + \frac{p_{\mathbf{y}}(y_2)}{|x'(y_2)|}$$

where
$$x'(y) = \frac{dx}{dy} = -\sin(\omega_c t + y) = -\sqrt{1 - x^2}$$

Since
$$|x'(y_1)| = |x'(y_2)|$$

we deduce

$$p(x;\, t) = p_{\mathbf{x}}(t) = \begin{cases} \dfrac{1}{\pi\sqrt{1 - x^2}} & \text{for } |x| < 1 \\ 0 & \text{otherwise} \end{cases}$$

10.5 MOMENTS AND AUTOCORRELATION

The first-order statistics give the mean, mean square, and other moments of a random process at any instant t. From Sec. 10.2

$$E\{\mathbf{x}(t)\} = \int_{-\infty}^{\infty} x p(x;\, t)\, dx$$

$$E\{\mathbf{x}^2(t)\} = \int_{-\infty}^{\infty} x^2 p(x;\, t)\, dx$$

The second-order statistics give the *autocorrelation function* of a random process, which is defined as

$$r_{\mathbf{x}}(t_1, t_2) = E\{\mathbf{x}(t_1)\mathbf{x}(t_2)\} = \int_{-\infty}^{\infty}\int_{-\infty}^{\infty} x_1 x_2 p(x_1, x_2;\, t_1, t_2)\, dx_1\, dx_2$$

The autocorrelation is a measure of the interdependence between the instantaneous signal values at $t = t_1$ and those at $t = t_2$. This is the most important attribute of a random process, as it leads to a frequency-domain description of the process.

Example 10.4. (*a*) Find the mean, mean square, and autocorrelation for the process in Example 10.2. (*b*) Repeat part (*a*) for the process in Example 10.3.

Solution. (*a*) From part (*b*) of Example 10.1 and Example 10.2

$$E\{\mathbf{x}(t)\} = -2 \qquad E\{\mathbf{x}^2(t)\} = t^2 + 4$$

$$r_\mathbf{x}(t_1, t_2) = E\{(\mathbf{y}t_1 - 2)(\mathbf{y}t_2 - 2)\} = t_1 t_2 E\{\mathbf{y}^2\} - 2(t_1 + t_2)E\{\mathbf{y}\} + 4$$

and since **y** is gaussian with

$$E\{\mathbf{y}\} = 0 \qquad E\{\mathbf{y}^2\} = 1$$

we have

$$r_\mathbf{x}(t_1, t_2) = t_1 t_2 + 4$$

(*b*) From Example 10.3

$$E\{\mathbf{x}(t)\} = \frac{1}{\pi} \int_{-1}^{1} \frac{x}{\sqrt{1-x^2}} dx = 0 \qquad (10.11)$$

$$E\{\mathbf{x}^2(t)\} = \frac{1}{\pi} \int_{-1}^{1} \frac{x^2}{\sqrt{1-x^2}} dx = \tfrac{1}{2}$$

$$r_\mathbf{x}(t_1, t_2) = E\{\cos(\omega_c t_1 + \mathbf{y}) \cos(\omega_c t_2 + \mathbf{y})\}$$

$$= \tfrac{1}{2} \cos(\omega_c t_1 - \omega_c t_2) - \tfrac{1}{2} E\{\cos(\omega_c t_1 + \omega_c t_2 + 2\mathbf{y})\}$$

$$= \tfrac{1}{2} \cos[\omega_c(t_1 - t_2)] \qquad (10.12)$$

10.6 STATIONARY PROCESSES

A random process is said to be *strictly stationary* if $\mathbf{x}(t)$ and $\mathbf{x}(t + T)$ have the same statistics (all orders) for any value of T. If the mean of $\mathbf{x}(t)$ is constant and its autocorrelation depends only on $t_2 - t_1$, i.e.,

$$E\{\mathbf{x}(t)\} = \text{const.} \qquad E\{\mathbf{x}(t_1)\mathbf{x}(t_2)\} = r_\mathbf{x}(t_2 - t_1)$$

the process is called *wide-sense stationary*. A strictly stationary process is also stationary in the wide sense; however, the converse is not necessarily true. The process of part (*b*) Example 10.4 is wide-sense stationary, according to Eqs. (10.11) and (10.12); that of part (*a*) of Example 10.4, however, is not stationary.

10.7 FREQUENCY-DOMAIN REPRESENTATION

The frequency-domain representation of deterministic signals is normally in terms of amplitude, phase, and energy-density spectra (see Chap. 6). Although such representations are possible for random processes [1], they are avoided in practice because of the mathematical difficulties associated with infinite-energy signals (see Sec. 6.3). Usually, random processes are represented in terms of power-density spectra.

Consider a signal $x(t)$ and let

$$x_{T_0}(t) = \begin{cases} x(t) & \text{for } |t| \leq T_0 \\ 0 & \text{otherwise} \end{cases}$$

The average power of $x(t)$ over the interval $[-T_0, T_0]$ is

$$P_{T_0} = \frac{1}{2T_0} \int_{-T_0}^{T_0} x^2(t) dt = \frac{1}{2T_0} \int_{-\infty}^{\infty} x_{T_0}^2(t) dt$$

and by virtue of Parseval's formula (see Theorem 6.9)

$$P_{T_0} = \frac{1}{2T_0} \int_{-\infty}^{\infty} \frac{|X_{T_0}(j\omega)|^2}{2T_0} \frac{d\omega}{2\pi}$$

where

$$X_{T_0}(j\omega) = \mathcal{F} x_{T_0}(t)$$

Evidently, the elemental area in the above integral, namely,

$$\frac{|X_{T_0}(j\omega)|^2}{2T_0} \frac{d\omega}{2\pi} = \frac{|X_{T_0}(j\omega)|^2}{2T_0} df$$

represents average power (f is the frequency in hertz). Therefore, the quantity

$$\frac{|X_{T_0}(j\omega)|^2}{2T_0}$$

represents the average power per unit bandwidth (in hertz) and can be referred to as the *power spectral density* (PSD) of $x_{T_0}(t)$. If $x_{T_0}(t)$ and $x(t)$ are sample functions of random processes $\mathbf{x}_{T_0}(t)$ and $\mathbf{x}(t)$, respectively, we can define

$$\text{PSD of } \mathbf{x}_{T_0}(t) = E\left\{\frac{|X_{T_0}(j\omega)|^2}{2T_0}\right\}$$

and since $\mathbf{x}_{T_0}(t) \to \mathbf{x}(t)$ as $T_0 \to \infty$, we obtain

$$\text{PSD of } \mathbf{x}(t) = S_{\mathbf{x}}(\omega) = \lim_{T_0 \to \infty} E\left\{\frac{|X_{T_0}(j\omega)|^2}{2T_0}\right\} \tag{10.13}$$

The function $S_{\mathbf{x}}(\omega)$ is said to be the *power-density spectrum* of the process.

For stationary processes, the *PSD is the Fourier transform of the autocorrelation function*, as we shall now demonstrate. From Eq. (10.13)

$$S_{\mathbf{x}}(\omega) = \lim_{T_0 \to \infty} E\left\{\frac{X_{T_0}(j\omega) X_{T_0}^*(j\omega)}{2T_0}\right\}$$

$$= \lim_{T_0 \to \infty} \frac{1}{2T_0} E\left\{\int_{-T_0}^{T_0} \mathbf{x}(t_2) e^{-j\omega t_2} dt_2 \int_{-T_0}^{T_0} \mathbf{x}(t_1) e^{j\omega t_1} dt_1\right\}$$

$$= \lim_{T_0 \to \infty} \frac{1}{2T_0} \int_{-T_0}^{T_0} \int_{-T_0}^{T_0} E\{\mathbf{x}(t_1)\mathbf{x}(t_2)\} e^{-j\omega(t_2-t_1)} dt_1 dt_2$$

For a wide-sense-stationary process

$$E\{x(t_1)x(t_2)\} = r_x(t_2 - t_1)$$

and hence we can write

$$S_x(\omega) = \lim_{T_0 \to \infty} \frac{1}{2T_0} \int_{-T_0}^{T_0} \int_{-T_0}^{T_0} f(t_2 - t_1) dt_1 dt_2 \tag{10.14}$$

where

$$f(t_2 - t_1) = r_x(t_2 - t_1) e^{-j\omega(t_2 - t_1)} \tag{10.15}$$

The preceding double integral represents the volume under the surface $y = f(t_2 - t_1)$ and above the square region in Fig. 10.4. Since $f(t_2 - t_1)$ is constant on any line of the form

$$t_2 = t_1 + c$$

the volume over the elemental area bounded by the square region and the lines

$$t_2 = t_1 + \tau \quad \text{and} \quad t_2 = t_1 + \tau + d\tau$$

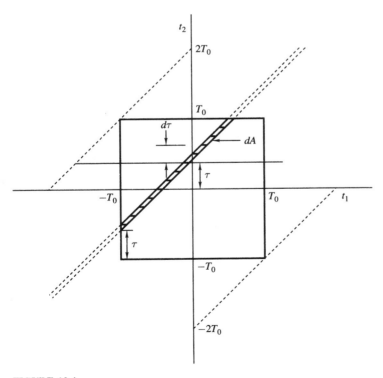

FIGURE 10.4
Domain of $y = f(t_2 - t_1)$.

is approximately constant. From the geometry of Fig. 10.4, we note that the elemental area dA is the difference between the areas of two overlapping equilateral right-angled triangles. For $\tau \geq 0$, the sides of the larger and smaller triangles are $2T_0 - \tau$ and $2T_0 - (\tau + d\tau)$, respectively, and hence

$$dA = \frac{1}{2}(2T_0 - \tau)^2 - \frac{1}{2}[2T_0 - (\tau + d\tau)]^2$$

$$= (2T_0 - \tau)d\tau + \frac{1}{2}(d\tau)^2$$

$$\approx (2T_0 - \tau)d\tau$$

Similarly, for $\tau < 0$

$$dA \approx (2T_0 + \tau)d\tau$$

and in general, as $d\tau \to 0$, we can write

$$dA = (2T_0 - |\tau|)d\tau$$

Hence the elemental volume for $t_2 - t_1 = \tau$ is

$$dV = f(\tau)(2T_0 - |\tau|)d\tau$$

In order to obtain the entire volume under the surface $y = f(t_2 - t_1)$ and above the square region in Fig. 10.4, τ must be increased from $-2T_0$ to $+2T_0$; thus Eq. (10.14) can be expressed as

$$S_\mathbf{x}(\omega) = \lim_{T_0 \to \infty} \frac{1}{2T_0} \int_{-2T_0}^{2T_0} f(\tau)(2T_0 - |\tau|)d\tau$$

$$= \int_{-\infty}^{\infty} f(\tau) \lim_{T_0 \to \infty} \left(1 - \frac{|\tau|}{2T_0}\right) d\tau = \int_{-\infty}^{\infty} f(\tau)d\tau$$

Therefore, from Eq. (10.15)

$$S_\mathbf{x}(\omega) = \int_{-\infty}^{\infty} r_\mathbf{x}(\tau)e^{-j\omega\tau}\, d\tau \qquad (10.16)$$

and if

$$\int_{-\infty}^{\infty} |r_\mathbf{x}(\tau)|\, d\tau < \infty$$

we can write

$$r_\mathbf{x}(\tau) = E\{\mathbf{x}(t)\mathbf{x}(t+\tau)\} = \frac{1}{2\pi} \int_{-\infty}^{\infty} S_\mathbf{x}(\omega)e^{j\omega\tau}d\omega \qquad (10.17)$$

i.e.,

$$r_\mathbf{x}(\tau) \leftrightarrow S_\mathbf{x}(\omega)$$

by virtue of Theorem 6.1. The formula in Eq. (10.16) is known as the *Wiener-Khinchine relation*.

Example 10.5. Find the PSD of the process in Example 10.3.

Solution. From part (b) of Example 10.4 [Eq. (10.12)] the autocorrelation of the process can be expressed as

$$r(\tau) = \frac{1}{2} \cos \omega_c \tau$$

Hence from Eq. (10.16) and Table 6.2

$$S_x(\omega) = \frac{\pi}{2}[\delta(\omega - \omega_c) + \delta(\omega + \omega_c)]$$

The autocorrelation is an even function of τ, i.e.,

$$r_x(\tau) = r_x(-\tau)$$

as can be easily shown, and $S_x(\omega)$ is an even function of ω by definition. Equations (10.16) and (10.17) can thus be written as

$$S_x(\omega) = \int_{-\infty}^{\infty} r_x(\tau) \cos(\omega\tau) \, d\tau$$

$$r_x(\tau) = \frac{1}{2\pi} \int_{-\infty}^{\infty} S_x(\omega) \cos(\omega\tau) \, d\omega$$

If $\omega = 0$,

$$S_x(0) = \int_{-\infty}^{\infty} r_x(\tau) d\tau$$

i.e., *the total area under the autocorrelation function equals the PSD at zero frequency.* The average power of $x(t)$ is given by

$$\text{Average power} = E\{x^2(t)\} = r_x(0) = \int_{-\infty}^{\infty} S_x(\omega) \frac{d\omega}{2\pi}$$

as is to be expected.

A random process whose PSD is constant at all frequencies is said to be a *white-noise process.* If

$$S_x(\omega) = K$$

we have

$$r_x(\tau) = K\delta(\tau)$$

i.e., *the autocorrelation of a white-noise process is an impulse at the origin.*

10.8 DISCRETE-TIME RANDOM PROCESS

The concept of a random process can be readily extended to discrete-time random signals by simply assigning discrete-time waveforms to the possible outcomes of an experiment. The *mean, mean square,* and *autocorrelation* of a discrete-time process $\mathbf{x}(nT)$ can be expressed as

$$E\{\mathbf{x}(nT)\} = \int_{-\infty}^{\infty} x p(x; nT) dx$$

$$E\{\mathbf{x}^2(nT)\} = \int_{-\infty}^{\infty} x^2 p(x; nT) dx$$

$$r_{\mathbf{x}}(kT) = E\{\mathbf{x}(nT)\mathbf{x}(nT+kT)\}$$

A *frequency-domain* representation for a discrete-time process can be deduced by using the interrelations between the z transform and the Fourier transform (see Sec. 6.6). We can write

$$\mathcal{Z} r_{\mathbf{x}}(kT) = \sum_{k=-\infty}^{\infty} r(kT) z^{-k} = R_{\mathbf{x}}(z)$$

and from Eq. (6.33)

$$R_{\mathbf{x}}(e^{j\omega T}) = \mathcal{F}\hat{r}_{\mathbf{x}}(\tau) = \hat{S}_{\mathbf{x}}(\omega) \qquad (10.18)$$

where

$$\hat{r}_{\mathbf{x}}(\tau) = E\{\hat{\mathbf{x}}(t)\hat{\mathbf{x}}(t+\tau)\}$$

$$\hat{\mathbf{x}}(t) = \sum_{n=-\infty}^{\infty} \mathbf{x}(nT)\delta(t-nT)$$

$$\tau = kT$$

Therefore, from Eqs. (10.13) and (10.18)

$$R_{\mathbf{x}}(e^{j\omega T}) = \lim_{T_0 \to \infty} E\left\{ \frac{|\hat{X}_{T_0}(j\omega)|^2}{2T_0} \right\}$$

where

$$\hat{X}_{T_0}(j\omega) = \mathcal{F}\hat{\mathbf{x}}_{T_0}(t)$$

and

$$\hat{\mathbf{x}}_{T_0}(t) = \begin{cases} \hat{\mathbf{x}}(t) & \text{for } |t| \leq T_0 \\ 0 & \text{otherwise} \end{cases}$$

In effect, the *z transform of the autocorrelation of discrete-time process $\mathbf{x}(nT)$ evaluated on the unit circle $|z|=1$ is numerically equal to the PSD of sampled process $\hat{\mathbf{x}}(t)$.* This quantity can be referred to as the PSD of discrete-time process $\mathbf{x}(nT)$ and can be represented by $S_{\mathbf{x}}(e^{j\omega T})$ by analogy with the PSD of continuous-time process $\mathbf{x}(t)$ which is represented by $S_{\mathbf{x}}(\omega)$. Consequently, we can write

$$\mathcal{Z} r_{\mathbf{x}}(kT) = S_{\mathbf{x}}(z)$$

where
$$r_x(kT) = \frac{1}{2\pi j} \oint_\Gamma S_x(z) z^{k-1} dz \qquad (10.19a)$$

by virtue of Eq. (2.8). If $x(t)$ were a voltage or current waveform, then $E\{x^2(t)\}$ would represent the average energy that would be delivered in a 1-Ω resistor. Consequently, the quantity $E\{x^2(nT)\}$ is said to be the *power* in $x(nT)$. It can be obtained by evaluating the autocorrelation function at $k = 0$, i.e.,

$$E\{x^2(nT)\} = r_x(0) = \frac{1}{2\pi j} \oint_\Gamma S_x(z) z^{-1} dz \qquad (10.19b)$$

10.9 FILTERING OF DISCRETE-TIME RANDOM SIGNALS

If a discrete-time random signal is passed through a digital filter, we expect the PSD of the output signal to be related to that of the input signal. This indeed is the case, as will now be shown.

Consider a filter characterized by $H(z)$, and let $x(n)$ and $y(n)$ be the input and output processes, respectively. From the convolution summation (see Sec. 1.7)

$$y(i) = \sum_{p=-\infty}^{\infty} h(p)x(i-p) \qquad y(j) = \sum_{q=-\infty}^{\infty} h(q)x(j-q)$$

and hence
$$E\{y(i)y(j)\} = E\left\{\sum_{q=-\infty}^{\infty} \sum_{p=-\infty}^{\infty} h(p)h(q)x(i-p)x(j-q)\right\}$$

With $j = i + k$ and $q = p + n$ we have

$$r_y(k) = \sum_{n=-\infty}^{\infty} \sum_{p=-\infty}^{\infty} h(p)h(p+n) E\{x(i-p)x(i-p+k-n)\}$$

or
$$r_y(k) = \sum_{n=-\infty}^{\infty} g(n) r_x(k-n)$$

where
$$g(n) = \sum_{p=-\infty}^{\infty} h(p)h(p+n)$$

The use of Theorem 2.9 gives
$$S_y(z) = \mathcal{Z} r_y(k) = \mathcal{Z} g(k) \mathcal{Z} r_x(k) = G(z) S_x(z) \qquad (10.20)$$

Now
$$G(z) = \mathcal{Z} \sum_{p=-\infty}^{\infty} h(p)h(p+n) = \sum_{n=-\infty}^{\infty} \sum_{p=-\infty}^{\infty} h(p)h(p+n) z^{-n}$$

and with $n = k - p$

$$G(z) = \sum_{k=-\infty}^{\infty} h(k) z^{-k} \sum_{p=-\infty}^{\infty} h(p)(z^{-1})^{-p} = H(z)H(z^{-1}) \qquad (10.21)$$

Therefore, from Eqs.(10.20) and (10.21)

$$S_y(z) = H(z)H(z^{-1})S_x(z) \qquad (10.22)$$

or
$$S_y(e^{j\omega T}) = |H(e^{j\omega T})|^2 S_x(e^{j\omega T})$$

i.e., the *PSD of the output process is equal to the squared amplitude response of the filter times the PSD of the input process.*

Example 10.6. The output of a digital filter is given by

$$y(n) = x(n) + 0.8 y(n-1)$$

The input of the filter is a random signal with zero mean and variance σ_x^2; successive values of $x(n)$ are statistically independent. (a) Find the output PSD of the filter. (b) Obtain an expression for the average output power.

Solution. (a) The autocorrelation of the input signal is

$$r_x(k) = E\{\mathbf{x}(n)\mathbf{x}(n+k)\}$$

For $k = 0$
$$r_x(k) = E\{\mathbf{x}^2(n)\} = \sigma_x^2$$

For $k \neq 0$ the use of Eq. (10.4) gives

$$r_x(k) = E\{\mathbf{x}(n)\}E\{\mathbf{x}(n+k)\} = 0$$

Hence
$$r_x(k) = \sigma_x^2 \delta(k) \quad \text{and} \quad S_x(z) = \sigma_x^2$$

Now from Eq. (10.22)
$$S_y(z) = \sigma_x^2 H(z) H(z^{-1})$$

where
$$H(z) = \frac{z}{z - 0.8}$$

(b) From Eq. (10.19)

$$\text{Output power} = E\{\mathbf{y}^2(n)\} = r_y(0) = \frac{1}{2\pi j} \oint_\Gamma \sigma_x^2 H(z) H(z^{-1}) z^{-1} \, dz$$

and if Γ is taken to be the unit circle $|z| = 1$, we can let $z = e^{j\omega T}$, in which case

$$\text{Output power} = \frac{1}{\omega_s} \int_0^{\omega_s} \sigma_x^2 H(e^{j\omega T}) H(e^{-j\omega T}) \, d\omega$$

A simple numerical method for the evaluation of the output power can be found in [5].

REFERENCES

1. A. Papoulis, *Probability, Random Variables, and Stochastic Processes*, McGraw-Hill, New York, 1991.
2. W. B. Davenport, Jr. and W. L. Root, *Random Signals and Noise*, McGraw-Hill, New York, 1958.
3. B. P. Lathi, *An Introduction to Random Signals and Communication Theory*, International Textbook, Scranton, 1968.
4. G. R. Cooper and C. D. McGillem, *Probabilistic Methods of Signal and System Analysis*, Holt Rinehart and Winston, New York, 1971.
5. K. J. Åström, E. I. Jury, and R. G. Agniel, "A Numerical Method for the Evaluation of Complex Integrals," *IEEE Trans. Automatic Control*, vol. AC-15, pp. 468–471, August 1970.

PROBLEMS

10.1. A random variable **x** has a probability-density function

$$p_\mathbf{x}(x) = \begin{cases} Ke^{-x} & \text{for } 1 \leq x \leq \infty \\ 0 & \text{otherwise} \end{cases}$$

(a) Find K.
(b) Find $\Pr[0 \leq \mathbf{x} \leq 2]$.

10.2. A random variable **x** has a probability-density function

$$p_\mathbf{x}(x) = \begin{cases} \frac{1}{q} & 0 \leq x \leq q \\ 0 & \text{otherwise} \end{cases}$$

Find its mean, mean square, and variance.

10.3. Find the mean, mean square, and variance for the random variable of Prob. 10.1.

10.4. Demonstrate the validity of Eq. (10.7).

10.5. A gaussian random variable **x** has a mean η and a variance σ^2. Show that

$$P_\mathbf{x}(x_1 - \eta) = 1 - P_\mathbf{x}(\eta - x_1)$$

where $P_\mathbf{x}(x)$ is the probability-distribution function of a gaussian random variable with zero mean.

10.6. A gaussian random variable **x** has $\eta = 0$ and $\sigma = 2$.
(a) Find $\Pr[\mathbf{x} \geq 2]$.
(b) Find $\Pr[|\mathbf{x}| \geq 2]$.
(c) Find x_1 if $\Pr[|\mathbf{x}| \leq x_1] = 0.95$.

10.7. The random variable of Prob. 10.5 satisfies the relations

$$\Pr[\mathbf{x} \leq 60] = 0.2 \qquad \Pr[\mathbf{x} \geq 90] = 0.1$$

Find η and σ^2.

10.8. A random variable **x** has a Rayleigh probability-density function given by

$$p_\mathbf{x}(x) = \begin{cases} \dfrac{xe^{-x^2/2\alpha^2}}{\alpha^2} & \text{for } 0 \leq x \leq \infty \\ 0 & \text{otherwise} \end{cases}$$

Show that
(a) $E\{\mathbf{x}\} = \alpha\sqrt{\frac{\pi}{2}}$
(b) $E\{\mathbf{x}^2\} = 2\alpha^2$
(c) $\sigma_\mathbf{x}^2 = \left(2 - \frac{\pi}{2}\right)\alpha^2$

10.9. A random process is given by
$$x(t) = ye^{-t}u(t-z)$$
where y and z are random variables uniformly distributed in the range $(-1, 1)$. Sketch five sample functions.

10.10. A random process is given by
$$x(t) = 2 + \frac{yt}{\sqrt{2}}$$
where y is a random variable with a probability-density function
$$p_y(y) = \frac{1}{\sqrt{2\pi}}e^{-y^2/2} \qquad -\infty \leq y \leq \infty$$
Find the first-order probability-density function of $x(t)$.

10.11. A random process is given by
$$x(t) = z\cos(\omega_0 t + y)$$
Find the first-order probability-density function of $x(t)$.
(a) If z is a random variable distributed uniformly in the range $(-1, 1)$ and y is a constant.
(b) If y is a random variable distributed uniformly in the range $(-\pi, \pi)$ and z is a constant.

10.12. Find the mean, mean square, and autocorrelation for the process in Prob. 10.10. Is the process stationary?

10.13. Repeat Prob. 10.12 for the processes in Prob. 10.11.

10.14. A stationary discrete-time random process is given by
$$x(nT) = E\{x(nT)\} + x_0(nT)$$
where $x_0(nT)$ is a zero-mean process. Show that
(a) $r_x(0) = E\{x^2(nT)\}$
(b) $r_x(-kT) = r_x(kT)$
(c) $r_x(0) \geq |r_x(kT)|$
(d) $r_x(kT) = [E\{x(nT)\}]^2 + r_{x_0}(kT)$

10.15. Explain the physical significance of
(a) $E\{x(nT)\}$
(b) $E^2\{x(nT)\}$
(c) $E\{x^2(nT)\}$
(d) $\sigma_x^2 = E\{x^2(nT)\} - [E\{x(nT)\}]^2$

10.16. A discrete-time random process is given by
$$x(nT) = 3 + 4nTy$$
where y is a random variable with a probability-density function
$$p_y(y) = \frac{1}{2\sqrt{2\pi}}e^{-(y-4)^2/8} \qquad -\infty \leq y \leq \infty$$
Find its mean, mean square, and autocorrelation.

10.17. A discrete-time random process is given by
$$x(nT) = z\cos\left(\omega_0 nT + \frac{\pi}{8}\right)$$
where z is a random variable distributed uniformly in the range (0, 1). Find the mean, mean square, and autocorrelation of $x(nT)$. Is the process stationary?

10.18. A discrete-time random process is given by
$$x(nT) = \sqrt{2}\cos(\omega_0 nT + y)$$
where y is a random variable uniformly distributed in the range $(-\pi, \pi)$.
(a) Find the mean, mean square, and autocorrelation of $x(nT)$.
(b) Show that the process is wide-sense-stationary.
(c) Find the PSD of $x(nT)$.

10.19. The random process of Prob. 10.18 is passed through a digital filter characterized by
$$H(e^{j\omega T}) = \begin{cases} 1 & \text{for } |\omega| \le \omega_c \\ 0 & \text{otherwise} \end{cases}$$
Sketch the input and output power-density spectra if $\omega_0 \le \omega_c$.

10.20. A random process $x(nT)$ with a probability-density function
$$p_x(x; nT) = \begin{cases} 1 & \text{for } \tfrac{1}{2} \le x \le \tfrac{1}{2} \\ 0 & \text{otherwise} \end{cases}$$
is applied at the input of the filter depicted in Fig. P10.20. Find the output PSD if $x(nT)$ and $x(kT)$ ($n \ne k$) are statistically independent.

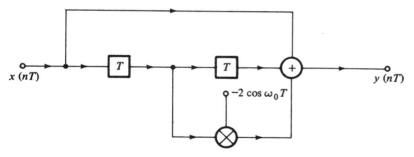

FIGURE P10.20

CHAPTER 11

EFFECTS OF FINITE WORD LENGTH IN DIGITAL FILTERS

11.1 INTRODUCTION

In software as well as hardware digital-filter implementations, numbers are stored in finite-length registers. Consequently, if coefficients and signal values cannot be accommodated in the available registers, they must be quantized before they can be stored. Number quantization gives rise to three types of errors:

1. Coefficient-quantization errors
2. Product-quantization errors
3. Input-quantization errors

The transfer-function coefficients are normally evaluated to a high degree of precision during the approximation step. If *coefficient quantization* is applied, the frequency response of the resulting filter may differ appreciably from the desired response, and if the quantization step is coarse, the filter may actually fail to meet the desired specifications.

Product-quantization errors arise at the outputs of multipliers. Each time a signal represented by b_1 digits is multiplied by a coefficient represented by b_2 digits, a product having as many as $b_1 + b_2$ digits is generated. Since a uniform register length must, in practice, be used throughout the filter, each multiplier output must be rounded or truncated before processing can continue. These errors tend to propagate through the filter and give rise to output noise commonly referred to as output *roundoff noise*.

Input-quantization errors arise in applications where digital filters are used to process continuous-time signals. These are the errors inherent in the analog-to-digital conversion process (see Sec. 6.9.1).

This chapter begins with a review of the various number systems and types of arithmetic that can be used in digital-filter implementations. It then describes various methods of analysis and design that can be applied to quantify and minimize the effects of quantization. Section 11.3 deals with a method of analysis that can be used to evaluate the effect of coefficient quantization and Sec. 11.4 describes two families of filter structures that are relatively insensitive to coefficient quantization. Section 11.5 deals with methods by which the effect of product quantization can be evaluated, and Secs. 11.6 to 11.8 describe methods by which this effect can be reduced or minimized. In Sec. 11.9, two types of parasitic oscillations known as *quantization* and *overflow limit cycles* are considered in some detail and methods for their elimination are described.

11.2 NUMBER REPRESENTATION

The hardware implementation of digital filters, like the implementation of other digital hardware, is based on the binary-number representation.

11.2.1 Binary System

In general, any number N can be expressed as

$$N = \sum_{i=-m}^{n} b_i r^i \tag{11.1}$$

where
$$0 \leq b_i \leq r - 1$$

If distinct symbols are assigned to the permissible values of b_i, the number N can be represented by the notation

$$N = (b_n b_{n-1} \cdots b_0 . b_{-1} \cdots b_{-m})_r \tag{11.2}$$

The parameter r is said to be the *radix* of the representation, and the point separating N into two parts is called the *radix point*.

If $r = 10$, Eq. (11.2) becomes the decimal representation of N and the radix point is the decimal point. Similarly, if $r = 2$ Eq. (11.2) becomes the binary representation of N and the radix point is referred to as the *binary point*. The common symbols used to represent the two permissible values of b_i are 0 and 1. These are called *bits*.

A mixed decimal number can be converted into a binary number through the following steps:

1. Divide the integer part by 2 repeatedly and arrange the resulting remainders in the reverse order.
2. Multiply the fraction part by 2 and remove the resulting integer part; repeat as many times as necessary, and then arrange the integers obtained in the forward order.

A binary number can be converted into a decimal number by using Eq. (11.1).

Example 11.1. (*a*) Form the binary representation of $N = 18.375_{10}$. (*b*) Form the decimal representation of $N = 11.101_2$.

Solution. (*a*)

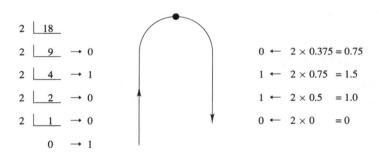

Hence
$$18.375_{10} = 10010.011_2$$

(*b*) From Eq. (11.1)
$$11.101_2 = 1(2^1) + 1(2^0) + 1(2^{-1}) + 0(2^{-2}) + 1(2^{-3}) = 3.625_{10}$$

The most basic electronic memory device is the *flip-flop*, which can be either in a low or a high state. By assigning a 0 to the low state and a 1 to the high state, a single-bit binary number can be stored. By arranging n flip-flops in juxtaposition, as in Fig. 11.1*a*, a register can be formed that will store an n-bit number.

A rudimentary 4-bit digital-filter implementation is shown in Fig. 11.1*b*. Registers R_y and R_b are used to store the past output $y(n-1)$ and the multiplier coefficient b, respectively. The output of the multiplier at steady state is $by(n-1)$. Once a new input sample is received, the adder goes into action to form the new output $y(n)$

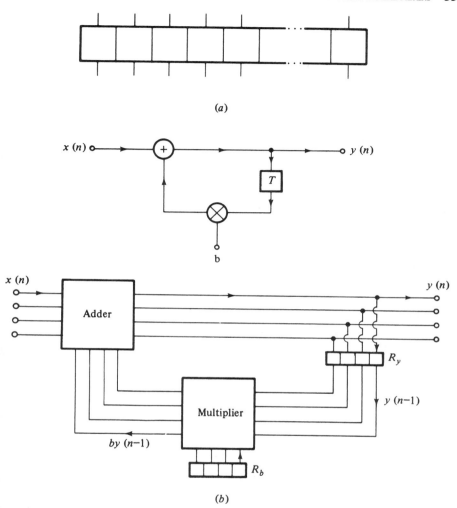

FIGURE 11.1
(a) Register, (b) rudimentary digital-filter implementation.

which is then used to update register R_y. Subsequently, the multiplier is triggered into operation and the product $by(n-1)$ is formed. The cycle is repeated when a new input sample is received.

A filter implementation like that in Fig. 11.1b can assume many forms, depending on the type of machine arithmetic used. The arithmetic can be of the *fixed-point* or *floating-point* type and in each case various conventions can be used for the representation of negative numbers. The two types of arithmetic differ in the way numbers are stored in registers and in the way by which they are manipulated by the digital hardware.

11.2.2 Fixed-Point Arithmetic

In fixed-point arithmetic, the numbers are usually assumed to be proper fractions. Integers and mixed numbers are avoided because (1) the number of bits representing an integer cannot be reduced by rounding or truncation without destroying the number and (2) mixed numbers are more difficult to multiply. For these reasons, the binary point is usually set between the first and second bit positions in the register, as depicted in Fig. 11.2a. The first position is reserved for the sign of the number.

Depending on the representation of negative numbers, fixed-point arithmetic can assume three forms:

1. Signed magnitude
2. One's complement
3. Two's complement

In the *signed-magnitude* arithmetic a fractional number

$$N = \pm 0.b_{-1}b_{-2}\cdots b_{-m}$$

is represented as

$$N_{sm} = \begin{cases} 0.b_{-1}b_{-2}\cdots b_{-m} & \text{for } N \geq 0 \\ 1.b_{-1}b_{-2}\cdots b_{-m} & \text{for } N \leq 0 \end{cases}$$

The most significant bit is said to be the *sign bit*; e.g., if $N = +0.1101$ or -0.1001, then $N_{sm} = 0.1101$ or 1.1001.

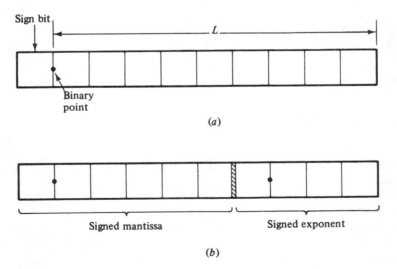

FIGURE 11.2
Storage of (a) fixed-point numbers, (b) floating-point numbers.

The *one's-complement* representation of a number N is defined as

$$N_1 = \begin{cases} N & \text{for } N \geq 0 \\ 2 - 2^{-L} - |N| & \text{for } N \leq 0 \end{cases} \quad (11.3)$$

where L, referred to as the *word length*, is the number of bit locations in the register to the right of the binary point. The binary form of $2 - 2^{-L}$ is a string of 1s filling the $L + 1$ locations of the register. Thus, the one's complement of a negative number can be deduced by representing the number by $L + 1$ bits, including zeros if necessary, and then complementing (changing 0s into 1s and 1s into 0s) all bits; e.g., if $N = -0.11010$, then $N_1 = 1.00101$ for $L = 5$ and $N_1 = 1.00101111$ for $L = 8$.

The *two's-complement* representation is similar. We now have

$$N_2 = \begin{cases} N & \text{for } N \geq 0 \\ 2 - |N| & \text{for } N < 0 \end{cases}$$

The two's complement of a negative number can be formed by adding 1 at the least significant position of the one's complement. Similarly, a negative number can be recovered from its two's complement by complementing and then adding 1 at the least significant position.

The possible numbers that can be stored in a 4-bit register together with their decimal equivalents are listed in Table 11.1. Some peculiarities of the three systems are evident. The signed-magnitude and the one's complement systems have two representations for zero whereas the two's complement system has only one. On the other hand, -1 is represented in the two's complement system but not in the other two.

TABLE 11.1
Decimal equivalents of numbers 0.000 to 1.111

Binary number	Decimal equivalent (eighths)		
	Signed magnitude	One's complement	Two's complement
0.000	0	0	0
0.001	1	1	1
0.010	2	2	2
0.011	3	3	3
0.100	4	4	4
0.101	5	5	5
0.110	6	6	6
0.111	7	7	7
1.000	-0	-7	-8
1.001	-1	-6	-7
1.010	-2	-5	-6
1.011	-3	-4	-5
1.100	-4	-3	-4
1.101	-5	-2	-3
1.110	-6	-1	-2
1.111	-7	-0	-1

The merits and demerits of the three types of arithmetic can be envisaged by examining how arithmetic operations are performed in each case.

One's-complement addition of any two numbers is carried out by simply adding their one's complements bit by bit. A carry bit at the most significant position, if one is generated, is added at the least significant position (*end-around carry*). Two's-complement addition is exactly the same except that a carry bit at the most significant position is ignored. Signed-magnitude addition, on the other hand, is much more complicated as it involves sign checks as well as complementing and end-around carry [1].

In the one's- or two's-complement arithmetic, direct multiplication of the complements does not always yield the product, and as a consequence special algorithms must be employed. By contrast, signed-magnitude multiplication is accomplished by simply multiplying the magnitudes of the two numbers bit by bit and then adjusting the sign bit of the product.

Example 11.2. Form the sum $0.53125 + (-0.40625)$ using the one's- and two's-complement additions assuming a word length of 5 bits.

Solution

$$0.53125_{10} = 0.10001_2$$

$$0.40625_{10} = 0.01101_2$$

	One's complement	Two's complement
0.53125	0.10001	0.10001
−0.40625	1.10010	1.10011
0.12500	↓ 0.00011	1 ← 0.00100
	→ 1	
	0.00100	

An important feature of the one's- or two's-complement addition is that a machine-representable sum $S = n_1 + n_2 + \cdots + n_i + \cdots$ will always be evaluated correctly, even if overflow does occur in the evaluation of partial sums.

Example 11.3. Form the sum $\frac{7}{8} + \frac{4}{8} + (-\frac{6}{8})$ using the two's-complement addition. Assume $L = 3$.

Solution. From Table 11.1

	7/8	0.111	
+	4/8	0.100	
	11/8	1.011	incorrect partial sum
−	6/8 +	1.010	
	5/8	0.101	correct sum

11.2.3 Floating-Point Arithmetic

There are two basic disadvantages in a fixed-point arithmetic: (1) The range of numbers that can be handled is small; e.g., in the two's-complement representation the smallest number is -1 and the largest is $1 - 2^{-L}$. (2) The percentage error produced by truncation or rounding tends to increase as the magnitude of the number is decreased. For example, if numbers 0.11011010 and 0.000110101 are both truncated such that only 4 bits are retained to the right of the binary point, the respective errors will be 4.59 and 39.6 percent.

These problems can be alleviated to a large extent by using a floating-point arithmetic. In this type of arithmetic, a number N is expressed as

$$N = M \times 2^e \tag{11.4}$$

where e is an integer and

$$\frac{1}{2} \le M < 1$$

M and e are referred to as the *mantissa* and *exponent*, respectively. For example, numbers 0.00110101 and 1001.11 are represented by 0.110101×2^{-2} and 0.100111×2^4, respectively. Negative numbers are handled in the same way as in fixed-point arithmetic.

Floating-point numbers are stored in registers, as depicted in Fig. 11.2b. The register is subdivided into two segments, one for the signed mantissa and one for the signed exponent.

Floating-point addition is carried out by shifting the mantissa of the smaller number to the right and increasing the exponent until the exponents of the two numbers are equal. The mantissas are then added to form the sum, which is subsequently put back into the normalized representation of Eq. (11.4). Multiplication is accomplished by multiplying mantissas, adding exponents, and then readjusting the product.

Floating-point arithmetic, as implied above, leads to increased dynamic range and improved precision of processing. Unfortunately, it also leads to increased cost of hardware and to reduced speed of processing. The reason is that hardware is in a sense duplicated since both the mantissa and exponent have to be manipulated. For software non-real-time implementations on general purpose digital computers, floating-point arithmetic is always preferred since neither the cost of hardware nor the speed of processing is a significant factor.

11.2.4 Number Quantization

Once the register length in a fixed-point implementation is assigned, the set of machine representable numbers is fixed. If the word length is L bits (excluding the sign bit), the smallest number variation that can be represented is a 1 at the least significant register position, which corresponds to 2^{-L}. Therefore, any number consisting of B bits (excluding the sign bit), where $B > L$, must be quantized. This can be accomplished (1) by *truncating* all bits that cannot be accommodated in the register, and (2) by *rounding* the number to the nearest machine-representable number.

Obviously, if a number x is quantized, an error ε will be introduced given by

$$\varepsilon = x - Q[x] \tag{11.5}$$

where $Q[x]$ denotes the quantized value of x. The range of ε tends to depend on the type of number representation and also on the type of quantization. Let us examine the various possibilities, starting with truncation.

As can be seen in Table 11.1, the representation of positive numbers is identical in all three fixed-point representations. Since truncation can only reduce a positive number, ε is positive. Its maximum value occurs when all disregarded bits are 1s, in which case

$$0 \leq \varepsilon_T \leq 2^{-L} - 2^{-B} \quad \text{for } x \geq 0$$

For negative numbers the three representations must be considered individually. For the signed-magnitude representation, truncation will decrease the magnitude of the number or increase its signed value, and hence $Q[x] > x$ or

$$-(2^{-L} - 2^{-B}) \leq \varepsilon_T \leq 0 \quad \text{for } x < 0$$

The one's-complement representation of a negative number

$$x = -\sum_{i=1}^{B} b_{-i} 2^{-i} \tag{11.6}$$

(where $b_{-i} = 0$ or 1) is obtained from Eq. (11.3) as

$$x_1 = 2 - 2^{-L} - \sum_{i=1}^{B} b_{-i} 2^{-i}$$

If all the disregarded bits are 0s, obviously $\varepsilon = 0$. At the other extreme if all the disregarded bits are 1s, we have

$$Q[x_1] = 2 - 2^{-L} - \sum_{i=1}^{B} b_{-i} 2^{-i} - (2^{-L} - 2^{-B})$$

Consequently, the decimal equivalent of $Q[x_1]$ is

$$Q[x] = -\left[\sum_{i=1}^{B} b_{-i} 2^{-i} + (2^{-L} - 2^{-B})\right] \tag{11.7}$$

and, therefore, from Eqs. (11.5) to (11.7)

$$0 \leq \varepsilon_T \leq 2^{-L} - 2^{-B} \quad \text{for } x < 0$$

The same inequality holds for two's-complement numbers, as can easily be shown. In summary, for signed-magnitude numbers

$$-q < \varepsilon_T < q$$

where $q = 2^{-L}$ is the quantization step, whereas for one's- or two's-complement numbers

$$0 \le \varepsilon_T < q$$

Evidently, quantization errors can be kept as low as desired by using a sufficiently large value of L.

For rounding, the quantization error can be positive as well as negative by definition, and its maximum value is $q/2$. If numbers lying halfway between quantization levels are rounded up, we have

$$-\frac{q}{2} \le \varepsilon_R < \frac{q}{2} \tag{11.8}$$

Rounding can be effected, in practice, by adding 1 at position $L+1$ and then truncating the number to L bits.

A convenient way of visualizing the process of quantization is to imagine a quantizer with input x and output $Q[x]$. Depending on the type of quantization, the transfer characteristic of the device can assume one of the forms illustrated in Fig. 11.3.

The range of quantization error in floating-point arithmetic can be evaluated by a similar approach.

11.3 COEFFICIENT QUANTIZATION

Coefficient-quantization errors introduce perturbations in the zeros and poles of the transfer function, which in turn manifest themselves as errors in the frequency response. Product-quantization errors, on the other hand, can be regarded as noise sources which give rise to output roundoff noise. Since the importance of the two types of errors can vary considerably from application to application, it is frequently advantageous to use different word lengths for the coefficient and signal values. The coefficient word length can be chosen to satisfy prescribed frequency-response specifications, whereas the signal word length can be chosen to satisfy a signal-to-noise ratio specification.

Consider a digital filter characterized by $H(z)$ and let

$M(\omega) = |H(e^{j\omega T})|$ = amplitude response without quantization

$M_Q(\omega)$ = amplitude response with quantization

$M_I(\omega)$ = ideal amplitude response

δ_p (δ_a) = passband (stopband) tolerance on amplitude response

These quantities are illustrated in Fig. 11.4.

The effect of coefficient quantization is to introduce an error ΔM in $M(\omega)$ given by

$$\Delta M = M(\omega) - M_Q(\omega)$$

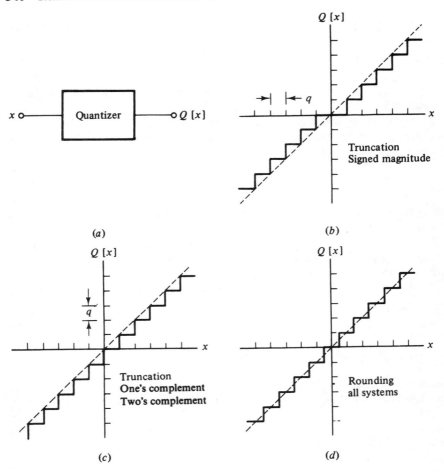

FIGURE 11.3
Number quantization: (a) quantizer, (b) to (d) $Q[x]$ versus x.

The maximum permissible value of $|\Delta M|$, denoted by $\Delta M_{\max}(\omega)$, can be deduced from Fig. 11.4 as

$$\Delta M_{\max}(\omega) = \begin{cases} \delta_p - |M(\omega) - M_I(\omega)| & \text{for } \omega \leq \omega_p \\ \delta_a - |M(\omega) - M_I(\omega)| & \text{for } \omega \geq \omega_a \end{cases}$$

and if

$$|\Delta M| \leq \Delta M_{\max}(\omega) \tag{11.9}$$

for $0 \leq \omega \leq \omega_p$ and $\omega_a \leq \omega \leq \omega_s/2$, the desired specification will be met. The *optimum* word length can thus be determined exactly by evaluating $|\Delta M|$ as a function of frequency for successively larger values of the word length until Eq. (11.9) is

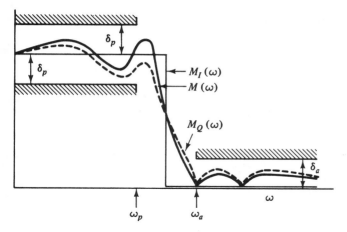

FIGURE 11.4
Coefficient quantization.

satisfied. Evidently, this is a trial-and-error approach and may entail considerable computation.

An alternative approach is to employ a statistical method proposed by Avenhaus [2] and later modified by Crochiere [3]. This method yields a fairly accurate estimate of the required word length and is, in general, more efficient than the exact method outlined above. Its details follow.

Consider a fixed-point implementation and assume that quantization is carried out by rounding. From Eq. (11.8) the error in coefficient c_i ($i = 1, 2, \cdots, m$), denoted by Δc_i, can assume any value in the range $-q/2$ to $+q/2$; that is, Δc_i is a random variable. If the probability density of Δc_i is assumed to be uniform, i.e.,

$$p(\Delta c_i) = \begin{cases} \frac{1}{q} & \text{for } -\frac{q}{2} \leq \Delta c_i \leq \frac{q}{2} \\ 0 & \text{otherwise} \end{cases}$$

then from Eqs. (10.8) and (10.10)

$$E\{\Delta c_i\} = 0 \tag{11.10}$$

$$\sigma^2_{\Delta c_i} = \frac{q^2}{12} \tag{11.11}$$

The variation ΔM in $M(\omega)$ is also a random variable. By virtue of Taylor's theorem we can write

$$\Delta M = \sum_{i=1}^{m} \Delta c_i S^M_{c_i}$$

where

$$S^M_{c_i} = \frac{\partial M(\omega)}{\partial c_i}$$

is the sensitivity of the amplitude response $M(\omega)$ with respect to variations in coefficient c_i. Evidently,

$$E\{\Delta M\} = \sum_{i=1}^{m} S_{c_i}^{M} E\{\Delta c_i\} = 0$$

according to Eq. (11.10). If Δc_i and Δc_j ($i \neq j$) are assumed to be statistically independent, then from Eq. (10.7)

$$\sigma_{\Delta M}^2 = \sum_{i=1}^{m} \sigma_{c_i}^2 (S_{c_i}^{M})^2$$

and, therefore, from Eq. (11.11)

$$\sigma_{\Delta M}^2 = \frac{q^2 S_T^2}{12} \tag{11.12}$$

where

$$S_T^2 = \sum_{i=1}^{m} (S_{c_i}^{M})^2 \tag{11.13}$$

For a large value of m, ΔM is approximately gaussian by virtue of the *central-limit theorem* [4], and since $E\{\Delta M\} = 0$, Eq. (10.1) gives

$$p(\Delta M) = \frac{1}{\sigma_{\Delta M} \sqrt{2\pi}} e^{-\Delta M^2 / 2\sigma_{\Delta M}^2} \qquad -\infty \leq \Delta M \leq \infty$$

Consequently, ΔM will be in some range $-\Delta M_1 \leq \Delta M \leq \Delta M_1$ with a probability y given by

$$y = \Pr[|\Delta M| \leq \Delta M_1] = \frac{2}{\sigma_{\Delta M} \sqrt{2\pi}} \int_0^{\Delta M_1} e^{-\Delta M^2 / 2\sigma_{\Delta M}^2} d(\Delta M) \tag{11.14}$$

With the variable transformation

$$\Delta M = x \sigma_{\Delta M} \qquad \Delta M_1 = x_1 \sigma_{\Delta M} \tag{11.15}$$

Eq. (11.14) can be put in the standard form

$$y = \frac{2}{\sqrt{2\pi}} \int_0^{x_1} e^{-x^2/2} \, dx$$

Once an acceptable confidence factor y is selected, the corresponding value of x_1 can be obtained from published tables or by using a numerical method. The quantity ΔM_1 is essentially a statistical bound on ΔM, and if the word length is chosen such that

$$\Delta M_1 \leq \Delta M_{\max}(\omega) \tag{11.16}$$

the desired specifications will be satisfied to within a confidence factor y. The resulting word length can be referred to as the *statistical word length*. A statistical bound on the quantization step can be deduced from Eqs. (11.12), (11.15), and (11.16) as

$$q \leq \frac{\sqrt{12} \Delta M_{\max}(\omega)}{x_1 S_T} \tag{11.17}$$

The register length should be sufficiently large to accommodate the quantized value of the largest coefficient; so let

$$Q[\max c_i] = \sum_{i=-K}^{J} b_i 2^i$$

where b_J and $b_{-K} \neq 0$. The required word length must be

$$L = 1 + J + K \qquad (11.18)$$

and since $q = 2^{-K}$ or

$$K = \log_2 \frac{1}{q} \qquad (11.19)$$

Eqs. (11.17) to (11.19) give the desired result as

$$L \geq L(\omega) = 1 + J + \log_2 \frac{x_1 S_T}{\sqrt{12} \Delta M_{\max}(\omega)}$$

A reasonable agreement between the statistical and exact word lengths is achieved by using $x_1 = 2$ [3, 5]. This value of x_1 corresponds to a confidence factor of 0.95.

The sensitivities $S_{c_i}^M$ in Eq. (11.13) can be computed efficiently by using the transpose or adjoint approach described in Sec. 4.10.5. This approach gives

$$S_{c_i}^H(e^{j\omega T}) = \frac{\partial H(e^{j\omega T})}{\partial c_i} = \text{Re}\,[S_{c_i}^H(e^{j\omega T})] + j\,\text{Im}\,[S_{c_i}^H(e^{j\omega T})]$$

and if

$$H(e^{j\omega T}) = M(\omega) e^{j\theta(\omega)}$$

we can show that

$$\text{Re}\,[S_{c_i}^H(e^{j\omega T})] = [\cos\theta(\omega)]\frac{\partial M(\omega)}{\partial c_i} - M(\omega)[\sin\theta(\omega)]\frac{\partial \theta(\omega)}{\partial c_i}$$

$$\text{Im}\,[S_{c_i}^H(e^{j\omega T})] = [\sin\theta(\omega)]\frac{\partial M(\omega)}{\partial c_i} + M(\omega)[\cos\theta(\omega)]\frac{\partial \theta(\omega)}{\partial c_i}$$

Therefore

$$S_{c_i}^M = \frac{\partial M(\omega)}{\partial c_i} = [\cos\theta(\omega)]\,\text{Re}\,S_{c_i}^H(e^{j\omega T}) + [\sin\theta(\omega)]\,\text{Im}\,S_{c_i}^H(e^{j\omega T})$$

and

$$S_{c_i}^\theta = \frac{\partial \theta(\omega)}{\partial c_i} = \frac{1}{M(\omega)}\{[\cos\theta(\omega)]\,\text{Im}\,S_{c_i}^H(e^{j\omega T}) - [\sin\theta(\omega)]\,\text{Re}\,S_{c_i}^H(e^{j\omega T})\}$$

where $S_{c_i}^\theta$ is the sensitivity of the phase response $\theta(\omega)$ with respect to coefficient c_i.

The statistical word length is a convenient figure of merit of a specific filter structure. It can serve as a sensitivity measure in studies where a general comparison of various structures is desired. It can also be used as an objective function in word-length optimization algorithms [3].

A different approach for the study of quantization effects was proposed by Jenkins and Leon [6]. In this approach a computer-aided analysis scheme is used to generate confidence-interval error bounds on the time-domain response of the filter. The method can be used to study the effects of coefficient or product quantization in fixed-point or floating-point implementations. Furthermore, the quantization can be by rounding or truncation.

11.4 LOW-SENSITIVITY STRUCTURES

The effects of coefficient quantization are most serious in applications where the poles of the transfer function are located close to the unit circle $|z| = 1$. In such applications, small changes in the coefficients can cause large changes in the frequency response of the filter, and in extreme cases they can actually cause the filter to become unstable. In this section, we show that second-order structures can be derived whose sensitivity to coefficient quantization is much lower than that of the standard direct realizations described in Chap. 4. These structures can be used in the cascade or parallel realizations for the design of high-selectivity or narrow-band filters.

Let $M(\omega)$ be the amplitude response of a digital-filter structure and assume that b is a multiplier constant. Now let $\Delta M(\omega)$ be the change in $M(\omega)$ due to a quantization error Δb in b. The *normalized sensitivity* of $M(\omega)$ with respect to b is defined as

$$\bar{S}_b^M = \lim_{\Delta b \to 0} \frac{\frac{\Delta M(\omega)}{M(\omega)}}{\frac{\Delta b}{b}}$$

$$= \frac{b}{M(\omega)} \frac{\partial M(\omega)}{\partial b} \qquad (11.20)$$

and for small values of Δb, we have

$$\frac{\Delta M(\omega)}{M(\omega)} \approx \frac{\Delta b}{b} \bar{S}_b^M \qquad (11.21)$$

The normalized sensitivity can be used to compare different structures.

Consider the direct realization of Fig. 11.5a. Straightforward analysis gives the transfer function

$$H(z) = \frac{1}{z^2 + b_1 z + b_0}$$

and hence the amplitude response of the realization can be readily obtained as

$$M(\omega) = \frac{1}{[1 + b_0^2 + b_1^2 + 2b_1(1 + b_0)\cos \omega T + 2b_0 \cos 2\omega T]^{1/2}} \qquad (11.22)$$

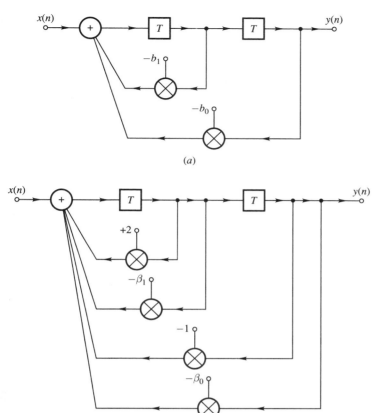

FIGURE 11.5
(a) Second-order direct realization, (b) corresponding low-sensitivity realization.

Using Eqs. (11.20) and (11.22), the normalized sensitivities of $M(\omega)$ with respect to b_0 and b_1 can be obtained as

$$\bar{S}_{b_0}^M = -b_0(b_0 + b_1 \cos \omega T + \cos 2\omega T)[M(\omega)]^2$$

$$\bar{S}_{b_1}^M = -b_1[b_1 + (1 + b_0) \cos \omega T][M(\omega)]^2$$

A modified version of the structure in Fig. 11.5a can be obtained by replacing each of the multipliers by two multipliers in parallel, as shown in Fig. 11.5b, as suggested by Agarwal and Burrus [7]. The transfer function of the original structure will be maintained in the new structure if

$$b_0 = 1 + \beta_0 \quad \text{and} \quad b_1 = \beta_1 - 2$$

and from Eq. (11.20)

$$\bar{S}^M_{\beta_0} = \frac{\beta_0}{M(\omega)} \frac{\partial M(\omega)}{\partial \beta_0} = \frac{\beta_0}{b_0} \frac{\partial b_0}{\partial \beta_0} \times \frac{b_0}{M(\omega)} \frac{\partial M(\omega)}{\partial b_0}$$

$$= \frac{\beta_0}{1+\beta_0} \bar{S}^M_{b_0} \tag{11.23}$$

and

$$\bar{S}^M_{\beta_1} = \frac{\beta_1}{M(\omega)} \frac{\partial M(\omega)}{\partial \beta_1} = \frac{\beta_1}{b_1} \frac{\partial b_1}{\partial \beta_1} \times \frac{b_1}{M(\omega)} \frac{\partial M(\omega)}{\partial b_1}$$

$$= \frac{\beta_1}{2+\beta_1} \bar{S}^M_{b_1} \tag{11.24}$$

Now if the poles of the transfer function are located close to the point $z = 1$, as may be the case in a narrow-band lowpass filter of high selectivity, then $b_0 \approx 1$ and $b_1 \approx -2$. As a consequence, β_0 and β_1 will be small and, therefore, from Eqs. (11.23) and (11.24)

$$\bar{S}^M_{\beta_0} \ll \bar{S}^M_{b_0} \quad \text{and} \quad \bar{S}^M_{\beta_1} \ll \bar{S}^M_{b_1}$$

In effect, if coefficients β_0 and β_1 are represented to the same degree of precision as coefficients b_0 and b_1, then the use of the structure in Fig. 11.5b instead of that in Fig. 11.5a leads to a significant reduction in the sensitivity to quantization errors, as can be seen from Eq. (11.21). The same degree of precision in the representation of the coefficients can be achieved by using either floating-point or fixed-point arithmetic. In the latter case, each multiplier coefficient should be scaled up to eliminate any zeros between the binary point and the most significant nonzero bit and the product scaled down by a corresponding shift operation.

The structure of Fig. 11.5b, like other structures in which all the outputs of multipliers are inputs to one and the same adder, has the advantage that the quantization of products can be carried out using one quantizer at the output of the adder instead of one quantizer at the output of each multiplier. Structures of this type are suitable for the application of *error-spectrum shaping*, which is a technique for the reduction of roundoff noise (see Sec. 11.8).

The disadvantage of the structure of Fig. 11.5b is that the low-sensitivity property can be achieved only if the poles of the transfer function are close to point $z = 1$. A family of structures that are suitable for the application of error-spectrum shaping and simultaneously lead to low sensitivity for a variety of pole locations close to the unit circle $|z| = 1$ can be obtained from the general second-order configuration depicted in Fig. 11.6 by using a method reported by Diniz and Antoniou [8]. In this configuration, branches A, B, C, D, and E represent unit delays or machine-representable multiplier constants, e.g., 0, ± 1, or ± 2.

The structure of Fig. 11.6 realizes the transfer function

$$H(z) = \frac{N(z)}{D(z)} \tag{11.25}$$

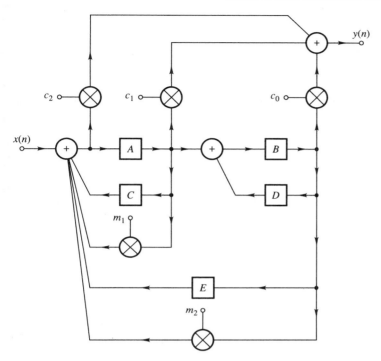

FIGURE 11.6
General second-order direct realization.

where $N(z)$ depends on the choice of multiplier coefficients c_0 to c_2 and

$$D(z) = z^2(1 - BD - AC - m_1 A + ABE \\
+ m_2 AB + ABCD + m_1 ABD) \tag{11.26}$$

Assuming that $H(z)$ is of the form

$$H(z) = \frac{a_2 z^2 + a_1 z + a_0}{z^2 + b_1 z + b_0} \tag{11.27}$$

and then comparing Eq. (11.25) with Eq. (11.27), a number of second-order structures can be deduced. In order to avoid delay-free loops and keep the number of delays to the minimum of two, the constraints

$$A = z^{-1} \quad \text{and} \quad B \quad \text{or} \quad D = z^{-1}$$

must be satisfied. Therefore, two cases are possible, namely, Case I where $A = B = z^{-1}$ and Case II where $A = D = z^{-1}$.

11.4.1 Case I

For Case I, polynomial $D(z)$ of Eq. (11.26) assumes the form

$$D(z) = z^2 - z(C + D + m_1) + CD + m_1 D + m_2 + E$$

and in order to achieve low sensitivity, multipliers C, D, and E must be chosen as

$$C + D = I_R[-b_1] \quad \text{and} \quad E = I_R[b_0 + b_1 D + D^2] \quad (11.28)$$

where $I_R[x]$ is the closest integer to x. Equation (11.28) forces the values of m_1 and m_2 to be as low as possible and, as in the structure of Fig. 11.5b, low sensitivity is assured.

If the poles are close to point $z = 1$, then $b_1 \approx -2$ and $b_0 \approx 1$, and so

$$C + D = 2$$

We can thus assign

$$C = 1, \quad D = 1, \quad \text{and} \quad E = 0$$

This choice of coefficients gives the structure of Fig. 11.5b, which is suitable for values of b_1 in the range $-2.0 < b_1 < -1.5$. Proceeding in the same way, the 15 structures in Table 11.2 can be deduced [8]. Structure I-2, like I-1, was reported in [7].

TABLE 11.2
Structures for Case I

Structure	C	D	E	Range of b_1
I-1	1	1	0	$-2.0 < b_1 < -1.5$
I-2	2	0	1	
I-3	0	2	1	$-2.0 < b_1 < -1.75$
I-4	0	2	2	$-1.75 < b_1 < -1.5$
I-5	1	0	1	$-1.5 < b_1 < -0.5$
I-6	0	1	1	
I-7	0	0	1	
I-8	-1	1	2	$-0.5 < b_1 < 0.5$
I-9	1	-1	2	
I-10	-1	0	1	$0.5 < b_1 < 1.5$
I-11	0	-1	1	
I-12	0	-2	2	$1.5 < b_1 < 1.75$
I-13	-2	0	1	$1.5 < b_1 < 2.0$
I-14	-1	-1	0	
I-15	0	-2	1	$1.75 < b_1 < 2.0$

11.4.2 Case II

For Case II, polynomial $D(z)$ of Eq. (11.26) assumes the form

$$D(z) = z^2 - z(B + C + m_1 - m_2 B - BE) + BC + m_1 B$$

and in order to achieve low sensitivity, constants $B, C,$ and E must be chosen as

$$B = 1, \quad C = 1, \quad \text{and} \quad E = I_R[b_1 + b_0 + 1] \quad (11.29)$$

for poles with positive real part, and

$$B = -1, \quad C = -1, \quad \text{and} \quad E = -I_R[b_1 - b_0 - 1] \quad (11.30)$$

for poles with negative real part. Using Eqs. (11.29) and (11.30), the structures of Table 11.3 can be deduced [8]. Structure II-1 was reported by Nishimura, Hirano, and Pal [9].

Different biquadratic transfer functions can be realized by using the formulas in Table 11.4.

In the above approach, the poles of the transfer function have been assumed to be close to the unit circle of the z plane. An alternative approach for selecting the

TABLE 11.3
Structures for Case II

Structure	B	C	E	Range of b_1
II-1	1	1	0	$-2.0 < b_1 < -1.5$
II-2	1	1	1	$-1.5 < b_1 < -0.5$
II-3	1	1	2	$-0.5 < b_1 < 0$
II-4	-1	-1	2	$0 < b_1 < 0.5$
II-5	-1	-1	1	$0.5 < b_1 < 1.5$
II-6	-1	-1	0	$1.5 < b_1 < 2.0$

TABLE 11.4
Realization of biquadratic transfer functions

Multiplier constant	Case I	Case II
c_0	$a_0 + a_1 D + a_2 D^2$	$a_2 + \dfrac{a_1}{B} + \dfrac{a_0}{B^2}$
c_1	$a_1 + a_2 D$	$-\dfrac{a_0}{B}$
c_2	a_2	a_2
m_1	$-b_1 - C - D$	$\dfrac{b_0}{B} - C$
m_2	$b_0 + b_1 D + D^2 - E$	$1 + \dfrac{b_1}{B} + \dfrac{b_0}{B^2} - E$

optimum structure for a given transfer function, which is applicable for any pair of poles in the unit circle, was described by Ramana Rao and Eswaran [10].

11.5 PRODUCT QUANTIZATION

The output of a finite-word-length multiplier can be expressed as

$$Q[c_i x(n)] = c_i x(n) + e(n)$$

where $c_i x(n)$ and $e(n)$ are the exact product and quantization error, respectively. A machine multiplier can thus be represented by the model depicted in Fig. 11.7a, where $e(n)$ is a noise source.

Consider the filter structure of Fig. 11.7b and assume a fixed-point implementation. Each multiplier can be replaced by the model of Fig. 11.7a, as in Fig. 11.7c. If product quantization is carried out by rounding, each noise signal $e_i(n)$ can be

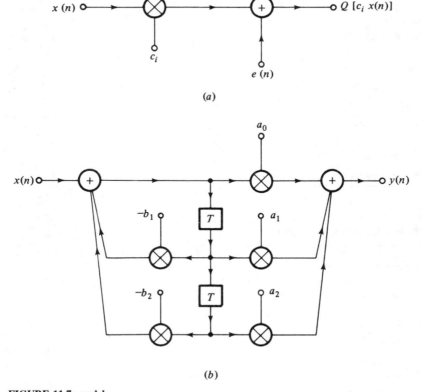

FIGURE 11.7a and b
Product quantization: (a) noise model for a multiplier, (b) second-order canonic section.

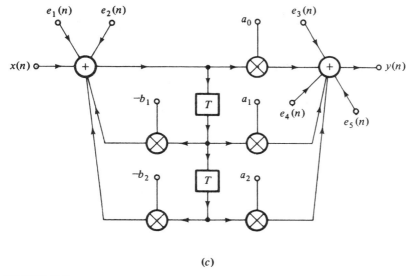

FIGURE 11.7c
Product quantization: (c) noise model for a second-order canonic section.

regarded as a random process with uniform probability density, i.e.,

$$p(e_i; n) = \begin{cases} \frac{1}{q} & \text{for } -\frac{q}{2} \le e_i(n) \le \frac{q}{2} \\ 0 & \text{otherwise} \end{cases}$$

Hence from Eqs. (10.8) and (10.9) and Sec. 10.8, we have

$$E\{e_i(n)\} = 0 \tag{11.31}$$

$$E\{e_i^2(n)\} = \frac{q^2}{12} \tag{11.32}$$

$$r_{e_i}(k) = E\{e_i(n)e_i(n+k)\} \tag{11.33}$$

If the signal levels throughout the filter are much larger than q, the following reasonable assumptions can be made: (1) $e_i(n)$ and $e_i(n+k)$ are statistically independent for any value of n ($k \neq 0$), and (2) $e_i(n)$ and $e_j(n+k)$ are statistically independent for any value of n or k ($i \neq j$). Let us examine the implications of these assumptions starting with the first assumption. From Eqs. (11.31) to (11.33) and (10.4)

$$r_{e_i}(0) = E\{e_i^2(n)\} = \frac{q^2}{12}$$

and

$$r_{e_i}(k)\Big|_{k \neq 0} = E\{e_i(n)\}E\{e_i(n+k)\} = 0$$

i.e.,

$$r_{e_i}(k) = \frac{q^2}{12}\delta(k)$$

where $\delta(k)$ is the impulse function. Therefore, the power spectral density (PSD) of $e_i(n)$ is

$$S_{e_i}(z) = \mathcal{Z} r_{e_i}(k) = \frac{q^2}{12} \tag{11.34}$$

that is, $e_i(n)$ *is a white-noise process.*

Let us now consider the implications of the second assumption. The autocorrelation of sum $e_i(n) + e_j(n)$ is

$$\begin{aligned} r_{e_i+e_j}(k) &= E\{[e_i(n) + e_j(n)][e_i(n+k) + e_j(n+k)]\} \\ &= E\{e_i(n)e_i(n+k)\} + E\{e_i(n)\}E\{e_j(n+k)\} \\ &\quad + E\{e_j(n)\}E\{e_i(n+k)\} + E\{e_j(n)e_j(n+k)\} \end{aligned}$$

or $\qquad r_{e_i+e_j}(k) = r_{e_i}(k) + r_{e_j}(k)$

Therefore

$$S_{e_i+e_j}(z) = \mathcal{Z}[r_{e_i}(k) + r_{e_j}(k)] = S_{e_i}(z) + S_{e_j}(z)$$

i.e., *the PSD of a sum of two statistically independent processes is equal to the sum of their respective PSDs.* In effect, superposition can be employed.

Now from Fig. 11.7c and Eq. (10.22)

$$S_y(z) = H(z)H(z^{-1}) \sum_{i=1}^{2} S_{e_i}(z) + \sum_{i=3}^{5} S_{e_i}(z)$$

where $H(z)$ is the transfer function of the filter, and hence from Eq. (11.34) the output PSD is given by

$$S_y(z) = \frac{q^2}{6} H(z)H(z^{-1}) + \frac{q^2}{4}$$

The above approach is applicable to any filter structure. Furthermore, it can be used to study the effects of input quantization.

11.6 SIGNAL SCALING

If the amplitude of any internal signal in a fixed-point implementation is allowed to exceed the dynamic range, *overflow* will occur and the output signal will be severely distorted. On the other hand, if all the signal amplitudes throughout the filter are unduly low, the filter will be operating inefficiently and *the signal-to-noise ratio will be poor.* Therefore, for optimum filter performance suitable *signal scaling* must be employed to adjust the various signal levels.

A scaling technique applicable to one's- or two's-complement implementations was proposed by Jackson [11]. In this technique a scaling multiplier is used at the input of a filter section, as in Fig. 11.8, with its constant λ chosen such that amplitudes of multiplier inputs are bounded by M if $|x(n)| \leq M$. Under these circumstances, adder outputs are also bounded by M and cannot overflow. This is due to the fact

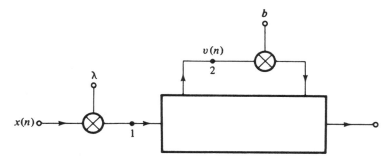

FIGURE 11.8
Signal scaling.

that a machine-representable sum is always evaluated correctly in one's- or two's-complement arithmetic, even if overflow does occur in one of the partial sums (see Example 11.3). There are two methods for the determination of λ, as follows.

11.6.1 Method A

Consider the filter section of Fig. 11.8, where $v(n)$ is a multiplier input. The transfer function between nodes 1 and 2 can be denoted by $F(z)$. From the convolution summation

$$v(n) = \sum_{k=0}^{\infty} \lambda f(k) x(n-k) \qquad (11.35)$$

where

$$f(n) = \mathcal{Z}^{-1} F(z)$$

Evidently

$$|v(n)| \le \sum_{k=0}^{\infty} |\lambda f(k)| \cdot |x(n-k)|$$

and if

$$|x(n)| \le M$$

then

$$|v(n)| \le M \sum_{k=0}^{\infty} |\lambda f(k)|$$

Thus a sufficient condition for $|v(n)| \le M$ is

$$\sum_{k=0}^{\infty} |\lambda f(k)| \le 1$$

or

$$\lambda \le \frac{1}{\sum_{k=0}^{\infty} |f(k)|} \qquad (11.36)$$

Now consider the specific signal

$$x(n-k) = \begin{cases} M & \text{for } \lambda f(k) > 0 \\ -M & \text{for } \lambda f(k) < 0 \end{cases}$$

where $M > 0$. From Eq. (11.35)

$$v(n) = M \sum_{k=0}^{\infty} |\lambda f(k)|$$

and therefore $|v(n)| \leq M$ if and only if Eq. (11.36) holds. Signal scaling can be applied by calculating the infinite sum of the magnitude of the impulse response from the input of the filter to the input of each multiplier and then evaluating λ using the largest sum so obtained in Eq. (11.36).

The above method guarantees that overflow will never occur as long as the input is bounded as prescribed. Unfortunately, the signal levels at the various nodes can be quite low and since quantization errors are independent of the signal level, a reduced signal-to-noise ratio may result. In addition, the computation of the sum in Eq. (11.36) is not usually straightforward.

11.6.2 Method B

The second and more efficient method for the evaluation of λ is based on L_p-*norm notation*. The L_p norm of an arbitrary periodic function $A(e^{j\omega T})$ with period ω_s is defined as

$$\|A\|_p = \left[\frac{1}{\omega_s} \int_0^{\omega_s} |A(e^{j\omega T})|^p d\omega \right]^{1/p}$$

where $p \geq 1$. It exists if

$$\int_0^{\omega_s} |A(e^{j\omega T})|^p d\omega < \infty$$

and if $A(e^{j\omega T})$ is continuous, then the limit

$$\lim_{p \to \infty} \|A\|_p = \|A\|_\infty = \max_{0 \leq \omega \leq \omega_s} |A(e^{j\omega T})| \qquad (11.37)$$

exists, as can be easily demonstrated (see Prob. 11.22). Usually, $A(e^{j\omega T})$ is obtained by evaluating function $A(z)$ on the unit circle $z = e^{j\omega T}$ and $\|A\|_p$ is often referred to as the L_p norm of either $A(e^{j\omega T})$ or $A(z)$.

Now let

$$X(z) = \sum_{n=-\infty}^{\infty} x(n) z^{-n} \qquad a < |z| < b$$

$$F(z) = \sum_{n=-\infty}^{\infty} f(n) z^{-n} \qquad c < |z| < b$$

where $c < 1$ for a stable filter and $b > 1$. From Eq. (11.35)
$$V(z) = \lambda F(z)X(z) \qquad d < |z| < b$$
where $d = \max(a, c)$. The inverse z transform of $V(z)$ is
$$v(n) = \frac{1}{2\pi j} \oint_\Gamma \lambda F(z)X(z)z^{n-1}\, dz \tag{11.38}$$
where Γ is a contour in the annulus of convergence. If $a < 1$, Γ can be taken to be the unit circle $|z| = 1$. With $z = e^{j\omega T}$ Eq. (11.38) becomes
$$v(n) = \frac{1}{\omega_s} \int_0^{\omega_s} \lambda F(e^{j\omega T}) X(e^{j\omega T}) e^{jn\omega T}\, d\omega$$
We can thus write
$$|v(n)| \le \left[\max_{0 \le \omega \le \omega_s} |X(e^{j\omega T})|\right] \frac{1}{\omega_s} \int_0^{\omega_s} |\lambda F(e^{j\omega T})|\, d\omega \tag{11.39}$$
or
$$|v(n)| \le \left[\max_{0 \le \omega \le \omega_s} |\lambda F(e^{j\omega T})|\right] \frac{1}{\omega_s} \int_0^{\omega_s} |X(e^{j\omega T})|\, d\omega \tag{11.40}$$
and by virtue of the *Schwarz inequality* [11]
$$|v(n)| \le \left[\frac{1}{\omega_s} \int_0^{\omega_s} |\lambda F(e^{j\omega T})|^2 d\omega\right]^{1/2} \left[\frac{1}{\omega_s} \int_0^{\omega_s} |X(e^{j\Omega T})|^2\, d\Omega\right]^{1/2} \tag{11.41}$$
When L_p-norm notation is used, Eqs. (11.39) to (11.41) can be put in the compact form
$$|v(n)| \le \|X\|_\infty \|\lambda F\|_1 \qquad |v(n)| \le \|X\|_1 \|\lambda F\|_\infty \qquad |v(n)| \le \|X\|_2 \|\lambda F\|_2$$
In fact these inequalities are particular cases of the *Holder inequality* [11, 12]
$$|v(n)| \le \|X\|_q \|\lambda F\|_p \tag{11.42}$$
where the relation
$$p = \frac{q}{q-1} \tag{11.43}$$
must hold.

Equation (11.42) is valid for any transfer function $\lambda F(z)$ including $\lambda F(z) = 1$, in which case $v(n) = x(n)$ and $\|1\|_p = 1$ for all $p \ge 1$. Consequently, from Eq. (11.42)
$$|x(n)| \le \|X\|_q \qquad \text{for all } q \ge 1$$
Now if
$$|x(n)| \le \|X\|_q \le M$$
Eq. (11.42) gives
$$|v(n)| \le M\|\lambda F\|_p$$
Therefore
$$|v(n)| \le M$$

provided that

$$\|\lambda F\|_p \leq 1$$

or
$$\lambda \leq \frac{1}{\|F\|_p} \quad \text{for } \|X\|_q \leq M \quad (11.44)$$

where Eq. (11.43) must hold.

11.6.3 Types of Scaling

Depending on the values of p and q, two types of scaling can be identified, namely, L_2 scaling if $p = q = 2$ and L_∞ scaling if $p = \infty$ and $q = 1$.

From the definition of the L_p norm and Eq. (11.37), we have

$$\|F\|_2 = \left[\frac{1}{\omega_s}\int_0^{\omega_s} |F(e^{j\omega T})|^2 d\omega\right]^{1/2} \leq \left\{\frac{1}{\omega_s}\int_0^{\omega_s}\left[\max_{0\leq\omega\leq\omega_s} |F(e^{j\omega T})|\right]^2 d\omega\right\}^{1/2}$$

$$\leq \left(\frac{1}{\omega_s}\int_0^{\omega_s} \|F\|_\infty^2 d\omega\right)^{1/2}$$

$$\leq \|F\|_\infty$$

or

$$\frac{1}{\|F\|_2} \geq \frac{1}{\|F\|_\infty}$$

As a consequence, L_2 scaling usually yields larger scaling constants than L_∞ scaling. This means that the signal levels at the various nodes are usually larger, and thus a better signal-to-noise ratio can be achieved. However, L_2 scaling is more likely to cause overflow than L_∞ scaling. The circumstances in which these two types of scaling are applicable are examined below.

If $x(n)$ is obtained by sampling a random or deterministic, finite-energy, band-limited, continuous-time signal $x(t)$ such that

$$X_A(j\omega) = \mathcal{F}x(t) = 0 \quad \text{for } |\omega| \geq \omega_s/2 \quad (11.45)$$

we can write

$$\|X\|_2 = \left[\frac{1}{\omega_s}\int_0^{\omega_s} |X(e^{j\omega T})|^2 d\omega\right]^{1/2}$$

$$= \left[\frac{1}{\omega_s}\int_{-\omega_s/2}^{\omega_s/2} |X(e^{j\omega T})|^2 d\omega\right]^{1/2}$$

where $X(z) = \mathcal{Z}x(n)$. From Eq. (6.36), we have

$$X(e^{j\omega T}) = \frac{1}{T}X_A(j\omega) \quad \text{for } |\omega| < \omega_s/2$$

and hence

$$\|X\|_2 = \left[\frac{1}{2\pi T}\int_{-\omega_s/2}^{\omega_s/2}|X_A(j\omega)|^2 d\omega\right]^{1/2}$$

$$= \left[\frac{1}{2\pi T}\int_{-\infty}^{\infty}|X_A(j\omega)|^2 d\omega\right]^{1/2}$$

On using Parseval's formula (see Theorem 6.9), we obtain

$$\|X\|_2 = \left[\frac{1}{T}\int_{-\infty}^{\infty}|x(t)|^2 dt\right]^{1/2} \tag{11.46}$$

For a finite-energy signal, the above integral converges. Therefore, Eq. (11.42) holds with $p = 2$ and $q = 2$, and L_2 scaling is applicable.

If $x(n)$ is obtained by sampling a continuous-time signal $x(t)$ whose energy content is not finite (e.g., a sinusoidal signal) the integral in Eq. (11.46) does not converge, $\|X\|_2$ does not exist, and L_2 scaling is not applicable; therefore, if such a signal is applied to a structure incorporating L_2 scaling, then signal overflow may occur. If $x(t)$ is bounded and bandlimited, Eq. (11.45) is satisfied, and hence we can write

$$\|X\|_1 = \frac{1}{\omega_s}\int_{-\omega_s/2}^{\omega_s/2}|X(e^{j\omega T})| d\omega$$

$$= \frac{1}{2\pi}\int_{-\omega_s/2}^{\omega_s/2}|X_A(j\omega)| d\omega \tag{11.47}$$

i.e., $\|X\|_1$ exists and Eq. (11.42) holds with $p = \infty$ and $q = 1$, and L_∞ scaling is applicable. The amplitude spectrum of $x(t)$ may become unbounded if $x(t)$ is a sinusoidal signal, in which case $X_A(j\omega)$ has poles on the $j\omega$ axis, or if $x(t)$ is constant, in which case $X_A(j\omega)$ is an impulse function. However, in both of these cases $\|X\|_1$ exists, as will now be demonstrated.

If $x(t) = M \cos \omega_0 nT$, where $0 \le \omega_0 \le \omega_s/2$, we have

$$X_A(j\omega) = \pi M[\delta(\omega - \omega_0) + \delta(\omega + \omega_0)]$$

and Eq. (11.47) gives

$$\|X\|_1 = \frac{1}{\omega_s}\int_{-\omega_s/2}^{\omega_s/2}|\pi M[\delta(\omega - \omega_0) + \delta(\omega + \omega_0)]| d\omega$$

$$= M$$

On the other hand, if $x(t) = M$, then

$$X_A(j\omega) = 2\pi M \delta(\omega)$$

and

$$\|X\|_1 = \frac{1}{2\pi} \int_{-\omega_s/2}^{\omega_s/2} |2\pi M \delta(\omega)| d\omega$$
$$= M$$

Therefore, if we select λ such that

$$\|\lambda F\|_\infty = \max_{0 \leq \omega \leq \omega_s} |\lambda F(e^{j\omega T})| \leq 1$$

then

$$v(n) \leq M$$

This result is to be expected in the case of a sinusoidal input. With the gain between the input and node 2 in Fig. 11.8 equal to or less than unity, the signal at node 2 will be a sinusoid with an amplitude equal to or less than M.

11.6.4 Application of Scaling

If there are m multipliers in the filter of Fig. 11.8, then $|v_i(n)| \leq M$ provided that

$$\lambda_i \leq \frac{1}{\|F_i\|_p}$$

for $i = 1, 2, \ldots, m$. Therefore, in order to ensure that all multiplier inputs are bounded by M we must assign

$$\lambda = \min(\lambda_1, \lambda_2, \ldots, \lambda_m)$$

or

$$\lambda = \frac{1}{\max(\|F_1\|_p, \|F_2\|_p, \ldots, \|F_m\|_p)} \quad (11.48)$$

In the case of parallel or cascade realizations, efficient scaling can be accomplished by using one scaling multiplier per section.

Example 11.4. Deduce the scaling formulation for the cascade filter of Fig. 11.9a assuming $p = \infty$ and $q = 1$.

Solution. The only critical signals are $y'_j(n)$ and $y_j(n)$ since the inputs of the feedback multipliers are delayed versions of $y'_j(n)$. The filter can be represented by the signal flow graph of Fig. 11.9b, where

$$F'_j(z) = \frac{z^2}{z^2 + b_{1j}z + b_{2j}} \qquad F_j(z) = \frac{(z+1)^2}{z^2 + b_{1j}z + b_{2j}}$$

By using Eq. (11.48) we obtain

$$\lambda_0 = \frac{1}{\max(\|F'_1\|_\infty, \|F_1\|_\infty)} \qquad \lambda_1 = \frac{1}{\lambda_0 \max(\|F_1 F'_2\|_\infty, \|F_1 F_2\|_\infty)}$$

$$\lambda_2 = \frac{1}{\lambda_0 \lambda_1 \max(\|F_1 F_2 F'_3\|_\infty, \|F_1 F_2 F_3\|_\infty)}$$

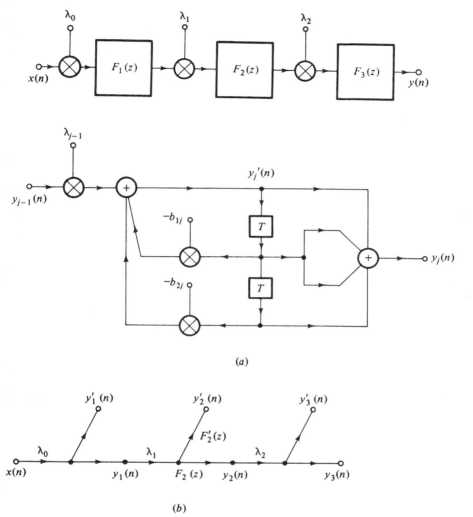

FIGURE 11.9
(a) Cascade filter, (b) signal flow-graph representation.

The scaling constants can be evaluated by noting that

$$\|\Pi F_i\|_\infty = \max_{0 \leq \omega \leq \omega_s} |\Pi F_i(e^{j\omega T})|$$

according to Eq. (11.37).

The scaling constants are usually chosen to be the *nearest powers of 2* satisfying the overflow constraints. In this way scaling multiplications can be reduced to simple data shifts.

In cascade filters, the *ordering of sections* has an influence on scaling, which in turn has an influence on the output noise. Analytical techniques for determining the optimum sequential ordering have not yet been devised. Nevertheless, some guidelines suggested by Jackson [13] lead to a good ordering.

11.7 MINIMIZATION OF OUTPUT ROUNDOFF NOISE

The level of output roundoff noise in fixed-point implementations can be reduced by increasing the word length. An alternative approach is to assume a general structure and vary its topology or parameters in such a way as to minimize the output roundoff noise. A method of this type that leads to *optimal* state-space structures was proposed by Mullis and Roberts [14]. The method is based on a state-space noise formulation reported by these authors and Hwang [15] at approximately the same time, and the principles involved are detailed below. The method is applicable to the general Nth-order realization but for the sake of simplicity it will be presented in terms of the second-order case.

A second-order state-space realization can be represented by the signal flow graph in Fig. 11.10 where $e_i(n)$ for $i = 1, 2,$ and 3 are noise sources due to the quantization of products. From Sec. 1.9.1, the filter can be represented by the equations

$$\mathbf{q}(n+1) = \mathbf{A}\mathbf{q}(\mathbf{n}) + \mathbf{b}x(n) + \mathbf{e}(n) \quad (11.49a)$$

$$y(n) = \mathbf{c}^T \mathbf{q}(\mathbf{n}) + dx(n) + e_3(n) \quad (11.49b)$$

where $\mathbf{e}^T(n) = [e_1(n)\, e_2(n)]$. Let $F_1(z)$, $F_2(z)$ and $G_1(z)$, $G_2(z)$ be the transfer functions from the input to nodes $q_1(n)$, $q_2(n)$ and from nodes $e_1(n)$, $e_2(n)$ to the output,

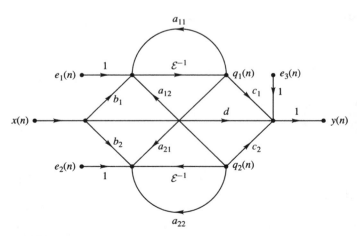

FIGURE 11.10
Second-order state-space realization.

respectively. In terms of this notation, the column vectors $\mathbf{f}(z)$ and $\mathbf{g}(z)$ can be formed as

$$\mathbf{f}^T(z) = [F_1(z)\ F_2(z)] \quad \text{and} \quad \mathbf{g}^T(z) = [G_1(z)\ G_2(z)] \tag{11.50}$$

and from Eq. (11.49), we obtain

$$\mathbf{f}(z) = (z\mathbf{I} - \mathbf{A})^{-1}\mathbf{b} \quad \text{and} \quad \mathbf{g}(z) = (z\mathbf{I} - \mathbf{A}^T)^{-1}\mathbf{c} \tag{11.51}$$

(see Prob. 11.23).

Now if the realization of Fig. 11.10 is represented by the set $\{\mathbf{A}, \mathbf{b}, \mathbf{c}^T, d\}$ and the state vector $\mathbf{q}(n)$ is subjected to a transformation of the form $\tilde{\mathbf{q}}(n) = \mathbf{T}\mathbf{q}(n)$, a new realization $\{\tilde{\mathbf{A}}, \tilde{\mathbf{b}}, \tilde{\mathbf{c}}^T, \tilde{d}\}$ is obtained where

$$\tilde{\mathbf{A}} = \mathbf{T}\mathbf{A}\mathbf{T}^{-1}, \quad \tilde{\mathbf{b}} = \mathbf{T}\mathbf{b}, \quad \tilde{\mathbf{c}}^T = \mathbf{c}^T\mathbf{T}^{-1}, \quad \tilde{d} = d \tag{11.52}$$

and from Eq. (11.49), one can show that

$$\tilde{\mathbf{f}}(z) = \mathbf{T}\mathbf{f}(z) \quad \text{and} \quad \tilde{\mathbf{g}}(z) = \mathbf{T}^{-1}\mathbf{g}(z) \tag{11.53}$$

(see Prob. 11.24). The realization $\{\tilde{\mathbf{A}}, \tilde{\mathbf{b}}, \tilde{\mathbf{c}}^T, \tilde{d}\}$ has minimum output roundoff noise subject to L_2-norm scaling if and only if

$$\tilde{\mathbf{W}} = \mathbf{D}\tilde{\mathbf{K}}\mathbf{D} \tag{11.54}$$

and

$$\tilde{K}_{ii}\tilde{W}_{ii} = \tilde{K}_{jj}\tilde{W}_{jj} \quad \text{for all } i, j \tag{11.55}$$

where $\mathbf{D}$ is a diagonal matrix and $\tilde{\mathbf{K}} = \{\tilde{K}_{ij}\}$ and $\tilde{\mathbf{W}} = \{\tilde{W}_{ij}\}$ are the matrices given by

$$\tilde{\mathbf{K}} = \frac{1}{2\pi j} \oint_\Gamma \tilde{\mathbf{f}}(z)\tilde{\mathbf{f}}^T(z^{-1})z^{-1}dz \tag{11.56}$$

and

$$\tilde{\mathbf{W}} = \frac{1}{2\pi j} \oint_\Gamma \tilde{\mathbf{g}}(z)\tilde{\mathbf{g}}^T(z^{-1})z^{-1}dz \tag{11.57}$$

respectively [14]. Matrices $\mathbf{K}$ and $\mathbf{W}$ are known as the *reachability* and *observability gramians*, respectively.

From Eq. (11.44), L_2 scaling can be applied by ensuring that

$$\|\tilde{F}_i\|_2 = 1 \quad \text{for all } i \tag{11.58}$$

and from Eqs. (11.56) and (11.58), we have

$$\tilde{K}_{ii} = \frac{1}{2\pi j} \oint_\Gamma \tilde{F}_i(z)\tilde{F}_i(z^{-1})z^{-1}dz$$

$$= \frac{1}{\omega_s} \int_0^{\omega_s} |\tilde{F}_i(e^{j\omega T})|^2 d\omega$$

$$= \|\tilde{F}_i\|_2^2 = 1 \tag{11.59}$$

Therefore, the condition for minimum output roundoff noise in Eq. (11.55) assumes the form

$$\tilde{W}_{ii} = \tilde{W}_{jj} \quad \text{for all } i, j \tag{11.60}$$

and from Eq. (11.57), we have

$$\|G_i\|_2^2 = \|G_j\|_2^2 \quad \text{for all } i, j$$

In effect, *the output noise is minimum if the individual contributions due to the different noise sources are all equal*, as may be expected.

The application of the above method to the Nth-order general state-space realization would require $N^2 + 2N + 1$ multipliers, as opposed to $2N + 1$ in parallel or cascade canonic structures. That is, the method is uneconomical. Recognizing this problem, Mullis and Roberts applied their methodology to obtain so-called *block-optimal* parallel and cascade structures that require only $4N + 1$ and $9N/2$ multipliers, respectively. Unfortunately, in both cases the realization process is relatively complicated; in addition, in the latter case the pairing of zeros and poles into biquadratic transfer functions and the ordering of second-order sections are not optimized, and to be able to obtain a structure that is *fully* optimized the designer must undertake a large number of designs. A practical approach to this problem is to obtain second-order sections that are individually optimized and then use sections of this type in parallel or cascade for the realization of Nth-order transfer functions. Realizations so obtained are said to be *section-optimal*. This approach gives optimal parallel structures since in this case the output noise is independent of the pairing of zeros and poles and the ordering of sections; furthermore, as was shown by Jackson, Lindgren, and Kim [16], with some experience the approach gives suboptimal cascade structures that are nearly as good as corresponding block-optimal cascade structures.

Optimized second-order sections can be obtained by noting that Eq. (11.54) is satisfied if and only if $\mathbf{D} = \rho \mathbf{I}$, according to Eqs. (11.59) and (11.60); hence Eq. (11.54) can be expressed as

$$\tilde{\mathbf{W}} = \rho^2 \tilde{\mathbf{K}} \tag{11.61}$$

Since $\tilde{\mathbf{W}}$ and $\tilde{\mathbf{K}}$ are symmetric matrices with equal diagonal elements, Eq. (11.61) assumes the form

$$\tilde{\mathbf{W}} = \rho^2 \mathbf{J} \tilde{\mathbf{K}} \mathbf{J} \tag{11.62}$$

where

$$\mathbf{J} = \begin{bmatrix} 0 & 1 \\ 1 & 0 \end{bmatrix}$$

for a second-order realization. Eq. (11.62) is satisfied by a network in which

$$\tilde{\mathbf{A}}^T = \mathbf{J} \tilde{\mathbf{A}} \mathbf{J}$$

and

$$\tilde{\mathbf{c}} = \rho \mathbf{J} \tilde{\mathbf{b}}$$

If $\tilde{\mathbf{A}} = \{\tilde{a}_{ij}\}$, $\tilde{\mathbf{b}} = \{\tilde{b}_i\}$, and $\tilde{\mathbf{c}}^T = \{\tilde{c}_i\}$, then the above conditions yield

$$\tilde{a}_{11} = \tilde{a}_{22}$$

and

$$\frac{\tilde{b}_1}{\tilde{b}_2} = \frac{\tilde{c}_2}{\tilde{c}_1}$$

If $\{\hat{\mathbf{A}}, \hat{\mathbf{b}}, \hat{\mathbf{c}}^T, \hat{d}\}$ represents a specific realization that satisfies these conditions, then applying the scaling transformation

$$\mathbf{T} = \begin{bmatrix} \|\hat{F}_1\|_2 & 0 \\ 0 & \|\hat{F}_2\|_2 \end{bmatrix} \quad (11.63)$$

results in a structure that satisfies Eqs. (11.54), (11.59), and (11.60) simultaneously and, therefore, is optimal for L_2 scaling. It should be mentioned that if the transformation

$$\mathbf{T} = \begin{bmatrix} \|\hat{F}_1\|_\infty & 0 \\ 0 & \|\hat{F}_2\|_\infty \end{bmatrix} \quad (11.64)$$

is used instead, the structure obtained is not optimal for L_∞ scaling, although good results are usually obtained.

A biquadratic second-order transfer function with complex-conjugate poles can be expressed as

$$H(z) = \frac{\gamma_1 z + \gamma_0}{z^2 + \beta_1 z + \beta_0} + \delta \quad (11.65)$$

and on the basis of the above principles, Jackson et al. [16] obtained the following optimal state-space realization:

$$\hat{a}_{11} = \hat{a}_{22} = -\beta_1/2 \quad (11.66a)$$

$$\hat{a}_{12} = (1 + \gamma_0)(K_1 \pm K_2)/\gamma_1^2 \quad (11.66b)$$

$$\hat{a}_{21} = [K_1 \pm (-K_2)]^2/(1 + \gamma_0) \quad (11.66c)$$

$$\hat{b}_1 = \frac{1}{2}(1 + \gamma_0), \quad \hat{b}_2 = \frac{1}{2}\gamma_1 \quad (11.66d)$$

$$\hat{c}_1 = \frac{\gamma_1}{1 + \gamma_0}, \quad \hat{c}_2 = 1 \quad (11.66e)$$

$$\hat{d} = \delta \quad (11.66f)$$

$$K_1 = \gamma_0 - \frac{1}{2}\beta_1\gamma_1$$

$$K_2 = \sqrt{(\gamma_0^2 - \gamma_0\gamma_1\beta_1 + \beta_0\gamma_1^2)}$$

An arbitrary parallel or cascade design can be obtained by expressing the individual biquadratic transfer functions as in Eq. (11.65) and then using the scaling transformation

$$\mathbf{T} = \begin{bmatrix} \|F_{1i}\|_p & 0 \\ 0 & \|\hat{F}_{2i}\|_p \end{bmatrix}$$

with $p = 2$ or ∞ for each section, where $F_{1i}(z)$ and $F_{2i}(z)$ are the transfer functions between the input of the filter and the state-variable nodes 1 and 2, respectively, of the ith section.

11.8 APPLICATION OF ERROR-SPECTRUM SHAPING

An alternative approach for the reduction of output roundoff noise is through the application of a technique known as *error-spectrum shaping* [17, 18]. This technique involves the generation of a roundoff-error signal and the application of local feedback for the purpose of controlling and manipulating the output roundoff noise. The technique entails additional hardware which increases in direct proportion to the number of adders in the structure. Consequently, only structures in which the outputs of all multipliers are inputs to one and the same adder are suitable for the application of error-spectrum shaping. The most well-known structure of this type is the classical direct realization. Other structures of this type are the low-sensitivity structures described in Sec. 11.4.

The application of error-spectrum shaping to the direct realization of Fig. 11.11a is illustrated in Fig. 11.11b. Signals and coefficients are assumed to be in fixed-point format using L bits for the magnitude and one bit for the sign, and each of the two adders A_1 and A_2 can add products of $2L$ bits to produce a sum of $2L$ bits. Quantizer Q_1 rounds the output of adder A_1 to L bits and simultaneously generates a scaled-up version of the quantization error which is fed back to adder A_1 through the β subnetwork. Quantizer Q_2, on the other hand, scales down and rounds the output of adder A_2 to $2L$ bits. A suitable scaling factor for the β subnetwork is 2^L since the leading L bits of the quantization error are zeros. Constant λ is used to scale the input of quantizer Q_1. Assuming L_2 signal scaling, then

$$\lambda = \frac{1}{\|H\|_2^2} \tag{11.67}$$

where $H(z)$ is the transfer function of the structure in Fig. 11.11a.

A noise model for the configuration in Fig. 11.11b can be readily obtained as shown in Fig. 11.11c, where $-q_i/2 \leq e_i(n) \leq q_i/2$ with $q_1 = 2^{-L}$ and $q_2 = 2^{-2L}$. Hence the PSDs of signals $e_1(n)$ and $e_2(n)$ are given by

$$S_{e_i}(z) = \sigma_{e_i}^2 = \frac{q_i^2}{12}$$

As in Sec. 11.5, the PSD of the output noise can be obtained as

$$S_n(z) = \sum_{i=1}^{2} \frac{q_i^2}{12} H_i(z) H_i(z^{-1}) \tag{11.68}$$

where

$$H_1(z) = \frac{1}{\lambda} \left(\frac{z^2 + \beta_1 z + \beta_0}{z^2 + b_1 z + b_0} \right) \tag{11.69}$$

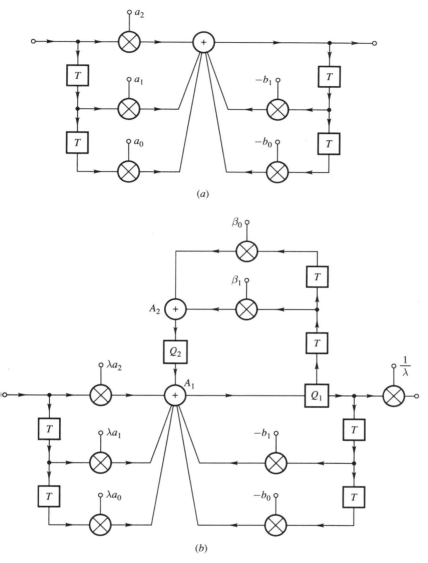

FIGURE 11.11a and b
(a) Second-order direct realization, (b) application of error-spectrum shaping.

and

$$H_2(z) = \frac{1}{\lambda(z^2 + b_1 z + b_0)} \tag{11.70}$$

are the transfer functions from noise sources $e_1(n)$ and $e_2(n)$ to the output, respectively. The output noise power is numerically equal to the autocorrelation of the output noise

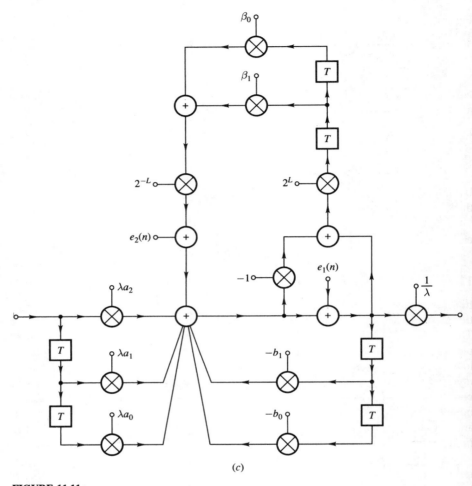

FIGURE 11.11c
(c) Noise model.

evaluated at $k = 0$ and from Eqs. (10.19) and (11.68), we obtain

$$r_n(0) = \sigma_n^2 = \frac{1}{2\pi j} \oint_\Gamma S_n(z)z^{-1}dz$$

$$= \frac{1}{2\pi j} \oint_\Gamma \sum_{i=1}^{2} \frac{q_i^2}{12} H_i(z)H_i(z^{-1})z^{-1}dz$$

$$= \sum_{i=1}^{2} \frac{q_i^2}{12} \|H_i\|_2^2 \tag{11.71}$$

For a random input signal whose amplitude is uniformly distributed in the range $(-1, 1)$, we have $r_x(k) = \sigma_x^2 = 1/3$; hence the output power due to the signal is given by

$$r_y(0) = \sigma_y^2 = \frac{1}{3}\|H\|_2^2 \qquad (11.72)$$

Now from Eqs. (11.67) and (11.69) to (11.72), the signal-to-noise ratio can be obtained as

$$\text{SNR} = \frac{\sigma_y^2}{\sigma_n^2} = \frac{4 \times 2^{2L}}{\left\|\dfrac{z^2 + z\beta_1 + \beta_0}{z^2 + zb_1 + b_0}\right\|_2^2 + 2^{-2L}\left\|\dfrac{1}{z^2 + zb_1 + b_0}\right\|_2^2}$$

If the parameters β_1 and β_2 are chosen to be equal to b_1 and b_2, respectively, then the signal-to-noise ratio is maximized, as demonstrated by Higgins and Munson [18].

Expressions for the coefficients of the error-spectrum shaping network for the case of cascade structures have been derived in [19].

11.9 LIMIT-CYCLE OSCILLATIONS

In the methods of analysis presented in Sec. 11.5, we made the fundamental assumption that signal levels are much larger than the quantization step throughout the filter. This allowed us to assume statistically independent noise signals from sample to sample and from source to source. On many occasions, signal levels can become very low or constant, at least for short periods of time, e.g., during pauses in speech and music signals. Under such circumstances, quantization errors tend to become highly correlated and can actually cause a filter to lock in an unstable mode whereby a steady output oscillation is generated. This phenomenon is known as the *deadband effect*, and the oscillation generated is commonly referred to as *quantization* or *granularity limit cycle*.

Quantization limit cycles are low-level oscillations whose amplitudes can be reduced by increasing the word length of the implementation. Another type of oscillation that can cause serious problems is sometimes brought about by overflow in the arithmetic devices used. Oscillations of this type are known as *overflow limit cycles* and their amplitudes can be quite large, sometimes as large as the maximum signal handling capacity of the hardware.

In this section, we examine the mechanisms by which quantization and overflow limit cycles can be generated and present methods for their elimination.

11.9.1 Quantization Limit Cycles

The deadband effect can be studied by using a technique developed by Jackson [20]. Consider the first-order filter of Fig. 11.12a. The transfer function and difference equation of the filter are given by

$$H(z) = \frac{H_0 z}{z - b}$$

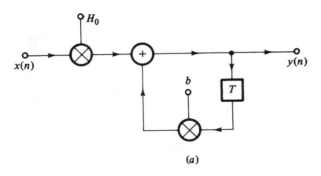

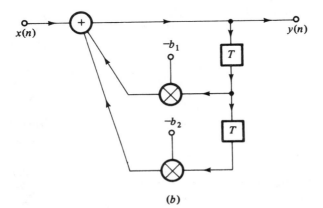

FIGURE 11.12
(a) First-order filter,
(b) second-order filter.

and

$$y(n) = H_0 x(n) + b y(n-1) \tag{11.73}$$

respectively. The impulse response is

$$h(n) = H_0(b)^n$$

If $b = 1$ or -1, the filter is unstable and has an impulse response

$$h(n) = \begin{cases} H_0 & \text{for } b = 1 \\ H_0(-1)^n & \text{for } b = -1 \end{cases}$$

With $H_0 = 10.0$ and $b = -0.9$, the exact impulse response given in the second column of Table 11.5 can be obtained.

Now assume that the filter is implemented using fixed-point decimal arithmetic where each product $by(n-1)$ is rounded to the nearest integer according to the rule

$$Q[|by(n-1)|] = \text{Int}\,[|by(n-1)| + 0.5] \tag{11.74}$$

With $H_0 = 10.0$ and $b = -0.9$, the response in the third column of Table 11.5 is obtained. As can be seen, for $n \geq 5$ the response oscillates between $+5$ and -5 and, in a sense, quantization has rendered the filter unstable.

TABLE 11.5
Impulse response of first-order filter

n	$h(n)$	$Q[h(n)]$
0	10.0	10.0
1	−9.0	−9.0
2	8.1	8.0
3	−7.29	−7.0
4	6.561	6.0
5	−5.9049	−5.0
6	5.31441	5.0
⋮	⋮	⋮
100	2.65614×10^{-4}	5.0

If Eq. (11.73) is assumed to hold during the unstable mode, the effective value of b must be 1 for $b > 0$ or -1 for $b < 0$. If this is the case

$$Q[|by(n-1)|] = |y(n-1)|$$

and from Eq. (11.74)

$$\text{Int}\,[|b| \cdot |y(n-1)| + 0.5] = |y(n-1)|$$

or

$$\text{Int}\,[|y(n-1)| - (1 - |b|)|y(n-1)| + 0.5] = |y(n-1)|$$

This equation can be satisfied if

$$0 \leq -(1 - |b|)|y(n-1)| + 0.5 < 1$$

and by using the left-hand inequality, we conclude that

$$|y(n-1)| \leq \frac{0.5}{1 - |b|} = k$$

Since $y(n-1)$ is an integer, instability cannot arise if $|b| < 0.5$. On the other hand, if $|b| \geq 0.5$, the response will tend to decay to zero once the input is removed, and eventually $y(n-1)$ will assume values in the so-called *deadband range* $[-k, k]$. When this happens, the filter will become unstable. Any tendency of $|y(n-1)|$ to exceed k will restore stability, but in the absence of an input signal the response will again decay to a value within the deadband. Thus the filter will lock into a limit cycle of amplitude equal to or less than k. Since the effective value of b is $+1$ for $0.5 \leq b < 1$ or -1 for $-1 < b \leq -0.5$, the frequency of the limit cycle will be 0 or $\omega_s/2$.

For the second-order filter of Fig. 11.12b, we have

$$H(z) = \frac{z^2}{z^2 + b_1 z + b_0}$$

and

$$y(n) = x(n) - b_1 y(n-1) - b_0 y(n-2) \qquad (11.75)$$

If the poles are complex, then

$$h(n) = \frac{r^n}{\sin\theta} \sin[(n+1)\theta]$$

where

$$r = \sqrt{b_0}$$

and

$$\theta = \cos^{-1} -\frac{b}{2\sqrt{b_0}}$$

For $b_0 = 1$, the impulse response is a sinusoid with constant amplitude and frequency

$$\omega_0 = \frac{1}{T}\cos^{-1} -\frac{b_1}{2} \qquad (11.76)$$

This is sometimes referred to as the *resonant frequency* of the filter.

In second-order filters, there are two distinct limit-cycle modes. In one mode, a limit cycle with frequency 0 or $\omega_s/2$ is generated, and a limit cycle whose frequency is related to the resonant frequency ω_0 is generated in the other.

If the filter is implemented using fixed-point decimal arithmetic and each of the products $b_1 y(n-1)$ and $b_0 y(n-2)$ is rounded to the nearest integer according to the rule in Eq. (11.74), then Eq. (11.75) yields

$$y(n) = x(n) - Q[b_1 y(n-1)] - Q[b_0 y(n-2)]$$

The filter can sustain a zero-input limit cycle of amplitude y_0 ($y_0 > 0$) and frequency 0 or $\omega_s/2$ if

$$y_0 = \pm Q[b_1 y_0] - Q[b_0 y_0] \qquad (11.77)$$

where the plus sign applies for limit cycles of frequency $\omega_s/2$ (see Prob. 11.29). Regions of the (b_0, b_1) plane that satisfy this equation and the corresponding values of y_0 are shown in Fig. 11.13a. The domain inside the triangle represents stable filters as can be easily shown [see Eqs. (11.91a) to (11.91c)].

If $e_1(n)$ and $e_2(n)$ are the quantization errors in products $b_1 y_0$ and $b_0 y_0$, respectively, then Eq. (11.77) gives

$$\pm b_1 = \frac{y_0 \pm e_1(n) \pm e_2(n)}{y_0} + b_0$$

and since $-0.5 < e_i(n) \leq 0.5$, a necessary but not sufficient condition for the existence of a limit cycle of frequency 0 or $\omega_s/2$ is obtained as

$$|b_1| \geq \frac{y_0 - 1}{y_0} + b_0$$

The second limit-cycle mode involves the quantization of product $b_0 y(n-2)$. If

$$Q[|b_0 y(n-2)|] = |y(n-2)|$$

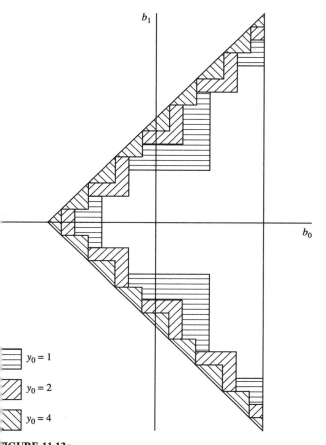

FIGURE 11.13a
Regions of the (b_0, b_1) plane that yield quantization limit cycles: (a) regions that satisfy Eq. (11.77),

then the effective value of b_0 is unity and, as in the first-order case, a condition for the existence of limit cycles can be deduced as

$$|y(n-2)| \leq \frac{0.5}{1 - |b_0|} = k \tag{11.78}$$

With k an integer, values of b_0 in the ranges

$$0.5 \leq |b_0| \leq 0.75$$
$$0.75 \leq |b_0| \leq 0.833$$
$$\cdots\cdots\cdots\cdots\cdots\cdots$$
$$\frac{2k-1}{2k} \leq |b_0| \leq \frac{2k+1}{2(k+1)}$$
$$\cdots\cdots\cdots\cdots\cdots\cdots$$

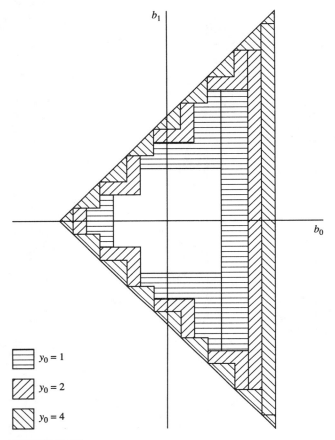

FIGURE 11.13b
Regions of the (b_0, b_1) plane that yield quantization limit cycles: (b) regions that satisfy Eqs. (11.77) and (11.78).

will yield deadbands $[-1, 1], [-2, 2], \ldots, [-k, k], \ldots$, respectively. Regions of the (b_0, b_1) plane that satisfy both Eqs. (11.77) and (11.78) are depicted in Fig. 11.13b.

If the poles are close to the unit circle, the limit cycle is approximately sinusoidal with a frequency close to the resonant frequency given by Eq. (11.76).

For signed-magnitude binary arithmetic, Eq. (11.78) becomes

$$|y(n-2)| \leq \frac{q}{2(1 - |b_0|)}$$

where q is the quantization step.

11.9.2 Overflow Limit Cycles

In one's- or two's- complement fixed-point implementations, the transfer characteristic of adders is periodic, as illustrated in Fig. 11.14a; as a consequence, if the inputs to an

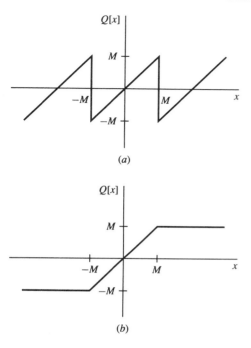

FIGURE 11.14
(a) Transfer characteristic of one's or two's complement fixed-point adder, (b) transfer characteristic of adder incorporating saturation mechanism.

adder are sufficiently large to cause overflow, unexpected results can occur. Under certain circumstances, oscillations of large amplitude can be sustained, which are known as *overflow limit-cycle oscillations*. These were identified and studied quite early in the development of digital filters by Ebert, Mazo, and Taylor [21]. The generation of overflow limit cycles is demonstrated by the following example.

> **Example 11.5.** A second-order digital filter characterized by Eq. (11.75) with $b_1 = -1.375$ and $b_0 = 0.625$ is implemented in terms of two's complement fixed-point arithmetic using a word length of 6 bits, excluding the sign bit. The quantization of products is carried out by rounding. Show that if $x(n) = 0$, $y(-2) = -43/64$, and $y(-1) = 43/64$, the filter will sustain an overflow limit cycle.

Solution. Using the difference equation, output $y(n)$ given in column 2 of Table 11.6 can be readily computed. Evidently, $y(4) = y(-2)$ and $y(5) = y(-1)$ and, therefore, a sustained oscillation of amplitude 43/64 and frequency $\omega_s/2$ will be generated.

11.9.3 Elimination of Quantization Limit Cycles

Quantization limit-cycle oscillations received considerable attention from researchers in the past, and two general approaches for minimizing or eliminating their effects have evolved. One approach entails the use of a sufficiently large signal word length to ensure that the amplitude of the limit-cycle is small enough to meet some system specification imposed by the application. Bounds on the limit-cycle amplitude that can be

TABLE 11.6
Overflow limit cycle in second-order filter

n	$64y(n)$	$64\tilde{y}(n)$
−2	−43	−43
−1	43	43
0	−42	63
1	43	60
2	−43	44
3	42	23
4	−43	4
5	43	−8
6	−42	−14
7	43	−14
8	−43	−10
9	42	−5
10	−43	−3
11	43	−1
12	−42	1
13	43	2
14	−43	2
15	42	2

used in this approach have been deduced by Sandberg and Kaiser [22], Long and Trick [23], and Green and Turner [24]. The other approach entails the elimination of limit cycles altogether. Quantization limit cycles can be eliminated by using appropriate signal quantization schemes in specific structures, whereas overflow limit cycles can be eliminated by incorporating suitable saturation mechanisms in arithmetic devices.

An important method for the elimination of zero-input limit cycles was proposed by Meerkötter [25] and was later used by Mills, Mullis, and Roberts [26], and Vaidyanathan and Liu [27] to show that there are several realizations that support the elimination of limit-cycle oscillations. In this method, a Lyapunov function related to the stored power is constructed and is then used to demonstrate that under certain conditions limit cycles cannot be sustained. The principles involved are as follows.

Consider the digital filter shown in Fig. 11.15 and assume that block A is a linear subnetwork containing adders, multipliers, and interconnections but no unit delays. Further, assume that signal quantization and overflow control are carried out by quantizers Q_k for $k = 1, 2, \ldots, N$ placed at the inputs of the unit delays as shown. The state-space characterization of the filter can be expressed as

$$\mathbf{v}(n) = \mathbf{A}\mathbf{q}(n) + \mathbf{b}x(n)$$
$$y(n) = \mathbf{c}^T \mathbf{q}(n) + dx(n)$$

and if $x(n) = 0$, we can write

$$\mathbf{v}(n) = \mathbf{A}\mathbf{q}(n) \tag{11.79}$$
$$\mathbf{q}(n+1) = \tilde{\mathbf{v}}(n) \tag{11.80}$$

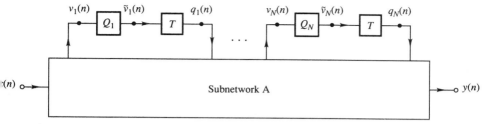

FIGURE 11.15
Nth-order digital filter incorporating nonlinearities.

where $\mathbf{A} = \{a_{ij}\}$ and $\tilde{v}_k(n)$ is related to $v_k(n)$ by some nonlinear and possibly time-varying functional relation of the form

$$\tilde{v}_k(n) = Q_k[v_k(n)] \qquad \text{for } k = 1, 2, \ldots, N \tag{11.81}$$

The quadratic form

$$p[\mathbf{q}(n)] = \mathbf{q}^T(n)\mathbf{D}\mathbf{q}(n) \tag{11.82}$$

where $\mathbf{D}$ is an $N \times N$ positive definite diagonal matrix, is related to the power stored in the unit delays at instant nT, and changes in this quantity can provide information about the stability of the filter under zero-input conditions. The increase in $p[\mathbf{q}(n)]$ in one filter cycle can be expressed as

$$\Delta p[\mathbf{q}(n)] = p[\mathbf{q}(n+1)] - p[\mathbf{q}(n)] \tag{11.83}$$

and from Eqs. (11.80), (11.82), and (11.83), we have

$$\Delta p[\mathbf{q}(n)] = -\mathbf{q}^T(n)\mathbf{D}\mathbf{q}(n) + \tilde{\mathbf{v}}^T(n)\mathbf{D}\tilde{\mathbf{v}}(n) \tag{11.84}$$

Hence Eqs. (11.79) and (11.84) yield

$$\Delta p[\mathbf{q}(n)] = -\mathbf{q}^T(n)\mathbf{D}\mathbf{q}(n) + \tilde{\mathbf{v}}^T(n)\mathbf{D}\tilde{\mathbf{v}}(n) + [\mathbf{A}\mathbf{q}(n)]^T\mathbf{D}[\mathbf{A}\mathbf{q}(n)] - \mathbf{v}^T(n)\mathbf{D}\mathbf{v}(n)$$
$$= -\mathbf{q}^T(n)(\mathbf{D} - \mathbf{A}^T\mathbf{D}\mathbf{A})\mathbf{q}(n) - \sum_{k=1}^{N}[v_k^2(n) - \tilde{v}_k^2(n)]d_{kk} \tag{11.85}$$

where d_{kk} for $k = 1, 2, \ldots, N$ are the diagonal elements of $\mathbf{D}$.

Now if

$$\mathbf{q}^T(n)(\mathbf{D} - \mathbf{A}^T\mathbf{D}\mathbf{A})\mathbf{q}(n) \geq 0 \tag{11.86}$$

and signals $v_k(n)$ are quantized such that

$$|\tilde{v}_k(n)| \leq |v_k(n)| \qquad \text{for } k = 1, 2, \ldots, N \tag{11.87}$$

then Eq. (11.85) yields

$$\Delta p[\mathbf{q}(n)] \leq 0 \qquad (11.88)$$

that is, the power stored in the unit delays cannot increase. Since a digital filter is a finite-state machine, signals $q_k(n)$ must after a finite number of filter cycles either become permanently zero or oscillate periodically. In the first case, there are no limit cycle oscillations. In the second case, at least one $q_k(n)$, say $q_l(n)$, must oscillate periodically. However, from Eq. (11.88), we conclude that the amplitude of the oscillation must decrease with each filter cycle by some fixed amount until $q_l(n)$ becomes permanently zero after a finite number of filter cycles. Therefore, Eq. (11.86) in conjunction with the conditions in Eq. (11.87) constitute a sufficient set of conditions for the elimination of limit cycles. A realization satisfying Eq. (11.86) is said to *support the elimination of zero-input limit cycles*. The conditions in Eq. (11.87) can be imposed by quantizing the state variables using magnitude truncation.

For a stable filter, the magnitudes of the eigenvalues of $\mathbf{A}$ are less than unity and Eq. (11.86) is satisfied if a positive definite diagonal matrix $\mathbf{D}$ can be found such that matrix $\mathbf{D} - \mathbf{A}^T \mathbf{D} \mathbf{A}$ is positive semidefinite [26, 27]. For second-order filters, this condition is satisfied if

$$a_{12} a_{21} \geq 0 \qquad (11.89a)$$

or

$$a_{12} a_{21} < 0 \quad \text{and} \quad |a_{11} - a_{22}| + \det(\mathbf{A}) \leq 1 \qquad (11.89b)$$

There are quite a few realizations that support the elimination of zero-input limit cycles. Some examples are: normal state-space structures in which

$$\mathbf{A} = \begin{bmatrix} \alpha & -\beta \\ -\beta & \alpha \end{bmatrix}$$

with $\beta > 0$ [28–30]; realizations that minimize the output roundoff noise such as those in [14, 16] (see Sec. 11.7); lattice realizations [27, 30, 31], etc.

Example 11.6. The structure shown in Fig. 11.16 realizes the biquadratic transfer function

$$H(z) = \frac{z^2 + a_1 z + a_0}{z^2 + b_1 z + b_0}$$

where

$$a_1 = -(\alpha_1 + \alpha_2) \qquad (11.90a)$$

$$a_0 = 1 + \alpha_1 - \alpha_2 \qquad (11.90b)$$

$$b_1 = -(\beta_1 + \beta_2) \qquad (11.90c)$$

$$b_0 = 1 + \beta_1 - \beta_2 \qquad (11.90d)$$

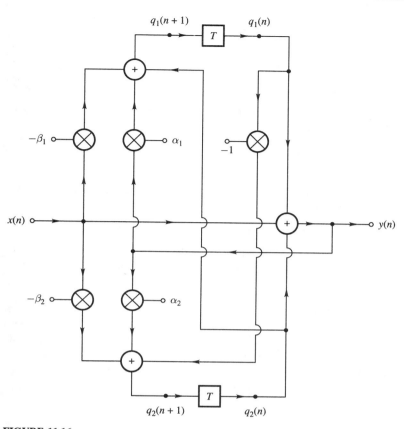

FIGURE 11.16
Biquadratic realization due to Meerkötter.

and is due to Meerkötter [25]. Show that the structure supports the elimination of zero-input limit cycles.

Solution. Straightforward analysis gives the state-space characterization of the structure as

$$\mathbf{q}(n+1) = \mathbf{A}\mathbf{q}(n) + \mathbf{b}x(n)$$

$$y(n) = \mathbf{c}^T\mathbf{q}(n) + dx(n)$$

where

$$\mathbf{A} = \begin{bmatrix} a_{11} & a_{12} \\ a_{21} & a_{22} \end{bmatrix} = \begin{bmatrix} \beta_1 & (\beta_1+1) \\ (\beta_2-1) & \beta_2 \end{bmatrix},$$

$$\mathbf{b} = \begin{bmatrix} (\beta_1-\alpha_1) \\ (\beta_2-\alpha_2) \end{bmatrix}, \quad \mathbf{c} = \begin{bmatrix} 1 \\ 1 \end{bmatrix}, \quad \text{and} \quad d = 1$$

The filter is stable if and only if

$$1 - b_0 > 0 \tag{11.91a}$$

$$1 + b_1 + b_0 > 0 \tag{11.91b}$$

$$1 - b_1 + b_0 > 0 \tag{11.91c}$$

as can be easily shown by using the Jury-Marden stability criterion (see Sec. 3.3.7). From Eq. (11.90), we can show that

$$a_{12}a_{21} = (\beta_1 + 1)(\beta_2 - 1) = \frac{1}{4}[b_1^2 - (1 + b_0)^2]$$

and since $1 + b_0 > b_1$, according to Eq. (11.91c), we conclude that $a_{12}a_{21} < 0$. Hence, zero-input limit cycles can be eliminated by using magnitude truncation only if the condition in Eq. (11.89b) is satisfied. Simple manipulation now yields

$$|a_{11} - a_{22}| + \det(\mathbf{A}) = |b_0 - 1| + b_0 - 1 + 1 = 1$$

since $b_0 - 1$ is negative according to Eq. (11.91a); that is, Eq. (11.89b) is satisfied with the equal sign and, therefore, the structure supports the elimination of zero-input limit cycles.

Limit cycles can also be generated if the input assumes a constant value for a certain period of time. Limit cycles of this type, which include zero-input limit cycles as a special case, are referred to as *constant-input limit cycles*; they can be eliminated by using techniques described by Verkroost [32], Turner [33], and Diniz and Antoniou [34]. A state-space realization of the transfer function in Eq. (11.65) that supports the elimination of zero- and constant-input limit cycles is illustrated in Fig. 11.17, where

$$a_{11} = a_{22} = -\beta_1/2 \tag{11.92a}$$

$$a_{12} = -\zeta/\sigma, \qquad a_{21} = \sigma\zeta \tag{11.92b}$$

$$c_1 = \frac{\gamma_1 + \gamma_0}{1 + \beta_1 + \beta_0}, \qquad c_2 = \frac{(2 + \beta_1)\gamma_0 - (\beta_1 + 2\beta_0)\gamma_1}{2\sigma\zeta(1 + \beta_1 + \beta_0)} \tag{11.92c}$$

$$d = \delta \tag{11.92d}$$

$$\zeta = \sqrt{\left(\beta_0 - \frac{\beta_1^2}{4}\right)}$$

Constant σ can be used to achieve optimal scaling. This structure is optimal or nearly optimal with respect to roundoff noise and is, in addition, slightly more economical than the state-space realization given by Eqs. (11.66a) to (11.66f) (see [34] for more details).

11.9.4 Elimination of Overflow Limit Cycles

Overflow limit cycles can be avoided to a large extent by applying strict scaling rules, e.g., using scaling method A in Sec. 11.6.1, to as far as possible prevent overflow from occurring. The problem with this approach is that signal levels throughout the filter

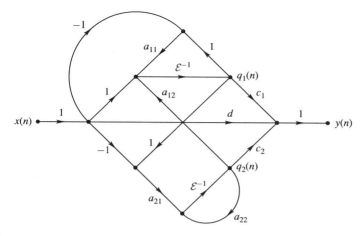

FIGURE 11.17
Second-order state-space realization that supports the elimination of zero- and constant-input limit cycles.

are low; as a result, a poor signal-to-noise ratio is achieved. The preferred solution is to allow overflow on occasion but prevent the limit-cycle oscillations from occurring. A solution of this type reported in [21] involves incorporating a saturation mechanism in the design of adders so as to achieve a transfer characteristic of the type depicted in Fig. 11.14b where

$$Q[x] = \begin{cases} x & \text{if } |x| < M \\ M & \text{if } |x| \geq M \end{cases}$$

If this type of adder is used in the filter of Example 11.5, output $\tilde{y}(n)$ given in column 3 of Table 11.6 will be obtained. Evidently, the overflow limit cycle will be eliminated but a quantization limit cycle of amplitude 2/64 and frequency 0 will be present. This is due to the fact that this amplitude satisfies Eq. (11.77), as can be easily verified.

A concept that is closely related to overflow oscillations is the stability of the forced response of a nonlinear system or filter. If $\tilde{v}(n)$ and $v(n)$ are the state variables in Fig. 11.15, first with and then without the quantizers installed, the *forced response* of the filter is said to be stable if

$$\lim_{n \to \infty} [\tilde{v}(n) - v(n)] = 0$$

In practical terms, the stability of the forced response implies that transients due to overflow effects tend to die out once the cause of the overflow has been removed.

Claasen, Mecklenbräuker, and Peek [35] have shown that if a filter incorporating certain nonlinearities, e.g., overflow nonlinearities, is stable under zero-input conditions, then the forced response is also stable with respect to a corresponding set of nonlinearities. On the basis of this equivalence, if a digital filter of the type shown in Fig. 11.15 is stable under zero-input conditions, i.e., it satisfies Eq. (11.86) subject to the conditions in Eq. (11.87), then the forced response is also stable provided that

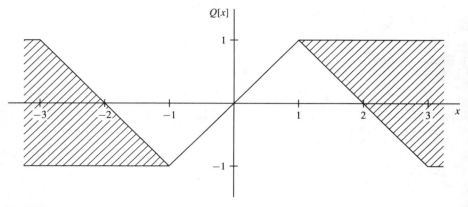

FIGURE 11.18
Transfer characteristic that guarantees the stability of the forced response.

the nonlinearities in Eq. (11.81) satisfy the conditions

$$2 - x < Q_k \leq 1 \quad \text{for} \quad 1 < x < 3$$
$$-2 - x > Q_k \geq -1 \quad \text{for} \quad -3 < x < -1$$
$$-1 \leq Q_k \leq 1 \quad \text{for} \quad |x| \geq 3$$

for $k = 1, 2, \ldots, N$, as illustrated in Fig. 11.18.

The stability of the forced response implies freedom from overflow limit cycles. It should be mentioned, however, that Claasen et al. deduced the above equivalence on the assumption that there is an infinite time separation between successive occurrences of overflow. Consequently, the above conditions may not guarantee the absence of overflow limit cycles if overflow occurs while the filter is recovering from a previous overflow.

REFERENCES

1. K. Hwang, *Computer Arithmetic, Principles, Architecture, and Design*, Wiley, New York, 1979.
2. E. Avenhaus, "On the Design of Digital Filters with Coefficients of Limited Word Length," *IEEE Trans. Audio Electroacoust.*, vol. AU-20, pp. 206–212, August 1972.
3. R. E. Crochiere, "A New Statistical Approach to the Coefficient Word Length Problem for Digital Filters," *IEEE Trans. Circuits Syst.*, vol. CAS-22, pp. 190–196, March 1975.
4. A. Papoulis, *Probability, Random Variables, and Stochastic Processes*, McGraw-Hill, New York, 1991.
5. R. E. Crochiere and A. V. Oppenheim, "Analysis of Linear Digital Networks," *Proc. IEEE*, vol. 63, pp. 581–595, April 1975.
6. W. K. Jenkins and B. J. Leon, "An Analysis of Quantization Error in Digital Filters Based on Interval Algebras," *IEEE Trans. Circuits Syst.*, vol. CAS-22, pp. 223–232, March 1975.
7. R. C. Agarwal and C. S. Burrus, "New Recursive Digital Filter Structures Having Very Low Sensitivity and Roundoff Noise," *IEEE Trans. Circuits Syst.*, vol. CAS-22, pp. 921–927, December 1975.
8. P. S. R. Diniz and A. Antoniou, "Low-Sensitivity Digital-Filter Structures which are Amenable to Error-Spectrum Shaping," *IEEE Trans. Circuits Syst.*, vol. CAS-32, pp. 1000–1007, October 1985.

9. S. Nishimura, K. Hirano, and R. N. Pal, "A New Class of Very Low Sensitivity and Low Roundoff Noise Recursive Digital Filter Structures," *IEEE Trans. Circuits Syst.*, vol. CAS-28, pp. 1152–1158, December 1981.
10. Y. V. Ramana Rao and C. Eswaran, "A Pole-Sensitivity Based Method for the Design of Digital Filters for Error-Spectrum Shaping," *IEEE Trans. Circuits Syst.*, vol. CAS-36, pp. 1017–1020, July 1989.
11. L. B. Jackson, "On the Interaction of Roundoff Noise and Dynamic Range in Digital Filters," *Bell Syst. Tech. J.*, vol. 49, pp. 159–184, February 1970.
12. G. Bachman and L. Naria, *Functional Analysis*, Academic, New York, 1966.
13. L. B. Jackson, "Roundoff-Noise Analysis for Fixed-Point Digital Filters Realized in Cascade or Parallel Form," *IEEE Trans. Audio Electroacoust.*, vol. AU-18, pp. 107–122, June 1970.
14. C. T. Mullis and R. A. Roberts, "Synthesis of Minimum Roundoff Noise Fixed Point Digital Filters," *IEEE Trans. Circuits Syst.*, vol. CAS-23, pp. 551–562, September 1976.
15. S. Y. Hwang, "Roundoff Noise in State-Space Digital Filtering: A General Analysis," *IEEE Trans. Acoust., Speech, Signal Process.*, vol. ASSP-24, pp. 256–262, June 1976.
16. L. B. Jackson, A. G. Lindgren, and Y. Kim, "Optimal Synthesis of Second-Order State-Space Structures for Digital Filters," *IEEE Trans. Circuits Syst.*, vol. CAS-26, pp. 149–153, March 1979.
17. T. Thong and B. Liu, "Error Spectrum Shaping in Narrow-Band Recursive Filters," *IEEE Trans. Acoust., Speech, Signal Process.*, vol. ASSP-25, pp. 200–203, April 1977.
18. W. E. Higgins and D. C. Munson, Jr., "Noise Reduction Strategies for Digital Filters: Error Spectrum Shaping Versus the Optimal Linear State-Space Formulation," *IEEE Trans. Acoust., Speech, Signal Process.*, vol. ASSP-30, pp. 963–973, December 1982.
19. W. E. Higgins and D. C. Munson, Jr., "Optimal and Suboptimal Error Spectrum Shaping for Cascade-Form Digital Filters," *IEEE Trans. Circuits Syst.*, vol. CAS-31, pp. 429–437, May 1984.
20. L. B. Jackson, "An Analysis of Limit Cycles Due to Multiplication Rounding in Recursive Digital Filters," *Proc. 7th Annu. Allerton Conf. Circuit Syst. Theory*, pp. 69–78, 1969.
21. P. M. Ebert, J. E. Mazo, and M. G. Taylor, "Overflow Oscillations in Digital Filters," *Bell Syst. Tech. J.*, vol. 48, pp. 2999–3020, November 1969.
22. I. W. Sandberg and J. F. Kaiser, "A Bound on Limit Cycles in Fixed-Point Implementations of Digital Filters," *IEEE Trans. Audio Electroacoust.*, vol. AU-20, pp. 110–114, June 1972.
23. J. L. Long and T. N. Trick, "An Absolute Bound on Limit Cycles Due to Roundoff Errors in Digital Filters," *IEEE Trans. Audio Electroacoust.*, vol. AU-21, pp. 27–30, February 1973.
24. B. D. Green and L. E. Turner, "New Limit Cycle Bounds for Digital Filters," *IEEE Trans. Circuits Syst.*, vol. CAS-35, pp. 365–374, April 1988.
25. K. Meerkötter, "Realization of Limit Cycle-Free Second-Order Digital Filters," *Proc. 1976 IEEE Int. Symp. Circuits Syst.*, pp. 295–298.
26. W. L. Mills, C. T. Mullis, and R. A. Roberts, "Digital Filter Realizations without Overflow Oscillations," *IEEE Trans. Acoust., Speech, Signal Process.*, vol. ASSP-26, pp. 334–338, August 1978.
27. P. P. Vaidyanathan and V. Liu, "An Improved Sufficient Condition for Absence of Limit Cycles in Digital Filters," *IEEE Trans. Circuits Syst.*, vol. CAS-34, pp. 319–322, March 1987.
28. C. M. Rader and B. Gold, "Effects of Parameter Quantization on the Poles of a Digital Filter," *Proc. IEEE*, vol. 55, pp. 688–689, May 1967.
29. C. W. Barnes and A. T. Fam, "Minimum Norm Recursive Digital Filters that Are Free of Overflow Limit Cycles," *IEEE Trans. Circuits Syst.*, vol. CAS-24, pp. 569–574, October 1977.
30. A. H. Gray, Jr. and J. D. Markel, "Digital Lattice and Ladder Filter Synthesis," *IEEE Trans. Audio Electroacoust.*, vol. AU-21, pp. 491–500, December 1973.
31. A. H. Gray, Jr., "Passive Cascaded Lattice Digital Filters," *IEEE Trans. Circuits Syst.*, vol. CAS-27, pp. 337–344, May 1980.
32. G. Verkroost, "A General Second-Order Digital Filter with Controlled Rounding to Exclude Limit Cycles for Constant Input Signals," *IEEE Trans. Circuits Syst.*, vol. CAS-24, pp. 428–431, August 1977.
33. L. E. Turner, "Elimination of Constant-Input Limit Cycles in Recursive Digital Filters Using a Generalised Minimum Norm," *IEE Proc.*, vol. 130, pt. G, pp. 69–77, June 1983.

34. P. S. R. Diniz and A. Antoniou, "More Economical State-Space Digital-Filter Structures Which are Free of Constant-Input Limit Cycles," *IEEE Trans. Acoust., Speech, Signal Process.*, vol. ASSP-34, pp. 807–815, August 1986.
35. T. A. C. M. Claasen, W. F. G. Mecklenbräuker, and J. B. H. Peek, "On the Stability of the Forced Response of Digital Filters with Overflow Nonlinearities," *IEEE Trans. Circuits Syst.*, vol. CAS-22, pp. 692–696, August 1975.

ADDITIONAL REFERENCES

Abu-El-Haija, A. I., and A. M. Peterson: "An Approach to Eliminate Roundoff Errors in Digital Filters," *IEEE Trans. Acoust., Speech, Signal Process.*, vol. ASSP-27, pp. 195–198, April 1979.

Barnes, C. W., and S. Shinnaka: "Finite Word Effects in Block-State Realizations of Fixed-Point Digital Filters," *IEEE Trans. Circuits Syst.*, vol. CAS-27, pp. 345–349, May 1980.

———: "Computationally Efficient Second-Order Digital Filter Sections with Low Roundoff Noise Gain," *IEEE Trans. Circuits Syst.*, vol. CAS-31, pp. 841–847, October 1984.

Bomar, B. W., and J. C. Hung: "Minimum Roundoff Noise Digital Filters with Some Power-of-Two Coefficients," *IEEE Trans. Circuits Syst.*, vol. CAS-31, pp. 833–840, October 1984.

Butterweck, H. J.,: "Suppression of Parasitic Oscillations in Second-Order Digital Filters by Means of a Controlled-Rounding Arithmetic," *Arch. Elektron. Übertragung.*, vol. 29, pp. 371–374, 1975.

———, A. C. P. van Meer, and G. Verkroost: "New Second-Order Digital Filter Sections without Limit Cycles," *IEEE Trans. Circuits Syst.*, vol. CAS-31, pp. 141–146, February 1984.

Chang, T. L.: "Suppression of Limit Cycles in Digital Filters with One Magnitude-Truncation Quantizer," *IEEE Trans. Circuits Syst.*, vol. CAS-28, pp. 107–111, February 1981.

Charalambous, C. and M. J. Best: "Optimization of Recursive Digital Filters with Finite Word Lengths," *IEEE Trans. Acoust., Speech, Signal Process.*, vol. ASSP-22, pp. 424–431, December 1974.

Claasen, T. A. C. M., W. F. G. Mecklenbräuker, and J. B. H. Peek: "Frequency Domain Criteria for the Absence of Zero-Input Limit Cycles in Nonlinear Discrete-Time Systems, with Applications to Digital Filters," *IEEE Trans. Circuits Syst.*, vol. CAS-22, pp. 232–239, May 1975.

———, ———, and ———: "Effects of Quantization and Overflow in Recursive Digital Filters," *IEEE Trans. Acoust., Speech, Signal Process.*, vol. ASSP-24, pp. 517–529, December 1976.

Hwang, S. Y.: "Minimum Uncorrelated Unit Noise in State-Space Digital Filtering," *IEEE Trans. Acoust., Speech, Signal Process.*, vol. ASSP-25, pp. 256–262, August 1977.

———: "Dynamic Range Constraint in State-Space Digital Filtering," *IEEE Trans. Acoust., Speech, Signal Process.*, vol. ASSP-23, pp. 591–593, December 1975.

Jackson, L. B.: "Limit Cycles in State-Space Structures for Digital Filters," *IEEE Trans. Circuits Syst.*, vol. CAS-26, pp. 67–68, January 1979.

Kawamata, M., and T. Higuchi: "A Systematic Approach to Synthesis of Limit Cycle-Free Digital Filters," *IEEE Trans. Acoust., Speech, Signal Process.*, vol. ASSP-31, pp. 212–214, February 1983.

Macedo Jr., T. C., T. Laakso, P. S. R. Diniz, and I. Hartimo: "Reformulation of Chang's Criterion for the Absence of Limit Cycles using Bilinear Transform," *Proc. 1991 IEEE Int. Symp. Circuits Syst.*, pp. 388–391.

Meerkötter, K., and W. Wegener: "A New Second-Order Digital Filter without Parasitic Oscillations," *Arch. Elektron. Übertragung.*, vol. 29, pp. 312–314, 1975.

Mitra, S. K., K. Hirano, and H. Sakaguchi: "A Simple Method of Computing the Input Quantization and Multiplication Roundoff Errors in a Digital Filter," *IEEE Trans. Acoust., Speech, Signal Process.*, vol. ASSP-22, pp. 326–329, October 1974.

Mullis, C. T., and R. A. Roberts: "Roundoff Noise in Digital Filters: Frequency Transformations and Invariants," *IEEE Trans. Acoust., Speech, Signal Process.*, vol. ASSP-24, pp. 538–550, December 1976.

Munson, D. C., and B. Liu: "Narrow-Band Recursive Filters with Error Spectrum Shaping," *IEEE Trans. Circuits Syst.*, vol. CAS-28, pp. 160–163, February 1981.

Nishihara, A., and K. Sugahara: "A Synthesis of Digital Filters with Minimum Pole Sensitivity," *Proc. 1982 IEEE Int. Symp. Circuits Syst.*, pp. 507–510.

Parker S. R., and S. F. Hess: "Limit-Cycle Oscillations in Digital Filters," *IEEE Trans. Circuit Theory*, vol. CT-18, pp. 687–697, November 1971.

Turner, L. E., and L. T. Bruton: "Elimination of Granularity and Overflow Limit Cycles in Minimum Norm Recursive Digital Filters," *IEEE Trans. Circuits Syst.*, vol. CAS-27, pp. 50–53, January 1980.

———: "Second-Order Recursive Digital Filter that is Free from All Constant-Input Limit Cycles," *Electron. Lett.*, vol. 18, pp. 743–745, August 1982.

Willson Jr., A. N.: "Limit Cycles Due to Adder Overflow in Digital Filters," *IEEE Trans. Circuit Theory*, vol. CT-19, pp. 342–346, July 1972.

PROBLEMS

11.1. (a) Convert the decimal numbers

$$730.796875 \quad \text{and} \quad -3521.8828125$$

into binary representation.

(b) Convert the binary numbers

$$11011101.011101 \quad \text{and} \quad -100011100.1001101$$

into decimal representation.

11.2. Deduce the signed-magnitude, one's-complement, and two's-complement representations of (a) 0.810546875 and (b) -0.9462890625. Assume a word length $L = 10$.

11.3. The two's complement of a number x can be designated as

$$\tilde{x} = x_0.x_1x_2\cdots x_L$$

(a) Show that

$$x = -x_0 + \sum_{i=1}^{L} x_i 2^{-i}$$

(b) Find x if $\tilde{x} = 0.1110001011$.
(c) Find x if $\tilde{x} = 1.1001110010$.

11.4. Perform the following operations by using the one's- and two's-complement additions.
(a) $0.6015625 - 0.4218750$
(b) $-0.359375 + (-0.218750)$
Assume that $L = 7$.

11.5. The two's complement of x is given by

$$\tilde{x} = x_0.x_1x_2\cdots x_L$$

(a) Show that

$$\text{Two's complement } (2^{-1}x) = \begin{cases} 2^{-1}\tilde{x} & \text{if } x_0 = 0 \\ 1 + 2^{-1}\tilde{x} & \text{if } x_0 = 1 \end{cases}$$

(b) Find the two's complement of $2^{-4}x$ if $\tilde{x} = 1.00110$.

11.6. (a) The register length in a fixed-point digital-filter implementation is 9 bits (including the sign bit), and the arithmetic is of the two's-complement type. Find the largest and smallest machine-representable decimal numbers.

(b) Show that the addition $0.8125 + 0.65625$ will cause overflow.

(c) Show that the addition $0.8125 + 0.65625 + (-0.890625)$ will be evaluated correctly despite the overflow in the first partial sum.

11.7. The mantissa and exponent register segments in a floating-point implementation are 8 and 4 bits long, respectively.
   (a) Deduce the register contents for -0.0234375, -5.0, 0.359375, and 11.5.
   (b) Determine the dynamic range of the implementation.
   Both mantissa and exponent are stored in signed-magnitude form.

11.8. A floating-point number

$$x = M \times 2^e \quad \text{where} \quad M = \sum_{i=1}^{B} b_{-i} 2^{-i}$$

is to be stored in a register whose mantissa and exponent segments comprise $L+1$ and $e+1$ bits, respectively. Assuming signed-magnitude representation and quantization by rounding, find the range of the quantization error.

11.9. A filter section is characterized by the transfer function

$$H(z) = H_0 \frac{(z+1)^2}{z^2 + b_1 z + b_0}$$

where $H_0 = -0.01903425$ $\quad b_0 = 0.8638557$ $\quad b_1 = -0.5596596$

   (a) Find the quantization error for each coefficient if signed-magnitude fixed-point arithmetic is to be used. Assume quantization by truncation and a word length $L = 6$ bits.
   (b) Repeat part (a) if the quantization is to be by rounding.

11.10. (a) Realize the transfer function of Prob. 11.9 by using a canonic structure.
   (b) The filter obtained in part (a) is implemented by using the arithmetic described in Prob. 11.9a. Plot the amplitude-response error versus frequency for $10 \leq \omega \leq 30$ rad/s. The sampling frequency is 100 rad/s.
   (c) Repeat part (b), assuming quantization by rounding.
   (d) Compare the results obtained in parts (b) and (c).

11.11. (a) The transfer function

$$H(z) = \frac{z^2 + 2z + 1}{z^2 + b_1 z + b_0} \quad \text{where} \quad \begin{array}{l} b_1 = -r\sqrt{2} \\ b_0 = r^2 \end{array}$$

is to be realized by using the canonic structure of Fig. 11.7b. Find the sensitivities $S_{b_1}^H(z)$ and $S_{b_0}^H(z)$.
   (b) The section is to be implemented by using fixed-point arithmetic, and the coefficient quantization is to be by rounding. Compute the statistical word length $L(\omega)$ for $0.7 \leq r \leq 0.95$ in steps of 0.05. Assume that $\Delta M_{\max}(\omega) = 0.02$, $x_1 = 2$ (see Sec. 11.3).
   (c) Plot the statistical word length versus r and discuss the results achieved.

11.12. (a) Using Tables 11.2 and 11.3, obtain all possible low-sensitivity direct realizations of the transfer function in Prob. 11.9.
   (b) The realizations in part (a) are to be implemented in terms of signed-magnitude fixed-point arithmetic using a word length $L = 6$, and quantization is to be by rounding. The sampling frequency is 100 rad/s. Plot the amplitude-response error versus frequency for $10 \leq \omega \leq 30$ rad/s for each realization.

(c) On the basis of the results in part (b), select the least sensitive of the possible realizations.
(d) Compare the realization selected in part (c) with the canonic realization obtained in Prob. 11.10.

11.13. The transfer function

$$H(z) = \prod_{i=1}^{3} \frac{a_i(z+1)^2}{z^2 + b_{1i}z + b_{0i}}$$

where a_i, b_{0i}, and b_{1i} are given in Table P11.13, represents a lowpass Butterworth filter.

TABLE P11.13

i	a_i	b_{0i}	b_{1i}
1	0.165765	0.735915	−1.404385
2	0.134910	0.412801	−1.142981
3	0.121819	0.275708	−1.032070

(a) Realize the transfer function using three canonic sections in cascade.
(b) The realization in part (a) is to be implemented in terms of fixed-point signed-magnitude arithmetic using a word length $L = 8$ bits, and coefficient quantization is to be by rounding. The sampling frequency is 10^4 rad/s. Plot the amplitude-response error versus frequency for $0 \leq \omega \leq 10^3$ rad/s.

11.14. (a) Realize the transfer function in Prob. 11.13 using structure II-2 of Table 11.3.
(b) The realization in part (a) is to be implemented as in part (b) of Prob. 11.13. Plot the amplitude-response error versus frequency for $0 \leq \omega \leq 10^3$ rad/s.
(c) Compare the realization in part (a) with the cascade canonic realization of Prob. 11.13 with respect to sensitivity and the number of arithmetic operations.

11.15. The response of an A/D converter to a signal $x(t)$ is given by

$$y(n) = x(n) + e(n)$$

where $x(n)$ and $e(n)$ are random variables uniformly distributed in the ranges $-1 \leq x(n) \leq 1$ and $-2^{-(L+1)} \leq e(n) \leq 2^{-(L+1)}$, respectively.
(a) Find the signal-to-noise ratio. This is defined as

$$\text{SNR} = 10 \log \frac{\text{average signal power}}{\text{average noise power}}$$

(b) Find the PSD of $y(n)$ if $x(n)$, $e(n)$, $x(k)$, and $e(k)$ are statistically independent.

11.16. The filter section of Prob. 11.9 is to be scaled using the scheme in Fig. 11.8.
(a) Find λ for L_∞ scaling.
(b) Find λ for L_2 scaling using a frequency-domain method.
(c) Find λ for L_2 scaling using a time-domain method. (Hint: Use Theorem 2.11.)
(d) Compare the methods in parts (b) and (c).
(e) Compare the values of λ obtained with L_∞ and L_2 scaling and comment on the advantages and disadvantages of the two types of scaling.

11.17. The canonic realization of Prob. 11.13 is to be scaled according to the scheme in Fig. 11.9 using the L_∞ norm.
(a) Find the scaling constants λ_0, λ_1, and λ_2.

(b) The scaled realization is to be implemented in terms of fixed-point arithmetic and product quantization is to be by rounding. Plot the relative, output-noise PSD versus frequency. This is defined as

$$\text{RPSD} = 10 \log \frac{S_y(e^{j\omega T})}{S_e(e^{j\omega T})}$$

where $S_y(e^{j\omega T})$ is the PSD of output noise and $S_e(e^{j\omega T})$ is the PSD of a single noise source. The sampling frequency is 10^4 rad/s.

11.18. Repeat Prob. 11.17 using L_2 scaling and compare the results with those obtained in Prob. 11.17.

11.19. The low-sensitivity realization of Prob. 11.14 is to be scaled according to the scheme in Fig. 11.9 using the L_2 norm.
(a) Find the scaling constants λ_0, λ_1, and λ_2.
(b) The scaled realization is to be implemented in terms of fixed-point arithmetic and product quantization is to be by rounding. Plot the relative, output-noise PSD versus frequency.

11.20. The transfer function

$$H(z) = \prod_{i=1}^{3} \frac{a_{0i} z^2 + a_{1i} z + a_{0i}}{z^2 + b_{1i} z + b_{0i}}$$

where a_{0i}, a_{1i}, b_{0i} and b_{1i} are given in Table P11.16 represents a bandstop elliptic filter.

TABLE P11.16

i	a_{0i}	a_{1i}	b_{0i}	b_{1i}
1	4.623281×10^{-1}	7.859900×10^{-9}	-7.534381×10^{-2}	7.859900×10^{-9}
2	4.879171×10^{-1}	5.904108×10^{-2}	8.051571×10^{-1}	8.883641×10^{-1}
3	1.269926	-1.536691×10^{-1}	8.051571×10^{-1}	-8.883640×10^{-1}

(a) Realize the transfer function using three canonic sections in cascade.
(b) Determine the scaling constants. Assume the section order implied by the transfer function and use L_∞ scaling. The sampling frequency is 18 rad/s.
(c) Plot the relative output-noise PSD versus frequency.

11.21. The transfer function

$$H(z) = \prod_{i=1}^{3} \frac{a_{0i} z^2 + a_{1i} z + 1}{z^2 + a_{1i} z + a_{0i}}$$

where a_{0i} and a_{1i} are given in Table P11.17, represents a digital equalizer. Repeat parts (a) to (c) of Prob. 11.16. The sampling frequency is 2.4π rad/s.

TABLE P11.17

i	a_{0i}	a_{1i}
1	0.973061	-1.323711
2	0.979157	-1.316309
3	0.981551	-1.345605

11.22. Demonstrate the validity of Eq. (11.37).

11.23. Show that the column vectors $\mathbf{f}(z)$ and $\mathbf{g}(z)$ defined in Eq. (11.50) are given by the expressions in Eq. (11.51).

11.24. The vector $\mathbf{q}(n)$ in the state-space realization $\{\mathbf{A},\mathbf{b},\mathbf{c}^T,d\}$ is subjected to the transformation $\tilde{\mathbf{q}}(n) = \mathbf{T}\mathbf{q}(n)$.
 (a) Show that the transformed realization $\{\tilde{\mathbf{A}}, \tilde{\mathbf{b}}, \tilde{\mathbf{c}}^T, \tilde{d}\}$ is given by Eq. (11.52).
 (b) Show that the transformed vectors $\tilde{\mathbf{f}}(z)$ and $\tilde{\mathbf{g}}(z)$ are given by Eq. (11.53).

11.25. (a) Obtain a state-space section-optimal realization of the lowpass filter in Prob. 11.13.
 (b) Apply L_2 scaling to the realization.
 (c) The scaled realization is to be implemented in terms of fixed-point arithmetic and product quantization is to be by rounding. Plot the relative, output-noise PSD versus frequency.
 (d) Compare the results with those obtained in the case of the direct canonic realization in Prob. 11.18.

11.26. (a) Apply error-spectrum shaping to the scaled, low-sensitivity realization obtained in Prob. 11.19.
 (b) The modified realization is to be implemented in terms of fixed-point arithmetic and product quantization is to be by rounding. Plot the relative, output-noise PSD versus frequency.
 (c) Compare the results with those obtained without error-spectrum shaping in Prob. 11.19.

11.27. A second-order filter characterized by Eq. (11.75) with $b_1 = -1.343503$ and $b_0 = 0.9025$ is to be implemented using signed-magnitude decimal arithmetic. Quantization is to be performed by rounding each product to the nearest integer, and $\omega_s = 2\pi$ rad/s.
 (a) Estimate the peak-to-peak amplitude and frequency of the limit cycle by using Jackson's approach.
 (b) Determine the actual amplitude and frequency of the limit cycle by simulation.
 (c) Compare the results obtained in parts (a) and (b).

11.28. Repeat Prob. 11.27 for the coefficients $b_1 = -1.8$ and $b_0 = 0.99$.

11.29. A second-order filter represented by Eq. (11.75) is implemented in terms of fixed-point decimal arithmetic.
 (a) Show that the filter can sustain zero-input limit cycles of amplitude y_0 ($y_0 > 0$) and frequency 0 or $\omega_s/2$ if Eq. (11.77) is satisfied.
 (b) Find y_0 if $b_1 = -1.375$ and $b_0 = 0.625$.

11.30. The second-order realization shown in Fig. 1.5b can under certain conditions support the elimination of zero-input limit cycles. Deduce these conditions.

11.31. Show that the state-space realization of Eqs. (11.92a) to (11.92d) supports the elimination of zero-input limit cycles.

11.32. Realize the lowpass filter of Prob. 11.13 using Meerkötter's structure shown in Fig. 11.16.

11.33. Design a sinusoidal oscillator by using a digital filter in cascade with a bandpass filter. The frequency of oscillation is required to be $\omega_s/10$.

CHAPTER 12

WAVE DIGITAL FILTERS

12.1 INTRODUCTION

The effects of coefficient quantization in digital filters can be kept small by realizing the transfer function in terms of a cascade or parallel arrangement of second-order filter sections of the type described in Sec. 11.4.

Alternative *low-sensitivity* structures can be obtained by using a synthesis methodology advanced by Fettweis [1–3] and developed further by Fettweis, Sedlmeyer, and others [4–6]. In this approach, an *equally terminated LC* filter satisfying prescribed specifications is first designed. Then by replacing analog elements by appropriate digital realizations, the LC filter is transformed into a topologically equivalent digital filter. The synthesis is based on the wave network characterization, and for this reason the resulting structures are referred to collectively as *wave digital filters*. The low sensitivity comes about because equally terminated LC filters are inherently low-sensitivity structures.

The chapter begins with a qualitative justification of the low-sensitivity attribute of equally terminated LC filters. It then proceeds to the derivation of digital realizations for the various analog elements and then to the design and analysis details of structures based on LC lattice and ladder filters. An important advantage of wave digital filters is that they can be designed to be free of zero-input and overflow limit-cycle oscillations. This property of wave digital filters is demonstrated in Sec. 12.10. Later, in Sec. 12.12, an alternative cascade synthesis is developed by using the concept of the generalized-immittance converter. This approach yields filters with improved in-band signal-to-noise ratio and leads to digital biquadratic multiple-output realizations that are amenable to VLSI implementation.

12.2 SENSITIVITY CONSIDERATIONS

An equally terminated LC filter like that in Fig. 12.1a can be characterized in terms of its *insertion loss*, which is defined as

$$L(\omega) = 10 \log \frac{P_m(\omega)}{P(\omega)}$$

$P(\omega)$ is the actual output power, and $P_m(\omega)$ is the maximum output power under perfect matching conditions. Since the LC 2-port is a passive lossless network, $P(\omega) \leq P_m(\omega)$ and thus $L(\omega) \geq 0$. Now let us assume that $L(\omega_i) = 0$ for $i = 1, 2, \ldots$, as depicted in Fig. 12.1b, as in the case of an elliptic characteristic. At frequency ω_i the

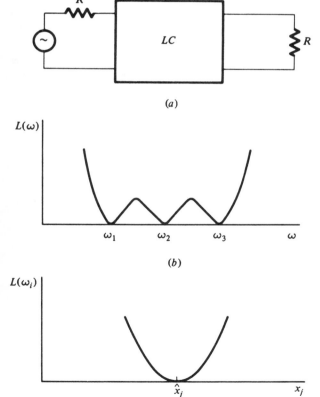

FIGURE 12.1
(a) Equally terminated LC filter, (b) equiripple loss characteristic, (c) $L(\omega_i)$ versus x_j.

filter delivers the maximum available power, and if any lossless element x_j is increased above or decreased below its nominal value $\hat{x}_j$, $L(\omega_i)$ must necessarily increase above zero as illustrated in Fig. 12.1c. Clearly

$$\lim_{x_j \to \hat{x}_j} \frac{\Delta L(\omega_i)}{\Delta x_j} = \frac{dL(\omega_i)}{dx_j} = 0$$

for $i = 1, 2, \ldots$ and $j = 1, 2, \ldots$ independently of the order of the filter [7]. Consequently, the sensitivity of the passband loss to element variations in equally terminated LC filters is inherently low. Therefore, by simulating filters of this type digitally, low-sensitivity digital-filter structures can be obtained.

In Chap. 7, we have shown that a discrete-time transfer function $H_D(z)$ can be readily obtained by applying the bilinear transformation

$$s = \frac{2}{T}\left(\frac{z-1}{z+1}\right)$$

to a continuous-time transfer function $H_A(s)$. It would, therefore, appear that one should be able to obtain low-sensitivity digital structures by simply applying the bilinear transformation to signal flow graphs of equally terminated LC filters. Unfortunately, this approach leads to flow graphs with *delay-free loops* which are not realizable (see Sec. 1.9.1). The problem is due to the fact that any realization of the bilinear transformation has a direct delay-free path between input and output.

The problem of delay-free loops can be avoided in a somewhat circuitous manner by using the wave network characterization, as will be demonstrated in the next and subsequent sections.

12.3 WAVE NETWORK CHARACTERIZATION

An analog N-port network like that in Fig. 12.2a can be represented by the set of equations

$$\left.\begin{array}{l} A_k = V_k + I_k R_k \\ B_k = V_k - I_k R_k \end{array}\right\} k = 1, 2, \ldots, N \qquad (12.1)$$

The parameters A_k and B_k are referred to as the *incident* and *reflected wave quantities*, respectively, and R_k is the *port resistance*. The representation can be either in the time or frequency domain except that lower-case symbols are usually used for the time-domain representation.

If two N-ports are cascaded as in Fig. 12.2b, it is necessary to assign

$$R_j = R_k$$

so that $\qquad A_k = B_j \qquad A_j = B_k$

i.e., a common resistance must be assigned to two interconnected ports to maintain continuity in the wave flow. Otherwise, R_k can be assigned on an arbitrary basis.

An LC filter can be regarded as a conglomerate of a number of impedances (R, sL, or $1/sC$), a source (voltage or current), and a number of 3-port series and

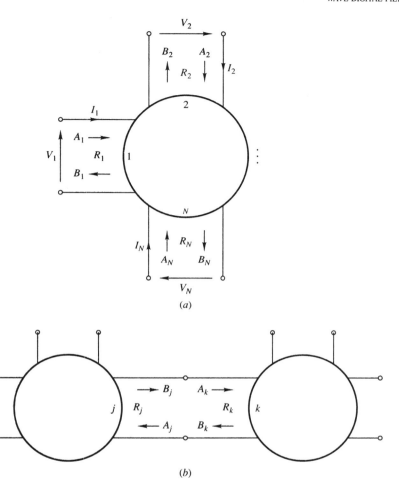

FIGURE 12.2
(a) Analog N-port network, (b) interconnected N-ports.

parallel wire interconnections like those depicted in Fig. 12.3c and d. By realizing these elements digitally and subsequently replacing analog elements in LC filters by their digital realizations, wave digital filters can be synthetized.

12.4 ELEMENT REALIZATIONS

Digital realizations for analog elements can be derived by using the following procedure:

1. Represent the element in terms of the wave characterization.

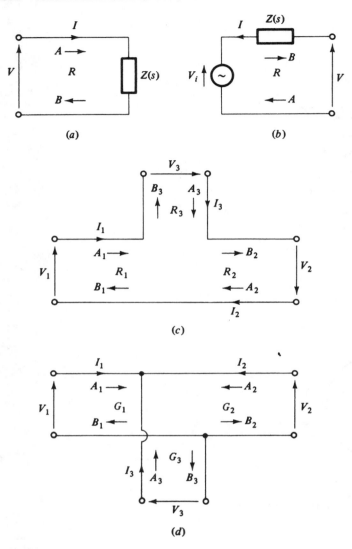

FIGURE 12.3
(a) Impedance, (b) voltage source, (c) series wire interconnection, (d) parallel wire interconnection.

2. Eliminate variables V_k, I_k, and s, using the loop and node equations and the bilinear transformation

$$s = \frac{2}{T}\left(\frac{z-1}{z+1}\right)$$

3. Express the reflected wave quantities as functions of the incident wave quantities.

4. Realize the resulting set of equations using unit delays, adders, inverters, and multipliers.

12.4.1 Impedances

Consider an *impedance*

$$Z(s) = s^\lambda R_x \qquad (12.2)$$

where R_x is a positive constant and $\lambda = -1$ for a capacitance, $\lambda = 0$ for a resistance, and $\lambda = 1$ for an inductance. From Eq. (12.1) and Fig. 12.3a

$$A = V + IR \qquad (12.3a)$$
$$B = V - IR \qquad (12.3b)$$

where

$$V = IZ(s)$$

and if the bilinear transformation is applied to continuous-time variables A, B, V, and I, i.e.,

$$Q|_{s=\frac{2}{T}\left(\frac{z-1}{z+1}\right)} \to Q \qquad (12.4)$$

for $Q = A, B, V$, and I, we obtain

$$B = f(z)A \qquad (12.5)$$

where

$$f(z) = \left.\frac{Z(s) - R}{Z(s) + R}\right|_{s=\frac{2}{T}\left(\frac{z-1}{z+1}\right)} \qquad (12.6)$$

Now on choosing

$$R = \left(\frac{2}{T}\right)^\lambda R_x \qquad (12.7)$$

and then using Eqs. (12.2) and (12.6) we have

$$f(z) = \begin{cases} z^{-1} & \text{for } \lambda = -1 \\ 0 & \text{for } \lambda = 0 \\ -z^{-1} & \text{for } \lambda = 1 \end{cases}$$

Hence Eq. (12.5) results in the element realizations of Fig. 12.4; that is, a resistance translates into a digital sink, a capacitance into a unit delay, and an inductance into a unit delay in cascade with an inverter.

12.4.2 Voltage Sources

For the *voltage source* of Fig. 12.3b, where

$$Z(s) = s^\lambda R_x$$

Element	R	Realization	Symbol
 R	R		
 C	$\dfrac{T}{2C}$	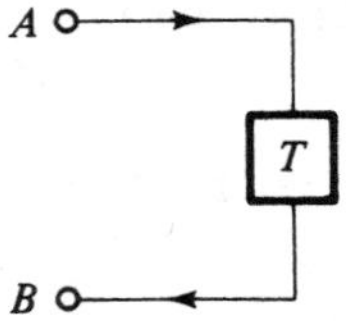	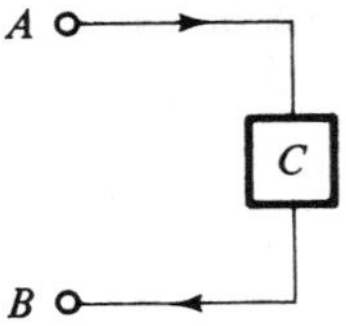
 L	$\dfrac{2L}{T}$	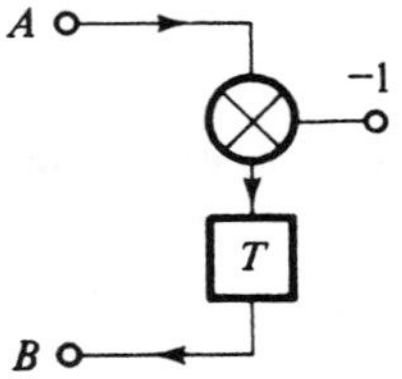	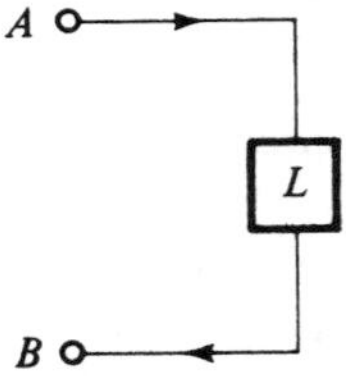

FIGURE 12.4
Digital realization of impedances.

we can write

$$A = V + IR \qquad B = V - IR \qquad V = IZ(s) + V_i$$

and on eliminating V, I, and s we deduce

$$B = f_1(z)V_i + f_2(z)A \tag{12.8}$$

where $\quad f_1(z) = \left.\dfrac{2R}{R + Z(s)}\right|_{s=\frac{2}{T}\left(\frac{z-1}{z+1}\right)} \qquad f_2(z) = \left.\dfrac{Z(s) - R}{Z(s) + R}\right|_{s=\frac{2}{T}\left(\frac{z-1}{z+1}\right)}$

With

$$R = \left(\dfrac{2}{T}\right)^\lambda R_x$$

$f_1(z)$ and $f_2(z)$ simplify to

$$f_1(z) = \begin{cases} 1 - z^{-1} & \text{for } \lambda = -1 \\ 1 & \text{for } \lambda = 0 \\ 1 + z^{-1} & \text{for } \lambda = 1 \end{cases} \quad \text{and} \quad f_2(z) = \begin{cases} z^{-1} & \text{for } \lambda = -1 \\ 0 & \text{for } \lambda = 0 \\ -z^{-1} & \text{for } \lambda = 1 \end{cases}$$

Hence Eq. (12.8) yields realizations for capacitive, resistive, and inductive sources, as depicted in Fig. 12.5.

12.4.3 Series Wire Interconnection

The preceding approach can be readily extended to *wire interconnections*. For the *series* interconnection of Fig. 12.3c

$$I_1 = I_2 = I_3 \qquad V_1 + V_2 + V_3 = 0$$

and on eliminating voltages and currents in Eq. (12.1) we can show that

$$\mathbf{B} = (\mathbf{I} - \mathbf{M_s})\mathbf{A} \tag{12.9}$$

where $\mathbf{I}$ is the 3×3 unity matrix, $\mathbf{A}$ and $\mathbf{B}$ are column vectors

$$\mathbf{M_s} = \begin{bmatrix} m_{s1} & m_{s1} & m_{s1} \\ m_{s2} & m_{s2} & m_{s2} \\ m_{s3} & m_{s3} & m_{s3} \end{bmatrix} \qquad m_{s3} = 2 - m_{s1} - m_{s2}$$

and

$$m_{sk} = \frac{2R_k}{R_1 + R_2 + R_3} \qquad \text{for } k = 1, 2 \tag{12.10}$$

A realization of Eq. (12.9) is shown in Fig. 12.6a. This can be referred to as type S2 adaptor, i.e., series 2-multiplier adaptor.

With R_2 unspecified, one can choose

$$R_2 = R_1 + R_3$$

so that

$$m_{s1} = \frac{R_1}{R_2} \qquad m_{s2} = 1$$

according to Eq. (12.10). As a consequence, the above adaptor can be simplified to the series 1-multiplier adaptor (type S1) of Fig. 12.6b.

12.4.4 Parallel Wire Interconnection

Similarly, for the *parallel* wire interconnection of Fig. 12.3d

$$V_1 = V_2 = V_3 \qquad I_1 + I_2 + I_3 = 0$$

and from Eq. (12.1)

$$\mathbf{B} = (\mathbf{M_p} - \mathbf{I})\mathbf{A} \tag{12.11}$$

where

$$\mathbf{M_p} = \begin{bmatrix} m_{p1} & m_{p2} & m_{p3} \\ m_{p1} & m_{p2} & m_{p3} \\ m_{p1} & m_{p2} & m_{p3} \end{bmatrix} \qquad m_{p3} = 2 - m_{p1} - m_{p2}$$

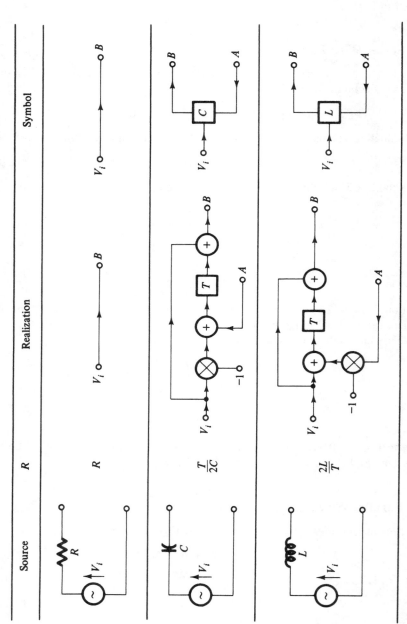

FIGURE 12.5
Digital realization of voltage sources.

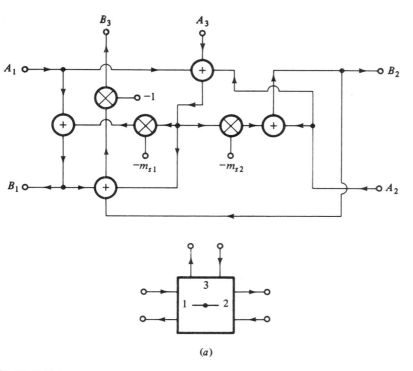

FIGURE 12.6a
Series adaptors: (a) type S2 adaptor.

and
$$m_{pk} = \frac{2G_k}{G_1 + G_2 + G_3} \quad \text{for } k = 1, 2 \tag{12.12}$$

G_k is the *port conductance*. A realization of Eq. (12.11), refered to here as type P2 adaptor, is shown in Fig. 12.7a. The corresponding 1-multiplier realization (type P1 adaptor) shown in Fig. 12.7b is obtained by choosing the conductance at port 2 as

$$G_2 = G_1 + G_3$$

so that

$$m_{p1} = \frac{G_1}{G_2} \qquad m_{p2} = 1$$

Adaptors S2 and P2 are said to be *unconstrained* since their port resistances can be assigned arbitrary values.

12.4.5 2-Port Adaptors

Unconstrained 2-port adaptors can be obtained by modifying series or parallel 3-port adaptors. By letting $A_3 = G_3 = 0$ and deleting the terminal for B_3 in the parallel

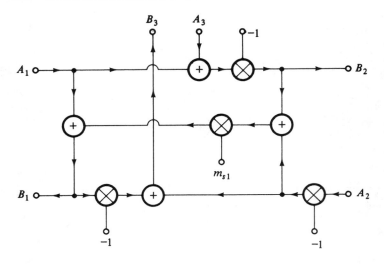

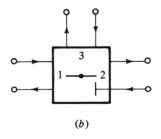

(b)

FIGURE 12.6b
Series adaptors: (b) type S1 adaptor.

adaptor of Fig. 12.7a, the 2-port adaptor depicted in Fig. 12.8a can be obtained. From Eq. (12.12)

$$m_{p1} - 1 = -(m_{p2} - 1)$$

and hence the number of multipliers in the adaptor can be reduced to one, as shown in Fig. 12.8b. The value of multiplier constant μ is given by

$$\mu = m_{p2} - 1 = \frac{G_2 - G_1}{G_2 + G_1}$$

Alternative adaptor configurations can be found in [8, 9].

12.4.6 Transformers

The above principles can be used for the derivation of digital equivalent networks for 2-port devices such as transformers, gyrators, circulators, etc. For example, the *ideal*

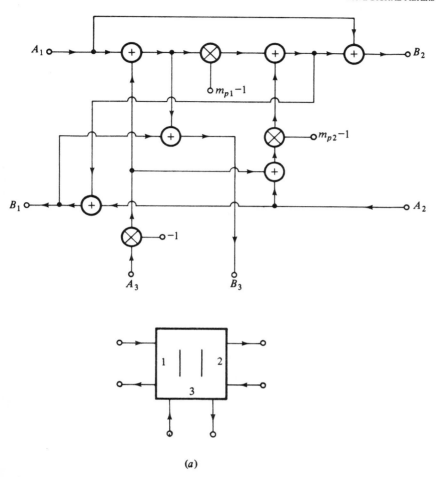

FIGURE 12.7a
Parallel adaptors: (*a*) type P2 adaptor.

transformer of Fig. 12.9a can be represented by the relations

$$V_2 = kV_1 \qquad I_2 = -I_1/k$$

where k is the turns ratio. On assigning resistances R_1 and R_2 to ports 1 and 2, respectively, we can show that

$$\mathbf{B} = \mathbf{M}_T \mathbf{A} \qquad (12.13)$$

where

$$\mathbf{M}_T = \begin{bmatrix} m_{11} & m_{12} \\ m_{21} & m_{22} \end{bmatrix}$$

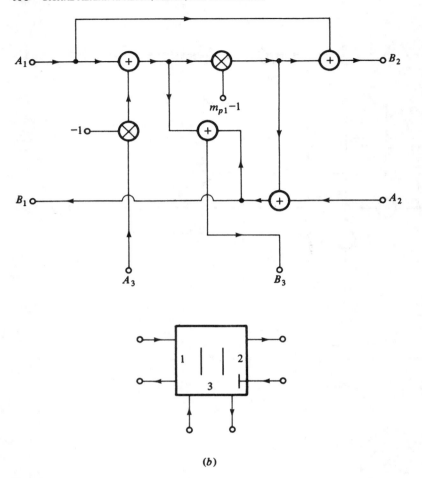

FIGURE 12.7b
Parallel adaptors: (b) type P1 adaptor.

with

$$m_{11} = \frac{R_2 - k^2 R_1}{R_2 + k^2 R_1}, \qquad m_{12} = \frac{2k R_1}{R_2 + k^2 R_1}$$

$$m_{21} = \frac{2k R_2}{R_2 + k^2 R_1}, \qquad m_{22} = -\frac{R_2 - k^2 R_1}{R_2 + k^2 R_1}$$

A realization of Eq. (12.13) is shown in Fig. 12.9b. If we assign $R_2 = k^2 R_1$, then $m_{11} = 0$, $m_{12} = 1/k$, $m_{21} = k$, and $m_{22} = 0$ and the simplified realization of Fig. 12.9c is obtained.

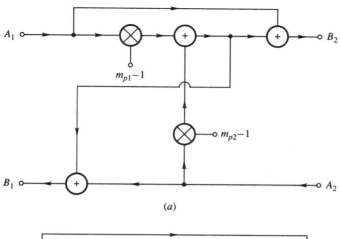

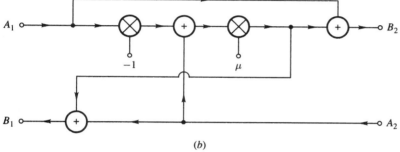

FIGURE 12.8
(a) 2-port adaptor, (b) 1-multiplier version.

12.4.7 Unit Elements

Another 2-port element which is often quite useful is the so-called *unit element*. This device simulates a transmission line of delay $T/2$ and *characteristic impedance R*, and it has been used extensively in the design of microwave filters. It can be represented by the symbol of Fig. 12.10a and is characterized by the equations

$$V_1 = k_{11}V_2 - k_{12}I_2 \tag{12.14a}$$

$$I_1 = k_{21}V_2 - k_{22}I_2 \tag{12.14b}$$

where

$$\mathbf{K} = \begin{bmatrix} k_{11} & k_{12} \\ k_{21} & k_{22} \end{bmatrix} = \frac{1}{\sqrt{1-\hat{s}^2}} \begin{bmatrix} 1 & \hat{s}R \\ \hat{s}/R & 1 \end{bmatrix} \tag{12.15}$$

with $\hat{s} = sT/2$. **K** is said to be the *chain matrix* of the 2-port. On assigning port resistances

$$R_1 = R_2 = R$$

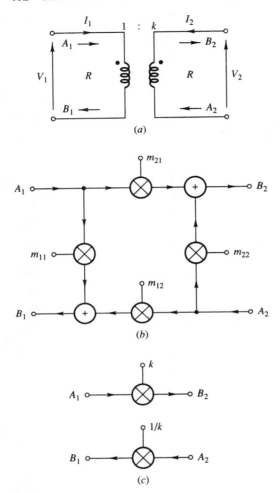

FIGURE 12.9
(a) Ideal transformer, (b) digital realization, (c) simplified version.

and then expressing B_1 and B_2 in terms of A_1 and A_2 using Eqs. (12.1), (12.4), and (12.14), we obtain

$$B_1 = z^{-\frac{1}{2}} A_2$$

$$B_2 = z^{-\frac{1}{2}} A_1$$

Therefore, a digital realization of the unit element can be obtained as shown in Fig. 12.10b.

12.4.8 Circulators

Circulators are N-port devices with $N \geq 3$ in which the reflected wave at a given port is equal to the incident wave at the adjacent port, say in the counterclockwise

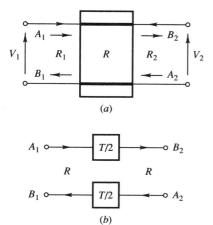

FIGURE 12.10
(a) Unit element, (b) digital realization.

direction, i.e.,

$$B_1 = A_N, B_2 = A_1, \ldots, B_N = A_{N-1}$$

A 3-port circulator and its digital realization are illustrated in Fig. 12.11a and b. Circulators, like unit elements, are used in the design of microwave circuits and filters but, as will be shown in Sec. 12.5, they can also be used in the design of an important class of wave digital filters known as *lattice filters*.

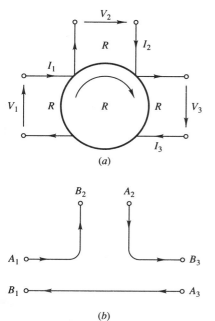

FIGURE 12.11
(a) 3-port circulator, (b) digital realization.

12.4.9 Resonant Circuits

Wave digital filters are sometimes designed as simple interconnections of *series* or *parallel resonant circuits*. Digital realizations for circuits of this type can be readily obtained by using the principles described so far. For example, a series resonant circuit comprising a capacitor C and an inductor L can be drawn as shown in Fig. 12.12a and on assigning port resistances

$$R_1 = R, \qquad R_2 = T/2C, \qquad \text{and} \qquad R_3 = 2L/T$$

the digital realization of Fig. 12.12b is obtained.

An alternative realization of a series resonant circuit can be derived by using a unit element terminated by a capacitor, as depicted in Fig. 12.13a. From Eqs. (12.14) and (12.15), the input impedance of the circuit can be deduced as

$$Z_i = \frac{1}{K_3}\left(sK_1 + \frac{1}{sK_2}\right) \tag{12.16}$$

where

$$K_1 = \frac{1}{2}R_0 T, \qquad K_2 = C_0, \qquad \text{and} \qquad K_3 = 1 + \frac{T}{2R_0 C_0}$$

Now if we assign

$$R_0 = \frac{4LC + T^2}{2TC} \qquad \text{and} \qquad C_0 = \frac{4LC^2}{4LC + T^2}$$

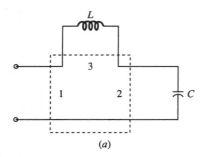

(a)

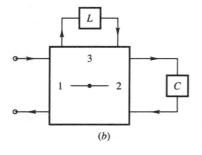

(b)

FIGURE 12.12
(a) Series resonant circuit, (b) digital realization using a 3-port adaptor.

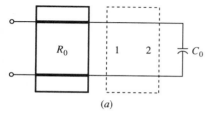

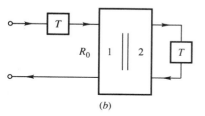

FIGURE 12.13
(*a*) Series resonant circuit using a unit element and a capacitor, (*b*) digital realization.

Eq. (12.16) assumes the form

$$Z_i = sL + \frac{1}{sC}$$

i.e., the network realizes a series resonant circuit. On assigning conductances $G_1 = 1/R_0$ and $G_2 = 2C_0/T$ to ports 1 and 2 of the wire interconnection and then replacing the analog elements by their digital counterparts, the digital realization of Fig. 12.13*b* is obtained; the multiplier constant of the 2-port adaptor can be expressed as

$$\mu = \frac{G_2 - G_1}{G_2 + G_1} = \frac{4LC - T^2}{4LC + T^2}$$

12.4.10 Realizability Constraint

Digital networks containing delay-free loops are said to be unrealizable because certain node signals in such networks cannot be computed (see Sec. 1.9.1). The networks derived so far do not contain delay-free loops. However, such can arise if adaptor ports with direct paths are interconnected. The only adaptor port without direct paths is port 2 in adaptors S1 and P1, as can be seen in Figs. 12.6*b* and 12.7*b*. Therefore, for the sake of realizability, every direct connection between adaptor ports *must necessarily involve port 2 of either an S1 or a P1 adaptor*.

12.5 LATTICE WAVE DIGITAL FILTERS

With digital realizations available for the various analog elements, several families of wave digital filters can be obtained by converting classical lattice, ladder, microwave, and other types of analog filters into digital filters [1, 4-6, 9, 10, 30, 34]. In this and the next section, we examine the realization of *lattice* and *ladder* digital filters.

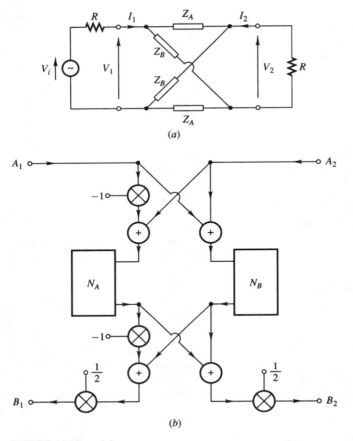

FIGURE 12.14a and b
(a) Analog lattice network, (b) alternative realization based on wave characterization.

The class of wave lattice filters is based on the lattice network of Fig. 12.14a where Z_A and Z_B are usually canonic, lossless, LC impedances.

12.5.1 Analysis

Like any other 2-port network, the lattice network of Fig. 12.14a can be represented by the wave characterization of Eq. (12.1). Applying Kirchhoff's laws to the network yields

$$I_1 = \frac{1}{Z_B}(V_1 + V_2) - I_2$$

$$I_2 = \frac{2V_2 + (Z_A - Z_B)I_1}{Z_A + Z_B}$$

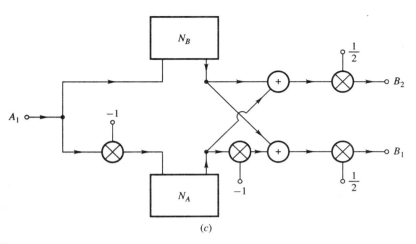

FIGURE 12.14c
(c) Simplified configuration.

and on assigning port resistances $R_1 = R_2 = R$ and then eliminating I_1, I_2, V_1, and V_2 in Eq. (12.1), we obtain

$$\mathbf{B} = \mathbf{SA} \tag{12.17}$$

where

$$\mathbf{S} = \begin{bmatrix} S_{11} & S_{12} \\ S_{21} & S_{22} \end{bmatrix} \tag{12.18}$$

with

$$S_{11} = S_{22} = (S_B + S_A)/2 \tag{12.19a}$$

$$S_{12} = S_{21} = (S_B - S_A)/2 \tag{12.19b}$$

$$S_A = \frac{Z_A - R}{Z_A + R} \tag{12.20a}$$

$$S_B = \frac{Z_B - R}{Z_B + R} \tag{12.20b}$$

12.5.2 Alternative Lattice Configuration

From Eqs. (12.17)–(12.19), we can write

$$B_1 = \frac{1}{2}[S_A(A_1 - A_2) + S_B(A_1 + A_2)]$$

$$B_2 = \frac{1}{2}[S_A(A_2 - A_1) + S_B(A_1 + A_2)]$$

Thus an alternative analog realization for the lattice network can be derived, as depicted in Fig. 12.14b, where networks N_A and N_B realize S_A and S_B, respectively.

With port 2 of the network in Fig. 12.14a terminated by a resistance R, we have $A_2 = 0$ and hence, the simplified configuration of Fig. 12.14c is obtained. The transfer functions of the configuration from input to outputs B_1 and B_2 are given by

$$\tilde{H}_A(s) = \frac{B_1}{A_1} = S_{11} = \frac{1}{2}(S_B + S_A) \qquad (12.21a)$$

$$H_A(s) = \frac{B_2}{A_1} = S_{21} = \frac{1}{2}(S_B - S_A) \qquad (12.21b)$$

If Z_A and Z_B are assumed to be lossless LC impedances in order to achieve the low-sensitivity property described in Sec. 12.2, then for $s = j\omega$ they assume imaginary values and, therefore, from Eq. (12.20)

$$|S_A| = |S_B| = 1 \qquad \text{for all } \omega$$

i.e., S_A and S_B are *allpass* transfer functions.

An analog lattice filter can be designed by expressing its transfer function as the difference of allpass transfer functions S_A and S_B, as in Eq. (12.21b), and then realizing these transfer functions. Below we consider these tasks in the reverse order.

An allpass transfer function of the form

$$S = \frac{B}{A} = \frac{Z(s) - R}{Z(s) + R}$$

can be realized either directly as an allpass network or indirectly by realizing impedance $Z(s)$ and then applying the wave characterization, as can be seen from Eq. (12.3). An arbitrary LC impedance can be realized in terms of the classical *Foster* or *Cauer forms* or a combination of the two or by a cascade arrangement of unit elements [9]. Consequently, a large variety of realizations are possible for S_A and S_B.

The design effort can be kept to a minimum by realizing S_A and S_B in terms of cascade arrangements of first- and second-order allpass sections which can be realized by simple reactances or resonant circuits. Thus, an arbitrary allpass network represented by

$$S = \prod_{i=1}^{K} S_i$$

can be realized by the cascade arrangement illustrated in Fig. 12.15a. Now if the return path from A_2 to B_1 is included as shown and the port resistances are assumed to be R throughout, an LC network realizing S can be obtained as illustrated in Fig. 12.15b.

Let us now consider the decomposition of the required transfer function into a difference of two allpass transfer functions, as in Eq. (12.21b). If we let

$$S_A = -\frac{d_A(-s)}{d_A(s)} \qquad (12.22a)$$

$$S_B = \frac{d_B(-s)}{d_B(s)} \qquad (12.22b)$$

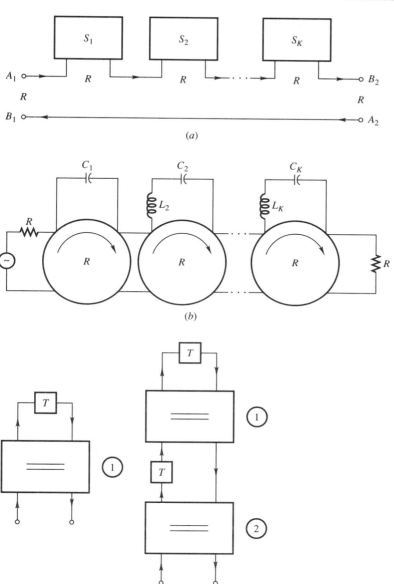

FIGURE 12.15
(a) Realization of transfer function S, (b) corresponding LC network, (c) digital realization of first-order allpass network, (d) digital realization of second-order allpass network.

$$S_{11} = \frac{1}{2}(S_B + S_A) = \frac{M(s)}{D(s)} \tag{12.23a}$$

$$S_{21} = \frac{1}{2}(S_B - S_A) = \frac{N(s)}{D(s)} \tag{12.23b}$$

where $d_A(s)$ and $d_B(s)$ are *Hurwitz polynomials* of degree N_1 and N_2, respectively, with N_1 odd and N_2 even, then Eqs. (12.22) and (12.23) give

$$D(s) = d_A(s)d_B(s)$$

$$\frac{1}{2}[d_A(s)d_B(-s) - d_A(-s)d_B(s)] = M(s)$$

and

$$\frac{1}{2}[d_A(s)d_B(-s) + d_A(-s)d_B(s)] = N(s)$$

where $D(s)$ is of degree N with N odd. Hence

$$d_A(s)d_B(-s) = M(s) + N(s) \tag{12.24}$$

For Butterworth, Chebyshev, inverse-Chebyshev, and elliptic lowpass filters, the factorization of Eq. (12.24) exists. Furthermore, it can be easily obtained by using a technique due to Rhodes [11], as will now be shown. If the denominator of the transfer function is expressed as

$$D(s) = (s + p_1) \prod_{i=2}^{(N+1)/2} (s + p_i)(s + p_i^*)$$

where p_1 is real and Im p_i > Im p_{i-1} > 0 for $i = 2, 3, \ldots, (N+1)/2$, then the factorization in Eq. (12.24) can be carried out by assigning poles with odd index to $d_A(s)$ and poles with even index to $d_B(s)$, i.e.,

$$d_A(s) = (s + p_1)(s + p_3)(s + p_3^*) \cdots \tag{12.25a}$$

$$d_B(s) = (s + p_2)(s + p_2^*)(s + p_4)(s + p_4^*) \cdots \tag{12.25b}$$

For the aforementioned types of filters, the transfer functions $\tilde{H}_A(s)$ and $H_A(s)$ satisfy the so-called *Feldkeller equation* [10] given by

$$|\tilde{H}_A(j\omega)|^2 + |H_A(j\omega)|^2 = 1$$

Consequently, if $\tilde{H}_A(s)$ represents a lowpass filter such that

$$\tilde{H}_A(e^{j\omega T}) \approx \begin{cases} 1 & \text{for } 0 < |\omega| < \omega_p \\ 0 & \text{otherwise} \end{cases}$$

then

$$H_A(e^{j\omega T}) \approx \begin{cases} 1 & \text{for } \omega_p < |\omega| < \omega_s/2 \\ 0 & \text{otherwise} \end{cases} \tag{12.26}$$

and vice versa. In effect, the configuration of Fig. 12.14c realizes simultaneously a lowpass and a highpass filter whose frequency responses are complementary. The structure finds applications in the design of quadrature mirror-image filter banks (see Sec. 16.3).

12.5.3 Digital Realization

Given a lowpass transfer function of the form

$$H_A(s) = \frac{N(s)}{(s + b_{01}) \prod_{i=2}^{(N+1)/2}(s^2 + b_{1i}s + b_{0i})} \quad (12.27)$$

allpass sections characterized by

$$H_1(s) = \frac{-s + b_{01}}{s + b_{01}} \quad \text{and} \quad H_i(s) = \frac{s^2 - b_{1i}s + b_{0i}}{s^2 + b_{1i}s + b_{0i}}$$

are required. $H_1(s)$ can be expressed as

$$H_1(s) = \frac{Z(s) - R}{Z(s) + R} \quad (12.28)$$

where $Z(s) = Rb_{01}/s$ is the impedance of a capacitor. It can, therefore, be realized by a 2-port adaptor with a multiplier constant

$$\mu_{11} = \frac{2 - Tb_{01}}{2 + Tb_{01}} \quad (12.29)$$

terminated by a unit delay, as depicted in Fig. 12.15c. Similarly, $H_i(s)$ can be put in the form of Eq. (12.28) with

$$Z(s) = \frac{Rs}{b_{1i}} + \frac{Rb_{0i}}{sb_{1i}}$$

In this case, $Z(s)$ represents a series resonant circuit with $L = R/b_{1i}$ and $C = b_{1i}/Rb_{0i}$. A realization of $H_i(s)$ can, therefore, be obtained as shown in Fig. 12.15d by using the structure of Fig. 12.13b. The bottom adaptor is used to match the port resistance of the resonant circuit to that of the circulator. The multiplier constants of the two adaptors are given by

$$\mu_{1i} = \frac{4 - T^2 b_{0i}}{4 + T^2 b_{0i}} \quad (12.30a)$$

and

$$\mu_{2i} = \frac{2Tb_{1i} - T^2 b_{0i} - 4}{2Tb_{1i} + T^2 b_{0i} + 4} \quad (12.30b)$$

for $i = 2, 3, \ldots, (N+1)/2$.

A transfer function of the form given by Eq. (12.27) can be realized in terms of a wave lattice structure by using the following procedure:

1. Carry out the decomposition in Eq. (12.24) as in Eq. (12.25) using the technique described.
2. Form the allpass transfer functions S_A and S_B as in Eq. (16.22).
3. Realize the allpass sections obtained in (2) using the structure in Fig. 12.15c for the first-order section and the structure of Fig. 12.15d for second-order sections.
4. Form the digital realizations of networks N_A and N_B and connect them as in Fig. 12.14c.

The transfer function of the digital filter obtained is given by Eq. (12.21b) as

$$H_D(z) = H_A(s)|_{s=\frac{2}{T}(\frac{z-1}{z+1})}$$

Example 12.1. Obtain a lattice realization for the 5th-order Butterworth lowpass transfer function

$$H_A(s) = \frac{1}{(s+1)(s^2 + 0.618034s + 1)(s^2 + 1.618034s + 1)}$$

Assume that $\omega_s = 2\pi$.

Solution. The polynomials $d_A(s)$ and $d_B(s)$ are given by

$$d_A(s) = (s+1)(s^2 + 0.618034s + 1)$$

$$d_B(s) = s^2 + 1.618034s + 1$$

Hence

$$S_A(s) = -\frac{(-s+1)(s^2 - 0.618034s + 1)}{(s+1)(s^2 + 0.618034s + 1)}$$

$$S_B(s) = \frac{s^2 - 1.618034s + 1}{s^2 + 1.618034s + 1}$$

With $\omega_s = 2\pi$, we have $T = 1s$; on using Eqs. (12.28)–(12.30), the multiplier constants of the adaptors can be computed as $\mu_{11} = 0.333333$, $\mu_{12} = 0.60$, $\mu_{22} = -0.214172$, $\mu_{13} = 0.60$, and $\mu_{23} = -0.603575$. The realization obtained is shown in Fig. 12.16.

12.6 LADDER WAVE DIGITAL FILTERS

Although lattice filters are easy to design and have low sensitivity to coefficient quantization at passband frequencies, the sensitivity at stopband frequencies can be quite high. This is due to the fact that transmission zeros (frequencies of zero gain) are achieved by the exact cancellation of the signals through networks N_A and N_B. A class of wave digital filters in which the sensitivity is low at passband as well as stopband frequencies can be obtained by realizing the 2-port in Fig. 12.1a in terms of an *LC ladder* network [11–15] and then applying the wave characterization. The following design procedure can be employed.

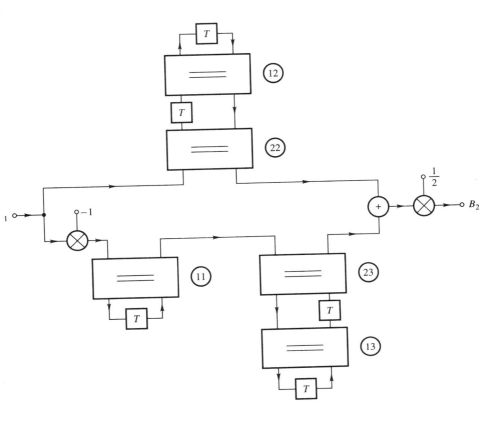

FIGURE 12.16
Lattice realization of Butterworth lowpass filter (Example 12.1).

1. Identify the various series and parallel wire interconnections in the LC filter and number the ports such that every direct connection between wire-interconnection ports involves a port 2.
2. Assign port resistances to the wire-interconnection ports. For a port terminated by an impedance $s^\lambda R_x$ or by a voltage source with an internal impedance $s^\lambda R_x$ assign a port resistance $(2/T)^\lambda R_x$. Then choose the unspecified port resistances to give as far as possible type S1 and P1 adaptors, ensuring that a common resistance is assigned to any two interconnected ports.
3. Calculate the multiplier constants for the various adaptors.
4. Replace each analog element in the LC filter by its digital realization.

The transfer function of the filter obtained is given by

$$H_D(z) = \frac{B_2}{A_1} = \frac{A_o}{B_i} \tag{12.31}$$

where A_o is the incident wave quantity for the output resistance and B_i is the reflected wave quantity for the input source. From Eqs. (12.3)–(12.6), and (12.8), we obtain $B_i = V_i$ and $A_o = 2V_o$ and so Eq. (12.31) yields

$$H_D(z) = 2H_A(s)\Big|_{s=\frac{2}{T}\left(\frac{z-1}{z+1}\right)}$$

where $H_A(s)$ is the transfer function of the analog filter.

Example 12.2. Figure 12.17a represents an elliptic lowpass filter satisfying the following specifications:

Maximum passband ripple: 1 dB   Minimum stopband loss: 34.5 dB

Passband edge: $\sqrt{0.5}$ rad/s   Stopband edge: $1/\sqrt{0.5}$ rad/s

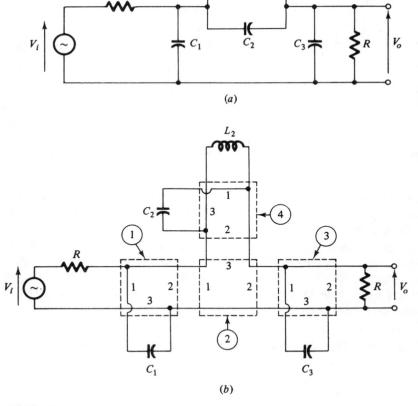

FIGURE 12.17a and b
(a) Elliptic lowpass filter (Example 12.2), (b) identification of wire interconnections.

The element values of the filter are

$$C_1 = C_3 = 2.6189 \text{ F} \qquad C_2 = 0.31946 \text{ F} \qquad L_2 = 1.2149 \text{ H} \qquad R = 1 \text{ }\Omega$$

Derive a corresponding wave digital filter using a sampling frequency of 10 rad/s.

Solution. The wire interconnections can be identified as illustrated in Fig. 12.17b. Let G_{jk} (R_{jk}) represent the port conductance (resistance) assigned to the jth port of the kth wire interconnection. From step 2 of the above procedure the following assignments can be made:

Interconnection 1:

$$G_{11} = \frac{1}{R} \qquad G_{31} = \frac{2C_1}{T} \qquad G_{21} = G_{11} + G_{31} \qquad m_{p1} = 0.107110$$

Interconnection 4:

$$G_{14} = \frac{T}{2L_2} \qquad G_{34} = \frac{2C_2}{T} \qquad G_{24} = G_{14} + G_{34} \qquad m_{p1} = 0.202741$$

Interconnection 2:

$$R_{12} = \frac{1}{G_{21}} \qquad R_{32} = \frac{1}{G_{24}} \qquad R_{22} = R_{12} + R_{32} \qquad m_{s1} = 0.120194$$

Interconnection 3:

$$G_{13} = \frac{1}{R_{22}} \qquad G_{23} = \frac{1}{R} \qquad G_{33} = \frac{2C_3}{T} \qquad \begin{matrix} m_{p1} = 0.214595 \\ m_{p2} = 0.191234 \end{matrix}$$

Interconnections 1, 2, 3, and 4 result in P1, S1, P2, and P1 adaptors, respectively, as depicted in Fig. 12.17c. The multiplier coefficients can be computed as shown above by using Eqs. (12.10) and (12.12).

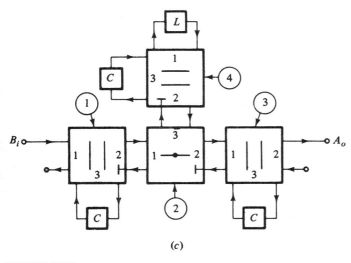

FIGURE 12.17c
(c) Wave digital filter.

The realization of Fig. 12.17c is one of a large number of possibilities since the assignment of port resistances to wire interconnections is not unique and one may realize the resonant circuits using unit elements and capacitors or inductors instead of 3-port adaptors, capacitors, and inductors (see Figs. 12.12 and 12.13). Furthermore, alternative realizations can be obtained by viewing the 2-port ladder network as a cascade connection of elemental 2-ports which can be realized individually [16–18]. While many of these structures may appear to be equivalent in terms of computational complexity, frequently there are important practical considerations that may favor one or the other type of structure, depending on the application at hand. Topological constraints necessitate that one of the adaptors in a ladder digital filter be of the unconstrained type (i.e., of type S2 or P2). Increased speed of operation and signal-to-noise ratio can often be achieved by assigning port resistances such that an unconstrained adaptor is obtained at the center of the structure, e.g., for interconnection 2 of the filter in Fig. 12.17b. There are many other issues involved in the design of wave digital filters, like their VLSI implementation and the compensation for parasitic delays that arise in practical circuits. These issues and many others are considered in some detail by Fettweis [9][†].

It should be mentioned that the numbers of delays and adders in ladder wave digital filters can sometimes be reduced somewhat by employing a pair of impedance transformations first used by Bruton [19] in the domain of active filters. In these transformations, each impedance $Z(s)$ in the analog filter is replaced by $sZ(s)$ or $Z(s)/s$. In the first case, impedances R_x, R_x/s, and sR_x are replaced by impedances sR_x, R_x, and $s^2 R_x$, i.e., resistances translate into inductances, capacitances into resistances, and inductances into s^2-impedance elements. With $s = j\omega$, we have $s^2 = -\omega^2$, i.e., an s^2-impedance element behaves as a frequency-dependent negative resistance and for this reason the transformed network is often referred to as an *FDNR network*. With this transformation, filters with a large number of capacitances and a small number of inductances, e.g., minimum inductance lowpass and highpass filters, translate into filters with a large number of resistances and a small number of s^2-impedance elements. Since the digital equivalents of resistances are simple sinks, the FDNR filter leads to a somewhat more economical digital design [20].

12.7 FILTERS SATISFYING PRESCRIBED SPECIFICATIONS

In lattice as well as ladder digital filters, the wave quantities are transformed using the bilinear transformation (see Eq. 12.4) and, therefore, the filters obtained, like other filters based on the bilinear transformation, are subject to the warping effect discussed in Sec. 7.6.3.

Wave digital filters satisfying prescribed specifications can be designed by using the prewarping techniques of Chap. 8. A detailed design procedure is as follows:

[†]This paper includes, in addition, a fairly long list of references on the subject.

1. Using the specifications, derive an appropriate normalized lowpass transfer function according to steps 1 to 3 in Sec. 8.4.
2. Realize the transfer function derived in step 1 as an equally terminated LC lattice or ladder filter.
3. Transform the lowpass filter realized in step 2 using the appropriate formula in Table 8.1.
4. Form the desired digital filter using the procedure in Sec. 12.5 or 12.6.

For the case of ladder filters, step 2 can be carried out by using filter-design tables like those found in Skwirzynski [12], Saal [13], and Zverev [14] or filter-synthesis programs like that described by Szentirmai in [21].

Example 12.3. Design a wave bandpass digital filter satisfying the following specifications:

Maximum passband ripple : 1 dB   Minimum stopband loss : 35 dB

Lower and upper passband edges : 2.0, 3.0 rad/s

Lower and upper stopband edges : 1.5, 3.5 rad/s

Sampling frequency : 10 rad/s.

Solution. On choosing an elliptic approximation and then using the procedure in Sec. 8.4, we obtain

$$n = 3 \quad k = 0.4472136 \quad \omega_0 = 3.183099 \quad B = 3.093133$$

where n and k are the order and selectivity factor of the normalized lowpass filter, respectively, and ω_0 and B are the parameters in the transformation

$$s = \frac{1}{B}\left(\bar{s} + \frac{\omega_0^2}{\bar{s}}\right)$$

A normalized lowpass LC filter with $n = 3$ and $k = 0.45$ can be obtained from [12] as depicted in Fig. 12.17a, where we now have

$$C_1 = C_3 = 2.8130 \text{ F} \quad C_2 = 0.26242 \text{ F} \quad L_2 = 1.3217 \text{ H} \quad R = 1 \; \Omega$$

On applying the above lowpass-to-bandpass transformation, the bandpass filter of Fig. 12.18a can be formed, where

$$L_1' = L_4' = 0.108525 \text{ H} \quad C_1' = C_4' = 0.909434 \text{ F}$$

$$L_2' = 0.427301 \text{ H} \quad C_2' = 0.230975 \text{ F}$$

$$L_3' = 1.16333 \text{ H} \quad C_3' = 0.0848396 \text{ F}$$

$$R = 1 \; \Omega$$

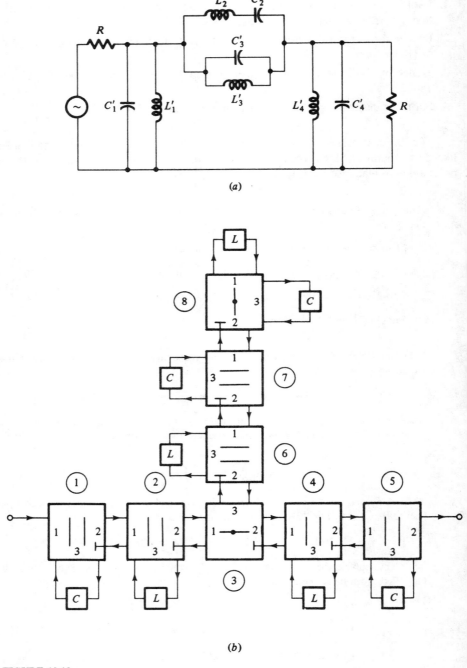

FIGURE 12.18
(a) Bandpass elliptic filter (Example 12.3), (b) wave digital filter.

TABLE 12.1
Multiplier constants (Example 12.3)

Adaptor	Type	k	m_{pk} or m_{sk}
1	P1	1	0.256751
2	P1	1	0.573642
3	S1	1	0.117926
4	P1	1	0.216662
5	P2	1	0.973738
		2	0.263494
6	P1	1	0.702492
7	P1	1	0.576495
8	S1	1	0.500000

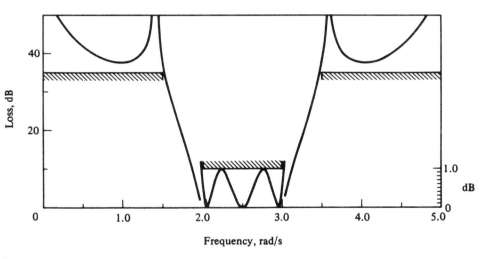

FIGURE 12.19
Loss characteristic of bandpass filter (Example 12.3).

Subsequently, on using the procedure of Sec. 12.6, the wave digital filter of Fig. 12.18b can be derived. The resulting multiplier constants are given in Table 12.1. The loss characteristic achieved is plotted in Fig. 12.19.

12.8 FREQUENCY-DOMAIN ANALYSIS

Once a wave digital filter is designed, a frequency-domain analysis is often necessary to study quantization effects or simply to verify the design. Such an analysis will now be described.

Consider the network in Fig. 12.20a, where adaptor q is terminated by subnetworks N_p, N_r, and N_s. Adaptor q can be characterized by

$$H_q(z) = \frac{B_{2q}}{A_{1q}} \qquad F_{1q} = \frac{B_{1q}}{A_{1q}} \qquad F_{2q} = \frac{B_{2q}}{A_{2q}}$$

$H_q(z)$ is the transfer function of the terminated adaptor, and F_{1q} and F_{2q} are its *input functions* at ports 1 and 2, respectively, Similarly, subnetworks N_p, N_r, and N_s can be characterized by the input functions

$$F_p = \frac{B_p}{A_p} = \frac{A_{1q}}{B_{1q}} \qquad F_r = \frac{B_r}{A_r} = \frac{A_{2q}}{B_{2q}} \qquad F_s = \frac{B_s}{A_s} = \frac{A_{3q}}{B_{3q}}$$

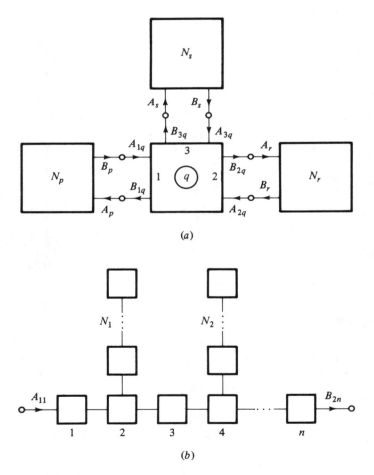

FIGURE 12.20
Analysis of wave digital filters: (a) embedded adaptor, (b) block diagram.

Expressions for $H_q(z)$, F_{1q}, and F_{2q} in terms of F_p, F_r, and F_s for series and parallel adaptors can be derived from Eqs. (12.9) and (12.11). For the S2 adaptor we have

$$H_q(z) = \frac{m_{s2}(F_s - 1)}{D_1} \qquad F_{1q} = \frac{-C_1 + C_3 F_r - C_2 F_s - F_r F_s}{D_1}$$

$$F_{2q} = \frac{-C_2 + C_3 F_p - C_1 F_s - F_p F_s}{D_2}$$

where

$$D_1 = 1 + C_2 F_r - C_3 F_s + C_1 F_r F_s \qquad D_2 = 1 + C_1 F_p - C_3 F_s + C_2 F_p F_s$$

$$C_1 = m_{s1} - 1 \qquad C_2 = m_{s2} - 1 \qquad C_3 = m_{s1} + m_{s2} - 1$$

Similarly, for the P2 adaptor

$$H_q(z) = \frac{m_{p1}(1 + F_s)}{D_1} \qquad F_{1q} = \frac{C_1 + C_3 F_r - C_2 F_s + F_r F_s}{D_1}$$

$$F_{2q} = \frac{C_2 + C_3 F_p - C_1 F_s + F_p F_s}{D_2}$$

where

$$D_1 = 1 - C_2 F_r + C_3 F_s + C_1 F_r F_s \qquad D_2 = 1 - C_1 F_p + C_3 F_s + C_2 F_p F_s$$

$$C_1 = m_{p1} - 1 \qquad C_2 = m_{p2} - 1 \qquad C_3 = m_{p1} + m_{p2} - 1$$

These relations apply to S1 and P1 adaptors except that $m_{s2} = 1$ and $m_{p2} = 1$.

Now consider the filter of Fig. 12.20b and assume that the adaptors of the main path are numbered consecutively from input to output. The overall transfer function of the filter is

$$H(z) = \frac{B_{2n}}{A_{11}}$$

Since the reflected and incident wave quantities at the output of adaptor q become the incident and reflected wave quantities at the input of adaptor $q + 1$, respectively, we can write

$$H(z) = \frac{B_{21}}{A_{11}} \frac{B_{22}}{B_{21}} \cdots \frac{B_{2n}}{B_{2(n-1)}}$$

$$= \frac{B_{21}}{A_{11}} \frac{B_{22}}{A_{12}} \cdots \frac{B_{2n}}{A_{1n}}$$

Therefore

$$H(z) = \prod_{q=1}^{n} H_q(z) \qquad (12.32)$$

For the connection of Fig. 12.20a, $H_q(z)$ and F_{1q} depend on F_s and F_r, as was shown earlier. If N_s (N_r) comprises a cascade of adaptors, F_s (F_r) will depend on the input function of the second adaptor in the cascade, which will in turn depend on the input function of the third adaptor, and so on. Consequently, for the filter in Fig. 12.20b, the input functions of branches $N_1, N_2, \ldots$ must be evaluated first, starting with the last adaptor and proceeding to the branch input in each case. Subsequently, the main-path adaptors should be analyzed, starting with the output adaptor and proceeding to the filter input. With the frequency responses of the individual main-path adaptors known, the overall response of the filter can be evaluated by using Eq. (12.32).

12.9 SCALING

The signal-to-noise ratio in wave digital filters can be improved by applying signal scaling, as in other types of digital filters. This can be achieved by scaling the incident and reflected wave quantities at port 1 of the ith adaptor by factors λ_i and $1/\lambda_i$, respectively, as depicted in Fig. 12.21. The first multiplier scales down the inputs of adaptor multipliers in order to avoid overflow, whereas the second one ensures that the input functions at ports 1 and 2 of the adaptor remain unchanged after the application of signal scaling. Note that the two multipliers form a 2-port which is equivalent to an ideal transformer in the analog network with a turns ratio of λ, as can be seen in Fig. 12.9c.

If the top and bottom multiplier constants are not exactly the reciprocal of each other, then frequency-response errors similar to coefficient-quantization errors will

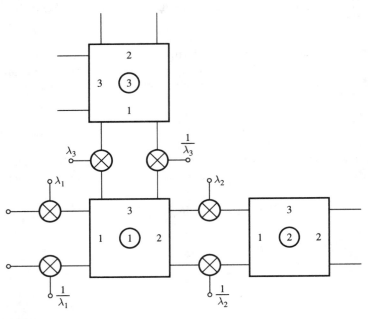

FIGURE 12.21
Application of signal scaling to wave digital filters.

occur. This problem can be avoided in practice by choosing λ to be a power of 2. This choice of constant has the additional advantage that both scaling multiplications become simple data shifts which are easy to implement. Unfortunately, however, with this choice of λ the signal-to-noise ratio cannot be optimal.

12.10 ELIMINATION OF LIMIT-CYCLE OSCILLATIONS

In Sec. 11.9.3, it was shown that in certain second-order structures zero-input limit-cycle oscillations can be eliminated by carrying out the quantization of signals in terms of magnitude truncation. A similar approach is applicable for the class of wave digital filters, as was demonstrated by Fettweis and Meerkötter [22]. The details of this approach are as follows.

Consider the wave digital filter of Fig. 12.22, where block B is a linear subnetwork containing adders, multipliers, and interconnections but no unit delays or delay-free loops. Further, assume that signal quantization is carried out by using quantizers Q_k for $k = 3, 4, \ldots, N$, as shown, and let the block enclosed by dashed lines be referred to as block $\tilde{B}$. The quantity

$$p_k(n) = G_k a_k^2(n)$$

where G_k is the conductance assigned to the kth port, represents the power[†] stored in the kth unit delay at instant nT. Since port conductances can be assigned on

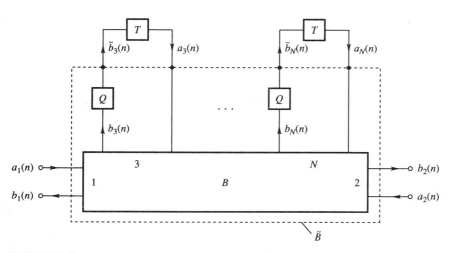

FIGURE 12.22
Elimination of zero-input limit-cycle oscillations in wave digital filters.

[†]Fettweis refers to this quantity as *pseudopower* to distinguish it from its analog counterpart which is actual power.

an arbitrary basis as long as pairs of interconnected ports are assigned the same conductances (see Sec. 12.3), G_k for $k = 3, 4, \ldots, N$ can be assumed to be positive without any loss of generality. Hence the *total power stored* in all the unit delays at instant nT can be obtained as

$$p_D(n) = \sum_{k=3}^{N} G_k a_k^2(n)$$

The increase in $p_D(n)$ after one filter cycle is given by

$$\Delta p_D(n) = p_D(n+1) - p_D(n)$$
$$= \sum_{k=3}^{N} G_k [a_k^2(n+1) - a_k^2(n)]$$

and since

$$a_k(n+1) = \tilde{b}_k(n)$$

we have

$$\Delta p_D(n) = \sum_{k=3}^{N} G_k [\tilde{b}_k^2(n) - a_k^2(n)] \tag{12.33}$$

The total power absorbed by block $\tilde{B}$ at instant nT is given by [3]

$$\tilde{p}_N(n) = \sum_{k=1}^{2} G_k [a_k^2(n) - b_k^2(n)] + \sum_{k=3}^{N} G_k [a_k^2(n) - \tilde{b}_k^2(n)] \tag{12.34}$$

Hence, Eqs. (12.33) and (12.34) give

$$\Delta p_D(n) = -\tilde{p}_N(n) + \sum_{k=1}^{2} G_k [a_k^2(n) - b_k^2(n)]$$

and under zero-input conditions such that $a_1(n) = a_2(n) = 0$, we have

$$\Delta p_D(n) = -\tilde{p}_N(n) - G_1 b_1^2(n) - G_2 b_2^2(n) \tag{12.35}$$

If the reflected quantities at ports 3 to N are quantized such that

$$|\tilde{b}_k(n)| \leq |b_k(n)| \quad \text{for } k = 3, 4, \ldots, N \tag{12.36}$$

then from Eqs. (12.35) and (12.36), the power absorbed by block $\tilde{B}$ can be expressed as

$$\tilde{p}_N(n) = \sum_{k=1}^{2} G_k [a_k^2(n) - b_k^2(n)] + \sum_{k=3}^{N} G_k [a_k^2(n) - \tilde{b}_k^2(n)]$$
$$\geq \sum_{k=1}^{N} G_k [a_k^2(n) - b_k^2(n)] = p_N(n) \tag{12.37}$$

where $p_N(n)$ is the power absorbed by block B. Now if block B represents a wave digital filter derived from a passive network, we have [3]

$$p_N(n) \geq 0 \qquad (12.38)$$

and from Eqs. (12.37) and (12.38), we conclude that

$$\tilde{p}_N(n) \geq 0$$

Hence Eq. (12.35) yields

$$\Delta p_D(n) \leq 0$$

Under these circumstances, the total power stored in the unit delays cannot increase and, for the reasons stated immediately after Eq. (11.88) on p. 376, the wave digital filter cannot sustain zero-input limit-cycle oscillations. Therefore, wave digital filters obtained from passive networks support the elimination of zero-input limit cycles.

Overflow oscillations can be eliminated as in other realizations that support the elimination of zero-input limit cycles, as described in Sec. 11.9.4 (see p. 378 and [35] of Chap. 11).

12.11 RELATED SYNTHESIS METHODS

An alternative but closely related methodology for the design of low-sensitivity digital filters has been developed by Vaidyanathan and Mitra and others [23, 24]. This methodology encompasses concepts that are analogous to those found in classical network synthesis (e.g., passivity, positive real functions, 2-port networks, extraction of elements, etc.) and provides a framework for the realization to be carried out entirely in the z domain without recourse to LC prototype filters. The methodology can be used for the realization of a variety of types of filters, including wave ladder and lattice filters as well as low-sensitivity nonrecursive filters.

Yet another class of low-sensitivity filters is the class of *lossless-discrete-integrator (LDI) ladder* filters, proposed by Bruton [25] and developed further by Bruton and Vaughan-Pope, and others [26–28]. As in wave digital filters, low sensitivity is achieved by emulating analog LC filters. In this approach, the problem of delay-free loops is avoided by replacing the bilinear transformation by the transformation

$$s = \frac{1}{T}\left(\frac{z-1}{z^{\frac{1}{2}}}\right)$$

In this way, the required digital filter is obtained directly from the analog filter without recourse to the wave characterization. The penalty paid is that the one-to-one correspondence between the imaginary axis of the s plane and the unit circle of the z plane is lost and, as a consequence, various techniques must be used to correct the distortion introduced in the amplitude response.

12.12 A CASCADE SYNTHESIS BASED ON THE WAVE CHARACTERIZATION

The wave characterization along with the concept of the *generalized-immittance converter* (GIC) [29] can be used to develop an alternative to the cascade synthesis of Sec. 4.7 [30, 31]. The details of this approach are as follows.

12.12.1 Generalized-Immittance Converters

A GIC is a 2-port whose input admittance Y_i is related to the load admittance Y_L by

$$Y_i = h(s)Y_L$$

where $h(s)$ is the *admittance conversion function* of the device. Two specific types of GIC can be identified, namely, voltage- and current-conversion GICs. The current-conversion GIC (CGIC) is characterized by the terminal relations

$$V_1 = V_2 \qquad I_1 = -h(s)I_2 \qquad (12.39)$$

This is usually represented by the symbol of Fig. 12.23a.

12.12.2 Analog G-CGIC Configuration

By interconnecting three conductances and two CGICs, we can construct the G-CGIC configuration of Fig. 12.24a [32]. If each CGIC is assumed to have a conversion function $h(s) = s$, straightforward analysis yields

$$\frac{V_o}{V_i} = \frac{k_0 G_0 + k_1 G_1 s + k_2 G_2 s^2}{G_0 + G_1 s + G_2 s^2}$$

and if $G_r = b_r$ and $k_r = a_r/b_r$ for $r = 0, 1, 2$, the network realizes the transfer function

$$H(s) = \frac{a_0 + a_1 s + a_2 s^2}{b_0 + b_1 s + b_2 s^2} \qquad (12.40)$$

By cascading a number of sections like the above any stable continuous-time transfer function can be realized.

12.12.3 Digital G-CGIC Configuration

Like an *LC* network, the G-CGIC network of Fig. 12.24a can be readily simulated by digital elements. We need only develop a digital realization for the CGIC by using the procedure outlined in Sec. 12.4.

On assigning wave quantities and conductances to the CGIC ports, as illustrated in Fig. 12.23a, and then using Eqs. (12.1), (12.4), and (12.39), we can show that

$$B_1 = A_2 + (A_1 - A_2)F(z) \qquad B_2 = A_1 + (A_1 - A_2)F(z)$$

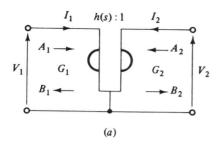

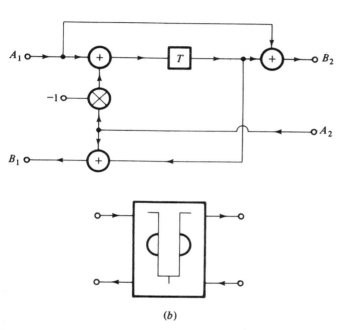

FIGURE 12.23
(a) Current-conversion generalized-immittance converter, (b) digital realization.

where
$$F(z) = \frac{G_1 - G_2 h(z)}{G_1 + G_2 h(z)} \tag{12.41}$$

$$h(z) = h(s)\big|_{s=\frac{2}{T}\left(\frac{z-1}{z+1}\right)}$$

Hence with $h(s) = s$, and $G_1 = 2G_2/T$ Eq. (12.41) reduces to

$$F(z) = z^{-1}$$

Therefore, a digital realization for the CGIC can be obtained, as depicted in Fig. 12.23b.

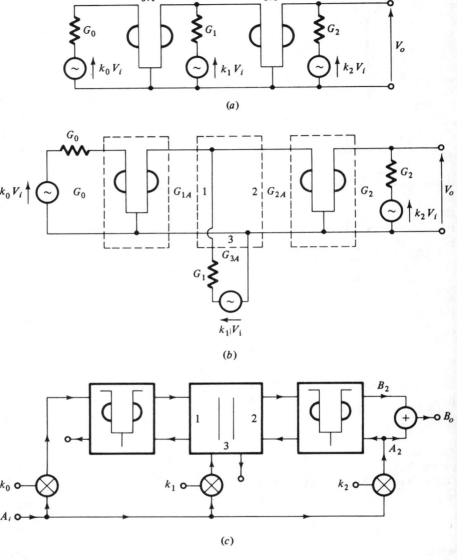

FIGURE 12.24
(a) Analog G-CGIC configuration, (b) identification of N-ports, (c) digital realization.

The individual N-ports of the G-CGIC configuration can now be identified, as indicated in Fig. 12.24b. On assigning the port conductances

$$G_{1A} = \frac{TG_0}{2} \qquad G_{2A} = \frac{2G_2}{T} \qquad G_{3A} = G_1$$

the general second-order digital section of Fig. 12.24c can be derived. An output proportional to V_o can be formed by using an adder at the input or output of any one of the CGICs, as in Fig. 12.24c, or at port 3 of the adaptor. This is permissible by virtue of Eq. (12.39).

The transfer function of the derived structure can be obtained from Eqs. (12.1) and (12.4) as

$$H_D(z) = \frac{B_o}{A_i} = \frac{B_2 + A_2}{A_i} = \frac{2V_o}{V_i} = 2H(s)\Big|_{s=\frac{2}{T}\left(\frac{z-1}{z+1}\right)}$$

12.12.4 Cascade Synthesis

Almost invariably recursive filters are designed by using Butterworth, Chebyshev, Bessel, or elliptic transfer functions which have zeros at the origin of the s plane, on the imaginary axis, or at infinity (see Chap. 5). Hence the continuous-time transfer function can be realized as a cascade connection of second-order sections characterized by transfer functions of the type

$$H_A(s) = \frac{N_A(s)}{b_0 + b_1 s + s^2}$$

where $N_A(s)$ can take the form b_0, s^2, $b_1 s$, or $a_0 + s^2$ for a lowpass (LP), highpass (HP), bandpass (BP), or notch (N) section, respectively. On the other hand, delay equalizers are designed by using allpass (AP) sections, in which

$$H_A(s) = \frac{b_0 - b_1 s + s^2}{b_0 + b_1 s + s^2}$$

Evidently, the above transfer functions are special cases of the transfer function in Eq. (12.40), and therefore they can all be readily realized by using the digital structure of Fig. 12.24c. The resulting structures are shown in Fig. 12.25, where

$$k_0 = \frac{a_0}{b_0} \tag{12.42}$$

$$m_1 = \frac{b_0 - (2/T)b_1 - (2/T)^2}{b_0 + (2/T)b_1 + (2/T)^2} \tag{12.43}$$

$$m_2 = -\frac{b_0 + (2/T)b_1 - (2/T)^2}{b_0 + (2/T)b_1 + (2/T)^2} \tag{12.44}$$

in each case.

With a set of universal sections available, any Butterworth, Chebyshev, Bessel, or elliptic digital filter satisfying prescribed specifications can be designed by using the following procedure:

1. Using the specifications, derive the appropriate normalized lowpass transfer function according to steps 1 to 3 in Sec. 8.4.
2. Apply the transformation in Eq. (8.1).
3. Select suitable sections from Fig. 12.25.

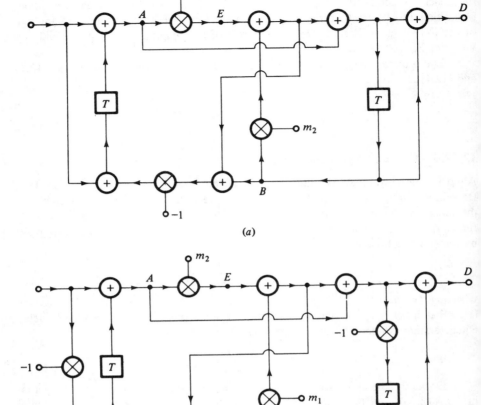

FIGURE 12.25a and b
Universal second-order CGIC sections: (a) lowpass, (b) highpass.

4. Calculate the multiplier constants using Eqs. (12.42) to (12.44).
5. Connect the various sections in cascade.

Example 12.4. A Butterworth lowpass filter is characterized by

$$H(s) = \prod_{j=1}^{3} \frac{b_{0j}}{b_{0j} + b_{1j}s + s^2}$$

where coefficients b_{ij} are given in Table 12.2. Design a corresponding digital filter by using the CGIC cascade synthesis. The sampling frequency is 10^4 rad/s.

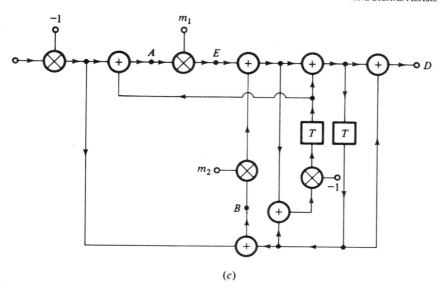

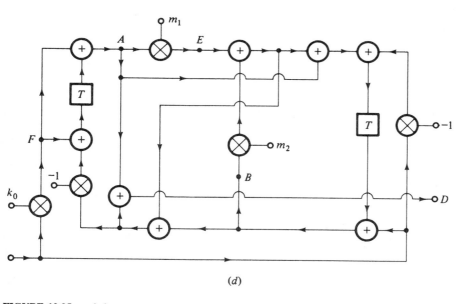

FIGURE 12.25c and d
Universal second-order CGIC sections: (c) bandpass, (d) notch.

Solution. The filter can be designed by cascading three LP sections of the type shown in Fig. 12.25a. The values of the multiplier constants can be readily evaluated as in columns 4 and 5 of Table 12.2.

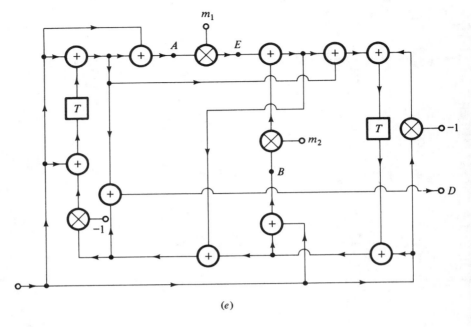

FIGURE 12.25e
Universal second-order CGIC sections: (e) allpass.

TABLE 12.2
Lowpass-filter parameters (Example 12.4)

j	b_{0j}	b_{1j}	m_{1j}	m_{2j}
1	1.069676×10^6	5.353680×10^2	-8.342350×10^{-1}	5.701500×10^{-1}
2	1.069676×10^6	1.462653×10^3	-8.650900×10^{-1}	2.778910×10^{-1}
3	1.069676×10^6	1.998021×10^3	-8.781810×10^{-1}	1.538890×10^{-1}

12.12.5 Signal Scaling

Assuming a fixed-point implementation, the CGIC sections of Fig. 12.25 can be scaled by using Jackson's technique (see Sec. 11.6.2). For this purpose each of the five sections can be represented by the signal flow graph of Fig. 12.26a, where

$$H_A(z) = \frac{N_A(z)}{D(z)}, \qquad H_B(z) = \frac{N_B(z)}{D(z)}, \quad \text{and} \quad H_D(z) = \frac{N_D(z)}{D(z)}$$

are the transfer functions between section input and nodes A, B, and D, respectively. The above polynomials are given in Table 12.3. The optimum value of λ, for L_∞ scaling, is given by

$$\lambda = \frac{1}{\max \left[\|H_A(e^{j\omega T})\|_\infty, \|H_B(e^{j\omega T})\|_\infty, \|H_D(e^{j\omega T})\|_\infty\right]}$$

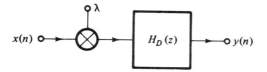

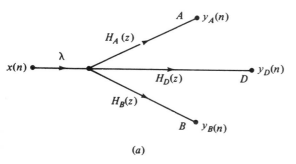

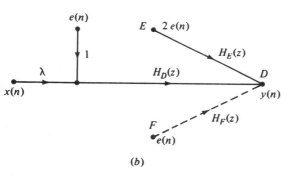

FIGURE 12.26
Universal CGIC sections: (a) scaling model, (b) noise model.

12.12.6 Output Noise

For the purpose of noise analysis, the five CGIC sections can be represented by the model of Fig. 12.26b, where $e(n)$ is the noise component generated by one multiplier and

$$H_E(z) = \frac{(z+1)(z-1)}{D(z)} \qquad H_F(z) = \frac{(1+m_1)(z+1)^2}{D(z)}$$

are the transfer functions between nodes E and F and section output. The dotted line in Fig. 12.26b applies to the N section only.

On using the approach of Sec. 11.5 the *output-noise PSD* can be deduced as

$$S_o(e^{j\omega T}) = [|H_D(e^{j\omega T})|^2 + 2|H_E(e^{j\omega T})|^2 + |H_F(e^{j\omega T})|^2]S_e(e^{j\omega T})$$

where
$$S_e(e^{j\omega T}) = \frac{q^2}{12} \qquad \text{and} \qquad H_F(e^{j\omega T}) = 0$$

TABLE 12.3
Polynomials in CGIC sections

Type	$N_A(s)$	$N_D(z)$	$N_A(z)$	$N_B(z)$
LP	b_0	$(1+m_1)(z+1)^2$	$(z-m_2)(z+1)$	$(1+m_1)(z+1)$
HP	s^2	$(1+m_2)(z-1)^2$	$(z+m_1)(z-1)$	$-(1+m_2)(z-1)$
BP	$b_1 s$	$(m_1+m_2)(1-z^2)$	$-(z^2-2m_2z+1)$	$-(z^2+2m_1z+1)$
N	a_0+s^2	$k_0(1+m_1)(z+1)^2$ $+(1+m_2)(z-1)^2$	$k_0(z-m_2)(z+1)$ $-(1+m_2)(z-1)$	$k_0(1+m_1)(z+1)$ $+(z+m_1)(z-1)$
AP	$b_0-b_1 s+s^2$	$2[(1+m_1+m_2)z^2$ $+(m_1-m_2)z+1]$	$2(z^2-2m_2z+1)$	$2(z^2+2m_1z+1)$

$D(z) = z^2 + (m_1 - m_2)z + (1 + m_1 + m_2)$

in all sections except for the N section in which

$$H_F(e^{j\omega T}) \neq 0$$

A useful property of the CGIC sections can be identified at this point. $H_E(z)$ is a bandpass transfer function in each of the five sections. As a consequence, noise generated by multipliers m_1 and m_2 will be attenuated at low as well as high frequencies, becoming zero at $\omega = 0$ as well as at $\omega = \omega_s/2$. By contrast, in the conventional canonic sections noise due to the multipliers is subjected to the same transfer function as the signal; e.g., in a lowpass section, the quantization noise is subjected to a lowpass transfer function. Because of this property, the CGIC synthesis tends to yield lowpass, highpass, and bandstop filters and also equalizers with improved in-band signal-to-noise ratio [30, 31]. In addition, like wave structures derived from passive networks, CGIC structures can be designed to be free of zero-input and overflow limit-cycle oscillations, as was demonstrated by Eswaran and Ganapathy [33].

Alternative biquadratic realizations (or *biquads*) can be obtained by applying the wave characterization to other types of analog-filter configurations. Interesting multiple-output biquads that realize all the standard second-order filter functions simultaneously can be found in [34] and [35]. Such structures could form the basis of a standard VLSI filter chip which could be used in a large variety of filter applications much like a standard digital signal processing chip.

12.13 CHOICE OF STRUCTURE

This chapter, like Chap. 4, has demonstrated that many distinct structures are possible for a given set of filter specifications. Hence one of the initial tasks of the filter designer is to choose a structure. The principal factors in this task are the sensitivity of the structure to coefficient quantization, the level of output roundoff noise, and the computational efficiency of the structure. As may be expected, these factors tend to depend to a large extent on the desired specifications; on the type of filter, i.e., lowpass, bandpass, etc.; on the type of approximation used, i.e., Butterworth, elliptic, etc.; on the type of arithmetic, i.e., fixed-point or floating-point; on the number system used, i.e., two's-complement, signed-magnitude, etc.; on the scaling norm used, i.e., L_2 or L_∞; and, in cascade structures, on the pairing of zeros and poles into second-order transfer functions and the ordering of sections and so on. Consequently, categorical statements about one or the other type of structure are difficult to make. Nevertheless, certain tendencies have been noted by researchers in the field, as follows:

1. Direct and continued-fraction ladder structures tend to be very sensitive to coefficient quantization [36, 37].
2. Direct and continued-fraction ladder structures tend to generate a high level of roundoff noise [37, 38].
3. Cascade, parallel, and wave structures tend to have similar sensitivities for fixed-point arithmetic [31, 36–38].
4. Parallel structures tend to generate a lower level of roundoff noise than cascade structures [37, 41, 42]. However, they tend to be more sensitive to coefficient quantization at stopband frequencies because the zeros can wander off the unit circle.
5. For filters with zeros on the unit circle of the z plane, cascade canonic structures involve the lowest number of arithmetic operations.
6. The direct second-order structures described in Sec. 11.4 lead to low sensitivity and are, in addition, suitable for the application of error-spectrum shaping which can reduce the level of output roundoff noise quite significantly. However, signals must be scaled before and after each multiplication (see [8] of Chap. 11).
7. State-space structures can be designed to have minimum roundoff noise but require a large number of multiplications (see [14] and [16] of Chap. 11).
8. Wave structures are significantly less sensitive than cascade structures for floating-point arithmetic [39, 40]. Similar results are expected for fixed-point arithmetic provided that scaling is applied before and after each multiplication. However, if the signal scaling constants are not powers of two, mismatch can arise at the inputs of adaptors, which can increase the sensitivity.
9. Lattice wave structures tend to be more sensitive to coefficient quantization at stopband frequencies than ladder wave structures.
10. CGIC cascade structures tend to yield improved in-band signal-to-noise ratio [30, 31].
11. State-space, wave, and CGIC structures can be designed to be free of limit-cycle oscillations.

It should be mentioned that the choice of structure involves many other issues besides the above, e.g., the suitability of the structure to the application at hand, the amenability of the structure to VLSI implementation (see Sec. 4.9), and the cost of the hardware. Also, in applications where very high sampling rates are employed, the degree of parallelism inherent in the various structures should be considered. In canonic structures, all multiplications can be performed simultaneously and, as a consequence, the time taken to complete the processing for one filter cycle can be nearly as short as the time taken to perform one multiplication. In wave structures, multiplications must be performed in sequence according to a certain hierarchy because of topological constraints, and hence the minimum time required to do the processing for one filter cycle can be much longer [31, 36].

REFERENCES

1. A. Fettweis, "Digital Filter Structures Related to Classical Filter Networks," *Arch. Elektron. Übertrag.*, vol. 25, pp. 79–89, 1971.
2. A. Fettweis, "Some Principles of Designing Digital Filters Imitating Classical Filter Structures," *IEEE Trans. Circuit Theory*, vol. CT-18, pp. 314–316, March 1971.
3. A. Fettweis, "Pseudopassivity, Sensitivity, and Stability of Wave-Digital Filters," *IEEE Trans. Circuit Theory*, vol. CT-19, pp. 668–673, November 1972.
4. A. Sedlmeyer and A. Fettweis, "Digital Filters with True Ladder Configuration," *Int. J. Circuit Theory Appl.*, vol. 1, pp. 5–10, March 1973.
5. R. Nouta, "The Jaumann Structure in Wave-Digital Filters," *Int. J. Circuit Theory Appl.*, vol. 2, pp. 163–174, June 1974.
6. A. Fettweis, H. Levin, and A. Sedlmeyer, "Wave Digital Lattice Filters," *Int. J. Circuit Theory Appl.*, vol. 2, pp. 203–211, June 1974.
7. H. J. Orchard, "Inductorless Filters," *Electron. Lett.*, vol. 2, pp. 224–225, June 1966.
8. A. Fettweis and K. Meerkötter, "On Adaptors for Wave Digital Filters," *IEEE Trans. Acoust., Speech, Signal Process.*, vol. ASSP-23, pp. 516–525, December 1975.
9. A. Fettweis, "Wave Digital Filters: Theory and Practice," *Proc. IEEE*, vol. 74, pp. 270–327, February 1986.
10. L. Gazsi, "Explicit Formulas for Lattice Wave Digital Filters," *IEEE Trans. Circuits Syst.*, vol. CAS-32, pp. 68–88, January 1985.
11. J. D. Rhodes, *Theory of Electrical Filters*, Wiley, London, 1976.
12. J. K. Skwirzynski, *Design Theory and Data for Electrical Filters*, Van Nostrand, London, 1965.
13. R. Saal, *Handbook of Filter Design*, AEG Telefunken, Backnang, 1979.
14. A. I. Zverev, *Handbook of Filter Synthesis*, Wiley, New York, 1967.
15. E. Chirlian, *LC Filters: Design, Testing, and Manufacturing*, Wiley, New York, 1983.
16. A. G. Constantinides, "Alternative Approach to Design of Wave Digital Filters," *Electron. Lett.*, vol. 10, pp. 59–60, March 1974.
17. M. N. S. Swamy and K. S. Thyagarajan, "A New Type of Wave Digital Filter," *J. Franklin Inst.*, vol. 300, pp. 41–58, July 1975.
18. A. G. Constantinides, "Design of Digital Filters from *LC* Ladder Networks," *IEE Proc.*, vol. 123, pp. 1307–1312, December 1976.
19. L. T. Bruton, "Network Transfer Functions Using Concept of Frequency-Dependent Negative Resistance," *IEEE Trans. Circuit Theory*, vol. CT-16, pp. 406–408, August 1969.
20. A. Fettweis, "Wave Digital Filters with Reduced Number of Delays," *Int. J. Circuit Theory Appl.*, vol. 2, pp. 319–330, December 1974.
21. G. Szentirmai, "FILSYN—A General Purpose Filter Synthesis Program," *Proc. IEEE*, vol. 65, pp. 1443–1458, October 1977.
22. A. Fettweis and K. Meerkötter, "Suppression of Parasitic Oscillations in Wave Digital Filters," *IEEE Trans. Circuits Syst.*, vol. CAS-22, pp. 239–246, March 1975.
23. P. P. Vaidyanathan and S. K. Mitra, "Low Passband Sensitivity Digital Filters: A Generalized Viewpoint and Synthesis Procedures," *Proc. IEEE*, vol. 72, pp. 404–423, April 1984.

24. P. P. Vaidyanathan, "A Unified Approach to Orthogonal Digital Filters and Wave Digital Filters, Based on LBR Two-Pair Extraction," *IEEE Trans. Circuits Syst.*, vol. CAS-32, pp. 673–686, July 1985.
25. L. T. Bruton, "Low-Sensitivity Digital Ladder Filters," *IEEE Trans. Circuits Syst.*, vol. CAS-22, pp. 168–176, March 1975.
26. L. T. Bruton and D. A. Vaughan-Pope, "Synthesis of Digital Ladder Filters from LC Filters," *IEEE Trans. Circuits Syst.*, vol. CAS-23, pp. 395–402, June 1976.
27. E. S. K. Liu, L. E. Turner, and L. T. Bruton, "Exact Synthesis of LDI and LDD Ladder Filters," *IEEE Trans. Circuits Syst.*, vol. CAS-31, pp. 369–381, April 1984.
28. B. D. Green and L. E. Turner, "Digital LDI Filters Using Lattice Equivalents and Wave Concepts," *IEEE Trans. Circuits Syst.*, vol. CAS-37, pp. 133–135, January 1990.
29. A. Antoniou, "Realisation of Gyrators Using Operational Amplifiers, and Their Use in RC-Active-Network Synthesis," *IEE Proc.*, vol. 116, pp. 1838–1850, November 1969.
30. A. Antoniou and M. G. Rezk, "Digital-Filter Synthesis Using Concept of Generalized-Immittance Convertor," *IEE J. Electron. Circuits Syst.*, vol. 1, pp. 207–216, November 1977 (see vol. 2, p. 88, May 1978 for errata).
31. A. Antoniou and M. G. Rezk, "A Comparison of Cascade and Wave Fixed-Point Digital-Filter Structures," *IEEE Trans. Circuits Syst.*, vol. CAS-27, pp. 1184–1194, December 1980.
32. A. Antoniou, "Novel RC-Active-Network Synthesis Using Generalized-Immittance Converters," *IEEE Trans. Circuit Theory*, vol. CT-17, pp. 212–217, May 1970.
33. C. Eswaran and V. Ganapathy, "On the Stability of Digital Filters Designed Using the Concept of Generalized-Immittance Convertor," *IEEE Trans. Circuits Syst.*, vol. CAS-28, pp. 745–747, July 1981.
34. C. Eswaran, V. Ganapathy, and A. Antoniou, "Wave Digital Biquads Derived from RC-Active Configurations," *IEEE Trans. Circuits Syst.*, vol. CAS-31, pp. 779–787, September 1984.
35. P. S. R. Diniz and A. Antoniou, "Digital-Filter Structures Based on the Concept of the Voltage-Conversion Generalized-Immittance Converter," *Can. J. Elect. & Comp. Eng.*, vol. 13, pp. 90–98, 1988.
36. R. E. Crochiere and A. V. Oppenheim, "Analysis of Linear Digital Networks," *Proc. IEEE*, vol. 63, pp. 581–595, April 1975.
37. W. K. Jenkins and B. J. Leon, "An Analysis of Quantization Error in Digital Filters Based on Interval Algebras," *IEEE Trans. Circuits Syst.*, vol. CAS-22, pp. 223–232, March 1975.
38. J. L. Long and T. N. Trick, "Sensitivity and Noise Comparison of Some Fixed-Point Recursive Digital Filter Structures," *Proc. 1975 IEEE Int. Symp. Circuits Syst.*, pp. 56–59.
39. R. E. Crochiere, "Digital Ladder Structures and Coefficient Sensitivity," *IEEE Trans. Audio Electroacoust.*, vol. AU-20, pp. 240–246, October 1972.
40. W. H. Ku and S.-M. Ng, "Floating-Point Coefficient Sensitivity and Roundoff Noise of Recursive Digital Filters Realized in Ladder Structures," *IEEE Trans. Circuits Syst.*, vol. CAS-22, pp. 927–936, December 1975.
41. L. B. Jackson, "Roundoff-Noise Analysis for Fixed-Point Digital Filters Realized in Cascade or Parallel Form," *IEEE Trans. Audio Electroacoust.*, vol. AU-18, pp. 107–122, June 1970.
42. L. B. Jackson, "Roundoff Noise Bounds Derived from Coefficient Sensitivities for Digital Filters," *IEEE Trans. Circuits Syst.*, vol. CAS-23, pp. 481–485, August 1976.

ADDITIONAL REFERENCES

Eswaran, C., Y. V. Ramana Rao, and A. Antoniou: "Modified Series and Parallel Adaptors that Lead to Reduced Sensitivity in Wave Digital Filters," *Can. J. Elect. & Comp. Eng.*, vol. 16, pp. 47–52, 1991.

Fettweis, A.: "On the Connection Between Multiplier Word Length Limitation and Roundoff Noise in Digital Filters," *IEEE Trans. Circuit Theory*, vol. CT-19, pp. 486–491, September 1972.

———: "Roundoff Noise and Attenuation Sensitivity in Digital Filters with Fixed-Point Arithmetic," *IEEE Trans. Circuit Theory*, vol. CT-20, pp. 174–175, March 1973.

———: "On Sensitivity and Roundoff Noise in Wave Digital Filters," *IEEE Trans. Acoust., Speech, Signal Process.*, vol. ASSP-22, pp. 383–384, October 1974.

———: "On the Evaluation of Roundoff Noise in Digital Filters," *IEEE Trans. Circuits Syst.*, vol. CAS-22 p. 896, November 1975.

———, and K. Meerkötter: "On Parasitic Oscillations in Digital Filters Under Looped Conditions," *IEEE Trans. Circuits Syst.*, vol. CAS-24, pp. 475–481, September 1977.

———, and J. A. Nossek: "Sampling Rate Increase and Decrease in Wave Digital Filters," *IEEE Trans. Circuits Syst.*, vol. CAS-29, pp. 797–806, December 1982.

Gazsi, L.: "DSP-Based Implementation of a Transmultiplexer Using Wave Digital Filters," *IEEE Trans. Commun.*, vol. COM-30, pp. 1587–1597, July 1982.

———: "Single Chip Filter Bank with Wave Digital Filters," *IEEE Trans. Acoust., Speech, Signal Process.* vol. ASSP-30, pp. 709–718, October 1982.

Lawson, S. S.: "Implementation of Wave Digital Filter Structures," *IEE Proc.*, vol. 128, pt. G, pp. 224–226 August 1981.

———: "Wave Digital Filter Hardware Structure," *IEE Proc.*, vol. 128, pt. G, pp. 307–312, December 1981.

Manivannan, K., C. Eswaran, and A. Antoniou: "Design of Low-Sensitivity Universal Wave Digital Biquads," *IEEE Trans. Circuits Syst.*, vol. CAS-36, pp. 108–113, January 1989.

Owenier, K.-A.: "A General Method for the Efficient Computation of Sensitivities in Wave Digital Filters," *IEEE Trans. Circuits Syst.*, vol. CAS-30, pp. 750–757, October 1983.

Rao, S. K., and T. Kailath: "Orthogonal Digital Filters for VLSI Implementation," *IEEE Trans. Circuit. Syst.*, vol. CAS-31, pp. 933–945, November 1984.

Sikström, B.: "An Approach to the Hardware Implementation of Wave Digital Filters," *Signal Processing* vol. 1, pp. 259–263, October 1979.

Vaidyanathan, P. P., S. K. Mitra, and Y. Neuvo: "A New Approach to the Realization of Low-Sensitivity IIR Digital Filters," *IEEE Trans. Acoust., Speech, Signal Process.*, vol. 34, pp. 350–361, April 1986

Van Ginderdeuren, J. K. J., H. J. De Man, N. F. Goncalves, and W. A. M. Van Noije: "Compact NMOS Building Blocks and a Methodology for Dedicated Digital Filter Applications," *IEEE J. Solid-State Circuits*, vol. SC-18, pp. 306–316, June 1983.

Van Haften, D., and P. M. Chirlian: "An Analysis of Errors in Wave Digital Filters," *IEEE Trans. Circuits Syst.*, vol. CAS-28, pp. 154–160, February 1981.

Verkroost, G., and H. J. Butterweck: "Suppression of Parasitic Oscillations in Wave Digital Filters and Related Structures by Means of Controlled Rounding," *Arch. Elektron. Übertrag.*, vol. 30, pp. 181–186, May 1976.

Wegener, W.: "On the Design of Wave Digital Lattice Filters with Short Coefficient Word Lengths and Optimal Dynamic Range," *IEEE Trans. Circuits Syst.*, vol. CAS-25, pp. 1091–1098, December 1978.

———: "Wave Digital Directional Filters with Reduced Number of Multipliers and Adders," *Arch. Elektron Übertrag.*, vol. 33, pp. 239–243, June 1979.

PROBLEMS

12.1. Figure P12.1 depicts an equally terminated LC filter.
  (a) Obtain a signal flow graph for the filter by applying Kirchhoff's voltage and current laws.

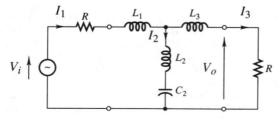

FIGURE P12.1

(b) Show that the application of the bilinear transformation to the signal flow graph of part (a) leads to at least one delay-free loop.

12.2. Figure P12.2 represents an independent current source with an internal impedance $Z(s) = s^\lambda R_x$. Obtain digital realizations for $\lambda = -1, 0,$ and 1 if $R = (2/T)^\lambda R_x$.

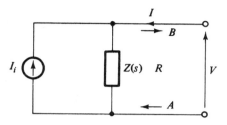

FIGURE P12.2

12.3. Analyze the series adaptors of Fig. 12.6a and b.

12.4. Analyze the parallel adaptors of Fig. 12.7a and b.

12.5. Show that a 2-port series adaptor and a 2-port parallel adaptor are interrelated by the equivalence of Fig. P12.5.

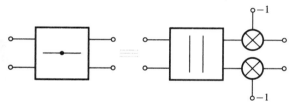

FIGURE P12.5

12.6. (a) Analyze the 2-port parallel adaptor of Fig. 12.8b.
   (b) Obtain an alternative 1-multiplier 2-port parallel adaptor.

12.7. Show that a 2-port parallel adaptor is equivalent to the realization of a transformer with a turns ratio $k = 1$.

12.8. The 2-port of Fig. P12.8, where $V_1 = -RI_2$ and $V_2 = RI_1$, represents a gyrator circuit. Obtain a corresponding digital realization.

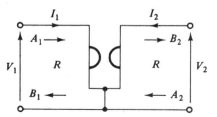

FIGURE P12.8

12.9. A 2-port in which the input impedance Z_i is related to the load impedance Z_L by $Z_i = -kZ_L$ is said to be a *negative-impedance converter* (NIC). The parameter k is referred to as the *impedance-conversion factor* of the device. Two types of NICs can be identified, namely, voltage-conversion NICs, in which

$$V_1 = -kV_2 \qquad I_1 = -I_2$$

and current-conversion NICs, in which

$$V_1 = V_2 \qquad I_1 = kI_2$$

Derive digital realizations for each case if port resistances R_1 and R_2 are assigned to the input and output ports, respectively.

12.10. (a) Obtain a digital realization of the parallel resonant circuit of Fig. P12.10a using a parallel adaptor.
(b) Show that the resonant circuit of Fig. P12.10a can be realized by a unit element terminated by an inductance L_0 as depicted in Fig. P12.10b. Obtain expressions for R_0 and L_0.
(c) Obtain a digital realization of the resonant circuit in part (b).

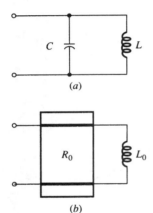

FIGURE P12.10a and b

12.11. Obtain a digital realization for the transfer function

$$H(s) = \frac{24s^4 - 24s^3 + 20s^2 - 2s + 1}{24s^4 + 24s^3 + 20s^2 + 2s + 1}$$

12.12. Derive the wave characterization of the lattice network given in Fig. 12.14a.

12.13. A 5th-order Chebyshev lowpass filter with passband edge $\omega_p = 1$ rad/s and passband ripple 1 dB is characterized by the transfer function

$$H(s) = \frac{0.1228}{(s + 0.2895)(s^2 + 0.4684s + 0.4293)(s^2 + 0.1789s + 0.9883)}$$

Assuming a sampling frequency of 8 rad/s, obtain a lattice realization for the filter.

12.14. An application calls for a highpass digital filter satisfying the specifications

$$A_p = 1.0 \text{ dB} \qquad A_a = 45.0 \text{ dB} \qquad \tilde{\Omega}_p = 3.5 \text{ rad/s} \qquad \tilde{\Omega}_a = 1.5 \text{ rad/s}$$

and the sampling frequency ω_s is to be 10 rad/s.
(a) Obtain the required transfer function using a Butterworth approximation.
(b) Realize the transfer function obtained in part (a) using a wave lattice structure.

12.15. Repeat parts (a) and (b) of Prob. 12.14 using an elliptic approximation.

12.16. Figure P12.16 shows an elliptic lowpass filter. Obtain a corresponding wave structure, assuming a sampling frequency $\omega_s = 10$ rad/s.

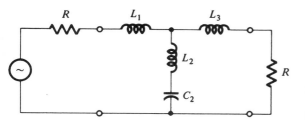

FIGURE P12.16
$L_1 = L_3 = 3.0316$ H, $L_2 = 0.21286$ H, $C_2 = 1.4396$ F, $R = 1$ Ω

12.17. Figure P12.17 shows an elliptic highpass filter in which

$A_p = 0.5$ dB $\quad A_a = 31.2$ dB $\quad \omega_p = 1/\sqrt{0.5}$ rad/s $\quad \omega_a = \sqrt{0.5}$ rad/s

(a) Obtain a corresponding wave digital filter, assuming that $\omega_s = 10$ rad/s.
(b) Determine the resulting passband and stopband edges.

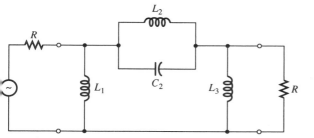

FIGURE P12.17
$L_1 = L_3 = 0.48948$ H, $L_2 = 3.4132$ H, $C_2 = 0.75489$ F, $R = 1$ Ω

12.18. An analog bandpass filter can be obtained by applying the lowpass-to-bandpass transformation

$$s = \frac{1}{10}\left(\bar{s} + \frac{625}{\bar{s}}\right)$$

to the lowpass filter of Fig. P12.16. Derive a corresponding wave digital filter if $\omega_s = 250$ rad/s.

12.19. Design a lowpass digital filter satisfying the specifications

$A_p = 1.0$ dB $\quad A_a \geq 60.0$ dB $\quad \tilde{\Omega}_p \approx 100$ rad/s $\quad \tilde{\Omega}_a \approx 200$ rad/s

assuming a sampling frequency of 1000 rad/s. Use an elliptic approximation and a wave ladder realization. (Hint: Use the tables in [12] or any other filter-design tables).

12.20. By applying the impedance transformation $Z(s) \rightarrow Z(s)/s$ to the filter of Fig. P12.16, derive a corresponding FDNR wave digital filter.

12.21. Repeat Prob. 12.20 using the highpass filter of Fig. P12.17 as a prototype.

12.22. The multiplier constants for the filter of Fig. 12.17c are given in Table P12.22. Compute the amplitude response of the filter if $\omega_s = 10$ rad/s.

TABLE P12.22

Adaptor	Multiplier constants
1	$m_{p1} = 1.341381 \times 10^{-1}$ $m_{p2} = 1.0$
2	$m_{s1} = 9.615504 \times 10^{-2}$ $m_{s2} = 1.0$
3	$m_{p1} = 1.720167 \times 10^{-1}$ $m_{p2} = 2.399664 \times 10^{-1}$
4	$m_{p1} = 2.436145 \times 10^{-1}$ $m_{p2} = 1.0$

12.23. Compute the amplitude response of the digital filter depicted in Fig. P12.23, assuming that $\omega_s = 10$ rad/s. The values of the multiplier constants are given in Table P12.23.

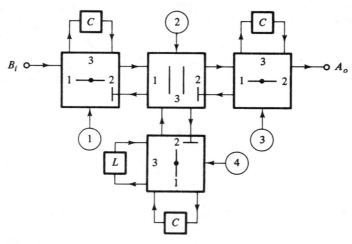

FIGURE P12.23

12.24. Derive Eq. (12.17).
12.25. (a) Obtain a digital realization for the 2-port of Fig. P12.25a, assuming that $R_1 = R_2 + L$ and $\omega_s = \pi$ rad/s.
  (b) Repeat part (a) for the 2-port of Fig. P12.25b if $G_1 = G_2 + C$ and $\omega_s = \pi$ rad/s.
12.26. (a) Show that the analog filter of Fig. P12.26, where $R = 1\ \Omega$, $C = 1$ F, and $L = 2$ H, represents a 3rd-order Butterworth filter.
  (b) Derive a corresponding digital filter using the realizations of the inductor and capacitor obtained in Prob. 12.25.
12.27. Analyze the configuration of Fig. 12.24a.
12.28. (a) Derive the lowpass section of Fig. 12.25a.
  (b) Derive the highpass section of Fig. 12.25b.

TABLE P12.23

Adaptor	Multiplier constants
1	$m_{s1} = 5.846557 \times 10^{-1}$ $m_{s2} = 1.0$
2	$m_{p1} = 4.307685 \times 10^{-1}$ $m_{p2} = 1.0$
3	$m_{s1} = 6.021498 \times 10^{-1}$ $m_{s2} = 8.172611 \times 10^{-1}$
4	$m_{s1} = 1.466151 \times 10^{-2}$ $m_{s2} = 1.0$

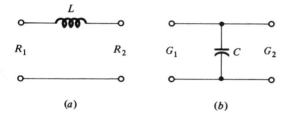

FIGURE P12.25

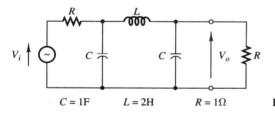

$C = 1F$ $L = 2H$ $R = 1\Omega$ **FIGURE P12.26**

12.29. An analog highpass filter is characterized by

$$H(s) = \prod_{j=1}^{3} \frac{s^2}{b_{0j} + b_{1j}s + s^2}$$

where $b_{01} = b_{02} = b_{03} = 31.15762$ $b_{11} = 10.78340$

$b_{12} = 7.8940$ $b_{13} = 2.889405$

Obtain a corresponding digital filter by using the CGIC synthesis, assuming that $\omega_s = 10$ rad/s.

12.30. Design a CGIC digital lowpass filter satisfying the following specifications:

$A_p = 0.5$ dB $A_a \geq 65$ dB $\tilde{\Omega}_p = 200$ rad/s

$\tilde{\Omega}_a = 300$ rad/s $\omega_s = 1000$ rad/s

Use an elliptic approximation.

CHAPTER 13

THE DISCRETE FOURIER TRANSFORM

13.1 INTRODUCTION

An important mathematical tool for the software implementation of digital filters is the *discrete Fourier transform* (DFT). This is closely related to the z transform on the one hand and to the continuous Fourier transform (CFT) on the other. Its importance arises because it can be efficiently computed by using some very powerful algorithms known collectively as the *fast-Fourier-transform* (FFT) *method* [1–4].

The DFT finds numerous applications in digital signal processing. However, in this chapter we focus our attention on some of its applications in the design and implementation of digital filters. In Sec. 13.8, we show that the DFT can serve as an alternative tool for the solution of the approximation problem in nonrecursive filters and in Sec. 13.12 we show that it leads to the efficient implementation of these filters.

13.2 DEFINITION

Given a finite-duration, real, discrete-time signal $x(nT)$, a corresponding *periodic signal* $x_p(nT)$ with period NT can be formed as

$$x_p(nT) = \sum_{r=-\infty}^{\infty} x(nT + rNT) \tag{13.1}$$

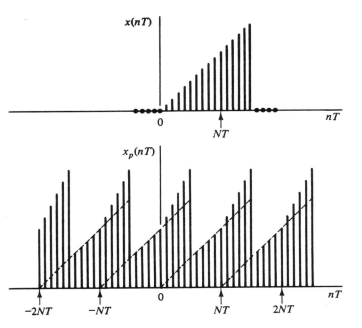

FIGURE 13.1
Signals $x(nT)$ and $x_p(nT)$.

(see Fig. 13.1). The DFT of $x_p(nT)$ is defined as

$$X_p(jk\Omega) = \sum_{n=0}^{N-1} x_p(nT) W^{-kn} = \mathcal{D} x_p(nT) \qquad (13.2)$$

where
$$W = e^{j2\pi/N} \qquad \Omega = \frac{\omega_s}{N} \qquad \omega_s = \frac{2\pi}{T}$$

In general, $X_p(jk\Omega)$ is complex and can be put in the form

$$X_p(jk\Omega) = A(k\Omega) e^{j\phi(k\Omega)}$$

where $\qquad A(k\Omega) = |X_p(jk\Omega)| \qquad$ and $\qquad \phi(k\Omega) = \arg X_p(jk\Omega)$

are discrete-frequency functions. The functions $A(k\Omega)$ and $\phi(k\Omega)$ are referred to as the *amplitude spectrum* and *phase spectrum* of $x_p(nT)$, respectively. They are analogous to the corresponding spectra of the CFT.

13.3 INVERSE DFT

The function $x_p(nT)$ is said to be the *inverse DFT* (IDFT) of $X_p(jk\Omega)$ and is given by

$$x_p(nT) = \frac{1}{N}\sum_{k=0}^{N-1} X_p(jk\Omega)W^{kn} = \mathcal{D}^{-1}X_p(jk\Omega) \qquad (13.3)$$

Proof. From the definition of the DFT

$$\frac{1}{N}\sum_{k=0}^{N-1} X_p(jk\Omega)W^{kn} = \frac{1}{N}\sum_{k=0}^{N-1}\left[\sum_{m=0}^{N-1} x_p(mT)W^{-km}\right]W^{kn}$$

$$= \frac{1}{N}\sum_{m=0}^{N-1} x_p(mT)\sum_{k=0}^{N-1} W^{k(n-m)}$$

where one can show that

$$\sum_{k=0}^{N-1} W^{k(n-m)} = \begin{cases} N & \text{for } m = n \\ 0 & \text{otherwise} \end{cases}$$

Therefore

$$\frac{1}{N}\sum_{k=0}^{N-1} X_p(jk\Omega)W^{kn} = x_p(nT)$$

13.4 PROPERTIES

13.4.1 Linearity

The DFT obeys the law of *linearity*; that is, for any two constants a and b,

$$\mathcal{D}[ax_p(nT) + by_p(nT)] = aX_p(jk\Omega) + bY_p(jk\Omega)$$

13.4.2 Periodicity

From Eq. (13.2)

$$X_p[j(k+rN)\Omega] = \sum_{n=0}^{N-1} x_p(nT)W^{-(k+rN)n} = \sum_{n=0}^{N-1} x_p(nT)W^{-kn}$$

$$= X_p(jk\Omega)$$

since $W^{-rnN} = 1$. In effect, $X_p(jk\Omega)$ is a *periodic* function of $k\Omega$ with period $N\Omega \,(= \omega_s)$.

13.4.3 Symmetry

The DFT has certain *symmetry* properties which are often useful. For example,

$$X_p[j(N-k)\Omega] = \sum_{n=0}^{N-1} x_p(nT) W^{-(N-k)n} = \sum_{n=0}^{N-1} x_p(nT) W^{kn}$$

$$= \left[\sum_{n=0}^{N-1} x_p(nT) W^{-kn} \right]^* = X_p^*(jk\Omega)$$

and as a result

$$\operatorname{Re} X_p[j(N-k)\Omega] = \operatorname{Re} X_p(jk\Omega) \qquad \operatorname{Im} X_p[j(N-k)\Omega] = -\operatorname{Im} X_p(jk\Omega)$$
$$A[(N-k)\Omega] = A(k\Omega) \qquad \phi[(N-k)\Omega] = -\phi(k\Omega) + 2\pi r$$

where r is any integer. If

$$x_p(nT) = \pm x_p[(N-n)T]$$

we obtain

$$X_p(jk\Omega) = \pm \sum_{n=0}^{N-1} x_p[(N-n)T] W^{-kn} = \pm \sum_{m=1}^{N} x_p(mT) W^{-k(N-m)}$$

$$= \pm \left[\sum_{n=0}^{N-1} x_p(nT) W^{-kn} \right]^* = \pm X_p^*(jk\Omega)$$

Thus if

$$x_p(nT) = x_p[(N-n)T]$$

we have

$$\operatorname{Im} X_p(jk\Omega) = 0$$

and if

$$x_p(nT) = -x_p[(N-n)T]$$

then

$$\operatorname{Re} X_p(jk\Omega) = 0$$

Example 13.1. Find the DFT of $x_p(nT)$ if

$$x_p(nT) = \begin{cases} 1 & \text{for } 2 \leq n \leq 6 \\ 0 & \text{for } n = 0, 1, 7, 8, 9 \end{cases}$$

assuming that $N = 10$.

Solution. From Eq. (13.2)

$$X_p(jk\Omega) = \sum_{n=2}^{6} W^{-kn} = \frac{W^{-2k} - W^{-7k}}{1 - W^{-k}}$$

$$= e^{-j4\pi k/5} \frac{\sin(\pi k/2)}{\sin(\pi k/10)}$$

The amplitude and phase spectra of $x_p(nT)$ are shown in Fig. 13.2.

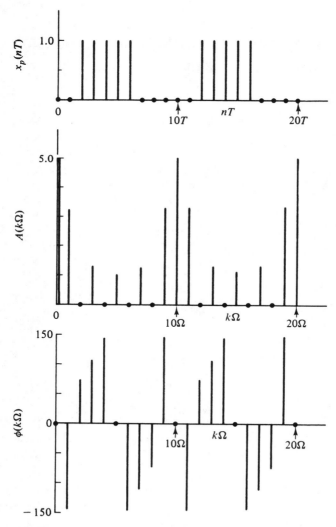

FIGURE 13.2
Amplitude and phase spectra of $x_p(nT)$ (Example 13.1).

13.5 INTERRELATION BETWEEN THE DFT AND THE z TRANSFORM

The DFT of $x_p(nT)$ can be derived from the z transform of $x(nT)$ as we shall now show.

From Eqs. (13.1) and (13.2)

$$X_p(jk\Omega) = \sum_{n=0}^{N-1} \sum_{r=-\infty}^{\infty} x(nT + rNT)W^{-kn} = \sum_{r=-\infty}^{\infty} \sum_{n=0}^{N-1} x(nT + rNT)W^{-kn}$$

and by letting $n = m - rN$ we have

$$X_p(jk\Omega) = \sum_{r=-\infty}^{\infty} \sum_{m=rN}^{rN+N-1} x(mT)W^{-k(m-rN)}$$

$$= \cdots + \sum_{m=-N}^{-1} x(mT)W^{-km} + \sum_{m=0}^{N-1} x(mT)W^{-km} + \sum_{m=N}^{2N-1} x(mT)W^{-km} + \cdots$$

$$= \sum_{m=-\infty}^{\infty} x(mT)W^{-km}$$

Alternatively, by replacing W by $e^{j2\pi/N}$ and m by n we have

$$X_p(jk\Omega) = \sum_{n=-\infty}^{\infty} x(nT)e^{-jk\Omega nT}$$

and therefore
$$X_p(jk\Omega) = X_D(e^{jk\Omega T}) \tag{13.4}$$

where
$$X_D(z) = \mathcal{F}x(nT)$$

In effect, *the DFT of $x_p(nT)$ is numerically equal to the discrete-frequency function obtained by sampling the z transform of $x(nT)$ on the unit circle $|z| = 1$*, as in Fig. 13.3.

13.5.1 Frequency-Domain Sampling Theorem

The application of the DFT in digital filtering is made possible by the *frequency-domain sampling theorem*, which is analogous to the time-domain sampling theorem considered in Chap. 6. It states that a z transform $X_D(z)$ for which

$$x(nT) = \mathcal{Z}^{-1}X_D(z) = 0 \quad \text{for} \quad n \geq N \quad \text{and} \quad n < 0 \tag{13.5}$$

can be uniquely determined from its values $X_D(e^{jk\Omega T})$, where $\Omega = \omega_s/N$. Equivalently, subject to the above condition, $X_D(z)$ can be determined from the DFT of $x_p(nT)$ by virtue of Eq. (13.4).

The validity of this theorem can be easily demonstrated. With Eq. (13.5) satisfied, $x_p(nT)$ as given by Eq. (13.1) is a periodic continuation of $x(nT)$. Hence

$$x(nT) = [u(nT) - u(nT - NT)]x_p(nT) \tag{13.6}$$

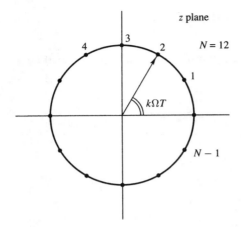

FIGURE 13.3
Relation between $X_p(jk\Omega)$ and $X_D(e^{jk\Omega T})$.

as depicted in Fig. 13.4, and so

$$X_D(z) = \mathcal{Z}\{[u(nT) - u(nT - NT)]x_p(nT)\}$$

Now
$$x_p(nT) = \mathcal{D}^{-1}X_p(jk\Omega)$$

where
$$X_p(jk\Omega) = X_D(e^{jk\Omega T})$$

and from Eq. (13.3)

$$X_D(z) = \mathcal{Z}\left\{[u(nT) - u(nT - NT)]\frac{1}{N}\sum_{k=0}^{N-1}X_p(jk\Omega)W^{kn}\right\}$$

$$= \frac{1}{N}\sum_{k=0}^{N-1}X_p(jk\Omega)\mathcal{Z}\{[u(nT) - u(nT - NT)]W^{kn}\}$$

Therefore, from Theorems 2.6 and 2.7

$$X_D(z) = \frac{1}{N}\sum_{k=0}^{N-1}X_p(jk\Omega)\frac{1-z^{-N}}{1-W^k z^{-1}} \tag{13.7}$$

since $W^{-kN} = 1$.

In summary, if $x(nT)$ is zero outside the range $0 \le nT \le (N-1)T$, then $x_p(nT)$ and $X_p(jk\Omega)$ can be obtained from $x(nT)$ and $X_D(z)$ by using Eqs. (13.1) and (13.4), respectively. Conversely, $x(nT)$ and $X_D(z)$ can be obtained from $x_p(nT)$ and $X_p(jk\Omega)$ by using Eqs. (13.6) and (13.7), respectively, as illustrated in Fig. 13.5. Therefore, $x(nT)$ can be represented by the DFT of $x_p(nT)$. As a result, any finite-duration discrete-time signal can be processed by employing FFT algorithms provided that a sufficiently large value of N is chosen.

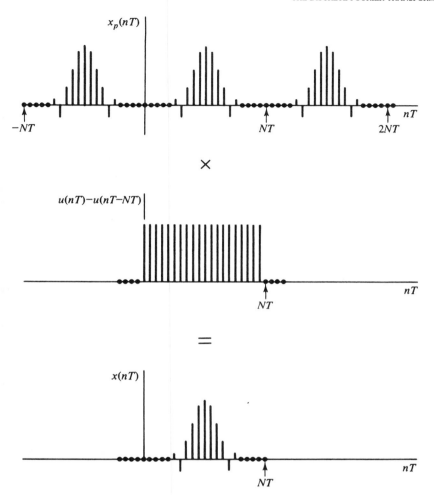

FIGURE 13.4
Derivation of $x(nT)$ from $x_p(nT)$.

13.5.2 Time-Domain Aliasing

If

$$x(nT) \neq 0 \quad \text{for} \quad n \geq N \quad \text{or} \quad n < 0$$

$x_p(nT)$ and $X_p(jk\Omega)$ can again be obtained from $x(nT)$ and $X_D(z)$ by using Eqs. (13.1) and (13.4), as depicted in Fig. 13.6. However, in this case $x(nT)$ cannot be recovered from $x_p(nT)$ by using Eq. (13.6), because of the inherent *time-domain aliasing*, and so Eq. (13.7) does not yield the z transform of $x(nT)$. Under these circumstances the DFT of $x_p(nT)$ is at best a distorted representation for $x(nT)$.

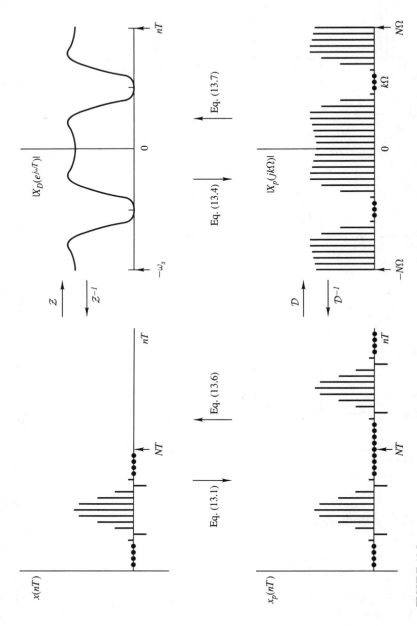

FIGURE 13.5
Interrelations between the DFT and the z transform.

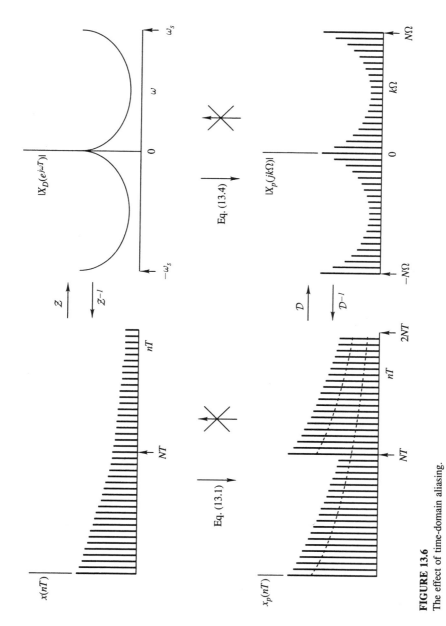

FIGURE 13.6
The effect of time-domain aliasing.

13.6 INTERRELATION BETWEEN THE DFT AND THE CFT

As is to be expected, a direct interrelation exists between the DFT and the CFT [5], which can be readily established by using the results of Secs. 13.5 and 6.6.

Let $X(j\omega)$ and $\hat{X}(j\omega)$ be the CFTs of $x(t)$ and $\hat{x}(t)$, respectively, where $\hat{x}(t)$ is the sampled (or impulse-modulated) version of $x(t)$. From Eqs. (6.33) and (13.4)

$$X_p(jk\Omega) = X_D(e^{jk\Omega T}) = \hat{X}(jk\Omega)$$

and therefore, from Eqs. (13.1) and (6.36), we have

$$\mathcal{D} \sum_{r=-\infty}^{\infty} x(nT + rNT) = \frac{1}{T} \sum_{r=-\infty}^{\infty} X(jk\Omega + jr\omega_s) \qquad (13.8)$$

Now if

$$x(t) = 0 \qquad \text{for } t < 0 \text{ and } t \geq NT \qquad (13.9)$$

and

$$X(j\omega) = 0 \qquad \text{for } |\omega| \geq \frac{\omega_s}{2} \qquad (13.10)$$

the left- and right-hand summations in the above relation become *periodic continuations* of $x(nT)$ and $X(jk\Omega)$, respectively, and as a result

$$x_p(nT) = x(nT) \qquad \text{for } 0 \leq nT \leq (N-1)T$$

$$X_p(jk\Omega) = \frac{1}{T}X(jk\Omega) \qquad \text{for } |k\Omega| < \frac{\omega_s}{2}$$

Hence $x_p(nT)$ and $X_p(jk\Omega)$ can be obtained from $x(t)$ and $X(j\omega)$ and conversely, as depicted in Fig. 13.7. That is, $x(t)$ can be represented by a DFT, and accordingly it can be processed by using the FFT method.

If Eq. (13.9) is violated, $x_p(nT)$ is no longer a periodic continuation of $x(nT)$ and the DFT of $x_p(nT)$ becomes an inaccurate representation for $x(t)$. However, for a bandlimited $x(t)$ this problem can often be overcome by using the window technique described in Sec. 9.4. We can form a truncated version of $x(t)$ [6] as

$$x'(t) = w(t)x(t)$$

where $w(t)$ is a continuous-time window function such that

$$w(t) = 0 \qquad \text{for } t < 0 \qquad \text{and} \qquad t \geq NT$$

The Fourier transform of $x'(t)$ is given by Theorem 6.8 as

$$X'(j\omega) = \frac{1}{2\pi} \int_{-\infty}^{\infty} X(jv)W(j\omega - jv)dv$$

Now by repeating the arguments of Sec. 9.4 we can demonstrate that

$$X'(j\omega) \approx X(j\omega) \qquad \text{for } |\omega| < \frac{\omega_s}{2}$$

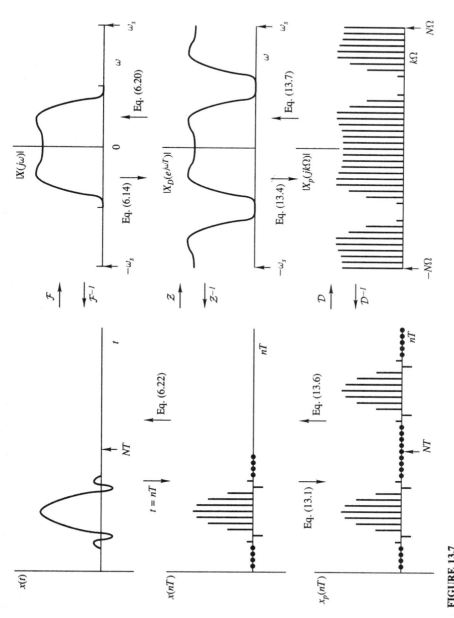

FIGURE 13.7
Interrelations between the DFT and the CFT.

and since

$$x'(t) \approx x(t) \quad \text{for } t < 0 \quad \text{and} \quad t \geq NT$$

the DFT of $x'(nT)$ is an approximate frequency-domain representation for $x(t)$. This technique is illustrated in Fig. 13.8.

The type of window function and the value of N can be chosen as in Sec. 9.4. The value of T ($= 2\pi/\omega_s$) must be chosen so that Eq. (13.10) is as far as possible satisfied. However, as the spectrum of $x(t)$ is usually unknown at the outset, one may be forced to carry out the computations for progressively smaller values of T until two successive sets of computations yield approximately the same results.

The above technique can also be used to obtain approximate DFT representations for discrete-time signals in which

$$x(nT) \neq 0 \quad \text{for } n < 0 \text{ or } n \geq N$$

13.7 INTERRELATION BETWEEN THE DFT AND THE FOURIER SERIES

The preceding results lead directly to a relationship between the DFT and the Fourier series [5].

A periodic signal $x_p(t)$ with a period T_0 can be expressed as

$$x_p(t) = \sum_{r=-\infty}^{\infty} x(t + rT_0) \tag{13.11}$$

where $x(t) = 0$ for $t < 0$ and $t \geq T_0$. Alternatively, by using the Fourier series (see Sec. 6.4), we have

$$x_p(t) = \sum_{k=-\infty}^{\infty} A(k) e^{jk\omega_0 t}$$

where $\omega_0 = 2\pi/T_0$ and

$$A(k) = \frac{1}{T_0} \int_0^{T_0} x(t) e^{-jk\omega_0 t} dt$$

Now with $t = nT$ and $T_0 = NT$, Eq. (13.11) becomes

$$x_p(nT) = \sum_{r=-\infty}^{\infty} x(nT + rNT)$$

and consequently Eq. (13.8) yields

$$X_p(jk\Omega) = \frac{1}{T} \sum_{r=-\infty}^{\infty} X(jk\Omega + jr\omega_s) \tag{13.12}$$

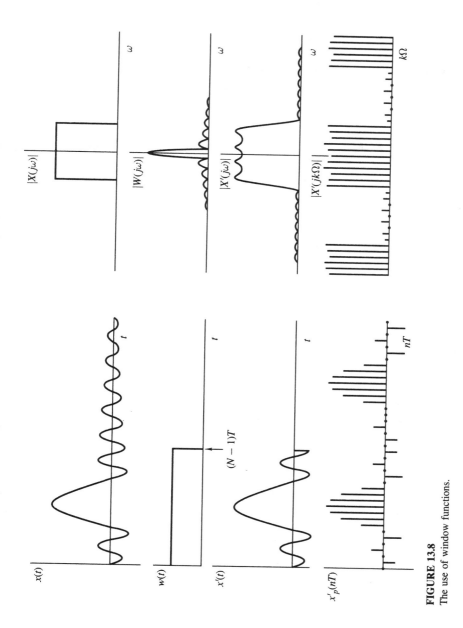

FIGURE 13.8
The use of window functions.

where
$$X(jk\Omega) = \mathcal{F}x(t)\Big|_{\omega=k\Omega} = \int_0^{T_0} x(t)e^{-jk\Omega t}\,dt$$

or
$$X(jk\Omega) = \int_0^{T_0} x(t)e^{-jk\omega_0 t}\,dt$$

since $\Omega = \omega_s/N = 2\pi/NT = 2\pi/T_0 = \omega_0$. Evidently
$$X(jk\Omega) = T_0 A(k)$$

and since $T_0 = NT$, Eq. (13.12) can be put in the form

$$X_p(jk\Omega) = \frac{1}{T}\sum_{r=-\infty}^{\infty} X[j(k+rN)\Omega] = N\sum_{r=-\infty}^{\infty} A(k+rN) \qquad (13.13)$$

In effect, *the DFT of $x_p(nT)$ can be expressed in terms of the Fourier-series coefficients of $x_p(t)$.*

Now with
$$A(k) \approx 0 \qquad \text{for } |k| \geq \frac{N}{2}$$

Eq. (13.13) gives
$$X_p(jk\Omega) \approx NA(k) \qquad \text{for } |k| < \frac{N}{2}$$

or
$$A(k) \approx \frac{1}{N} X_p(jk\Omega) \qquad \text{for } |k| < \frac{N}{2}$$

Thus the *Fourier-series coefficients* of $x_p(t)$ can be efficiently computed by using the FFT method.

13.8 NONRECURSIVE APPROXIMATIONS THROUGH THE USE OF THE DFT

The DFT can be used for the solution of the *approximation problem* in nonrecursive filters, as will now be demonstrated.

In Sec. 13.5.1 we have shown that a z transform $X_D(z)$ pertaining to a finite-duration signal can be deduced from its samples $X_D(e^{jk\Omega T})$ by using Eq. (13.7). Equivalently, if $H_p(jk\Omega)$ is a DFT, a corresponding transfer function can be formed as

$$H_D(z) = \frac{1}{N}\sum_{k=0}^{N-1} \frac{H_p(jk\Omega)(1-z^{-N})}{1-W^k z^{-1}}$$

After some manipulation, we can show that

$$H_D(e^{j\omega T}) = \frac{1}{N} e^{-j\omega(N-1)T/2} \sum_{k=0}^{N-1} \frac{H_p(jk\Omega)e^{-j\pi k/N}\sin(\omega NT/2)}{\sin(\omega T/2 - \pi k/N)}$$

and if
$$H_p(jk\Omega) = A(k\Omega)e^{j\theta(k\Omega)}$$

where $A(k\Omega) > 0$, then
$$|H_D(e^{jk\Omega T})| = A(k\Omega) \qquad \text{for } k = 0, 1, 2, \ldots, N-1$$
Furthermore, the above summation provides interpolation between samples, and as a consequence, given an amplitude response $A(\omega)$ and a phase response $\theta(\omega)$, a corresponding transfer function $H_D(z)$ can be obtained. These principles lead directly to the following approximation method:

1. Choose samples $A(k\Omega)$ and $\theta(k\Omega)$ according to the desired filter characteristic.
2. Find
$$h_p(nT) = \mathcal{D}^{-1} H_p(jk\Omega)$$
3. Form
$$h(nT) = [u(nT) - u(nT - NT)]h_p(nT)$$
4. Obtain the z transform
$$H_D(z) = \mathcal{Z} h(nT)$$

This approach is known as the *frequency-sampling approximation* method [7, 8] and is closely related to the Fourier-series method described in Sec. 9.3.

The samples of the phase response can be chosen such that a constant group delay is achieved. According to Sec. 9.2.1, constant-group-delay nonrecursive filters are characterized either by a symmetrical or an antisymmetrical impulse response such that
$$h(nT) = h[(N-1-n)T] \quad \text{or} \quad h(nT) = -h[(N-1-n)T]$$
for $0 \le n \le N-1$, and for each case N can be either odd or even. Hence there are four cases to consider.

13.8.1 Symmetrical Impulse Response

For a *symmetrical* and real impulse response, we can write
$$h_p(nT) = \frac{1}{N} \sum_{k=0}^{N-1} A(k\Omega) e^{j\theta(k\Omega)} W^{kn}$$
$$= h_p[(N-1-n)T] = h_p^*[(N-1-n)T]$$
$$= \frac{1}{N} \sum_{k=0}^{N-1} [A(k\Omega) e^{-j\theta(k\Omega)} W^{-(N-1)k}] W^{kn}$$
and hence a necessary condition for either odd or even N is obtained as
$$\theta(k\Omega) = \pi r - \frac{\pi k(N-1)}{N}$$

where r is some integer. By eliminating $\theta(k\Omega)$ in $h_p(nT)$, we can now determine the constraints on $A(k\Omega)$ and also the value (or values) of r that will guarantee a real impulse response.

For *odd N*, the constraints on $\theta(k\Omega)$ and $A(k\Omega)$ can be deduced as

$$\theta(k\Omega) = -\frac{\pi k(N-1)}{N} \quad \text{for } 0 \leq k \leq N-1$$

$$A(k\Omega) = A[(N-k)\Omega] \quad \text{for } 1 \leq k \leq N-1$$

in which case

$$h_p(nT) = \frac{1}{N}\left[A(0) + 2 \sum_{k=1}^{(N-1)/2} (-1)^k A(k\Omega) \cos\frac{\pi k(1+2n)}{N}\right]$$

Now from step 3 of the approximation method, the first formula in Table 13.1 follows. Similarly, for *even N* we can show that

$$\theta(k\Omega) = \begin{cases} -\dfrac{\pi k(N-1)}{N} & \text{for } 0 \leq k \leq \dfrac{N}{2} - 1 \\ \pi - \dfrac{\pi k(N-1)}{N} & \text{for } \dfrac{N}{2} + 1 \leq k \leq N-1 \end{cases}$$

and

$$A(k\Omega) = A[(N-k)\Omega] \quad \text{for } 1 \leq k \leq N-1$$

$$A\left(\frac{N\Omega}{2}\right) = 0$$

The corresponding impulse response is given by the second formula in Table 13.1.

13.8.2 Antisymmetrical Impulse Response

For an *antisymmetrical* and real impulse response

$$h_p(nT) = -h_p[(N-1-n)T] = -h_p^*[(N-1-n)T]$$

$$= \frac{1}{N}\sum_{k=0}^{N-1} A(k\Omega)[-e^{-j\theta(k\Omega)}W^{-(N-1)k}]W^{kn}$$

and hence $\theta(k\Omega)$ must in this case satisfy the relation

$$\theta(k\Omega) = \frac{(1+2r)\pi}{2} - \frac{\pi k(N-1)}{N}$$

where r is some integer.

TABLE 13.1
Formulas for frequency-sampling approximation method

Type	N	$h(nT)$ for $0 \leq n \leq N-1$	Restriction
Symmetrical	Odd	$\dfrac{1}{N}\left[A(0) + 2\displaystyle\sum_{k=1}^{(N-1)/2}(-1)^k A(k\Omega)\cos k\Lambda \right]$	
	Even	$\dfrac{1}{N}\left[A(0) + 2\displaystyle\sum_{k=1}^{N/2-1}(-1)^k A(k\Omega)\cos k\Lambda \right]$	$A\left(\dfrac{N\Omega}{2}\right)=0$
Antisymmetrical	Odd	$\dfrac{2}{N}\displaystyle\sum_{k=1}^{(N-1)/2}(-1)^{k+1} A(k\Omega)\sin k\Lambda$	$A(0)=0$
	Even	$\dfrac{1}{N}\left[(-1)^{N/2+n} A\left(\dfrac{N\Omega}{2}\right) + 2\displaystyle\sum_{k=1}^{N/2-1}(-1)^k A(k\Omega)\sin k\Lambda \right]$	$A(0)=0$

where $\Lambda = \dfrac{\pi(1+2n)}{N}$

For *odd* N, $\theta(k\Omega)$ is given by

$$\theta(k\Omega) = \begin{cases} \dfrac{\pi}{2} - \dfrac{\pi k(N-1)}{N} & \text{for } 1 \leq k \leq \dfrac{N-1}{2} \\ -\dfrac{\pi}{2} - \dfrac{\pi k(N-1)}{N} & \dfrac{N+1}{2} \leq k \leq N-1 \end{cases}$$

whereas for *even* N

$$\theta(k\Omega) = -\dfrac{\pi}{2} - \dfrac{\pi k(N-1)}{N} \qquad \text{for } 1 \leq k \leq N-1$$

For both these cases, $A(k\Omega)$ must satisfy the relations

$$A(k\Omega) = A[(N-k)\Omega] \qquad A(0) = 0$$

The corresponding impulse responses are given by the third and fourth formulas in Table 13.1.

The frequency responses for the four types of filters can be calculated by using the formulas given in Table 9.1.

Example 13.2. Design a lowpass filter with a frequency response

$$H(e^{j\omega T}) \approx \begin{cases} 1 & \text{for } 0 \leq |\omega| \leq 16 \text{ rad/s} \\ 0 & \text{for } 17 \leq |\omega| \leq 32.5 \text{ rad/s} \end{cases}$$

Assume that $\omega_s = 65$ rad/s and $N = 65$.

Solution. Since $\Omega = \omega_s/N = 1$, we can assign

$$A(k\Omega) = \begin{cases} 1 & \text{for } 0 \leq k \leq 16 \\ 0 & \text{for } 17 \leq k \leq 32 \end{cases}$$

By using the first formula in Table 13.1 and the corresponding formula in Table 9.1, the amplitude response shown as curve A in Fig. 13.9 is obtained.

The direct application of the frequency-sampling method usually yields unsatisfactory results; e.g., compare Figs. 13.9 and 9.4. However, by leaving some values of $A(k\Omega)$ as parameters and then using optimization methods of the type described in Chap. 14 some fairly good approximations can be obtained without a considerable amount of computation. This approach was used by Rabiner, Gold, and McGoneral [8] to develop tables of optimized filters. According to these tables, an optimized version of the above lowpass filter can be obtained by choosing

$$A(17\Omega) = 0.71742143 \qquad A(18\Omega) = 0.24385557 \qquad A(19\Omega) = 0.02368774$$

The corresponding amplitude response is shown as curve B in Fig. 13.9.

13.9 SIMPLIFIED NOTATION

The preceding somewhat complicated notation for the DFT was adopted to eliminate possible confusion between the various transforms. As we shall be dealing exclusively

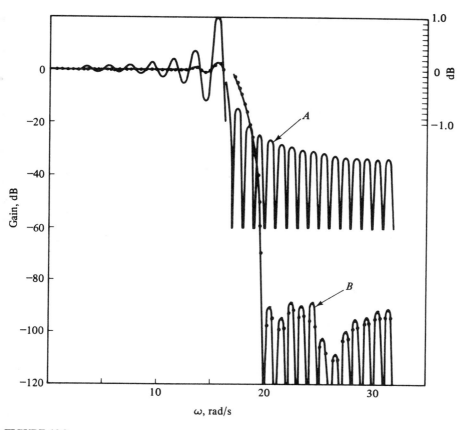

FIGURE 13.9
Amplitude response of lowpass filter (Example 13.2): curve A, without optimization, curve B, with optimization.

with the DFT from now on, we can write

$$X(k) = \sum_{n=0}^{N-1} x(n) W^{-kn} \qquad x(n) = \frac{1}{N} \sum_{k=0}^{N-1} X(k) W^{kn}$$

where $\qquad x(n) \equiv x_p(nT) \qquad X(k) \equiv X_p(jk\Omega)$

i.e., $x(n)$ and $X(k)$ are implicitly assumed to be periodic.

13.10 PERIODIC CONVOLUTIONS

The convolutions of the CFT, like those of the z transform, have been used extensively in previous chapters. Analogous and equally useful relations can be established for the DFT.

13.10.1 Time-Domain Convolution

The *time-domain convolution* of two periodic signals $x(n)$ and $h(n)$, each with period N, is defined as

$$y(n) = \sum_{m=0}^{N-1} x(m)h(n-m) = \sum_{m=0}^{N-1} x(n-m)h(m)$$

Like $x(n)$ or $h(n)$, $y(n)$ is a periodic function of n with period N.

Example 13.3. Find $y(5)$ if

$$x(n) = e^{-\alpha n} \quad \text{for } 0 \le n \le 9$$

and

$$h(n) = \begin{cases} 1 & \text{for } 3 \le n \le 6 \\ 0 & \text{for } 0 \le n \le 2 \\ 0 & \text{for } 7 \le n \le 9 \end{cases}$$

Each signal has a period of 10.

Solution. From the construction of Fig. 13.10b

$$y(5) = 1 + e^{-\alpha} + e^{-2\alpha} + e^{-9\alpha}$$

Owing to the periodicity of $x(n)$, $y(5)$ can also be calculated by using the construction of Fig. 13.10c. The DFT of $y(n)$ is

$$Y(k) = \sum_{n=0}^{N-1} \left[\sum_{m=0}^{N-1} h(m)x(n-m) \right] W^{-kn}$$

$$= \sum_{m=0}^{N-1} h(m) W^{-km} \sum_{n=0}^{N-1} x(n-m) W^{-k(n-m)}$$

and therefore

$$Y(k) = H(k)X(k) \tag{13.14}$$

13.10.2 Frequency-Domain Convolution

The *frequency-domain convolution* is defined as

$$Y(k) = \frac{1}{N} \sum_{m=0}^{N-1} X(m)H(k-m) = \frac{1}{N} \sum_{m=0}^{N-1} X(k-m)H(m)$$

and, as above, we can show that

$$y(n) = h(n)x(n)$$

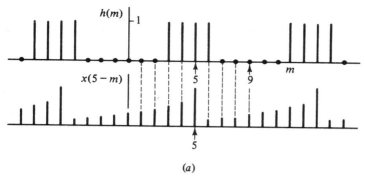

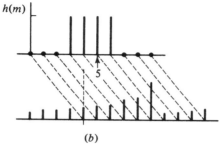

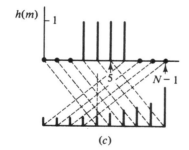

FIGURE 13.10
Time-domain, periodic convolution (Example 13.3).

13.11 FAST-FOURIER-TRANSFORM ALGORITHMS

The direct evaluation of the DFT involves N complex multiplications and $N - 1$ complex additions for each value of $X(k)$, and since there are N values to determine, N^2 multiplications and $N(N - 1)$ additions are necessary. Consequently, for large values of N, say in excess of 1000, direct evaluation involves a considerable amount of computation.

The efficient and up-to-date approach for the evaluation of the DFT is through the use of FFT algorithms. We describe here two, the so-called *decimation-in-time* and *decimation-in-frequency algorithms*.

13.11.1 Decimation-in-Time Algorithm

Let the desired DFT be

$$X(k) = \sum_{n=0}^{N-1} x(n) W_N^{-kn} \qquad \text{where } W_N = e^{j2\pi/N}$$

and assume that

$$N = 2^r$$

where r is an integer. The above summation can be split into two parts as

$$X(k) = \sum_{\substack{n=0 \\ n \text{ even}}}^{N-1} x(n) W_N^{-kn} + \sum_{\substack{n=0 \\ n \text{ odd}}}^{N-1} x(n) W_N^{-kn}$$

Alternatively

$$X(k) = \sum_{n=0}^{N/2-1} x_{10}(n) W_N^{-2kn} + W_N^{-k} \sum_{n=0}^{N/2-1} x_{11}(n) W_N^{-2kn} \qquad (13.15)$$

where

$$x_{10}(n) = x(2n) \qquad x_{11}(n) = x(2n+1) \qquad (13.16)$$

for $0 \le n \le N/2 - 1$. Since

$$W_N^{-2kn} = W_{N/2}^{-kn}$$

Eq. (13.15) can be expressed as

$$X(k) = \sum_{n=0}^{N/2-1} x_{10}(n) W_{N/2}^{-kn} + W_N^{-k} \sum_{n=0}^{N/2-1} x_{11}(n) W_{N/2}^{-kn}$$

Clearly

$$X(k) = X_{10}(k) + W_N^{-k} X_{11}(k) \qquad (13.17)$$

and since $X_{10}(k)$ and $X_{11}(k)$ are periodic, each with period $N/2$, we have

$$X\left(k + \frac{N}{2}\right) = X_{10}\left(k + \frac{N}{2}\right) + W_N^{-(k+N/2)} X_{11}\left(k + \frac{N}{2}\right) = X_{10}(k) - W_N^{-k} X_{11}(k)$$

$$(13.18)$$

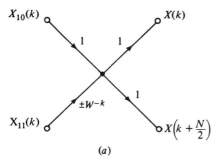

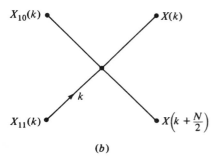

FIGURE 13.11
(a) Butterfly flow graph, (b) simplified diagram.

Equations (13.17) and (13.18) can be represented by the *butterfly* flow graph of Fig. 13.11a, where the minus sign in $\pm W_N^{-k}$ is pertinent in the computation of $X(k+N/2)$. This flow graph can be represented by the simplified diagram of Fig. 13.11b for convenience.

What we have accomplished so far is to express the desired N-element DFT as a function of two $(N/2)$-element DFTs. Assuming that the values of $X_{10}(k)$ and $X_{11}(k)$ are available in corresponding arrays, the values of $X(k)$ can be readily computed as depicted in Fig. 13.12a.

$X_{10}(k)$ and $X_{11}(k)$ can now be expressed in terms of $(N/4)$-element DFTs by repeating the above cycle. For $X_{10}(k)$, we can write

$$X_{10}(k) = \sum_{n=0}^{N/2-1} x_{10}(n) W_{N/2}^{-kn}$$

$$= \sum_{n=0}^{N/4-1} x_{10}(2n) W_{N/2}^{-2kn} + \sum_{n=0}^{N/4-1} x_{10}(2n+1) W_{N/2}^{-k(2n+1)}$$

$$= \sum_{n=0}^{N/4-1} x_{20}(n) W_{N/4}^{-kn} + W_N^{-2k} \sum_{n=0}^{N/4-1} x_{21}(n) W_{N/4}^{-kn} \qquad (13.19)$$

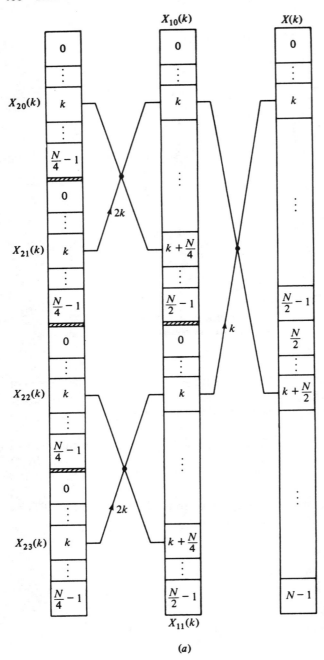

FIGURE 13.12a
Decimation-in-time FFT algorithm: (a) first and second cycles.

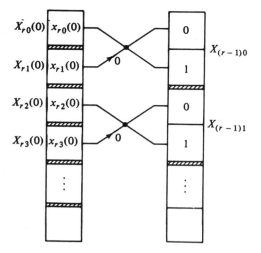

FIGURE 13.12b
Decimation-in-time FFT algorithm: (b) rth cycle.

and, similarly, for $X_{11}(k)$

$$X_{11}(k) = \sum_{n=0}^{N/2-1} x_{11}(n) W_{N/2}^{-kn}$$

$$= \sum_{n=0}^{N/4-1} x_{22}(n) W_{N/4}^{-kn} + W_N^{-2k} \sum_{n=0}^{N/4-1} x_{23}(n) W_{N/4}^{-kn} \quad (13.20)$$

where
$$x_{20}(n) = x_{10}(2n) \quad x_{21}(n) = x_{10}(2n+1) \quad (13.21a)$$
$$x_{22}(n) = x_{11}(2n) \quad x_{23}(n) = x_{11}(2n+1) \quad (13.21b)$$

for $0 \leq n \leq N/4 - 1$. Consequently, from Eqs. (13.19) and (13.20)

$$X_{10}(k) = X_{20}(k) + W_N^{-2k} X_{21}(k) \qquad X_{10}\left(k + \frac{N}{4}\right) = X_{20}(k) - W_N^{-2k} X_{21}(k)$$

$$X_{11}(k) = X_{22}(k) + W_N^{-2k} X_{23}(k) \qquad X_{11}\left(k + \frac{N}{4}\right) = X_{22}(k) - W_N^{-2k} X_{23}(k)$$

Thus if the values of $X_{20}(k)$, $X_{21}(k)$, $X_{22}(k)$, and $X_{23}(k)$ are available, those of $X_{10}(k)$ and $X_{11}(k)$ and in turn those of $X(k)$ can be computed, as illustrated in Fig. 13.12a.

In exactly the same way, the *mth cycle* of the procedure yields

$$X_{(m-1)0}(k) = X_{m0}(k) + W_N^{-2^{m-1}k} X_{m1}(k)$$

$$X_{(m-1)0}\left(k + \frac{N}{2^m}\right) = X_{m0}(k) - W_N^{-2^{m-1}k} X_{m1}(k)$$

$$X_{(m-1)1}(k) = X_{m2}(k) + W_N^{-2^{m-1}k} X_{m3}(k)$$

$$X_{(m-1)1}\left(k + \frac{N}{2^m}\right) = X_{m2}(k) - W_N^{-2^{m-1}k} X_{m3}(k)$$

....

where

$$x_{m0}(n) = x_{(m-1)0}(2n) \qquad (13.22a)$$

$$x_{m1}(n) = x_{(m-1)0}(2n+1) \qquad (13.22b)$$

$$x_{m2}(n) = x_{(m-1)1}(2n) \qquad (13.22c)$$

$$x_{m3}(n) = x_{(m-1)1}(2n+1) \qquad (13.22d)$$

....

for $0 \leq n \leq N/2^m - 1$. Clearly, the procedure terminates with the *r*th cycle ($N = 2^r$) since $x_{r0}(n), x_{r1}(n), \ldots$ reduce the one-element sequences, in which case

$$X_{ri}(0) = x_{ri}(0)$$

for $i = 0, 1, \ldots, N - 1$. The values of the penultimate DFTs can be obtained from the above equations as

$$X_{(r-1)0}(0) = x_{r0}(0) + W_N^0 x_{r1}(0)$$

$$X_{(r-1)0}(1) = x_{r0}(0) - W_N^0 x_{r1}(0)$$

$$X_{(r-1)1}(0) = x_{r2}(0) + W_N^0 x_{r3}(0)$$

$$X_{(r-1)1}(1) = x_{r2}(0) - W_N^0 x_{r3}(0)$$

....

Assuming that the sequence $\{x_{r0}(0), x_{r1}(0), \ldots\}$ is available in an array, the values of $X_{(r-1)i}(k)$ for $i = 0, 1, \ldots$ can be computed as in Fig. 13.12*b*. Then the values of $X_{(r-2)i}(k), X_{(r-3)i}(k), \ldots$ can be computed in sequence, and ultimately the values of $X(k)$ can be obtained.

The only remaining task at this point is to identify elements $x_{r0}(0), x_{r1}(0), \ldots$. Fortunately, this turns out to be easy. As can be shown, $x_{rp}(0)$ is given by

$$x_{rp}(0) = x(q)$$

where q is the r-bit binary representation of p reversed. For example, if $N = 16, r = 4$ and hence

$$x_{40}(0) = x(0)$$
$$x_{41}(0) = x(8)$$
$$x_{42}(0) = x(4)$$
$$\ldots\ldots\ldots\ldots$$
$$x_{4(15)}(0) = x(15)$$

In effect, *sequence* $\{x_{r0}(0), x_{r1}(0), \ldots\}$ *is a reordered version of sequence* $\{x(0), x(1), \ldots\}$.

In the above discussion, N has been assumed to be a power of 2. Nevertheless, the algorithm can be applied to any other finite-duration sequence by including a number of trailing zero elements in the given sequence.

Example 13.4. Construct the decimation-in-time algorithm for $N = 8$.

Solution. From Eq. (13.16)

$$\mathbf{x}_{10} = \{x(0), x(2), x(4), x(6)\}$$
$$\mathbf{x}_{11} = \{x(1), x(3), x(5), x(7)\}$$

and hence Eq. (13.21) gives

$$\mathbf{x}_{20} = \{x(0), x(4)\}$$
$$\mathbf{x}_{21} = \{x(2), x(6)\}$$
$$\mathbf{x}_{22} = \{x(1), x(5)\}$$
$$\mathbf{x}_{23} = \{x(3), x(7)\}$$

Finally, from Eq. (13.22)

$$\mathbf{x}_{30} = \{x(0)\}$$
$$\mathbf{x}_{31} = \{x(4)\}$$
$$\mathbf{x}_{32} = \{x(2)\}$$
$$\ldots\ldots\ldots\ldots$$
$$\mathbf{x}_{37} = \{x(7)\}$$

The complete algorithm is illustrated in Fig. 13.13.

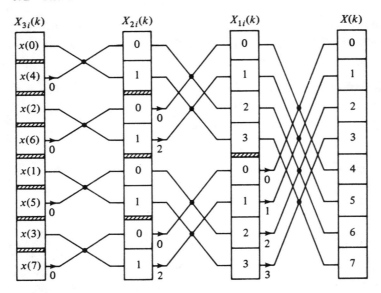

FIGURE 13.13
Decimation-in-time FFT algorithm for $N = 8$ (Example 13.4).

The algorithm can be easily programmed as computations can be carried out in place in a single array. As can be observed in Fig. 13.13, once the outputs of each input butterfly are computed, the input elements are no longer needed for further processing and can be replaced by the corresponding outputs. When we proceed in the same way from left to right, at the end of computation the input array will contain the elements of the desired DFT properly ordered. The input elements can be entered in the appropriate array locations by using a simple reordering subroutine.

In general, each cycle of the algorithm involves $N/2$ butterflies, as can be seen in Fig. 13.13, and each butterfly requires one (complex) multiplication. Since there are r cycles of computation and $r = \log_2 N$, the total number of multiplications is $(N/2) \log_2 N$ as opposed to N^2 in the case of direct evaluation. This constitutes a considerable saving in computation. For example, if $N \geq 512$, then the number of multiplications is reduced to a fraction of 1 percent of that required by direct evaluation.

13.11.2 Decimation-in-Frequency Algorithm

In the preceding algorithm the given sequence is split in two by separating the even- and odd-index elements. The same procedure is then applied repeatedly on each new sequence until one-element sequences are obtained. An alternative FFT algorithm, known as the *decimation-in-time* algorithm, can be developed by splitting the given sequence about its midpoint and then repeating the same for each resulting sequence.

We can write

$$X(k) = \sum_{n=0}^{N/2-1} x(n) W_N^{-kn} + \sum_{n=N/2}^{N-1} x(n) W_N^{-kn}$$

$$= \sum_{n=0}^{N/2-1} \left[x(n) + W_N^{-kN/2} x\left(n + \frac{N}{2}\right) \right] W_N^{-kn}$$

and on replacing k first by $2k$ and then by $2k+1$, we obtain

$$X(2k) = \sum_{n=0}^{N/2-1} x_{10}(n) W_{N/2}^{-kn}$$

$$X(2k+1) = \sum_{n=0}^{N/2-1} x_{11}(n) W_{N/2}^{-kn}$$

where

$$x_{10}(n) = x(n) + x\left(n + \frac{N}{2}\right) \tag{13.23}$$

$$x_{11}(n) = \left[x(n) - x\left(n + \frac{N}{2}\right) \right] W_N^{-n} \tag{13.24}$$

for $0 \le n \le N/2 - 1$. Thus the even- and odd-index values of $X(k)$ are given by the DFTs of $x_{10}(n)$ and $x_{11}(n)$, respectively. Assuming that the values of $x(n)$ are stored sequentially in an array, the values of $x_{10}(n)$ and $x_{11}(n)$ can be computed as illustrated in Fig. 13.14a, where the left-hand butterfly represents Eqs. (13.23) and (13.24).

The same cycle can now be applied to $x_{10}(n)$ and $x_{11}(n)$. For $x_{10}(n)$

$$X(2k) = \sum_{n=0}^{N/4-1} \left[x_{10}(n) + W_N^{-kN/2} x_{10}\left(n + \frac{N}{4}\right) \right] W_{N/2}^{-kn}$$

and, similarly, for $x_{11}(n)$

$$X(2k+1) = \sum_{n=0}^{N/4-1} \left[x_{11}(n) + W_N^{-kN/2} x_{11}\left(n + \frac{N}{4}\right) \right] W_{N/2}^{-kn}$$

Hence with k replaced first by $2k$ and then by $2k+1$, we have

$$X(4k) = \sum_{n=0}^{N/4-1} x_{20}(n) W_{N/4}^{-kn}$$

$$X(4k+2) = \sum_{n=0}^{N/4-1} x_{21}(n) W_{N/4}^{-kn}$$

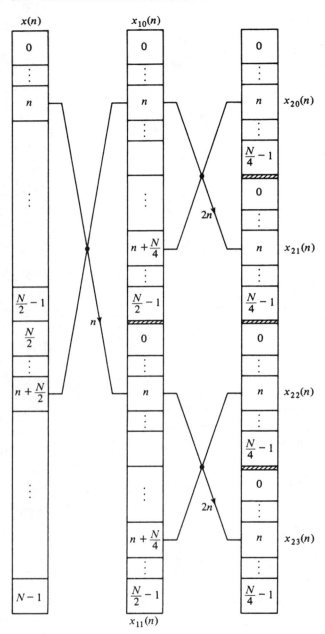

FIGURE 13.14a
Decimation-in-frequency FFT algorithm: (a) general case.

$$X(4k+1) = \sum_{n=0}^{N/4-1} x_{22}(n) W_{N/4}^{-kn}$$

$$X(4k+3) = \sum_{n=0}^{N/4-1} x_{23}(n) W_{N/4}^{-kn}$$

where

$$x_{20}(n) = x_{10}(n) + x_{10}\left(n + \frac{N}{4}\right)$$

$$x_{21}(n) = \left[x_{10}(n) - x_{10}\left(n + \frac{N}{4}\right)\right] W^{-2n}$$

$$x_{22}(n) = x_{11}(n) + x_{11}\left(n + \frac{N}{4}\right)$$

$$x_{23}(n) = \left[x_{11}(n) - x_{11}\left(n + \frac{N}{4}\right)\right] W^{-2n}$$

for $0 \leq n \leq (N/4 - 1)$. The values of $x_{20}(n), x_{21}(n), \ldots$ can be computed as in Fig. 13.14a. The DFT of each of these sequences gives one-quarter of the values of $X(k)$.

Similarly, the *mth cycle* yields

$$X(2^m k) = \sum_{n=0}^{N_1-1} x_{m0}(n) W_{N_1}^{-kn}$$

$$X(2^m k + 2^{m-1}) = \sum_{n=0}^{N_1-1} x_{m1}(n) W_{N_1}^{-kn}$$

....

where $N_1 = N/2^m$ and

$$x_{m0}(n) = x_{(m-1)0}(n) + x_{(m-1)0}\left(n + \frac{N}{2^m}\right)$$

$$x_{m1}(n) = \left[x_{(m-1)0}(n) - x_{(m-1)0}\left(n + \frac{N}{2^m}\right)\right] W_N^{-2^{m-1}n}$$

....

for $0 \leq n \leq N/2^m - 1$.

As in the previous case, the procedure terminates when $m = r$, at which time $x_{r0}(n), x_{r1}(n), \ldots$ reduce the one-element sequences each giving one value of the desired DFT, i.e.,

$$X(0) = x_{r0}(0)$$
$$X(2^{r-1}) = x_{r1}(0)$$
....

The complete algorithm for $N = 8$ is illustrated in Fig. 13.14b. As can be seen, the values of $X(k)$ appear disordered in the output array. However, reordering can be easily accomplished by reversing the r-bit binary representation of the location index at the end of computation. The advantage of this algorithm is that the values of $x(n)$ are entered in the input array sequentially.

13.11.3 Inverse DFT

Owing to the similarity between Eqs. (13.2) and (13.3), the preceding two algorithms can be readily employed for the computation of the IDFT. Equation (13.3) can be put in the form

$$x^*(n) = \left[\frac{1}{N}\sum_{k=0}^{N-1} X(k)W^{kn}\right]^* = \frac{1}{N}\sum_{k=0}^{N-1} X^*(k)W^{-kn}$$

or

$$x^*(n) = \mathcal{D}\left[\frac{1}{N}X^*(k)\right]$$

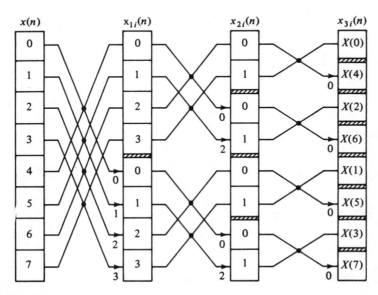

FIGURE 13.14b
Decimation-in-frequency FFT algorithm: (b) $N = 8$.

Thus if a program is available that can be used to compute the DFT of a complex signal $x(n)$, entering the complex conjugate of $X(k)/N$ as input will yield the complex conjugate of $x(n)$ as output.

13.12 DIGITAL-FILTER IMPLEMENTATION

The response of a nonrecursive filter to an excitation $x(n)$ is given by

$$y(n) = \sum_{m=-\infty}^{\infty} x(n-m)h(m)$$

and if

$$h(n) = 0 \quad \text{for } n < 0 \text{ and } n > N-1$$
$$x(n) = 0 \quad \text{for } n < 0 \text{ and } n > L-1$$

we have

$$y(n) = \sum_{m=0}^{N-1} x(n-m)h(m) \quad \text{for } 0 \leq n \leq N+L-2 \tag{13.25}$$

A software implementation for the filter can be readily obtained by programming Eq. (13.25) directly. However, this approach can involve a large amount of computation since N multiplications are necessary for each sample of the response. The alternative is to use the FFT method [9].

Let us define $(L+N-1)$-element DFTs for $h(n)$, $x(n)$, and $y(n)$, as in Sec. 13.2, which we can designate as $H(k)$, $X(k)$, and $Y(k)$, respectively. From Eqs. (13.14) and (13.25)

$$Y(k) = H(k)X(k)$$

and hence

$$y(n) = \mathcal{D}^{-1}[H(k)X(k)]$$

Therefore, the response of the filter can be computed by using the following procedure:

1. Compute the DFTs of $h(n)$ and $x(n)$ using an FFT algorithm.
2. Compute the product $H(k)X(k)$ for $k = 0, 1, \ldots$.
3. Compute the IDFT of $Y(k)$ using an FFT algorithm.

The evaluation of $H(k)$, $X(k)$, or $y(n)$ requires $[(L+N-1)/2]\log_2(L+N-1)$ complex multiplications, and step 2 above entails $L+N-1$ of the same. Since one complex multiplication corresponds to four real ones, the total number of real multiplications per output sample is $6\log_2(L+N-1)+4$, as opposed to N in the case of direct evaluation. Clearly, for large values of N, the FFT approach is the more efficient. For example, if $N = L = 512$, the number of multiplications is reduced to 12.5 percent of that required by direct evaluation.

13.12.1 Overlap-and-Add Method

In the above implementation, the entire input sequence must be available before the processing can start. Consequently, if the input sequence is long, a significant delay in computation is introduced, which is usually objectionable in *real-time* or *quasi-real-time* applications. For such applications, the input sequence can be *segmented*, and each segment or block of data can be processed individually, as will be shown below.

We can write

$$x(n) = \sum_{i=0}^{q} x_i(n)$$

for $0 \leq n \leq (q+1)L - 1$, where

$$x_i(n) = \{u(n - iL) - u[n - (i+1)L]\}x(n)$$

or
$$x_i(n) = \begin{cases} x(n) & \text{for } iL \leq n \leq (i+1)L - 1 \\ 0 & \text{otherwise} \end{cases}$$

as illustrated in Fig. 13.15. With this manipulation Eq. (13.25) becomes

$$y(n) = \sum_{m=0}^{N-1} \sum_{i=0}^{q} x_i(n-m)h(m)$$

and by interchanging the order of summation

$$y(n) = \sum_{i=0}^{q} c_i(n) \tag{13.26}$$

where
$$c_i(n) = \sum_{m=0}^{N-1} x_i(n-m)h(m) \tag{13.27}$$

or
$$c_i(n) = \sum_{m=0}^{N-1} \{u(n - m - iL) - u[n - m - (i+1)L]\}x(n-m)h(m) \tag{13.28}$$

In this way, $y(n)$ can be computed by evaluating a number of partial convolutions.
For $iL - 1 \leq n \leq (i+1)L + N - 1$, Eq. (13.28) gives

$$c_i(iL - 1) = 0$$
$$c_i(iL) = x(iL)h(0)$$
$$c_i(iL + 1) = x(iL + 1)h(0) + x(iL)h(1)$$
$$\cdots\cdots\cdots\cdots\cdots\cdots\cdots\cdots\cdots\cdots\cdots$$
$$c_i[(i+1)L + N - 2] = x[(i+1)L - 1]h(N - 1)$$
$$c_i[(i+1)L + N - 1] = 0$$

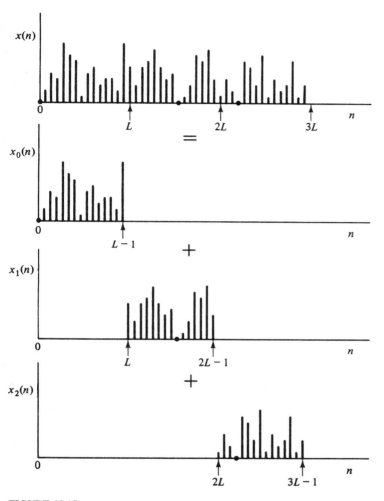

FIGURE 13.15
Segmentation of input sequence.

Evidently, the ith partial-convolution sequence has $L+N-1$ nonzero elements which can be stored in an array C_i, as demonstrated in Fig. 13.16. From Eq. (13.27) the elements of C_i can be computed as

$$c_i(n) = \mathcal{D}^{-1}[H(k)X_i(k)]$$

Now from Eq. (13.26) an array Y containing the values of $y(n)$ can be readily formed, as illustrated in Fig. 13.16, by *entering the elements of nonoverlapping segments in $C_0, C_1, \ldots$ and then adding the elements in overlapping adjacent segments*. As can be seen, processing can start as soon as L input elements are received, and the first batch

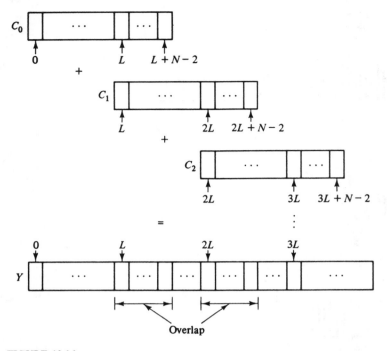

FIGURE 13.16
Overlap-and-add implementation.

of L output samples is available as soon as the first input segment is processed. In this way delays in computation can be minimized.

The above method is known as the *overlap-and-add method* for obvious reasons. An alternative scheme for the implementation of nonrecursive filters involves the so-called *overlap-and-save method* described below.

13.12.2 Overlap-and-Save Method

If $x(n) = 0$ for $n < 0$, as before, then the first L elements of $y(n)$, namely, elements 0 to $L-1$, are equal to the corresponding L elements of $c_0(n)$. This does not apply to the last $N-1$ elements of $y(n)$, i.e., elements L to $L+N-2$, owing to the overlap between $c_0(n)$ and $c_1(n)$, as can be seen in Fig. 13.16. However, if we redefine $x_1(n)$ such that

$$x_1(n) = \begin{cases} x(n) & \text{for } L-(N-1) \le n \le 2L-(N-1)-1 \\ 0 & \text{otherwise} \end{cases}$$

as in Fig. 13.17, then for $L \le n \le 2L-(N-1)-1$, we have

$$y(n) = \sum_{m=0}^{N-1} x(n-m)h(m) = \sum_{m=0}^{N-1} x_1(n-m)h(m) = c_1(n)$$

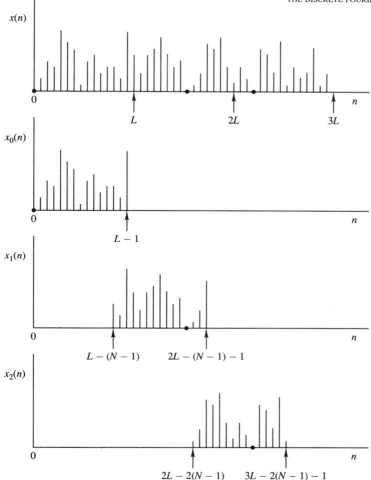

FIGURE 13.17
Alternative segmentation of input sequence.

Similarly, if

$$x_k(n) = \begin{cases} x(n) & \text{for } kL - k(N-1) \leq n \leq (k+1)L - k(N-1) - 1 \\ 0 & \text{otherwise} \end{cases}$$

then for $kL - (k-1)(N-1) \leq n \leq (k+1)2L - k(N-1) - 1$

$$y(n) = c_k(n) \tag{13.29}$$

for $k = 2, 3, \ldots$ (see Prob. 13.25). In effect, *the output of the filter can be evaluated by computing the first L elements of $c_0(n)$ and elements $N-1$ to $L-1$ of each of the subsequent partial convolutions and then concatenating the sequences obtained* as in Fig. 13.18. In the overlap-and-save method, the input rather than the output sequences are overlapped, as can be seen in Fig. 13.17, and the last $N-1$ elements of each

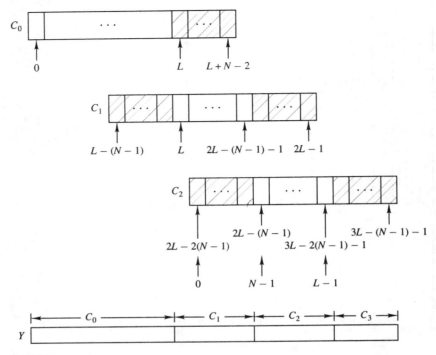

FIGURE 13.18
Overlap-and-save implementation.

input sequence are saved to be re-used for the computation of the subsequent partial convolution.

REFERENCES

1. J. W. Cooley and J. W. Tukey, "An Algorithm for the Machine Calculation of Complex Fourier Series," *Math Comp.*, vol. 19, pp. 297–301, April 1965.
2. W. T. Cochran, J. W. Cooley, D. L. Favin, H. D. Helms, R. A. Kaenel, W. W. Lang, G. C. Maling, D. E. Nelson, C. M. Rader, and P. D. Welch, "What Is the Fast Fourier Transform?," *IEEE Trans. Audio Electroacoust.*, vol. AU-15, pp. 45–55, June 1967.
3. G. D. Bergland, "A Guided Tour of the Fast Fourier Transform," *IEEE Spectrum*, vol. 6, pp. 41–52, July 1969.
4. J. W. Cooley, P. A. W. Lewis, and P. D. Welch, "Historical Notes on the Fast Fourier Transform," *IEEE Trans. Audio Electroacoust.*, vol. AU-15, pp. 76–79, June 1967.
5. J. W. Cooley, P. A. W. Lewis, and P. D. Welch, "Application of the Fast Fourier Transform to Computation of Fourier Integrals, Fourier Series and Convolution Integrals," *IEEE Trans. Audio Electroacoust.*, vol. AU-15, pp 79–84, June 1967.
6. H. Babic and G. C. Temes, "Optimum Low-Order Windows for Discrete Fourier Transform Systems," *IEEE Trans. Acoust., Speech, Signal Process.*, vol. ASSP-24, pp. 512–517, December 1976.
7. B. Gold and K. L. Jordan, Jr., "A Direct Search Procedure for Designing Finite Duration Impulse Response Filters," *IEEE Trans. Audio Electroacoust.*, vol. AU-17, pp. 33–36, March 1969.

8. L. R. Rabiner, B. Gold, and C. A. McGonegal, "An Approach to the Approximation Problem for Nonrecursive Digital Filters," *IEEE Trans. Audio Electroacoust.*, vol. AU-18, pp. 83–106, June 1970.
9. H. D. Helms, "Fast Fourier Transform Method of Computing Difference Equations and Simulating Filters," *IEEE Trans. Audio Electroacoust.*, vol. AU-15, pp. 85–90, June 1967.

ADDITIONAL REFERENCES

Agarwal, R. C., and C. S. Burrus: "Fast Convolution Using Fermat Number Transforms with Applications to Digital Filtering," *IEEE Trans. Acoust., Speech, Signal Process.*, vol. ASSP-22, pp. 87–97, April 1974.

—— and ——: "Number Theoretic Transforms to Implement Fast Digital Convolution," *Proc. IEEE*, vol. 63, pp. 550–560, April 1975.

Blahut, R. E.: *Fast Algorithms for Digital Signal Processing*, Addison-Wesley, Reading, MA, 1985.

Brigham, E. O.: *The Fast Fourier Transform*, Prentice-Hall, Englewood Cliffs, 1974.

Burrus, C. S., and T. W. Parks: *DFT/FFT and Convolution Algorithms*, Wiley, New York, 1985.

Cooley, J. W., P. A. W. Lewis, and P. D. Welch: "The Fast Fourier Transform Algorithm: Programming Considerations in the Calculation of Sine, Cosine and Laplace Transforms," *J. Sound Vib.*, vol. 12, pp. 315–337, July 1970.

——, ——, and ——: "The Finite Fourier Transform," *IEEE Trans. Audio Electroacoust.*, vol. AU-17, pp. 77–85, June 1969.

——: "How the FFT Gained Acceptance," *IEEE Signal Processing Magazine*, vol. 9, pp. 10–13, January 1992.

Duhamel, P.: "Implementation of 'Split-Radix' FFT Algorithms for Complex, Real, and Real-Symmetric Data," *IEEE Trans. Acoust., Speech, Signal Process.*, vol. ASSP-34, pp. 285–295, April 1986.

Elliot, D. F., and K. Ramamohan Rao: *Fast Transforms Algorithms, Analyses, Applications*, Academic Press, New York, 1982.

Johnson, H. W., and C. S. Burrus: "The Design of Optimal DFT Algorithms Using Dynamic Programming," *IEEE Trans. Acoust., Speech, Signal Process.*, vol. ASSP-31, pp. 378–387, April 1983.

Martens, J. B.: "Recursive Cyclotomic Factorization—A New Algorithm for Calculating the Discrete Fourier Transform," *IEEE Trans. Acoust., Speech, Signal Process.*, vol. ASSP-32, pp. 750–761, August 1984.

Nussbaumer, H. J.: "Relative Evaluation of Various Number Theoretic Transforms for Digital Filtering Applications," *IEEE Trans. Acoust., Speech, Signal Process.*, vol. ASSP-26, pp. 88–93, February 1978.

——: *Fast Fourier Transform and Convolution Algorithms*, Springer-Verlag, New York, 1982.

Oppenheim, A. V., and C. J. Weinstein: "Effects of Finite Register Length in Digital Filtering and the Fast Fourier Transform," *Proc. IEEE*, vol. 60, pp. 957–976, August 1972.

Rader, C. M.: "Discrete Fourier Transform When the Number of Data Samples Is Prime," *Proc. IEEE*, vol. 56, pp. 1107–1108, June 1968.

——: "Discrete Convolutions via Mersenne Transforms," *IEEE Trans. Comput.*, vol. C-21, pp. 1269–1273, December 1972.

Weinstein, C. J.: "Roundoff Noise in Floating Point Fast Fourier Transform Computation," *IEEE Trans. Audio Electroacoust.*, vol. AU-17, pp. 209–215, September 1969.

Welch, P. D.: "A Fixed-Point Fast Fourier Transform Error Analysis," *IEEE Trans. Audio Electroacoust.*, vol. AU-17, pp. 151–157, June 1969.

PROBLEMS

13.1. Show that

$$\sum_{k=0}^{N-1} W^{k(n-m)} = \begin{cases} N & \text{for } m = n \\ 0 & \text{otherwise} \end{cases}$$

13.2. Show that
(a) $\mathcal{D}x_p(nT + mT) = W^{km} X_p(jk\Omega)$
(b) $\mathcal{D}^{-1} X_p(jk\Omega + jl\Omega) = W^{-nl} x_p(nT)$

13.3. The definition of the DFT can be extended to include complex discrete-time signals. Show that
(a) $\mathcal{D}x_p^*(nT) = X_p^*(-jk\Omega)$
(b) $\mathcal{D}^{-1} X_p^*(jk\Omega) = x_p^*(-nT)$

13.4. (a) A complex discrete-time signal is given by

$$x_p(nT) = x_{p1}(nT) + j x_{p2}(nT)$$

where $x_{p1}(nT)$ and $x_{p2}(nT)$ are real. Show that

$$\operatorname{Re} X_{p1}(jk\Omega) = \frac{1}{2}\{\operatorname{Re} X_p(jk\Omega) + \operatorname{Re} X_p[j(N-k)\Omega]\}$$

$$\operatorname{Im} X_{p1}(jk\Omega) = \frac{1}{2}\{\operatorname{Im} X_p(jk\Omega) - \operatorname{Im} X_p[j(N-k)\Omega]\}$$

$$\operatorname{Re} X_{p2}(jk\Omega) = \frac{1}{2}\{\operatorname{Im} X_p(jk\Omega) + \operatorname{Im} X_p[j(N-k)\Omega]\}$$

$$\operatorname{Im} X_{p2}(jk\Omega) = -\frac{1}{2}\{\operatorname{Re} X_p(jk\Omega) - \operatorname{Re} X_p[j(N-k)\Omega]\}$$

(b) A DFT is given by

$$X_p(jk\Omega) = X_{p1}(jk\Omega) + j X_{p2}(jk\Omega)$$

where $X_{p1}(jk\Omega)$ and $X_{p2}(jk\Omega)$ are real DFTs. Show that

$$\operatorname{Re} x_{p1}(nT) = \frac{1}{2}\{\operatorname{Re} x_p(nT) + \operatorname{Re} x_p[(N-n)T]\}$$

$$\operatorname{Im} x_{p1}(nT) = \frac{1}{2}\{\operatorname{Im} x_p(nT) - \operatorname{Im} x_p[(N-n)T]\}$$

$$\operatorname{Re} x_{p2}(nT) = \frac{1}{2}\{\operatorname{Im} x_p(nT) + \operatorname{Im} x_p[(N-n)T]\}$$

$$\operatorname{Im} x_{p2}(nT) = -\frac{1}{2}\{\operatorname{Re} x_p(nT) - \operatorname{Re} x_p[(N-n)T]\}$$

13.5. Figure P13.5 shows four real discrete-time signals. Classify their DFTs as real, imaginary, or complex. Assume that $N = 10$ in each case.

13.6. Find the DFTs of the following periodic signals:

(a) $x_p(nT) = \begin{cases} 1 & \text{for } n = 3, 7 \\ 0 & \text{for } n = 0, 1, 2, 4, 5, 6, 8, 9 \end{cases}$

(b) $x_p(nT) = \begin{cases} 1 & \text{for } 0 \leq n \leq 5 \\ 2 & \text{for } 6 \leq n \leq 9 \end{cases}$

(c) $x_p(nT) = \begin{cases} n & \text{for } 0 \leq n \leq 2 \\ 0 & \text{for } 3 \leq n \leq 7 \\ -(10-n) & \text{for } n = 8, 9 \end{cases}$

The period is $10T$ in each case.

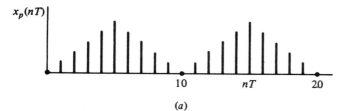

(a)

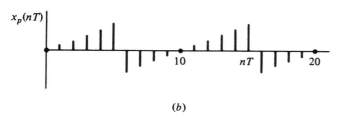

(b)

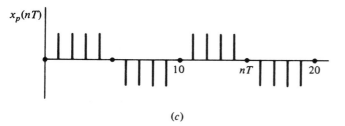

(c)

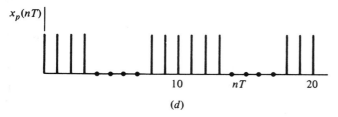

(d)

FIGURE P13.5

13.7. A periodic signal is given by

$$x_p(nT) = \sum_{r=-\infty}^{\infty} w_H(nT + rNT)$$

where
$$w_H(nT) = \begin{cases} \alpha + (1-\alpha)\cos\dfrac{2\pi n}{N-1} & \text{for } |n| \leq \dfrac{N-1}{2} \\ 0 & \text{otherwise} \end{cases}$$

Find $X_p(jk\Omega)$.

13.8. Obtain the IDFTs of the following:

(a) $X_p(jk\Omega) = (-1)^k \left(1 + 2\cos\dfrac{2\pi k}{10}\right)$

(b) $X_p(jk\Omega) = 1 + 2j(-1)^k \left(\sin\dfrac{3k\pi}{5} + \sin\dfrac{4k\pi}{5}\right)$

The value of N is 10.

13.9. Find the z transform of $x(nT)$ for the DFTs of Prob. 13.8. Assume that $x(nT) = 0$ outside the range $0 \le n \le 9$ in each case.

13.10. (a) Show that the first two formulas in Table 13.1 satisfy the relation

$$h(nT) = h[(N-1-n)T]$$

(b) Show that the last two formulas in Table 13.1 satisfy the relation

$$h(nT) = -h[(N-1-n)T]$$

13.11. (a) Derive the second formula in Table 13.1.

(b) Derive the fourth formula in Table 13.1.

13.12. Design a nonrecursive lowpass filter with a frequency response

$$H_D(e^{j\omega T}) \approx \begin{cases} 1 & \text{for } 0 \le |\omega| \le 12 \text{ rad/s} \\ 0 & \text{for } 13 \le |\omega| \le 32 \text{ rad/s} \end{cases}$$

Use the frequency-sampling method, and assume that $\omega_s = 64$ rad/s, $N = 64$.

13.13. (a) An optimized version of the filter in Prob. 13.12 can be obtained by assigning

$A(13\Omega) = 0.74040381 \qquad A(14\Omega) = 0.27101526 \qquad A(15\Omega) = 0.02996826$

(see [8]). Plot the impulse and amplitude responses of the filter.

(b) Compare this design with the direct design of Prob. 13.12.

13.14. (a) Design a highpass nonrecursive filter with a frequency response

$$H(e^{j\omega T}) \approx \begin{cases} 0 & \text{for } 0 \le |\omega| \le 20 \text{ rad/s} \\ 1 & \text{for } 21 \le |\omega| \le 32 \text{ rad/s} \end{cases}$$

Use the frequency-sampling method and assume that $\omega_s = 64$ rad/s, $N = 64$.

(b) Repeat part (a) if $\omega_s = 65$ rad/s and $N = 65$.

(c) Find the minimum stopband attenuation in each case.

13.15. (a) Show that a nonrecursive transfer function obtained through the frequency-sampling method can be realized by using a set of parallel, second-order recursive sections in cascade with an elementary Nth-order nonrecursive section.

(b) Using the approach in part (a), realize the transfer function of Prob. 13.12.

13.16. Periodic signals $x(n)$ and $h(n)$ are given by

$$x(n) = \begin{cases} 1 & \text{for } 0 \le n \le 4 \\ 2 & \text{for } 5 \le n \le 9 \end{cases}$$

$$h(n) = n \qquad \text{for } 0 \le n \le 9$$

Find

$$y(n) = \sum_{m=0}^{9} x(m) h(n-m)$$

for $0 \le n \le 9$.

13.17. Show that

$$\mathcal{D}[x(n)h(n)] = \frac{1}{N}\sum_{m=0}^{N-1} X(m)H(k-m)$$

where $X(k) = \mathcal{D}x(n)$ and $H(k) = \mathcal{D}h(n)$.

13.18. Construct the flow graph for a 16-element decimation-in-time FFT algorithm.

13.19. Construct the flow graph for a 16-element decimation-in-frequency FFT algorithm.

13.20. (a) Compute the Fourier-series coefficients for the periodic signal depicted in Fig. P13.20 by using a 32-element FFT algorithm.
(b) Repeat part (a) using a 64-element FFT algorithm.
(c) Repeat part (a) using an analytical method.
(d) Compare the results obtained.

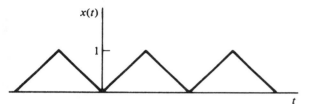

FIGURE P13.20

13.21. Repeat Prob. 13.20 for the signal of Fig. P13.21.

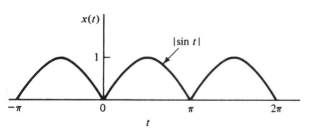

FIGURE P13.21

13.22. (a) Compute the Fourier transform of

$$x(t) = \begin{cases} \frac{1}{2}(1+\cos t) & \text{for } 0 \le |t| \le \pi \\ 0 & \text{otherwise} \end{cases}$$

by using a 64-element FFT algorithm. The desired resolution in the frequency domain is 0.5 rad/s.
(b) Repeat part (a) for a frequency domain resolution of 0.25 rad/s.
(c) Repeat part (a) by using an analytical method.
(d) Compare the results in parts (a) to (c).

13.23. Repeat Prob. 13.22 for the signal

$$x(t) = \begin{cases} 1-|t| & \text{for } |t| < 1 \\ 0 & \text{otherwise} \end{cases}$$

The desired frequency-domain resolutions for parts (*a*) and (*b*) are $\pi/4$ and $\pi/8$ rad/s, respectively.

13.24. An FFT program is available which allows for a maximum of 64 complex input elements. Show that this program can be used to process a real 128-element sequence.

13.25. Demonstrate the validity of Eq. (13.29).

CHAPTER 14

DESIGN OF RECURSIVE FILTERS USING OPTIMIZATION METHODS

14.1 INTRODUCTION

In Chaps. 7 and 8, several methods for the solution of the approximation problem in recursive filters have been described. These methods lead to a complete description of the transfer function in closed form, either in terms of its zeros and poles or its coefficients. They are, as a consequence, very efficient and lead to very precise designs. Their main disadvantage is that they are applicable only for the design of filters with piecewise-constant amplitude responses, i.e., filters whose passband and stopband gains are constant and zero, respectively, to within prescribed tolerances.

An alternative approach for the solution of the approximation problem in digital filters is through the application of *optimization methods* [1–5]. In these methods, a discrete-time transfer function is assumed and an error function is formulated on the basis of some desired amplitude and/or phase response. A norm of the error function is then minimized with respect to the transfer-function coefficients. As the value of the norm approaches zero, the resulting amplitude or phase response approaches the desired amplitude or phase response. These methods are *iterative* and, as a result, they usually involve a large amount of computation. However, unlike the closed-form methods of Chaps. 7 and 8, they are suitable for the design of filters having arbitrary amplitude or phase responses. Furthermore, they often yield superior designs.

In this chapter, the application of optimization methods for the design of recursive digital filters is considered. The chapter begins with an introductory section that deals with the formulation of the design problem as an optimization problem, and then proceeds with fairly detailed descriptions of algorithms that can be used to solve the optimization problem. The algorithms presented are based on the so-called *quasi-Newton* method which has been explored by Davidon, Fletcher, Powell, Broyden, and others. The exposition of the material begins with algorithms that are primarily of conceptual value and gradually proceeds to algorithms of increasing complexity and scope. It concludes with some highly sophisticated algorithms that are practical, flexible, efficient, and reliable. Throughout the chapter, emphasis is placed on the application of the algorithms rather than their theoretical foundation and convergence properties. Readers who are interested in a more mathematical treatment of the subject may consult one of the standard textbooks on optimization theory and practice [6–10].

14.2 PROBLEM FORMULATION

Assume that the amplitude response of a recursive filter is required to approach some specified amplitude response as closely as possible. Such a filter can be designed in two general steps, as follows:

1. An objective function which is dependent on the difference between the actual and specified amplitude response is formulated.
2. The objective function obtained is minimized with respect to the transfer-function coefficients.

An Nth-order recursive filter can be represented by the transfer function

$$H(z) = H_0 \prod_{j=1}^{J} \frac{a_{0j} + a_{1j}z + z^2}{b_{0j} + b_{1j}z + z^2} \tag{14.1}$$

where a_{ij} and b_{ij} are real coefficients, $J = N/2$, and H_0 is a positive multiplier constant. The amplitude response of the filter can be expressed as

$$M(\mathbf{x}, \omega) = |H(e^{j\omega T})| \tag{14.2}$$

where

$$\mathbf{x} = [a_{01} \quad a_{11} \quad b_{01} \quad b_{11} \quad \cdots \quad b_{1J} \quad H_0]^T$$

is a column vector with $4J + 1$ elements and ω is the frequency.

Let $M_0(\omega)$ be the specified amplitude response and, for the sake of exposition, assume that it is piecewise continuous, as illustrated in Fig. 14.1. The difference between $M(\mathbf{x}, \omega)$ and $M_0(\omega)$ is, in effect, the approximation error and can be expressed as

$$e(\mathbf{x}, \omega) = M(\mathbf{x}, \omega) - M_0(\omega) \tag{14.3}$$

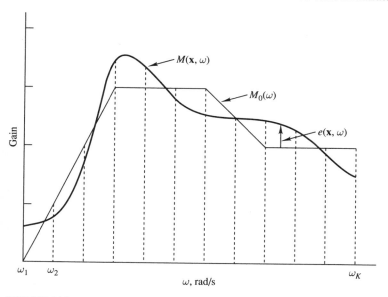

FIGURE 14.1
Formulation of error function.

By sampling $e(\mathbf{x}, \omega)$ at frequencies $\omega_1, \omega_2, \ldots, \omega_K$, as depicted in Fig. 14.1, the column vector

$$\mathbf{E}(\mathbf{x}) = [e_1(\mathbf{x}) \quad e_2(\mathbf{x}) \quad \ldots \quad e_K(\mathbf{x})]^T$$

can be formed where

$$e_i(\mathbf{x}) = e(\mathbf{x}, \omega_i) \tag{14.4}$$

for $i = 1, 2, \ldots, K$.

The approximation problem at hand can be solved by finding a point $\mathbf{x} = \check{\mathbf{x}}$ such that

$$e_i(\check{\mathbf{x}}) \approx 0$$

for $i = 1, 2, \ldots, K$. Assuming that a solution exists, a suitable *objective function* must first be formed which should satisfy a number of fundamental requirements. It should be a *scalar* quantity, and its minimization with respect to $\mathbf{x}$ should lead to the minimization of all the elements of $\mathbf{E}(\mathbf{x})$ in some sense. Further, it is highly desirable that it be *differentiable*. An objective function satisfying these requirements can be defined in terms of the L_p norm of $\mathbf{E}(\mathbf{x})$ as

$$\Psi(\mathbf{x}) = L_p = ||\mathbf{E}(\mathbf{x})||_p = \left[\sum_{i=1}^{K} |e_i(\mathbf{x})|^p\right]^{1/p}$$

where p is an integer.

Several special cases of the L_p norm are of particular interest. The L_1 *norm*, namely,

$$L_1 = \sum_{i=1}^{K} |e_i(\mathbf{x})|$$

is the sum of the magnitudes of the elements of $\mathbf{E}(\mathbf{x})$; the L_2 *norm* given by

$$L_2 = \left[\sum_{i=1}^{K} |e_i(\mathbf{x})|^2\right]^{1/2}$$

is the well-known *Euclidean norm*; and L_2^2 is the *sum of the squares* of the elements of $\mathbf{E}(\mathbf{x})$. In the case where $p = \infty$ and

$$\hat{E}(\mathbf{x}) = \max_{1 \leq i \leq K} |e_i(\mathbf{x})| \neq 0$$

we can write

$$L_\infty = \lim_{p \to \infty} \left\{\sum_{i=1}^{K} |e_i(\mathbf{x})|^p\right\}^{1/p}$$

$$= \hat{E}(\mathbf{x}) \lim_{p \to \infty} \left\{\sum_{i=1}^{K} \left[\frac{|e_i(\mathbf{x})|}{\hat{E}(\mathbf{x})}\right]^p\right\}^{1/p} \quad (14.5)$$

Since each of the terms in the above summation is equal to or less than unity, then as $p \to \infty$, we have

$$L_\infty = \hat{E}(\mathbf{x})$$

With an objective function available, the required design can be obtained by solving the optimization problem

$$\text{minimize}_\mathbf{x} \ \Psi(\mathbf{x}) \quad (14.6)$$

If $\Psi(\mathbf{x})$ is defined in terms of L_2^2, a *least-squares* solution is obtained; if the L_∞ norm is used, a so-called *minimax* solution is obtained, since in this case the largest element in $\mathbf{E}(\mathbf{x})$ is minimized.

In digital filters, the magnitude of the largest amplitude-response error is usually required to be as small as possible and, therefore, minimax solutions are preferred.

14.3 NEWTON'S METHOD

The optimization problem of Eq. (14.6) can be solved by using an *unconstrained* optimization algorithm. Various classes of these algorithms have been developed in recent years, ranging from *steepest-descent* to *conjugate-direction* algorithms [6–10]. An important class of optimization algorithms that have been found to be very effective for the design of digital filters is the class of *quasi-Newton* algorithms. These are based on Newton's method for finding the minimum in quadratic *convex* functions[†].

Consider a function $f(\mathbf{x})$ of n variables, where $\mathbf{x} = [x_1 \ x_2 \ \cdots \ x_n]^T$ is a column vector, and let $\boldsymbol{\delta} = [\delta_1 \ \delta_2 \ \cdots \ \delta_n]^T$ be a change in $\mathbf{x}$. If $f(\mathbf{x}) \in C^2$, i.e., $f(\mathbf{x})$ has continuous second derivatives, its Taylor series at point $\mathbf{x} + \boldsymbol{\delta}$ is given by

$$f(\mathbf{x} + \boldsymbol{\delta}) = f(\mathbf{x}) + \sum_{i=1}^{n} \frac{\partial f(\mathbf{x})}{\partial x_i} \delta_i$$

$$+ \frac{1}{2} \sum_{i=1}^{n} \sum_{j=1}^{n} \frac{\partial^2 f(\mathbf{x})}{\partial x_i \partial x_j} \delta_i \delta_j + o(\|\boldsymbol{\delta}\|_2^2) \quad (14.7)$$

where the remainder $o(\|\boldsymbol{\delta}\|_2^2)$ approaches zero faster than $\|\boldsymbol{\delta}\|_2^2$. If the remainder is negligible and a *stationary point* exists in the neighborhood of some point $\mathbf{x}$, it can be determined by differentiating $f(\mathbf{x} + \boldsymbol{\delta})$ with respect to elements δ_k for $k = 1, 2, \ldots, n$, and setting the result to zero. From Eq. (14.7), we obtain

$$\frac{\partial f(\mathbf{x} + \boldsymbol{\delta})}{\partial \delta_k} = \frac{\partial f(\mathbf{x})}{\partial x_k} + \sum_{i=1}^{n} \frac{\partial^2 f(\mathbf{x})}{\partial x_i \partial x_k} \delta_i = 0$$

for $k = 1, 2, \ldots, n$. This equation can be expressed in matrix form as

$$\mathbf{g} = -\mathbf{H}\boldsymbol{\delta} \quad (14.8)$$

where

$$\mathbf{g} = \nabla f(\mathbf{x}) = \left[\frac{\partial f(\mathbf{x})}{\partial x_1} \ \frac{\partial f(\mathbf{x})}{\partial x_2} \ \cdots \ \frac{\partial f(\mathbf{x})}{\partial x_n} \right]^T$$

and

$$\mathbf{H} = \begin{bmatrix} \frac{\partial^2 f(\mathbf{x})}{\partial x_1^2} & \frac{\partial^2 f(\mathbf{x})}{\partial x_1 \partial x_2} & \cdots & \frac{\partial^2 f(\mathbf{x})}{\partial x_1 \partial x_n} \\ \frac{\partial^2 f(\mathbf{x})}{\partial x_2 \partial x_1} & \frac{\partial^2 f(\mathbf{x})}{\partial x_2^2} & \cdots & \frac{\partial^2 f(\mathbf{x})}{\partial x_2 \partial x_n} \\ \vdots & \vdots & & \vdots \\ \frac{\partial^2 f(\mathbf{x})}{\partial x_n \partial x_1} & \frac{\partial^2 f(\mathbf{x})}{\partial x_n \partial x_2} & \cdots & \frac{\partial^2 f(\mathbf{x})}{\partial x_n^2} \end{bmatrix}$$

are the *gradient* vector and *Hessian* matrix (or simply the gradient and Hessian) of $f(\mathbf{x})$, respectively. Therefore, the value of $\boldsymbol{\delta}$ that yields the stationary point of $f(\mathbf{x})$ can be obtained from Eq. (14.8) as

$$\boldsymbol{\delta} = -\mathbf{H}^{-1}\mathbf{g} \quad (14.9)$$

[†] A two-variable convex function is one that represents a surface whose shape resembles a punch bowl.

This equation will give the solution if and only if the following two conditions hold:

(i) The remainder $o(||\delta||_2^2)$ in Eq. (14.7) can be neglected.
(ii) The Hessian is nonsingular.

If $f(\mathbf{x})$ is a *quadratic* function, its second partial derivatives are constants, i.e., $\mathbf{H}$ is a constant symmetric matrix, and its third and higher derivatives are zero. Therefore, condition (i) above holds. If $f(\mathbf{x})$ has a stationary point and the sufficiency conditions for a minimum hold at the stationary point, then the Hessian matrix is *positive definite* and, therefore, *nonsingular*. Under these circumstances, given an arbitrary point $\mathbf{x} \in E^n$†, the *minimum point* can be obtained as $\overset{\cup}{\mathbf{x}} = \mathbf{x} + \delta$ by using Eq. (14.9).

If $f(\mathbf{x})$ is a general nonquadratic convex function that has a minimum at point $\overset{\cup}{\mathbf{x}}$, then in the neighborhood $||\mathbf{x} - \overset{\cup}{\mathbf{x}}||_2 < \epsilon$ the remainder in Eq. (14.7) becomes negligible and the second partial derivatives of $f(\mathbf{x})$ become approximately constant. As a result, in this domain function $f(\mathbf{x})$ behaves as a quadratic function and the above two conditions are again satisfied. Therefore, for any point $\mathbf{x}$ such that $||\mathbf{x} - \overset{\cup}{\mathbf{x}}||_2 < \epsilon$, the use of Eq. (14.9) will yield an accurate estimate of the minimum point.

If a general function $f(\mathbf{x})$ is to be minimized and an arbitrary point $\mathbf{x} \in E^n$ is assumed, condition (i) and/or condition (ii) may be violated. If condition (i) is violated, then the use of Eq. (14.9) will not give the solution; if condition (ii) is violated, then Eq. (14.9) either has an infinite number of solutions or has no solutions at all. These problems can be overcome by using an iterative procedure in which the value of the function is progressively reduced by applying a series of corrections to $\mathbf{x}$ until a point in the neighborhood of the solution is obtained. When the remainder in Eq. (14.7) becomes negligible, an accurate estimate of the solution can be obtained by using Eq. (14.9). A suitable strategy to achieve this goal is based on the fundamental property that if $\mathbf{H}$ is positive definite, then $\mathbf{H}^{-1}$ is also positive definite. Furthermore, in such a case it can be shown through the use of the Taylor series that the direction pointed by the vector $-\mathbf{H}^{-1}\mathbf{g}$ of Eq. (14.9), which is known as the *Newton direction*, is a *descent direction* of $f(\mathbf{x})$. As a consequence, if at some initial point $\mathbf{x}$, $\mathbf{H}$ is positive definite, a reduction can be achieved in $f(\mathbf{x})$ by simply applying a correction of the form $\delta = \alpha \mathbf{d}$ to $\mathbf{x}$, where α is a positive factor and $\mathbf{d} = -\mathbf{H}^{-1}\mathbf{g}$. On the other hand, if $\mathbf{H}$ is not positive definite, it can be forced to become positive definite by means of some algebraic manipulation (e.g., it can be changed into the unity matrix) and, as before, a reduction can be achieved in $f(\mathbf{x})$. In either case, the largest possible reduction in $f(\mathbf{x})$ with respect to the direction $\mathbf{d}$ can be achieved by choosing variable α such that $f(\mathbf{x} + \alpha \mathbf{d})$ is minimized. This can be done by using one of many available one-dimensional minimization algorithms (also known as *line searches*) [6–10]. Repeating

---

† E^n represents the n-dimensional Euclidean space.

these steps a number of times will yield a value of **x** in the neighborhood of the solution and eventually the solution itself. An algorithm based on these principles, known as the *Newton algorithm*, is as follows:

Algorithm 1: Basic Newton algorithm

1. Input $\mathbf{x}_0$ and ε. Set $k = 0$.
2. Compute the gradient $\mathbf{g}_k$ and Hessian $\mathbf{H}_k$. If $\mathbf{H}_k$ is not positive definite, force it to become positive definite.
3. Compute $\mathbf{H}_k^{-1}$ and $\mathbf{d}_k = -\mathbf{H}_k^{-1} \mathbf{g}_k$.
4. Find α_k, the value of α that minimizes $f(\mathbf{x}_k + \alpha \mathbf{d}_k)$, using a line search.
5. Set $\mathbf{x}_{k+1} = \mathbf{x}_k + \boldsymbol{\delta}_k$, where $\boldsymbol{\delta}_k = \alpha_k \mathbf{d}_k$, and compute $f_{k+1} = f(\mathbf{x}_{k+1})$.
6. If $||\alpha_k \mathbf{d}_k||_2 < \varepsilon$, then output $\overset{\cup}{\mathbf{x}} = \mathbf{x}_{k+1}$, $f(\overset{\cup}{\mathbf{x}}) = f_{k+1}$, and stop.
   Else, set $k = k + 1$ and repeat from step 2.

The algorithm is terminated if the L_2 norm of $\alpha_k \mathbf{d}_k$, i.e., the magnitude of the change in **x**, is less than ε. The parameter ε is said to be the *termination tolerance* and is a small positive constant whose value is determined by the application under consideration.[†] In certain applications, a termination tolerance on the objective function itself, e.g., $|f_{k+1} - f_k| < \varepsilon$, may be preferable and sometimes termination tolerances may be imposed on the magnitudes of both the changes in **x** and the objective function.

So far, we have tacitly assumed that the optimization problem under consideration has a unique or *global* minimum. In practice, the problem may have more than one local minimum, sometimes a large number of minima, and on occasion a well-defined minimum may not even exist. We must, therefore, abandon the expectation that we shall always be able to obtain the best solution available. The best we can hope for is a solution that satisfies a number of the required specifications.

Example 14.1. (*a*) Show that the function

$$f(\mathbf{x}) = x_1^2 + 2x_1 x_2 + 2x_2^2 + 2x_1 + x_2$$

has a minimum. (*b*) Find the minimum of the function using Algorithm 1 with $\mathbf{x}_0 = [0 \ \ 0]^T$ as initial point.

Solution. (*a*) From basic calculus, the stationary points of a function are the points at which the gradient is equal to zero. If the Hessian at a specific stationary point is *positive definite, negative definite,* or *indefinite,* then the stationary point is a *minimum, maximum,* or *saddle point*; alternatively, if the Hessian is *positive* or *negative semidefinite,* then the stationary point can be either a *maximum* or a *minimum point.*

[†]Parameters ε, ε_1, and ε_2 represent termination tolerances throughout the chapter.

The partial derivatives of $f(\mathbf{x})$ are given by

$$\frac{\partial f}{\partial x_1} = 2x_1 + 2x_2 + 2 \quad \text{and} \quad \frac{\partial f}{\partial x_2} = 2x_1 + 4x_2 + 1$$

At a stationary point $\tilde{\mathbf{x}}$, the gradient $\mathbf{g}$ is zero; hence, we obtain $\tilde{\mathbf{x}} = [-1.5 \quad 0.5]^T$. The Hessian can be deduced as

$$\mathbf{H} = \begin{bmatrix} 2 & 2 \\ 2 & 4 \end{bmatrix}$$

Since the *principal minor determinants* of $\mathbf{H}$ are positive, the Hessian is positive definite (see page 75), and so $\tilde{\mathbf{x}}$ is a minimum point.

(b) The gradient at $\mathbf{x}_0 = [0 \quad 0]$ is $\mathbf{g}_0 = [2 \quad 1]^T$. The inverse of $\mathbf{H}_0$ is given by

$$\mathbf{H}_0^{-1} = \begin{bmatrix} 1 & -0.5 \\ -0.5 & 0.5 \end{bmatrix}$$

and hence the Newton direction can be obtained from step 3 of Algorithm 1 as $\mathbf{d}_0 = -\mathbf{H}_0^{-1}\mathbf{g}_0 = [-1.5 \quad 0.5]^T$. The function under consideration is quadratic and the solution can be obtained with $\alpha_0 = 1$. From step 5, $\mathbf{x}_1 = \overset{\cup}{\mathbf{x}} = [-1.5 \quad 0.5]^T$ and $f(\overset{\cup}{\mathbf{x}}) = f_1 = -1.25$. Note that Algorithm 1 will need two iterations to stop since the termination test in step 6 will not be satisfied until the second iteration.

14.4 QUASI-NEWTON ALGORITHMS

The Newton algorithm described in the preceding section has three major disadvantages. First, both the first and second partial derivatives of $f(\mathbf{x})$ must be computed in each iteration in order to construct the gradient and Hessian, respectively. Second, in each iteration the Hessian must be checked for positive definiteness and, if it is found to be nonpositive definite, it must be forced to become positive definite. Third, matrix inversion is required in each iteration. By contrast, in quasi-Newton algorithms only the first derivatives need to be computed, and it is unnecessary to manipulate or invert the Hessian. Consequently, for general problems other than convex quadratic problems, quasi-Newton algorithms are much more efficient and are preferred.

Quasi-Newton algorithms, like the Newton algorithm, are developed for the convex quadratic problem and are then extended to the general problem. The fundamental principle in these algorithms is that the direction of search is based on an $n \times n$ matrix $\mathbf{S}$ that serves the same purpose as the inverse Hessian in the Newton algorithm. This matrix is constructed using available data and is contrived to be an approximation of $\mathbf{H}^{-1}$. Furthermore, as the number of iterations is increased, $\mathbf{S}$ becomes progressively a more and more accurate representation of $\mathbf{H}^{-1}$. For convex quadratic objective functions, $\mathbf{S}$ becomes identical to $\mathbf{H}^{-1}$ in $n + 1$ iterations where n is the number of variables.

14.4.1 Basic Quasi-Newton Algorithm

Let the gradients of $f(\mathbf{x})$ at points $\mathbf{x}_k$ and $\mathbf{x}_{k+1}$ be $\mathbf{g}_k$ and $\mathbf{g}_{k+1}$, respectively. If

$$\mathbf{x}_{k+1} = \mathbf{x}_k + \boldsymbol{\delta}_k$$

then the Taylor series gives the elements of $\mathbf{g}_{k+1}$ as

$$g_{(k+1)m} = g_{km} + \sum_{i=1}^{n} \frac{\partial g_{km}}{\partial x_{ki}} \delta_{ki} + \frac{1}{2} \sum_{i=1}^{n} \sum_{j=1}^{n} \frac{\partial^2 g_{km}}{\partial x_{ki} \partial x_{kj}} \delta_{ki} \delta_{kj} + o(\|\delta\|_2^2)$$

for $m = 1, 2, \ldots, n$. Now if $f(\mathbf{x})$ is quadratic, the second and higher derivatives of $f(\mathbf{x})$ are constant and zero, respectively, and as a result the second and higher derivatives of g_{km} are zero. Thus

$$g_{(k+1)m} = g_{km} + \sum_{i=1}^{n} \frac{\partial g_{km}}{\partial x_{ki}} \delta_{ki}$$

and since

$$g_{km} = \frac{\partial f_k}{\partial x_{km}}$$

we have

$$g_{(k+1)m} = g_{km} + \sum_{i=1}^{n} \frac{\partial^2 f_k}{\partial x_{ki} \partial x_{km}} \delta_{ki}$$

for $m = 1, 2, \ldots, n$. Therefore, $\mathbf{g}_{k+1}$ is given by

$$\mathbf{g}_{k+1} = \mathbf{g}_k + \mathbf{H}\boldsymbol{\delta}_k$$

where $\mathbf{H}$ is the Hessian of $f(\mathbf{x})$. Alternatively, we can write

$$\boldsymbol{\gamma}_k = \mathbf{H}\boldsymbol{\delta}_k \tag{14.10}$$

where

$$\boldsymbol{\delta}_k = \mathbf{x}_{k+1} - \mathbf{x}_k$$

and

$$\boldsymbol{\gamma}_k = \mathbf{g}_{k+1} - \mathbf{g}_k$$

The above analysis has shown that, if the gradient of $f(\mathbf{x})$ is known at two points $\mathbf{x}_k$ and $\mathbf{x}_{k+1}$, a relation can be deduced that provides a certain amount of information about $\mathbf{H}$, namely, Eq. (14.10). Since $\mathbf{H}$ is a real symmetric matrix with $n \times (n+1)/2$ unknowns and Eq. (14.10) provides only n equations, $\mathbf{H}$ cannot be determined uniquely through the use of Eq. (14.10). This problem can be overcome by evaluating the gradient sequentially at $n+1$ points, say at $\mathbf{x}_0, \mathbf{x}_1, \ldots, \mathbf{x}_n$, such that the changes in $\mathbf{x}$, namely,

$$\begin{aligned} \boldsymbol{\delta}_0 &= \mathbf{x}_1 - \mathbf{x}_0 \\ \boldsymbol{\delta}_1 &= \mathbf{x}_2 - \mathbf{x}_1 \\ &\vdots \\ \boldsymbol{\delta}_{n-1} &= \mathbf{x}_n - \mathbf{x}_{n-1} \end{aligned}$$

form a set of linearly independent vectors. Under these circumstances, Eq. (14.10) yields

$$[\,\boldsymbol{\gamma}_0 \quad \boldsymbol{\gamma}_1 \quad \cdots \quad \boldsymbol{\gamma}_{n-1}\,] = \mathbf{H}[\,\boldsymbol{\delta}_0 \quad \boldsymbol{\delta}_1 \quad \cdots \quad \boldsymbol{\delta}_{n-1}\,]$$

Therefore, **H** can be uniquely determined as

$$\mathbf{H} = [\gamma_0 \quad \gamma_1 \quad \cdots \quad \gamma_{n-1}][\delta_0 \quad \delta_1 \quad \cdots \quad \delta_{n-1}]^{-1} \quad (14.11)$$

The above principles lead to the following algorithm:

Algorithm 2: Alternative Newton algorithm

1. Input $\mathbf{x}_{00}$ and ε. Input a set of n linearly independent vectors $\delta_0, \delta_1, \ldots, \delta_{n-1}$. Set $k = 0$.
2. Compute $\mathbf{g}_{k0}$.
3. For $i = 0$ to $n - 1$ do:
   a. Set $\mathbf{x}_{k(i+1)} = \mathbf{x}_{ki} + \delta_i$.
   b. Compute $\mathbf{g}_{k(i+1)}$.
   c. Set $\gamma_{ki} = \mathbf{g}_{k(i+1)} - \mathbf{g}_{ki}$.
4. Compute $\mathbf{H}_k$ using Eq. (14.11). If $\mathbf{H}_k$ is not positive definite, force it to become positive definite.
5. Determine $\mathbf{S}_k = \mathbf{H}_k^{-1}$.
6. Set $\mathbf{d}_k = -\mathbf{S}_k \mathbf{g}_{k0}$ and find α_k, the value of α that minimizes $f(\mathbf{x}_{k0} + \alpha \mathbf{d}_k)$, using a line search.
7. Set $\mathbf{x}_{(k+1)0} = \mathbf{x}_{k0} + \alpha_k \mathbf{d}_k$ and compute $f_{(k+1)0} = f(\mathbf{x}_{(k+1)0})$.
8. If $\|\alpha_k \mathbf{d}_k\|_2 < \varepsilon$, then output $\overset{\cup}{\mathbf{x}} = \mathbf{x}_{(k+1)0}$, $f(\overset{\cup}{\mathbf{x}}) = f_{(k+1)0}$, and stop.
   Else, set $k = k + 1$ and repeat from step 2.

The above algorithm is essentially an alternative implementation of the Newton method in which the generation of $\mathbf{H}^{-1}$ is accomplished using computed data instead of the second derivatives. However, as in Algorithm 1, for the general nonquadratic problem it is necessary to check, manipulate, and invert the Hessian in every iteration. In addition, we now need to provide a set of linearly independent vectors to the algorithm, namely, $\delta_0, \delta_1, \ldots, \delta_{n-1}$. In other words, though of considerable conceptual value, the algorithm is of little practical usefulness.

Further progress towards the development of the quasi-Newton method can be made by generating the matrix $\mathbf{H}^{-1}$ from computed data using a set of linearly independent vectors $\delta_0, \delta_1, \ldots, \delta_{n-1}$ that are themselves generated from available data. This objective can be accomplished by generating the vectors

$$\delta_k = -\mathbf{S}_k \mathbf{g}_k \quad (14.12)$$

$$\mathbf{x}_{k+1} = \mathbf{x}_k + \delta_k \quad (14.13)$$

and

$$\gamma_k = \mathbf{g}_{k+1} - \mathbf{g}_k$$

and then making an additive correction to $\mathbf{S}_k$ of the form

$$\mathbf{S}_{k+1} = \mathbf{S}_k + \mathbf{C}_k \quad (14.14)$$

for $k = 0, 1, \ldots, n-1$. If a correction matrix $\mathbf{C}_k$ can be found such that the conditions

$$\mathbf{S}_{k+1}\boldsymbol{\gamma}_i = \boldsymbol{\delta}_i \qquad \text{for } 0 \leq i \leq k \tag{14.15}$$

are satisfied and the vectors $\boldsymbol{\delta}_0, \boldsymbol{\delta}_1, \ldots, \boldsymbol{\delta}_{n-1}$ and $\boldsymbol{\gamma}_0, \boldsymbol{\gamma}_1, \ldots, \boldsymbol{\gamma}_{n-1}$ generated by this process are linearly independent, then for the case $k = n-1$ we can write

$$\mathbf{S}_n [\,\boldsymbol{\gamma}_0 \quad \boldsymbol{\gamma}_1 \quad \cdots \quad \boldsymbol{\gamma}_{n-1}\,] = [\,\boldsymbol{\delta}_0 \quad \boldsymbol{\delta}_1 \quad \cdots \quad \boldsymbol{\delta}_{n-1}\,]$$

or

$$\mathbf{S}_n = [\,\boldsymbol{\delta}_0 \quad \boldsymbol{\delta}_1 \quad \cdots \quad \boldsymbol{\delta}_{n-1}\,][\,\boldsymbol{\gamma}_0 \quad \boldsymbol{\gamma}_1 \quad \cdots \quad \boldsymbol{\gamma}_{n-1}\,]^{-1} \tag{14.16}$$

Now from Eqs. (14.11) and (14.16), we have

$$\mathbf{S}_n = \mathbf{H}^{-1} \tag{14.17}$$

and if $k = n$, Eqs. (14.12) and (14.17) yield the Newton direction

$$\boldsymbol{\delta}_n = -\mathbf{H}^{-1}\mathbf{g}_n \tag{14.18}$$

Therefore, subject to the conditions stated above, the solution of a convex quadratic problem can be obtained from Eqs. (14.13) and (14.18) as

$$\overset{\cup}{\mathbf{x}} = \mathbf{x}_{n+1} = \mathbf{x}_n - \mathbf{H}^{-1}\mathbf{g}_n$$

The above principles lead to the basic quasi-Newton algorithm detailed below.

Algorithm 3: Basic quasi-Newton algorithm

1. Input $\mathbf{x}_0$ and ε. Set $\mathbf{S}_0 = \mathbf{I}_n$ and $k = 0$. Compute $\mathbf{g}_0$.
2. Set $\mathbf{d}_k = -\mathbf{S}_k\mathbf{g}_k$ and find α_k, the value of α that minimizes $f(\mathbf{x}_k + \alpha\,\mathbf{d}_k)$, using a line search.
3. Set $\boldsymbol{\delta}_k = \alpha_k\mathbf{d}_k$ and $\mathbf{x}_{k+1} = \mathbf{x}_k + \boldsymbol{\delta}_k$, and compute $f_{k+1} = f(\mathbf{x}_{k+1})$.
4. If $\|\boldsymbol{\delta}_k\|_2 < \varepsilon$, then output $\overset{\cup}{\mathbf{x}} = \mathbf{x}_{k+1}$, $f(\overset{\cup}{\mathbf{x}}) = f_{k+1}$ and stop.
5. Compute $\mathbf{g}_{k+1}$ and set $\boldsymbol{\gamma}_k = \mathbf{g}_{k+1} - \mathbf{g}_k$.
6. Compute $\mathbf{S}_{k+1} = \mathbf{S}_k + \mathbf{C}_k$.
7. Check $\mathbf{S}_{k+1}$ for positive definiteness and if it is found to be nonpositive definite force it to become positive definite.
8. Set $k = k + 1$ and go to step 2.

In step 2, the vector $-\mathbf{S}_k\mathbf{g}_k$ is denoted as $\mathbf{d}_k$, instead of $\boldsymbol{\delta}_k$ as in Eq. (14.12), and $f(\mathbf{x}_k + \alpha\mathbf{d}_k)$ is minimized with respect to α. The purpose of this modification is to make the algorithm applicable to the general nonquadratic problem. Matrix $\mathbf{S}_k$ is required to be positive definite for each k to ensure that vector $\mathbf{d}_k$ is a descent direction in each iteration. To obtain a descent direction in the first iteration, $\mathbf{S}_0$ is assumed to be the $n \times n$ unity matrix in step 1. Vector $\boldsymbol{\gamma}_k$ in step 5 is required for the computation of correction matrix $\mathbf{C}_k$ in step 6, as will be demonstrated below.

Algorithm 3 eliminates the need to input a set of linearly independent vectors $\delta_0, \delta_1, \ldots, \delta_{n-1}$ and, in addition, the inversion of $\mathbf{H}_k$ is replaced by an additive correction to $\mathbf{S}_k$. However, matrices $\mathbf{S}_1, \mathbf{S}_2, \ldots$ need to be checked for positive definiteness and may need to be manipulated. This can be easily done in practice by diagonalizing $\mathbf{S}_{k+1}$ and then replacing any nonpositive diagonal elements by corresponding positive ones. However, this would increase the computational load quite significantly.

14.4.2 Updating Formulas for Matrix $\mathbf{S}_{k+1}$

The updating formula for matrix $\mathbf{S}_{k+1}$ of Eq. (14.14) must satisfy strict requirements in order to be useful in Algorithm 3. As was stated earlier, for a convex quadratic problem, Eq. (14.15) must be satisfied and the vectors $\delta_0, \delta_1, \ldots, \delta_{n-1}$ and $\gamma_0, \gamma_1, \ldots, \gamma_{n-1}$ must be linearly independent. The derivation and properties of updating formulas of this type have received considerable attention during the past thirty or so years, and several distinct formulas have appeared in the literature. Early in the development of the subject, the so-called *rank-one* formula was proposed, in which the correction matrix $\mathbf{C}_k$ is of rank one. This has largely been replaced in recent years by *rank-two* formulas, like the Davidon-Fletcher-Powell (DFP) and the Broyden-Fletcher-Goldfarb-Shanno (BFGS) formulas [6–10]. A very important property of these two formulas is that a positive definite matrix $\mathbf{S}_k$ yields a positive definite $\mathbf{S}_{k+1}$ not only for convex quadratic problems but also for the general nonquadratic problem, provided that the line search in step 2 of the algorithm is exact (see Fletcher [6] for proof). This property also holds in the case where an *inexact* line search is used in step 2, except that a scalar quantity inherent in the computation of $\mathbf{C}_k$ must be forced to remain positive. The usefulness of this property in Algorithm 3 is obvious: the checking and manipulation of $\mathbf{S}_{k+1}$ in step 7 of the algorithm become unnecessary, and hence a considerable amount of computation can be avoided.

The DFP and BFGS updating formulas are given by

$$\mathbf{S}_{k+1} = \mathbf{S}_k + \frac{\delta_k \delta_k^T}{\gamma_k^T \delta_k} - \frac{\mathbf{S}_k \gamma_k \gamma_k^T \mathbf{S}_k}{\gamma_k^T \mathbf{S}_k \gamma_k} \tag{14.19}$$

and

$$\mathbf{S}_{k+1} = \mathbf{S}_k + \left(1 + \frac{\gamma_k^T \mathbf{S}_k \gamma_k}{\gamma_k^T \delta_k}\right) \frac{\delta_k \delta_k^T}{\gamma_k^T \delta_k} - \frac{(\delta_k \gamma_k^T \mathbf{S}_k + \mathbf{S}_k \gamma_k \delta_k^T)}{\gamma_k^T \delta_k} \tag{14.20}$$

respectively. A condition that guarantees the positive definiteness of $\mathbf{S}_{k+1}$ in both formulas is

$$\delta_k^T \gamma_k = \delta_k^T \mathbf{g}_{k+1} - \delta_k^T \mathbf{g}_k > 0 \tag{14.21}$$

This will be put to good use in Algorithm 5.

14.4.3 Inexact Line Searches

In optimization algorithms in general, the bulk of the computational effort is spent executing line searches. Consequently, the amount of computation required to solve a problem tends to depend critically on the efficiency and precision of the line search used. If a high-precision line search is mandatory in a certain algorithm, then the algorithm can spend a considerable amount of computational effort minimizing the objective function with respect to scalar α. For this reason, low-precision or inexact line searches are usually preferable, provided of course that their use does not affect the convergence properties of the algorithm. Quasi-Newton algorithms have been found to be quite tolerant to line-search imprecision. As a result, inexact line searches are almost always used in these algorithms. An important line search of this type will now be examined.

Let

$$\mathbf{x}_{k+1} = \mathbf{x}_k + \alpha \, \mathbf{d}_k$$

where $\mathbf{d}_k$ is a given descent direction vector and α is an independent variable, and assume that $f(\mathbf{x}_{k+1})$ is a *unimodal function*[†] of α, with a minimum at some point $\alpha = \overset{\cup}{\alpha}$ where $\overset{\cup}{\alpha} > 0$, as depicted in Fig. 14.2a. The linear approximation of the Taylor series for $f(\mathbf{x}_{k+1})$ is of the form

$$f(\mathbf{x}_{k+1}) = f(\mathbf{x}_k) + \alpha \, \mathbf{g}_k^T \mathbf{d}_k \tag{14.22}$$

where

$$\mathbf{g}_k^T \mathbf{d}_k = \left. \frac{df(\mathbf{x}_k + \alpha \, \mathbf{d}_k)}{d\alpha} \right|_{\alpha=0}$$

is the slope at the origin of $f(\mathbf{x}_k + \alpha \, \mathbf{d}_k)$ as a function of α. Eq. (14.22) represents line A depicted in Fig. 14.2a. Similarly, the equation

$$f(\mathbf{x}_{k+1}) = f(\mathbf{x}_k) + \rho \alpha \, \mathbf{g}_k^T \mathbf{d}_k \tag{14.23}$$

where $0 < \rho < 0.5$ represents a line (line B in Fig. 14.2a) whose slope ranges from 0 to $0.5 \mathbf{g}_k^T \mathbf{d}_k$, depending on the value of ρ. Let us assume that this line intersects the curve in Fig. 14.2a at point $\alpha = \alpha_2$. On the other hand, the equation

$$\mathbf{g}_{k+1}^T \mathbf{d}_k = \sigma \, \mathbf{g}_k^T \mathbf{d}_k \tag{14.24}$$

where $0 < \sigma < 1$, and $\sigma \geq \rho$ relates the derivative of $f(\mathbf{x}_{k+1})$ at some point $\alpha = \alpha_1$ to the derivative of the function at $\alpha = 0$ and represents line C in Fig. 14.2a. Since $0 < \sigma < 1$, we have $0 < \alpha_1 < \overset{\cup}{\alpha}$.

Eqs. (14.23) and (14.24) define an interval $[\alpha_1, \alpha_2]$ which brackets the minimum point. Consequently, the two equations can be used as a termination criterion in a line search, much like a termination tolerance on $\mathbf{x}$ or $f(\mathbf{x})$. This possibility will now be examined.

[†]A unimodal function is one that has only one minimum.

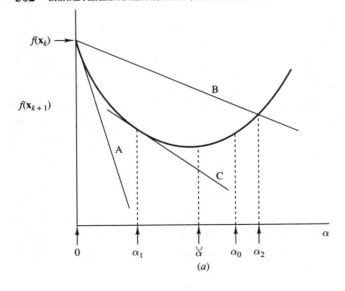

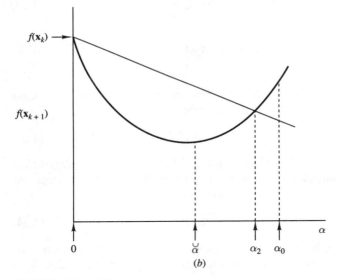

FIGURE 14.2a and b
Inexact line search: (a) case where the conditions in Eqs. (14.25) and (14.26) are both satisfied, (b) case where the condition in Eq. (14.25) is violated.

Let us assume that a mechanism is available by which an estimate of $\overset{\cup}{\alpha}$, say α_0, can be generated. If the actual value of $f(\mathbf{x}_{k+1})$ at $\alpha = \alpha_0$ is less than the value predicted by the linear approximation of Eq. (14.23), i.e.,

$$f(\mathbf{x}_{k+1}) \leq f(\mathbf{x}_k) + \rho\, \alpha_0 \mathbf{g}_k^T \mathbf{d}_k \qquad (14.25)$$

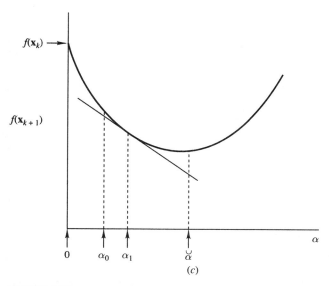

FIGURE 14.2c
Inexact line search: (c) case where the condition in Eq. (14.26) is violated.

then $\alpha_0 \leq \alpha_2$. On the other hand, if the actual slope at $\alpha = \alpha_0$ is more positive (less negative) than the slope of the line in Eq. (14.24), i.e.,

$$\mathbf{g}_{k+1}^T \mathbf{d}_k \geq \sigma \, \mathbf{g}_k^T \mathbf{d}_k \tag{14.26}$$

then $\alpha_1 \leq \alpha_0$. Under these circumstances, we have $\alpha_1 \leq \alpha_0 \leq \alpha_2$, as depicted in Fig. 14.2a, and a certain reduction in $f(\mathbf{x}_{k+1})$ is achieved, which can be considered to be acceptable. In other words, if both Eqs. (14.25) and (14.26) are satisfied, then α_0 can be accepted as a reasonable approximation of $\overset{\cup}{\alpha}$.

If either of the conditions in Eq. (14.25) and (14.26) is violated, then α_0 is outside the interval $[\alpha_1, \alpha_2]$ and the reduction in $f(\mathbf{x}_{k+1})$ can be considered to be unacceptable. If the condition in Eq. (14.25) is violated, then $\alpha_0 > \alpha_2$, as depicted in Fig. 14.2b; since $0 < \overset{\cup}{\alpha} < \alpha_0$, a better estimate for $\overset{\cup}{\alpha}$ (say $\overset{\cup}{\alpha}_0$) can be deduced by using some interpolation formula. If the condition in Eq. (14.26) is violated, then $0 < \alpha_0 < \alpha_1$, as depicted in Fig. 14.2c; in this case, a better estimate $\overset{\cup}{\alpha}_0$ can be deduced by using some extrapolation formula. With a new estimate for $\overset{\cup}{\alpha}$ available, the conditions in Eqs. (14.25) and (14.26) can be checked again and, if either of the two is not satisfied, the process is repeated. When an estimate of $\overset{\cup}{\alpha}$ is found that satisfies both Eqs. (14.25) and (14.26), the search is terminated. The precision of such a line search can be controlled by choosing the values of ρ and σ, since these parameters control the length of interval $[\alpha_1, \alpha_2]$.

Interpolation and *extrapolation* formulas that can be used in the above approach can be readily deduced by assuming a quadratic representation for $f(\mathbf{x}_k + \alpha \mathbf{d}_k)$. If the value of this function and its derivative with respect to α are known at two points, say at $\alpha = \alpha_L$ and $\alpha = \alpha_0$ where $\alpha_L < \alpha_0$, then for $\alpha_0 > \alpha_2$ we can show that

$$\breve{\alpha}_0 = \alpha_L + \frac{(\alpha_0 - \alpha_L)^2 f'_L}{2[f_L - f_0 + (\alpha_0 - \alpha_L) f'_L]} \tag{14.27}$$

and for $\alpha_0 < \alpha_1$

$$\breve{\alpha}_0 = \alpha_0 + \frac{(\alpha_0 - \alpha_L) f'_0}{(f'_L - f'_0)} \tag{14.28}$$

where

$$f_L = f(\mathbf{x}_k + \alpha_L \mathbf{d}_k)$$
$$f'_L = f'(\mathbf{x}_k + \alpha_L \mathbf{d}_k) = \mathbf{g}(\mathbf{x}_k + \alpha_L \mathbf{d}_k)^T \mathbf{d}_k$$
$$f_0 = f(\mathbf{x}_k + \alpha_0 \mathbf{d}_k)$$
$$f'_0 = f'(\mathbf{x}_k + \alpha_0 \mathbf{d}_k) = \mathbf{g}(\mathbf{x}_k + \alpha_0 \mathbf{d}_k)^T \mathbf{d}_k$$

An inexact line search based on the above principles due to Fletcher [6] is as follows:

Algorithm 4: Fletcher inexact line search

1. Input $\mathbf{x}_k$ and $\mathbf{d}_k$. Initialize algorithm parameters ρ, σ, τ, and χ. Set $\alpha_L = 0$ and $\alpha_U = 10^{99}$. Compute $\mathbf{g}_k$.
2. Compute $f_L = f(\mathbf{x}_k + \alpha_L \mathbf{d}_k)$ and $f'_L = \mathbf{g}(\mathbf{x}_k + \alpha_L \mathbf{d}_k)^T \mathbf{d}_k$.
3. Initialize α_0, say $\alpha_0 = 1$.
4. Compute $f_0 = f(\mathbf{x}_k + \alpha_0 \mathbf{d}_k)$.
5. (Interpolation)
   If $f_0 > f_L + \rho(\alpha_0 - \alpha_L) f'_L$, then do:
   a. If $\alpha_0 < \alpha_U$, then set $\alpha_U = \alpha_0$.
   b. Compute $\breve{\alpha}_0$ using Eq. (14.27).
   c. Compute $\breve{\alpha}_{0L} = \alpha_L + \tau(\alpha_U - \alpha_L)$; if $\breve{\alpha}_0 < \breve{\alpha}_{0L}$, then set $\breve{\alpha}_0 = \breve{\alpha}_{0L}$.
   d. Compute $\breve{\alpha}_{0U} = \alpha_U - \tau(\alpha_U - \alpha_L)$; if $\breve{\alpha}_0 > \breve{\alpha}_{0U}$, then set $\breve{\alpha}_0 = \breve{\alpha}_{0U}$.
   e. Set $\alpha_0 = \breve{\alpha}_0$ and go to step 4.
6. Compute $f'_0 = \mathbf{g}(\mathbf{x}_k + \alpha_0 \mathbf{d}_k)^T \mathbf{d}_k$.
7. (Extrapolation)
   If $f'_0 < \sigma f'_L$, then do:
   a. Compute $\Delta \alpha_0 = (\alpha_0 - \alpha_L) f'_0 / (f'_L - f'_0)$ [see Eq. (14.28)].
   b. If $\Delta \alpha_0 < \tau(\alpha_0 - \alpha_L)$, then set $\Delta \alpha_0 = \tau(\alpha_0 - \alpha_L)$.
   c. If $\Delta \alpha_0 > \chi(\alpha_0 - \alpha_L)$, then set $\Delta \alpha_0 = \chi(\alpha_0 - \alpha_L)$.
   d. Compute $\breve{\alpha}_0 = \alpha_0 + \Delta \alpha_0$.
   e. Set $\alpha_L = \alpha_0$, $\alpha_0 = \breve{\alpha}_0$, $f_L = f_0$, $f'_L = f'_0$ and go to step 4.
8. Output α_0 and f_0, and stop.

Assuming that $\mathbf{d}_k$ is a descent direction of $f(\mathbf{x})$ at point $\mathbf{x}_k$, the algorithm will carry out interpolations and/or extrapolations as necessary, which will progressively reduce the value of $f(\mathbf{x}_k + \alpha \mathbf{d}_k)$. When the conditions in Eqs. (14.25) and (14.26) are simultaneously satisfied, the algorithm terminates. The algorithm maintains a running bracket $[\alpha_L, \alpha_U]$ on the minimum point; if the interpolation formula yields a value of $\overset{\cup}{\alpha_0}$ outside this interval or very close to the lower or upper limit, a more reasonable value is assigned to $\overset{\cup}{\alpha_0}$ in step 5c or 5d. Similarly, if the value of $\Delta \alpha_0$ predicted in step 7a is negative, very small or very large, a more reasonable value is assigned to $\Delta \alpha_0$ in step 7b or 7c. The precision of the line search depends on the values of ρ and σ. Small values like $\rho = \sigma = 0.1$ yield a high-precision line search, whereas the values $\rho = 0.15$ and $\sigma = 0.9$ yield a somewhat imprecise one. Suitable values for τ and χ are 0.1 and 9, respectively. Further details about this line search can be found in the first edition of Fletcher [6]. A closely related inexact line search proposed by Al-Baali and Fletcher can be found in [11] (see also second edition of Fletcher [6]).

14.4.4 Practical Quasi-Newton Algorithm

A practical quasi-Newton algorithm that eliminates the problems associated with Algorithms 1 to 3 is detailed below. This is based on Algorithm 3 and uses a slightly modified version of Algorithm 4 as inexact line search. The algorithm is flexible, efficient, and very reliable, and is readily applicable to the design of digital filters and equalizers, as will be shown in Secs. 14.7 and 14.8.

Algorithm 5: Practical quasi-Newton algorithm

1. (Initialize algorithm)
    a. Input $\mathbf{x}_0$ and ε_1.
    b. Set $k = m = 0$.
    c. Set $\rho = 0.1$, $\sigma = 0.7$, $\tau = 0.1$, $\chi = 0.75$, $\hat{M} = 600$, and $\varepsilon_2 = 10^{-10}$.
    d. Set $\mathbf{S}_0 = \mathbf{I}_n$.
    e. Compute f_0 and $\mathbf{g}_0$, and set $m = m + 2$. Set $f_{00} = f_0$ and $\Delta f_0 = f_0$.

2. (Initialize line search)
    a. Set $\mathbf{d}_k = -\mathbf{S}_k \mathbf{g}_k$.
    b. Set $\alpha_L = 0$ and $\alpha_U = 10^{99}$.
    c. Set $f_L = f_0$ and compute $f'_L = \mathbf{g}(\mathbf{x}_k + \alpha_L \mathbf{d}_k)^T \mathbf{d}_k$.
    d. (Estimate α_0)
       If $|f'_L| > \varepsilon_2$, then compute $\alpha_0 = -2\Delta f_0 / f'_L$; else, set $\alpha_0 = 1$.
       If $\alpha_0 \leq 0$ or $\alpha_0 > 1$, then set $\alpha_0 = 1$.

3. Set $\boldsymbol{\delta}_k = \alpha_0 \mathbf{d}_k$ and compute $f_0 = f(\mathbf{x}_k + \boldsymbol{\delta}_k)$.
   Set $m = m + 1$.

4. (Interpolation)
   If $f_0 > f_L + \rho(\alpha_0 - \alpha_L) f'_L$ and $|(f_L - f_0)| > \varepsilon_2$ and $m < \hat{M}$, then do:
    a. If $\alpha_0 < \alpha_U$, then set $\alpha_U = \alpha_0$.

b. Compute $\overset{\cup}{\alpha}_0$ using Eq. (14.27).
   c. Compute $\overset{\cup}{\alpha}_{0L} = \alpha_L + \tau(\alpha_U - \alpha_L)$; if $\overset{\cup}{\alpha}_0 < \overset{\cup}{\alpha}_{0L}$, then set $\overset{\cup}{\alpha}_0 = \overset{\cup}{\alpha}_{0L}$.
   d. Compute $\overset{\cup}{\alpha}_{0U} = \alpha_U - \tau(\alpha_U - \alpha_L)$; if $\overset{\cup}{\alpha}_0 > \overset{\cup}{\alpha}_{0U}$, then set $\overset{\cup}{\alpha}_0 = \overset{\cup}{\alpha}_{0U}$.
   e. Set $\alpha_0 = \overset{\cup}{\alpha}_0$ and go to step 3.
5. Compute $f'_0 = \mathbf{g}(\mathbf{x}_k + \alpha_0 \mathbf{d}_k)^T \mathbf{d}_k$ and set $m = m + 1$.
6. (Extrapolation)

   If $f'_0 < \sigma f'_L$ and $|(f_L - f_0)| > \varepsilon_2$ and $m < \overset{\cap}{M}$, then do:
   a. Compute $\Delta \alpha_0 = (\alpha_0 - \alpha_L) f'_0 / (f'_L - f'_0)$ [see Eq. (14.28)].
   b. If $\Delta \alpha_0 \le 0$, then set $\overset{\cup}{\alpha}_0 = 2\alpha_0$; else, set $\overset{\cup}{\alpha}_0 = \alpha_0 + \Delta \alpha_0$.
   c. Compute $\overset{\cup}{\alpha}_{0U} = \alpha_0 + \chi(\alpha_U - \alpha_0)$; if $\overset{\cup}{\alpha}_0 > \overset{\cup}{\alpha}_{0U}$, then set $\overset{\cup}{\alpha}_0 = \overset{\cup}{\alpha}_{0U}$.
   d. Set $\alpha_L = \alpha_0$, $\alpha_0 = \overset{\cup}{\alpha}_0$, $f_L = f_0$, $f'_L = f'_0$ and go to step 3.
7. (Check termination criteria and output results)
   a. Set $\mathbf{x}_{k+1} = \mathbf{x}_k + \boldsymbol{\delta}_k$.
   b. Set $\Delta f_0 = f_{00} - f_0$.
   c. If $(\|\boldsymbol{\delta}_k\|_2 < \varepsilon_1$ and $|\Delta f_0| < \varepsilon_1)$ or $m > \overset{\cap}{M}$, then output $\overset{\cap}{\mathbf{x}} = \mathbf{x}_{k+1}$, $f(\overset{\cup}{\mathbf{x}}) = f_{k+1}$, and stop.
   d. Set $f_{00} = f_0$.
8. (Prepare for next iteration)
   a. Compute $\mathbf{g}_{k+1}$ and set $\boldsymbol{\gamma}_k = \mathbf{g}_{k+1} - \mathbf{g}_k$.
   b. Compute $D = \boldsymbol{\delta}_k^T \boldsymbol{\gamma}_k$; if $D \le 0$, then set $\mathbf{S}_{k+1} = \mathbf{I}_n$; else, compute $\mathbf{S}_{k+1}$ using Eq. (14.19) or (14.20).
   c. Set $k = k + 1$ and go to step 2.

Index m maintains a count of the number of function evaluations and is increased by one for each evaluation of f_0 or f'_0 in step 3 or 5, and $\overset{\cap}{M}$ is the maximum number of function evaluations allowed. When m becomes greater than $\overset{\cap}{M}$, the algorithm stops.

The estimate of α_0 in step 2d can be obtained by assuming that the function $f(\mathbf{x}_k + \alpha \mathbf{d}_k)$ can be represented by a quadratic polynomial of α and that the reduction achieved in $f(\mathbf{x}_k + \alpha \mathbf{d}_k)$ by changing α from 0 to α_0 is equal to Δf_0, the total reduction achieved in the previous iteration (see Prob. 14.11). This estimate can sometimes be quite inaccurate and may in certain circumstances become negative due to numerical ill-conditioning. For these reasons, if the estimate is equal to or less than zero or greater than unity, it is replaced by unity.

The quadratic extrapolation in step 6 of the algorithm may sometimes predict a maximum point at some negative value of α instead of a minimum point at some positive value of α (see Prob. 14.12). If such a case is identified in step 6b, the value of $2\alpha_0$ is assigned to $\overset{\cup}{\alpha}_0$ to ensure that α is changed in the direction of descent. If α_U is fixed by the interpolation, the minimum point cannot exceed this value; and, if extrapolation results in an unreasonably large value of $\overset{\cup}{\alpha}_0$, it is replaced by the value $\overset{\cup}{\alpha}_{0U}$ computed in step 6c.

While a positive definite matrix S_k will ensure that d_k is a direction of descent of function $f(x)$ at point x_k, in some rare occasions the function $f(x_k + \alpha d_k)$ may not have a well-defined minimum point. On the other hand, when the value of the function is very small, numerical ill-conditioning may arise occasionally due to roundoff errors. To avoid these problems, interpolation or extrapolation is carried out only if the expected reduction in the function $f(x_k + \alpha d_k)$ is larger than ε_2 and an upper limit in the number of function evaluations has not been exceeded.

If the DFP or BFGS updating formula is used in step 8b and the condition in Eq. (14.21) is satisfied, then a positive definite matrix S_k will result in a positive definite S_{k+1}, as was stated earlier. We will now demonstrate that if the Fletcher inexact line search is used and the search is not terminated until the inequality in Eq. (14.26) is satisfied, then Eq. (14.21) is, indeed, satisfied. When the search is terminated in the kth iteration, we have $\alpha_0 \equiv \alpha_k$ and from step 3 of the algorithm $\delta_k = \alpha_k d_k$. Now from Eqs. (14.21) and (14.26), we obtain

$$\delta_k^T \gamma_k = \delta_k^T g_{k+1} - \delta_k^T g_k$$
$$= \alpha_k (g_{k+1}^T d_k - g_k^T d_k)$$
$$> \alpha_k (\sigma - 1) g_k^T d_k$$

If d_k is a descent direction, then $g_k^T d_k < 0$ and $\alpha_k > 0$. Since $\sigma < 1$, we conclude that

$$\delta_k^T \gamma_k > 0$$

Under these circumstances, the positive definiteness of S_k is assured. In exceptional circumstances, the inexact line search in Algorithm 5 may not force the condition in Eq. (14.26) if the quantity $|(f_L - f_0)|$ is less than ε_2, and a nonpositive definite S_{k+1} matrix may on rare occasions arise. To safeguard against this possibility and ensure that a descent direction is achieved in every iteration, the quantity $\delta_k^T \gamma_k$ is checked in step 8b and if it is found to be negative or zero, the unity matrix I_n is assigned to S_{k+1}.

The DFP and BFGS updating formulas are very similar, and there are no clear theoretical advantages that apply to one and not the other. Indeed, the two formulas are interrelated in terms of a mathematical principle known as *duality*, which allows each of the two formulas to be derived from the other by simple algebraic manipulation. Nevertheless, extensive experimental results reported by Fletcher [6] show that the use of the BFGS formula tends to yield algorithms that are somewhat more tolerant to line-search imprecision. As a consequence, algorithms based on the BFGS formula are somewhat more efficient.

Example 14.2. In an application, the piecewise-continuous function

$$D(\omega) = \begin{cases} 2\omega & \text{for } 0 \leq \omega < 6 \\ 12 & \text{for } 6 \leq \omega < 12 \\ -\omega + 24 & \text{for } 12 \leq \omega < 16 \\ 8 & \text{for } 16 \leq \omega < 22 \end{cases}$$

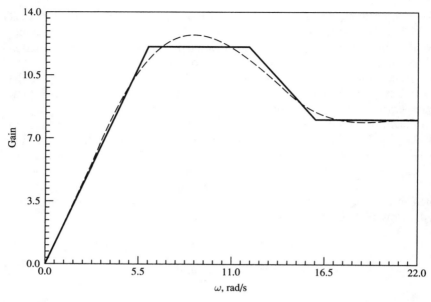

FIGURE 14.3
Plots of polynomials $D(\omega)$ and $P(\omega)$ (Example 14.2): ——— $D(\omega)$, - - - - - $P(\omega)$.

(see Fig. 14.3) has to be approximated by a polynomial of the form

$$P(\omega) = \sum_{k=0}^{5} a_k \omega^k$$

Using Algorithm 5, obtain a set of coefficients a_k for $k = 0, 1, \ldots, 5$ that minimizes the difference between $D(\omega)$ and $P(\omega)$ in the range $0 \leq \omega \leq 22$ in a least-squares sense.

Solution. A suitable objective function can be constructed as

$$\Psi(\mathbf{x}) = \frac{1}{2} L_2^2 = \frac{1}{2} \sum_{i=1}^{12} [D(\omega_i) - P(\omega_i)]^2$$

where $\omega_i = 2i - 2$ and

$$\mathbf{x} = [x_1 \quad x_2 \quad \ldots \quad x_6]^T = [a_0 \quad a_1 \quad \ldots \quad a_5]^T$$

by sampling the error $D(\omega) - P(\omega)$ at 12 points. The first partial derivatives of $\Psi(\mathbf{x})$ can be readily determined as

$$\frac{\partial \Psi(\mathbf{x})}{\partial x_k} = -\sum_{i=1}^{12} [D(\omega_i) - P(\omega_i)] \omega_i^k$$

TABLE 14.1
Coefficients of $P(\omega)$ (Example 14.2)

Coefficient	Value
a_0	-7.626758×10^{-2}
a_1	1.801233
a_2	2.389372×10^{-1}
a_3	-5.286809×10^{-2}
a_4	2.829081×10^{-3}
a_5	-4.791669×10^{-5}

TABLE 14.2
Progress of algorithm (Example 14.2)

k	Funct. evals.	$\Psi(\mathbf{x})$
0	7	5.060000×10^2
5	44	3.104894
10	87	1.017671
13	114	1.016952

for $k = 1, 2, \ldots, 6$. Using Algorithm 5 with an initial point $\mathbf{x}_0 = [0 \ 0 \ \cdots \ 0]^T$ and a termination tolerance $\varepsilon_1 = 10^{-6}$, the coefficients in Table 14.1 were obtained. The progress of the algorithm is illustrated in Table 14.2. The number of function evaluations is equal to the number of evaluations of the objective function $\Psi(\mathbf{x})$ plus the number of evaluations of the partial derivative function $\partial\Psi(\mathbf{x})/\partial x_k$. The polynomial $P(\omega)$ is compared with $D(\omega)$ in Fig. 14.3 and the error between the two is plotted versus frequency in Fig. 14.4. Note that the error is unevenly distributed with respect to the frequency. This is a common feature of least-squares solutions and is sometimes of concern.

14.5 MINIMAX ALGORITHMS

The design of digital filters can be accomplished by minimizing one of the norms described in Sec. 14.2. If the L_1 or L_2^2 norm is minimized, then the sum of the magnitudes or the sum of the squares of the elemental errors is minimized. The minimum error achieved usually turns out to be unevenly distributed with respect to the frequency and may exhibit large peaks (e.g., see the error achieved for Example 14.2 depicted in Fig. 14.4) which are often objectionable. If prescribed amplitude response specifications are to be met, the magnitude of the largest elemental error should be minimized and, therefore, the L_∞ norm of the error function should be used. Algorithms developed specifically for the minimization of the L_∞ norm are known as *minimax* algorithms and lead to designs in which the error is uniformly distributed with respect to frequency. The solutions obtained tend to be equiripple, much like the

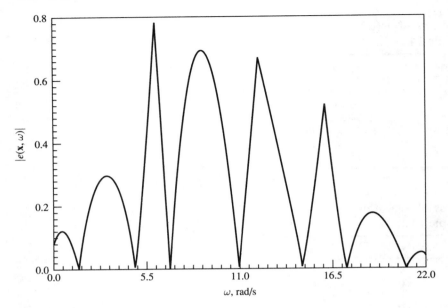

FIGURE 14.4
Error $|e(\mathbf{x}, \omega)|$ versus ω (Example 14.2).

solutions obtained by using the elliptic approximation of Chap. 5, which is, in effect, the minimax solution for filters with piecewise-constant amplitude responses.

The most fundamental minimax algorithm is the so-called *least-pth* algorithm, which involves minimizing an objective function of the type given in Eq. (14.5) for increasing values of p, say $p = 2, 4, 8, \ldots$, etc., and is as follows [12].

Algorithm 6: Least-pth minimax algorithm

1. Input $\overset{\cup}{\mathbf{x}}_0$ and ε_1. Set $k = 1$, $p = 2$, $\mu = 2$, $\overset{\cap}{E}_0 = 10^{99}$.
2. Initialize frequencies $\omega_1, \omega_2, \ldots, \omega_K$.
3. Using $\overset{\cup}{\mathbf{x}}_{k-1}$ as initial value, minimize

$$\Psi_k(\mathbf{x}) = \overset{\cap}{E}(\mathbf{x}) \left\{ \sum_{i=1}^{K} \left[\frac{|e_i(\mathbf{x})|}{\overset{\cap}{E}(\mathbf{x})} \right]^p \right\}^{1/p} \tag{14.29}$$

where

$$\overset{\cap}{E}(\mathbf{x}) = \max_{1 \leq i \leq K} |e_i(\mathbf{x})|$$

with respect to $\mathbf{x}$, to obtain $\overset{\cup}{\mathbf{x}}_k$. Set $\overset{\cap}{E}_k = \overset{\cap}{E}(\overset{\cup}{\mathbf{x}})$.
4. If $|\overset{\cap}{E}_{k-1} - \overset{\cap}{E}_k| < \varepsilon_1$, then output $\overset{\cup}{\mathbf{x}}_k$ and $\overset{\cap}{E}_k$, and stop. Else, set $p = \mu p$, $k = k + 1$ and go to step 3.

The underlying principle for the above algorithm is that the minimax problem is solved by solving a sequence of closely related problems whereby the solution of one renders the solution of the next one more tractable. Parameter μ in step 1, which must obviously be an integer, should not be too large in order to avoid numerical ill-conditioning. A value of 2 was found to give good results.

The minimization in step 3 can be carried out by using any unconstrained optimization algorithm; for example, Algorithm 5 described in the previous section. The gradient of $\Psi_k(\mathbf{x})$ is given by [12]

$$\nabla \Psi_k(\mathbf{x}) = \left\{ \sum_{i=1}^{K} \left[\frac{|e_i(\mathbf{x})|}{\hat{E}(\mathbf{x})} \right]^p \right\}^{(1/p)-1} \sum_{i=1}^{K} \left[\frac{|e_i(\mathbf{x})|}{\hat{E}(\mathbf{x})} \right]^{p-1} \nabla |e_i(\mathbf{x})| \qquad (14.30)$$

The preceding algorithm works very well, except that it requires a considerable amount of computation. An alternative and much more efficient minimax algorithm is one described in [13, 14]. This algorithm is based on principles developed by Charalambous [15] and involves the minimization of the objective function

$$\Psi(\mathbf{x}, \boldsymbol{\lambda}, \xi) = \sum_{i \in I_1} \frac{1}{2} \lambda_i [\phi_i(\mathbf{x}, \xi)]^2 + \sum_{i \in I_2} \frac{1}{2} [\phi_i(\mathbf{x}, \xi)]^2 \qquad (14.31)$$

where ξ and λ_i for $i = 1, 2, \ldots, K$ are constants

$$\phi_i(\mathbf{x}, \xi) = |e_i(\mathbf{x})| - \xi$$

$$I_1 = \{i : \phi_i(\mathbf{x}, \xi) > 0 \text{ and } \lambda_i > 0\} \qquad (14.32)$$

and

$$I_2 = \{i : \phi_i(\mathbf{x}, \xi) > 0 \text{ and } \lambda_i = 0\} \qquad (14.33)$$

The halves in Eq. (14.31) are included for the purpose of simplifying the gradient [see Eq. (14.34)].

If

(a) the second-order sufficiency conditions for a minimum of $\hat{E}(\mathbf{x})$ hold at $\check{\mathbf{x}}$,

(b) $\lambda_i = \check{\lambda}_i$ for $i = 1, 2, \ldots, K$ where $\check{\lambda}_i$ are the minimax multipliers corresponding to the minimum point $\check{\mathbf{x}}$ of $\hat{E}(\mathbf{x})$, and

(c) $\hat{E}(\check{\mathbf{x}}) - \xi$ is sufficiently small

then it can be proved that $\check{\mathbf{x}}$ is a *strong* local minimum point of function $\Psi(\mathbf{x}, \boldsymbol{\lambda}, \xi)$ given by Eq. (14.31) (see [15] for details). In practice, the conditions in (a) are satisfied for most practical problems. Consequently, if multipliers λ_i are forced to approach the minimax multipliers $\check{\lambda}_i$ and ξ is forced to approach $\hat{E}(\check{\mathbf{x}})$, then the minimization of $\hat{E}(\mathbf{x})$ can be accomplished by minimizing $\Psi(\mathbf{x}, \boldsymbol{\lambda}, \xi)$ with respect to $\mathbf{x}$. A minimax algorithm based on these principles is as follows:

Algorithm 7: Charalambous minimax algorithm

1. Input $\overset{U}{\mathbf{x}}_0$ and ε_1. Set $k = 1, \xi_1 = 0, \lambda_{11} = \lambda_{12} = \ldots = \lambda_{1K} = 1, \overset{\cap}{E}_0 = 10^{99}$.
2. Initialize frequencies $\omega_1, \omega_2, \ldots, \omega_K$.
3. Using $\overset{U}{\mathbf{x}}_{k-1}$ as initial value, minimize $\Psi(\mathbf{x}, \lambda_k, \xi_k)$ with respect to $\mathbf{x}$ to obtain $\overset{U}{\mathbf{x}}_k$. Set

$$\overset{\cap}{E}_k = \overset{\cap}{E}(\overset{U}{\mathbf{x}}_k) = \max_{1 \leq i \leq K} |e_i(\overset{U}{\mathbf{x}}_k)|$$

4. Compute

$$\Phi_k = \sum_{i \in I_1} \lambda_{ki} \phi_i(\overset{U}{\mathbf{x}}_k, \xi_k) + \sum_{i \in I_2} \phi_i(\overset{U}{\mathbf{x}}_k, \xi_k)$$

and update

$$\lambda_{(k+1)i} = \begin{cases} \lambda_{ki} \phi_i(\overset{U}{\mathbf{x}}_k, \xi_k)/\Phi_k & \text{for } i \in I_1 \\ \phi_i(\overset{U}{\mathbf{x}}_k, \xi_k)/\Phi_k & \text{for } i \in I_2 \\ 0 & \text{for } i \in I_3 \end{cases}$$

for $i = 1, 2, \ldots, K$ where

$$I_1 = \{i : \phi_i(\overset{U}{\mathbf{x}}_k, \xi_k) > 0 \text{ and } \lambda_{ki} > 0\}$$

$$I_2 = \{i : \phi_i(\overset{U}{\mathbf{x}}_k, \xi_k) > 0 \text{ and } \lambda_{ki} = 0\}$$

and

$$I_3 = \{i : \phi_i(\overset{U}{\mathbf{x}}_k, \xi_k) \leq 0\}$$

5. Compute

$$\xi_{k+1} = \sum_{i=1}^{K} \lambda_{(k+1)i} |e_i(\overset{U}{\mathbf{x}})|$$

6. If $|\overset{\cap}{E}_{k-1} - \overset{\cap}{E}_k| < \varepsilon_1$, then output $\overset{U}{\mathbf{x}}_k$ and $\overset{\cap}{E}_k$, and stop. Else, set $k = k+1$ and go to step 3.

The gradient of $\Psi(\mathbf{x}, \lambda_k, \xi_k)$, which is required in step 3 of the algorithm, is given by

$$\nabla \Psi(\mathbf{x}, \lambda_k, \xi_k) = \sum_{i \in I_1} \lambda_{ki} \phi_i(\mathbf{x}, \xi_k) \nabla |e_i(\mathbf{x})|$$

$$+ \sum_{i \in I_2} \phi_i(\mathbf{x}, \xi_k) \nabla |e_i(\mathbf{x})| \qquad (14.34)$$

Constant ξ is a lower bound of the minimum of $\overset{\cap}{E}(\mathbf{x})$ and as the algorithm progresses, it approaches $\overset{\cap}{E}(\overset{\cup}{\mathbf{x}})$ from below. Consequently, the number of functions $\phi_i(\mathbf{x}, \xi)$ that do not satisfy either Eq. (14.32) or Eq. (14.33) increases rapidly with the number of iterations. Since the derivatives of these functions are unnecessary in the minimization of $\Psi(\mathbf{x}, \lambda, \xi)$, they need not be evaluated. This increases the efficiency of the algorithm quite significantly.

As in Algorithm 6, the minimization in step 3 of Algorithm 7 can be carried out by using Algorithm 5.

14.6 IMPROVED MINIMAX ALGORITHMS

To achieve good results in the above minimax algorithms, the sampling of $e(\mathbf{x},\omega)$ with respect to ω must be dense; otherwise, the error function may develop spikes in the intervals between sampling points during the minimization. This problem is usually overcome by using a fairly large value of K of the order of three to six times the number of variables, e.g., if an eighth-order digital filter is to be designed, a value as high as 100 may be required. In such a case, each function evaluation in the minimization of the objective function would involve computing the gain of the filter as many as 100 times. A single optimization may sometimes necessitate 300 to 600 function evaluations, and a minimax algorithm like Algorithm 6 or 7 may require 5 to 10 unconstrained optimizations to converge. Consequently, the amount of computation required to complete a design is considerable.

A technique will now be described that can be used to suppress spikes in the error function without using a large value of K [16]. The technique entails the application of *nonuniform variable sampling* and involves the following steps:

1. Evaluate the error function of Eq. (14.3) with respect to a dense set of uniformly-spaced frequencies that span the frequency band of interest, say $\bar{\omega}_1, \bar{\omega}_2, \ldots, \bar{\omega}_L$, where L is fairly large, of the order of $10 \times K$.
2. Segment the frequency band of interest into K intervals.
3. For each of the K intervals, find the frequency that yields maximum error. Let these frequencies be $\overset{\cap}{\omega}_i$ for $i = 1, 2, \ldots, K$.
4. Use frequencies $\overset{\cap}{\omega}_i$ as sample frequencies in the evaluation of the objective function, i.e., set $\omega_i = \overset{\cap}{\omega}_i$ for $i = 1, 2, \ldots, K$.

By applying the above nonuniform sampling technique before the start of the second and subsequent optimizations, *frequency points at which spikes are beginning to form are located and are used as sample points in the next optimization*. In this way, the error at these frequencies is reduced and the formation of spikes is suppressed.

Assume that a digital filter is required to have a specified amplitude response with respect to a frequency band B which extends from $\bar{\omega}_1$ to $\bar{\omega}_L$, and let $\bar{\omega}_1, \bar{\omega}_2, \ldots, \bar{\omega}_L$ be uniformly-spaced frequencies such that

$$\bar{\omega}_i = \bar{\omega}_{i-1} + \Delta\omega$$

for $i = 2, 3, \ldots, L$ where

$$\Delta\omega = \frac{\bar{\omega}_L - \bar{\omega}_1}{L - 1} \tag{14.35}$$

These frequency points may be referred to as *virtual sample points*. Band B can be segmented into K intervals, say Ω_1 to Ω_K such that Ω_1 and Ω_K are of width $\Delta\omega/2$, Ω_2 and Ω_{K-1} are of width $l\Delta\omega$, and Ω_i for $i = 3, 4, \ldots, K - 2$ are of width $2l\Delta\omega$ where l is an integer. These requirements can be satisfied by letting

$$\Omega_1 = \left\{\omega : \bar{\omega}_1 \leq \omega < \bar{\omega}_1 + \frac{1}{2}\Delta\omega\right\}$$

$$\Omega_2 = \left\{\omega : \bar{\omega}_1 + \frac{1}{2}\Delta\omega \leq \omega < \bar{\omega}_1 + (l + \frac{1}{2})\Delta\omega\right\}$$

$$\Omega_i = \left\{\omega : \bar{\omega}_1 + \left[(2i - 5)l + \frac{1}{2}\right]\Delta\omega \leq \omega < \bar{\omega}_1 + \left[(2i - 3)l + \frac{1}{2}\right]\Delta\omega\right\}$$

for $i = 3, 4, \ldots, K - 2$

$$\Omega_{K-1} = \left\{\omega : \bar{\omega}_1 + \left[(2K - 7)l + \frac{1}{2}\right]\Delta\omega \leq \omega < \bar{\omega}_1 + \left[(2K - 6)l + \frac{1}{2}\right]\Delta\omega\right\}$$

and

$$\Omega_K = \left\{\omega : \bar{\omega}_1 + \left[(2K - 6)l + \frac{1}{2}\right]\Delta\omega \leq \omega \leq \bar{\omega}_L\right\}$$

where

$$\bar{\omega}_L = \bar{\omega}_1 + [(2K - 6)l + 1]\Delta\omega \tag{14.36}$$

The scheme is feasible if

$$L = (2K - 6)l + 2 \tag{14.37}$$

according to Eqs. (14.35) and (14.36), and is illustrated in Fig. 14.5 for the case where $K = 8$ and $l = 5$.

In the above segmentation scheme, there is only one sample in each of intervals Ω_1 and Ω_K, l samples in each of intervals Ω_2 and Ω_{K-1}, and $2l$ samples in each of intervals $\Omega_3, \Omega_4, \ldots, \Omega_{K-2}$, as can be seen in Fig. 14.5. Thus step 3 of the technique will yield $\hat{\omega}_1 = \bar{\omega}_1$ and $\hat{\omega}_K = \bar{\omega}_L$, i.e., the lower and upper band edges are forced to remain sample frequencies throughout the optimization. This strategy leads to two

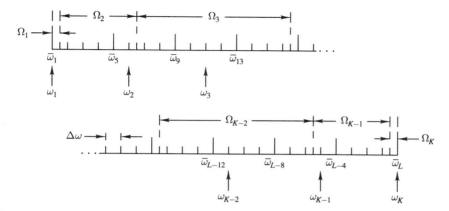

FIGURE 14.5
Segmentation of frequency axis.

advantages: (a) the error at the band edges is always minimized, and (b) a somewhat higher sampling density is maintained near the band edges where spikes are more likely to occur.

In the above technique, the required amplitude response needs to be specified with respect to a *dense* set of frequency points. This problem can be overcome through the use of interpolation. Let us assume that the amplitude response is specified at frequencies $\tilde{\omega}_1$ to $\tilde{\omega}_S$, where $\tilde{\omega}_1 = \bar{\omega}_1$ and $\tilde{\omega}_S = \bar{\omega}_L$. The required amplitude response for any frequency interval spanned by four successive specification points, say $\tilde{\omega}_j \leq \omega \leq \tilde{\omega}_{j+3}$, can be represented by a third-order polynomial of ω of the form

$$M_0(\omega) = a_{0j} + a_{1j}\omega + a_{2j}\omega^2 + a_{3j}\omega^3 \qquad (14.38)$$

and by varying j from 1 to $S - 3$, a set of $S - 3$ third-order polynomials can be obtained which can be used to interpolate the amplitude response to any desired degree of resolution. To achieve maximum interpolation accuracy, each of these polynomials should as far as possible be used only in the center of its frequency range of validity. Hence the first and last polynomials should be used for frequency ranges $\tilde{\omega}_1 \leq \omega < \tilde{\omega}_3$ and $\tilde{\omega}_{S-2} \leq \omega \leq \tilde{\omega}_S$, respectively, and the jth polynomial for $2 \leq j \leq S - 4$ should be used for the frequency range $\tilde{\omega}_{j+1} \leq \omega < \tilde{\omega}_{j+2}$.

Coefficients a_{ij} for $i = 0, 1, \ldots, 3$ and $j = 1$ to $S - 3$ can be determined by computing $\tilde{\omega}_m$, $(\tilde{\omega}_m)^2$, and $(\tilde{\omega}_m)^3$ for $m = j, j+1, \ldots, j+3$, and then constructing the system of simultaneous equations

$$\tilde{\Omega}_j \mathbf{a}_j = \mathbf{M}_{0j} \qquad (14.39)$$

where

$$\mathbf{a}_j = [a_{0j} \ldots a_{3j}] \qquad \text{and} \qquad \mathbf{M}_{0j} = [M_0(\tilde{\omega}_j) \ldots M_0(\tilde{\omega}_{j+3})]^T$$

are column vectors and Ω_j is the 4×4 matrix given by

$$\tilde{\Omega}_j = \begin{bmatrix} 1 & \tilde{\omega}_j & (\tilde{\omega}_j)^2 & (\tilde{\omega}_j)^3 \\ 1 & \tilde{\omega}_{j+1} & (\tilde{\omega}_{j+1})^2 & (\tilde{\omega}_{j+1})^3 \\ 1 & \tilde{\omega}_{j+2} & (\tilde{\omega}_{j+2})^2 & (\tilde{\omega}_{j+2})^3 \\ 1 & \tilde{\omega}_{j+3} & (\tilde{\omega}_{j+3})^2 & (\tilde{\omega}_{j+3})^3 \end{bmatrix}$$

Therefore, from Eq. (14.39) we have

$$\mathbf{a}_j = \tilde{\Omega}_j^{-1} \mathbf{M}_{0j} \qquad (14.40)$$

The above nonuniform sampling technique can be incorporated in Algorithm 6 by replacing steps 1, 2, and 4 by the modified steps 1A, 2A, and 4A listed below. The filter to be designed is assumed to be a single-band filter, for the sake of simplicity, although the technique is applicable to filters with an arbitrary number of bands.

1A. *a.* Input $\overset{\cup}{\mathbf{x}}_0$ and ε_1. Set $k = 1$, $p = 2$, $\mu = 2$, $\overset{\cap}{E}_0 = 10^{99}$. Initialize K.
  b. Input the required amplitude response $M_0(\tilde{\omega}_m)$ for $m = 1, 2, \ldots, S$.
  c. Compute L and $\Delta\omega$ using Eqs. (14.37) and (14.35), respectively.
  d. Compute coefficients a_{ij} for $i = 0, 1, \ldots, 3$ and $j = 1$ to $S - 3$ using Eq. (14.40).
  e. Compute the required amplitude response for $\bar{\omega}_1, \bar{\omega}_2, \ldots, \bar{\omega}_L$ using Eq. (14.38).

2A. Set $\omega_1 = \bar{\omega}_1$, $\omega_2 = \bar{\omega}_{1+l}$, $\omega_i = \bar{\omega}_{2(i-2)l+1}$ for $i = 3, 4, \ldots, K - 2$, $\omega_{K-1} = \bar{\omega}_{L-l}$, and $\omega_K = \bar{\omega}_L$.

4A. *a.* Compute $|e_i(\overset{\cup}{\mathbf{x}}_k)|$ for $i = 1, 2, \ldots, L$ using Eqs. (14.3) and (14.4).
  b. Determine frequencies $\overset{\cap}{\omega}_i$ for $i = 1, 2, \ldots, K$ and

$$\overset{\cap}{P}_k = \overset{\cap}{P}(\overset{\cup}{\mathbf{x}}_k) = \max_{1 \le i \le L} |e_i(\overset{\cup}{\mathbf{x}}_k)|$$

  c. Set $\omega_i = \overset{\cap}{\omega}_i$ for $i = 1, 2, \ldots, K$.
  d. If $|\overset{\cap}{E}_{k-1} - \overset{\cap}{E}_k| < \varepsilon_1$ and $|\overset{\cap}{P}_k - \overset{\cap}{E}_k| < \varepsilon_1$, then output $\overset{\cup}{\mathbf{x}}_k$ and $\overset{\cap}{E}_k$, and stop. Else, set $p = \mu p$, $k = k + 1$ and go to step 3.

Similarly, the technique can be applied to Algorithm 7, by replacing steps 1, 2, and 6 by the following modified steps:

1A. *a.* Input $\overset{\cup}{\mathbf{x}}_0$ and ε_1. Set $k = 1$, $\xi_1 = 0$, $\lambda_{11} = \lambda_{12} = \ldots = \lambda_{1K} = 1$, $\overset{\cap}{E}_0 = 10^{99}$. Initialize K.
  b. Input the required amplitude response $M_0(\tilde{\omega}_m)$ for $m = 1, 2, \ldots, S$.
  c. Compute L and $\Delta\omega$ using Eqs. (14.37) and (14.35), respectively.
  d. Compute coefficients a_{ij} for $i = 0, 1, \ldots, 3$ and $j = 1$ to $S - 3$ using Eq. (14.40).
  e. Compute the required amplitude response for $\bar{\omega}_1, \bar{\omega}_2, \ldots, \bar{\omega}_L$ using Eq. (14.38).

2A. Set $\omega_1 = \bar{\omega}_1$, $\omega_2 = \bar{\omega}_{1+l}$, $\omega_i = \bar{\omega}_{2(i-2)l+1}$ for $i = 3, 4, \ldots, K - 2$, $\omega_{K-1} = \bar{\omega}_{L-l}$, and $\omega_K = \bar{\omega}_L$.

6A. *a.* Compute $|e_i(\overset{\cup}{\mathbf{x}}_k)|$ for $i = 1, 2, \ldots, L$ using Eqs. (14.3) and (14.4).
  b. Determine frequencies $\hat{\omega}_i$ for $i = 1, 2, \ldots, K$ and
$$\hat{P}_k = \hat{P}(\overset{\cup}{\mathbf{x}}_k) = \max_{1 \leq i \leq L} |e_i(\overset{\cup}{\mathbf{x}}_k)|$$
  c. Set $\omega_i = \hat{\omega}_i$ for $i = 1, 2, \ldots, K$.
  d. If $|\hat{E}_{k-1} - \hat{E}_k| < \varepsilon_1$ and $|\hat{P}_k - \hat{E}_k| < \varepsilon_1$, then output $\overset{\cup}{\mathbf{x}}_k$ and $\hat{E}_k$, and stop. Else, set $k = k + 1$ and go to step 3.

In step 2A, the initial sample frequencies ω_1 and ω_K are assumed to be at the left-hand and right-hand band edges, respectively; ω_2 and ω_{K-1} are taken to be the last and first frequencies in intervals Ω_2 and Ω_{K-1}, respectively; and each of frequencies $\omega_3, \omega_4, \ldots, \omega_{K-2}$ is set near the center of each of intervals $\Omega_3, \Omega_4, \ldots, \Omega_{K-2}$. This assignment is illustrated in Fig. 14.5 for the case where $K = 8$ and $l = 5$.

Without the nonuniform sampling technique, the number of samples K should be chosen to be of the order of three to six times the number of variables, depending on the selectivity of the filter. While a value of 50 may be entirely satisfactory for an eighth-order lowpass filter with a wide transition band, a value of 100 may not be adequate for a highly selective narrow-band bandpass filter of the same order. With the technique, the number of virtual samples is approximately equal to $2l \times K$, according to Eq. (14.37). As l is increased above unity, the frequencies of maximum error $\hat{\omega}_i$ become progressively more precise, owing to the increased resolution; however, the amount of computation required in step 4A of Algorithm 6 or step 6A of Algorithm 7 is proportionally increased. Eventually, a situation of diminishing returns is reached whereby further increases in l bring about only slight improvements in the precision of the $\hat{\omega}_i$'s. The values $K = 35$ and $l = 5$, which correspond to 35 actual and 322 virtual sample points, were found to give excellent results for a diverse range of designs, including some complex 28th-order phase-equalizer designs (see Sec. 14.8).

14.7 DESIGN OF RECURSIVE FILTERS

The application of Algorithms 6 and 7 for the design of recursive digital filters can be readily accomplished by obtaining expressions for the objective functions $\Psi_k(\mathbf{x})$ and $\Psi(\mathbf{x}, \lambda_k, \xi_k)$ and their gradients.

The amplitude response of an Nth-order filter is given by Eqs. (14.1) and (14.2) as

$$M(\mathbf{x}, \omega) = H_0 \prod_{j=1}^{J} \frac{N_j(\omega)}{D_j(\omega)}$$

where

$$N_j(\omega) = [1 + a_{0j}^2 + a_{1j}^2 + 2a_{1j}(1 + a_{0j})\cos \omega T + 2a_{0j} \cos 2\omega T]^{\frac{1}{2}}$$

and

$$D_j(\omega) = [1 + b_{0j}^2 + b_{1j}^2 + 2b_{1j}(1 + b_{0j})\cos\omega T + 2b_{0j}\cos 2\omega T]^{\frac{1}{2}}$$

for $j = 1, 2, \ldots, J$. Hence Eqs. (14.3) and (14.4) yield

$$e_i(\mathbf{x}) = M(\mathbf{x}, \omega_i) - M_0(\omega_i)$$

and from Eqs. (14.29) and (14.31), $\Psi_k(\mathbf{x})$ and $\Psi(\mathbf{x}, \lambda_k, \xi_k)$ can be formed. Since $M_0(\omega_i)$ is a constant, we obtain

$$\frac{\partial e_i(\mathbf{x})}{\partial a_{0l}} = \frac{a_{0l} + a_{1l}\cos\omega_i T + \cos 2\omega_i T}{[N_l(\omega_i)]^2} \cdot M(\mathbf{x}, \omega_i)$$

$$\frac{\partial e_i(\mathbf{x})}{\partial a_{1l}} = \frac{a_{1l} + (1 + a_{0l})\cos\omega_i T}{[N_l(\omega_i)]^2} \cdot M(\mathbf{x}, \omega_i)$$

$$\frac{\partial e_i(\mathbf{x})}{\partial b_{0l}} = -\frac{b_{0l} + b_{1l}\cos\omega_i T + \cos 2\omega_i T}{[D_l(\omega_i)]^2} \cdot M(\mathbf{x}, \omega_i)$$

$$\frac{\partial e_i(\mathbf{x})}{\partial b_{1l}} = -\frac{b_{1l} + (1 + b_{0l})\cos\omega_i T}{[D_l(\omega_i)]^2} \cdot M(\mathbf{x}, \omega_i)$$

$$\frac{\partial e_i(\mathbf{x})}{\partial H_0} = \frac{1}{H_0} \cdot M(\mathbf{x}, \omega_i)$$

for $l = 1, 2, \ldots, J$ and $i = 1, 2, \ldots, K$. Hence the gradient of $e_i(\mathbf{x})$, namely, $\nabla e_i(\mathbf{x})$, can be formed, and since

$$\nabla |e_i(\mathbf{x})| = \text{sgn } e_i(\mathbf{x}) \nabla e_i(\mathbf{x})$$

where

$$\text{sgn } e_i(\mathbf{x}) = \begin{cases} 1 & \text{if } e_i(\mathbf{x}) \geq 0 \\ -1 & \text{otherwise} \end{cases}$$

$\nabla \Psi_k(\mathbf{x})$ and $\nabla \Psi(\mathbf{x}, \lambda_k, \xi_k)$ can be evaluated using Eqs. (14.30) and (14.34), respectively.

The minimax algorithms considered will yield filters which may or may not be stable, since the transfer function obtained may have poles outside the unit circle of the z plane. However, the problem can be easily eliminated by replacing the offending poles by their reciprocals and simultaneously adjusting the multiplier constant H_0 so as to compensate for the change in gain. This stabilization technique is described in Sec. 7.4 (see p. 226).

A problem associated with the design of filters with arbitrary amplitude and/or phase responses is that there are no known methods for the prediction of the filter order that will limit the approximation error to within prescribed bounds. However, satisfactory results can often be achieved on a cut-and-try basis by designing filters of increasing orders until the error is sufficiently small to satisfy the requirements.

Example 14.3. A lowpass digital filter is to be used in cascade with a D/A converter. The overall amplitude response from the input of the filter to the output of the D/A converter is required to be

$$M(\omega) = \begin{cases} 1.0 & \text{for } 0 \leq \omega \leq 4 \times 10^4 \text{rad/s} \\ 0.01 & \text{for } 4.5 \times 10^4 \leq \omega \leq 10^5 \end{cases}$$

and the amplitude response of the D/A converter is given by

$$\phi(\omega) = \left| \frac{\sin(\omega\tau/2)}{\omega\tau/2} \right|$$

where τ is the pulse duration at the output of the D/A converter (see Sec. 6.9). Design the lowpass filter using Algorithm 7 first without and then with the technique of Sec. 14.6 and compare the results obtained. Use an eighth-order transfer function and assume that $\omega_s = 2 \times 10^5$ rad/s and $\tau = T$.

Solution. The amplitude response of the filter must be modified as [17]

$$\tilde{M}(\omega) = \begin{cases} 1.0/\phi(\omega) & \text{for } 0 \leq \omega \leq 4 \times 10^4 \text{ rad/s} \\ 0.01/\phi(\omega) & \text{for } 4.5 \times 10^4 \leq \omega \leq 10^5 \end{cases}$$

in order to achieve the required amplitude response between the input of the filter and the output of the D/A converter.

The amount of computation required by optimization methods in general and the quality of the solution obtained tend to depend heavily on the initial solution assumed. If the initial point is close to the actual solution, the amount of computation tends to be low and the precision of the solution tends to be high. In this example, a good initial estimate of the solution can be obtained by designing an eighth-order lowpass filter with passband ripple $A_p = 0.1$ dB, minimum stopband attenuation $A_a = 50$ dB, passband edge $\omega_p = 4.0 \times 10^4$ rad/s, and stopband edge $\omega_a = 4.5 \times 10^4$ rad/s. A lowpass filter satisfying these specifications can readily be designed using the method of Chap. 8. The transfer-function coefficients of a design based on the elliptic approximation are given in Table 14.3.

Using Algorithm 7 with $K = 40$ (25 sample points in the passband and 15 in the stopband) first without and then with the technique of Sec. 14.6, designs A and B of Table 14.3 were obtained. The progress of the algorithm is illustrated in Table 14.4. The magnitude of the error function for each of the two designs is plotted in Fig. 14.6a and b. As can be seen, spikes are present in the error function of design A but are entirely eliminated in design B through the use of the technique in Sec. 14.6. The amplitude response achieved in design B is illustrated in Fig. 14.7.

14.06ab
14.07ab

Example 14.4. Through the application of the singular-value decomposition, the problem of designing two-dimensional digital filters (see Sec. 16.6) can be broken down into a problem of designing a set of one-dimensional digital filters [18]. The amplitude responses of the one-dimensional filters so obtained turn out to be quite irregular and, consequently, their design can be accomplished only through the use of optimization methods. The amplitude response of such a filter is specified at 21 frequency points, as in Table 14.5, and $\omega_s = 2$ rad/s. Obtain eighth-order designs using Algorithms 6 and 7 in conjunction with the technique of Sec. 14.6 in each case, and compare the results obtained. Assume that $K = 35$.

TABLE 14.3
Coefficients of $H(z)$ (Example 14.3)

	j	a_{0j}	a_{ij}	b_{0j}	b_{ij}
Initial filter	1	1.0	1.663591	2.964920×10^{-1}	-9.685886×10^{-1}
	2	1.0	4.646911×10^{-1}	5.578139×10^{-1}	-7.881536×10^{-1}
	3	1.0	-1.082131×10^{-1}	7.999954×10^{-1}	-6.314471×10^{-1}
	4	1.0	-2.936755×10^{-1}	9.452177×10^{-1}	-5.696838×10^{-1}
	$H_0 = 1.375814 \times 10^{-2}$				
Design A	1	1.201422	1.802335	1.826366×10^{-1}	-7.094977×10^{-1}
	2	1.023690	5.173944×10^{-1}	4.754965×10^{-1}	-6.411708×10^{-1}
	3	9.871557×10^{-1}	-9.208725×10^{-2}	7.562144×10^{-1}	-5.689563×10^{-1}
	4	9.970934×10^{-1}	-2.981699×10^{-1}	9.334971×10^{-1}	-5.428448×10^{-1}
	$H_0 = 1.987973 \times 10^{-2}$				
Design B	1	1.255164	1.822823	1.916731×10^{-1}	-7.257967×10^{-1}
	2	1.048624	4.620755×10^{-1}	4.905558×10^{-1}	-6.448195×10^{-1}
	3	1.003053	-1.109377×10^{-1}	7.657275×10^{-1}	-5.639906×10^{-1}
	4	1.000126	-2.939426×10^{-1}	9.349386×10^{-1}	-5.365697×10^{-1}
	$H_0 = 2.000669 \times 10^{-2}$				

TABLE 14.4
Progress of algorithm (Example 14.3)

	Design A		Design B	
k	ξ	$\Psi(x, \lambda_k, \xi_k)$	ξ	$\Psi(x, \lambda_k, \xi_k)$
1	0.0	7.509080×10^{-6}	0.0	7.509080×10^{-6}
2	7.510177×10^{-4}	5.098063×10^{-9}	7.510177×10^{-4}	6.274917×10^{-8}
3	8.854903×10^{-4}	6.874848×10^{-11}	1.158158×10^{-3}	3.401966×10^{-9}
4	9.013783×10^{-4}	2.732611×10^{-13}	1.250634×10^{-3}	3.311865×10^{-11}
5	9.023167×10^{-4}	2.371856×10^{-15}	1.260096×10^{-3}	4.468298×10^{-14}

Solution. Using an initial point

$$\mathbf{x} = [1 \quad 1 \quad 0.75 \quad 1 \quad 1 \quad 1 \quad 0.75 \quad 1 \quad 1 \quad -1 \quad 0.75 \quad -1 \quad 1 \quad -1 \quad 0.75 \quad -1 \quad 1]^T$$

designs A and B of Table 14.6 were obtained. The progress of each algorithm is illustrated in Table 14.7. The maximum amplitude-response errors in designs A and B were 3.2675×10^{-2} and 3.5292×10^{-2}, and the CPU times using an HP9000 series 310 workstation[†] were 4055 and 2259 seconds, respectively. Evidently, Algorithm 6 gave a somewhat better design, although the amount of computation time was nearly twice that required by Algorithm 7. The amplitude response achieved in design A is illustrated in Fig. 14.8.

14.8 DESIGN OF RECURSIVE DELAY EQUALIZERS

The minimax algorithms described can also be applied for the design of recursive delay equalizers, as will now be demonstrated. Consider a filter characterized by the transfer function

$$H_F(z) = H_0 \prod_{j=1}^{J} \frac{a_{0j} + a_{1j}z + a_{2j}z^2}{b_{0j} + b_{1j}z + b_{2j}z^2} \tag{14.41}$$

The group delay of the filter is given by

$$\tau_F(\omega) = -\frac{d\theta_F(\omega)}{d\omega} \tag{14.42}$$

where

$$\theta_F(\omega) = \arg H_F(e^{j\omega T}) \tag{14.43}$$

From Eqs. (14.41) and (14.42), we can show that

$$\tau_F(\omega) = -T \sum_{j=1}^{J} \frac{\tilde{N}_j(\omega)}{N_j(\omega)} + T \sum_{j=1}^{J} \frac{\tilde{D}_j(\omega)}{D_j(\omega)} \tag{14.44}$$

[†] A 68010-processor machine.

522 DIGITAL FILTERS: ANALYSIS, DESIGN, AND APPLICATIONS

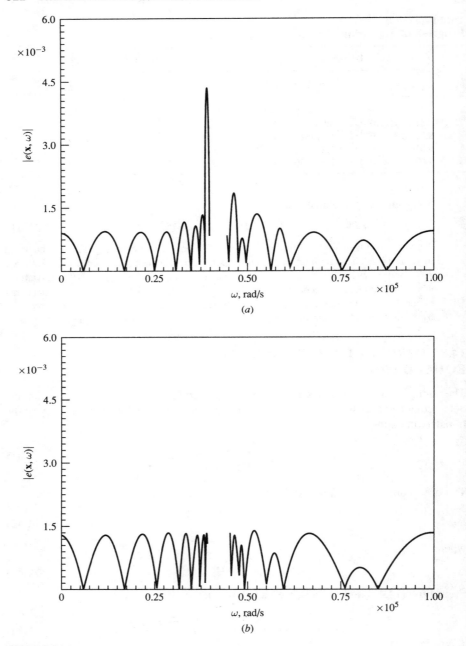

FIGURE 14.6
Error $|e(\mathbf{x}, \omega)|$ versus ω (Example 14.3): (a) without the technique of Sec. 14.6, (b) with the technique of Sec. 14.6.

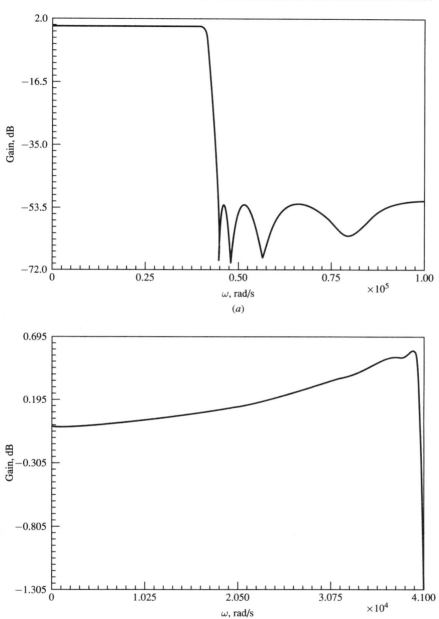

FIGURE 14.7
Amplitude response of lowpass filter (Example 14.3): (a) for $0 \leq \omega \leq 10^5$, (b) for $0 \leq \omega \leq 4.1 \times 10^4$.

TABLE 14.5
Specified amplitude response (Example 14.4)

ω	Gain	ω	Gain	ω	Gain
0.00	.1.0770	0.35	0.0304	0.70	0.7950
0.05	0.9863	0.40	0.1665	0.75	0.7950
0.10	0.9866	0.45	0.4402	0.80	0.7950
0.15	0.8428	0.50	0.6231	0.85	0.7950
0.20	0.8436	0.55	0.7471	0.90	0.7950
0.25	0.6466	0.60	0.7950	0.95	0.7950
0.30	0.3955	0.65	0.7950	1.00	0.7950

where

$$\tilde{N}_j(\omega) = a_{2j}^2 - a_{0j}^2 + a_{1j}(a_{2j} - a_{0j})\cos\omega T$$
$$N_j(\omega) = (a_{2j} - a_{0j})^2 + a_{1j}^2 + 2a_{1j}(a_{2j} + a_{0j})\cos\omega T + 4a_{0j}a_{2j}\cos^2\omega T$$
$$\tilde{D}_j(\omega) = b_{2j}^2 - b_{0j}^2 + b_{1j}(b_{2j} - b_{0j})\cos\omega T$$
$$D_j(\omega) = (b_{2j} - b_{0j})^2 + b_{1j}^2 + 2b_{1j}(b_{2j} + b_{0j})\cos\omega T + 4b_{0j}b_{2j}\cos^2\omega T$$

The group delay of the filter can be equalized with respect to a frequency range $\omega_1 \leq \omega \leq \omega_L$ by connecting an *allpass* delay equalizer in cascade with the filter, as described in Sec. 8.5.1. Let the transfer function of the equalizer be

$$H_E(z) = \prod_{j=1}^{M} \frac{1 + c_{1j}z + c_{0j}z^2}{c_{0j} + c_{1j}z + z^2}$$

The group delay of the equalizer can be obtained as

$$\tau_E(\mathbf{c}, \omega) = -\frac{d\theta_E(\omega)}{d\omega}$$

where

$$\theta_E(\mathbf{c}, \omega) = \arg H_E(e^{j\omega T})$$

Hence

$$\tau_E(\mathbf{c}, \omega) = 2T \sum_{j=1}^{M} \frac{\tilde{C}_j(\omega)}{C_j(\omega)} \qquad (14.45)$$

where

$$\tilde{C}_j(\omega) = 1 - c_{0j}^2 + c_{1j}(1 - c_{0j})\cos\omega T$$
$$C_j(\omega) = (1 - c_{0j})^2 + c_{1j}^2 + 2c_{1j}(1 + c_{0j})\cos\omega T + 4c_{0j}\cos^2\omega T$$

and

$$\mathbf{c} = [c_{01} \quad c_{11} \quad c_{02} \quad c_{12} \quad \cdots \quad c_{1M}]^T$$

The equalizer is stable if and only if the transfer function coefficients satisfy the relations

$$c_{0j} < 1, \qquad c_{1j} - c_{0j} < 1, \qquad c_{1j} + c_{0j} > -1$$

TABLE 14.6
Coefficients of $H(z)$ (Example 14.4)

	j	a_{0j}	a_{1j}	b_{0j}	b_{1j}
Design A	1	1.002238	2.482808	-4.716961×10^{-2}	-9.493371×10^{-1}
	2	-1.973023×10	1.880026×10	-2.123562×10^{-1}	3.655407×10^{-1}
	3	1.000000	-8.468213×10^{-1}	1.496466×10^{-1}	1.873191×10^{-2}
	4	1.830361	-2.033032	6.498825×10^{-1}	-1.155793
	$H_0 = 8.425338 \times 10^{-3}$				
Design B	1	-1.260454×10	3.977791×10	4.318101×10^{-1}	-1.055599
	2	2.377913	-2.490881	-1.831163×10^{-2}	-5.216264×10^{-1}
	3	9.849419×10^{-1}	-8.325620×10^{-1}	3.616646×10^{-1}	-2.230790×10^{-1}
	4	5.511632×10^{-1}	-9.021266×10^{-1}	6.733342×10^{-1}	-1.088983
	$H_0 = 6.418782 \times 10^{-3}$				

TABLE 14.7
Progress of algorithms (Example 14.4)

		Design A	Design B	
k	p	$\Psi(\mathbf{x})$	ξ	$\Psi(\mathbf{x}, \lambda_k, \xi_k)$
1	2	7.106816×10^{-2}	0.0	2.893164×10^{-4}
2	4	3.726389×10^{-2}	2.229626×10^{-2}	4.092217×10^{-5}
3	8	3.329217×10^{-2}	3.397612×10^{-2}	5.915443×10^{-7}
4	16	3.757264×10^{-2}	3.527249×10^{-2}	6.184436×10^{-20}
5	32	3.472619×10^{-2}	3.174503×10^{-2}	4.251311×10^{-5}
6	64	3.359927×10^{-2}	—	—
7	128	3.304717×10^{-2}	—	—

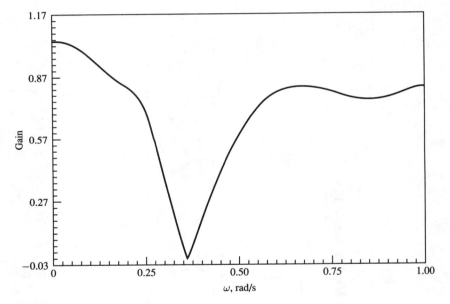

FIGURE 14.8
Amplitude response of one-dimensional filter (Design A, Example 14.4).

for $j = 1, 2, \ldots, M$ (see Sec. 3.3.7). The region of stability in the (c_0, c_1) plane is illustrated in Fig. 14.9. This may be referred to as the feasible region of the parameter space.

The group delay of the filter/equalizer combination can be expressed as

$$\tau_{FE}(\mathbf{c}, \omega) = \tau_F(\omega) + \tau_E(\mathbf{c}, \omega)$$

where $\tau_F(\omega)$ and $\tau_E(\mathbf{c}, \omega)$ are given by Eqs. (14.44) and (14.45), respectively.

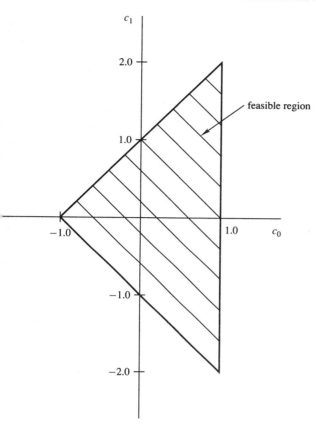

FIGURE 14.9
Feasible region of (c_0, c_1) plane.

The required equalizer can be designed by solving the optimization problem [13]

$$\text{minimize}_x \ \hat{E}(x)$$

where

$$\hat{E}(\mathbf{x}) = \max_{1 \leq i \leq K} |e_i(\mathbf{x})|$$

$$e_i(\mathbf{x}) = \frac{1}{T}\tau_{FE}(\mathbf{x}, \omega_i) - \tau_0$$

$$\mathbf{x} = [\ \mathbf{c}^T \quad \tau_0\ ]^T, \qquad \tau_0 = \frac{\tau}{T}$$

and

$$\omega_1 \leq \omega \leq \omega_L$$

The problem can be readily solved by using Algorithm 6 or 7. As the solution is approached, variable τ_0 approaches the average of τ_{FE}/T with respect to the frequency band of interest, i.e., τ approaches the average of τ_{FE}.

The gradient of $|e_i(\mathbf{x})|$, which is required for the evaluation of $\nabla \Psi(\mathbf{x}, \lambda_k, \xi_k)$, can be obtained, as in Sec. 14.7, by using the derivatives of $e_i(\mathbf{x})$, namely,

$$\frac{\partial e_i(\mathbf{x})}{\partial c_{0l}} = \frac{U_{0l} + U_{1l} \cos \omega_i T + U_{2l} \cos^2 \omega_i T + U_{3l} \cos^3 \omega_i T}{[C_l(\omega_i)]^2}$$

$$\frac{\partial e_i(\mathbf{x})}{\partial c_{1l}} = \frac{V_{0l} + V_{1l} \cos \omega_i T + V_{2l} \cos^2 \omega_i T + V_{3l} \cos^3 \omega_i T}{[C_l(\omega_i)]^2}$$

$$\frac{\partial e_i(\mathbf{x})}{\partial \tau_0} = -1$$

for $l = 1, 2, \ldots, M$ and $i = 1, 2, \ldots, K$, where

$$U_{0l} = 4[(1 - c_{0l})^2 - c_{0l} c_{1l}^2], \quad U_{1l} = -2c_{1l}(1 + 6c_{0l} + c_{0l}^2 + c_{1l}^2)$$

$$U_{2l} = -8(1 + c_{0l}^2 + c_{1l}^2), \quad U_{3l} = -8c_{1l}$$

$$V_{0l} = -4c_{1l}(1 - c_{0l})(1 + c_{0l}), \quad V_{1l} = -2(1 - c_{0l})(1 + 6c_{0l} + c_{0l}^2 + c_{1l}^2)$$

$$V_{2l} = 0, \quad V_{3l} = 8(1 - c_{0l})c_{0l}$$

The quality of an equalizer is inversely related to the maximum variation of τ_{FE} over the frequency band of interest. A measure that can be used to assess the quality of an equalizer design can, therefore, be defined as

$$Q = \frac{100(\hat{\tau}_{FE} - \check{\tau}_{FE})}{2\tilde{\tau}_{FE}} \tag{14.46}$$

where

$$\hat{\tau}_{FE} = \max_{\omega_1 \leq \omega \leq \omega_L} \tau_{FE}$$

$$\check{\tau}_{FE} = \min_{\omega_1 \leq \omega \leq \omega_L} \tau_{FE}$$

and

$$\tilde{\tau}_{FE} = \frac{1}{2}(\hat{\tau}_{FE} + \check{\tau}_{FE}) \tag{14.47}$$

Alternatively, from Eqs. (14.46) and (14.47),

$$Q = \frac{100(\hat{\tau}_{FE} - \check{\tau}_{FE})}{(\hat{\tau}_{FE} + \check{\tau}_{FE})} \tag{14.48}$$

As in the design of recursive filters, the application of Algorithm 6 or 7 for the design of equalizers may yield an unstable design. While it is possible to restore stability in such a design by replacing poles that are outside the unit circle of the z plane by their reciprocals, the group-delay characteristic of the equalizer will be changed and the resulting design will not be useful. A brute force approach to overcome this problem is to carry out several designs using different starting points, and then select the best design from the set of stable designs. An alternative and more methodical approach, which was found to give good results, is based on the following algorithm:

Algorithm 8: Design of equalizers

1. Compute $\tilde{\tau}_F = (\overset{\cap}{\tau}_F + \overset{\cup}{\tau}_F)/2$, where $\overset{\cap}{\tau}_F$ and $\overset{\cup}{\tau}_F$ are the maximum and minimum of the filter group delay, respectively. Assume a 1-section equalizer, and set $j = 1$ and $\tau_{01} = (1 + k_1)\tilde{\tau}_F/T$, where k_1 is a constant in the range $0 \le k_1 \le 0.5$. Carry out designs using points 1 to 8 in Table 14.8 for the initialization of the equalizer coefficients until a stable design is obtained; let the coefficients of the stable design be $\bar{c}_{01}$ and $\bar{c}_{11}$. Compute $\tilde{\tau}_{FE1}$ using Eq. (14.47).

2. *a.* Increase the number of equalizer sections to two; set $j = j + 1$ and $\tau_{02} = \tilde{\tau}_{FE1}/T$.†

   b. Carry out designs using point $(\bar{c}_{01}, \bar{c}_{11})$ for the initialization of the first section and each of the points

   $$P_{12} = [(1 - \varepsilon_1)\bar{c}_{01}, (1 - \varepsilon_1)\bar{c}_{11}]$$

   $$P_{22} = [(1 + \varepsilon_1)\bar{c}_{01}, (1 - \varepsilon_1)\bar{c}_{11}]$$

   $$P_{32} = [(1 + \varepsilon_1)\bar{c}_{01}, (1 + \varepsilon_1)\bar{c}_{11}]$$

   $$P_{42} = [(1 - \varepsilon_1)\bar{c}_{01}, (1 + \varepsilon_1)\bar{c}_{11}]$$

   in turn for the initialization of the second section (ε_1 is a small positive constant).

   c. Compute parameter Q using Eq. (14.48).

   d. If the design obtained is successful, i.e., it is stable and has a Q which is significantly lower than that of the 1-section design, compute $\tilde{\tau}_{FE2}$ and continue with step 3; otherwise, change ε_1 and repeat from step 2b.

3. *a.* Increase the number of equalizer sections by one. Set $j = j + 1$ and $\tau_{0j} = \tilde{\tau}_{FE(j-1)}/T$†, and carry out designs using the most recent successful design for the initialization of sections $1, 2, \ldots, j-1$ and point

   $$P_{0j} = \left[\frac{1}{2}(\overset{\cap}{c}_{0(j-1)} + \overset{\cup}{c}_{0(j-1)}), \frac{1}{2}(\overset{\cap}{c}_{1(j-1)} + \overset{\cup}{c}_{1(j-1)})\right]$$

---

† Randy K. Howell of the graduate digital-filter class of 1992 found that the amount of computation can be reduced by using $\tilde{\tau}_{FEj}/T$ instead of $\tilde{\tau}_{FE(j-1)}/T$ for τ_{0j} in steps 2 and 3; this modification can be readily incorporated in the algorithm by including the jth equalizer section in the calculation of $\tilde{\tau}_{FE}$ using the initial coefficient values for the jth section.

for the initialization of the jth section, where $\overset{\cap}{c}_{0(j-1)}$ ($\overset{\cap}{c}_{1(j-1)}$) and $\overset{U}{c}_{0(j-1)}$ ($\overset{U}{c}_{1(j-1)}$) are the largest and smallest coefficient c_0 (c_1), respectively, in the most recent successful design.

  b. If the design obtained in step 3a is unsuccessful, carry out designs using the most recent successful design for the initialization of sections $1, 2, \ldots, j-1$ and each of the points

$$P_{1j} = (\overset{U}{c}_{0(j-1)}, \overset{U}{c}_{1(j-1)})$$

$$P_{2j} = (\overset{\cap}{c}_{0(j-1)}, \overset{U}{c}_{1(j-1)})$$

$$P_{3j} = (\overset{\cap}{c}_{0(j-1)}, \overset{\cap}{c}_{1(j-1)})$$

$$P_{4j} = (\overset{U}{c}_{0(j-1)}, \overset{\cap}{c}_{1(j-1)})$$

in turn for the initialization of the jth section. If a successful design is obtained compute τ_{FEj} and proceed to step 4; otherwise, stop.

4. Compute Q; if $Q \leq Q_{\max}$, stop; otherwise, go to step 3a.

Extensive experimentation with Algorithm 8 has shown that for a given filter the solution points (c_{0j}, c_{1j}) tend to form a cluster in the (c_0, c_1) plane. Hence, once a stable 1-section design is obtained in step 1, the general domain of a multisection stable design is located. Consequently, as new sections are added in steps 2 and 3 one by one, a sequence of progressively improved stable designs are obtained. The logarithm of Q tends to decrease almost linearly with the number of equalizer sections at a rate that depends on the selectivity and passband width of the filter. In some examples, Q was found to reach a lower bound at some value less than 5 percent but the cause has not been identified.

The optimizations required in steps 1 to 3 can, in principle, be carried out using either Algorithm 6 or Algorithm 7. As in the design of recursive filters, Algorithm 7 tends to be much more efficient, while Algorithm 6 tends to yield better local minima (see Example 14.4). The advantages of the two algorithms can be combined by using Algorithm 6 in step 1, where a better design is highly desirable, and Algorithm 7 in steps 2 and 3, where computational efficiency is more important. Should Algorithm 7 fail to give a successful design in step 2 or 3, Algorithm 6 can be tried as an alternative.

At the solution, parameter τ_0 tends to approach the average of τ_{FE}/T. A fairly good estimate of this quantity for the 1-section design, which can be used to initialize τ_{01}, is obtained by letting $k_1 = 0.50$ in step 1. This value of k_1 was found to give good results.

For lowpass and highpass filters, points (c_{0j}, c_{1j}) tend to form clusters in the fourth and first quadrant of the feasible region, respectively. Hence only points 5 to 8 of Table 14.8 need be tried for lowpass filters and only points 1 to 4 need be tried for

TABLE 14.8
Initialization points in feasible region of (c_0, c_1) plane

No.	Point	No.	Point	No.	Point
1	(0.3, 0.3)	1A	(0.25, 0.50)	1B	(0.50, 0.25)
2	(0.7, 0.7)	2A	(0.50, 0.75)	2B	(0.75, 0.50)
3	(0.7, 1.3)	3A	(0.50, 1.25)	3B	(0.75, 1.50)
4	(−0.3, 0.3)	4A	(−0.25, 0.50)	4B	(−0.50, 0.25)
5	(0.3, −0.3)	5A	(0.25, −0.50)	5B	(0.50, −0.25)
6	(0.7, −0.7)	6A	(0.50, −0.75)	6B	(0.75, −0.50)
7	(0.7, −1.3)	7A	(0.50, −1.25)	7B	(0.75, −1.50)
8	(−0.3, −0.3)	8A	(−0.25, −0.50)	8B	(−0.50, −0.25)

highpass filters. In the unlikely situation where none of these points gives a solution, points 1A to 8A and 1B to 8B of Table 14.8 may be tried.

For filters with moderate or high selectivity, the value of ε_1 should be of the order of 0.01 or less; on the other hand, if the selectivity of the filter is low, a value as high as 0.1 may be necessary.

In steps 2b and 3b, a rectangular domain is established in the parameter space, which encloses points (c_{0i}, c_{1i}) for $i = 1, 2, \ldots j - 1$, and each of the corner points P_{1j} to P_{4j} is used for the initialization of the jth section. Occasionally, one or two of these points may be located outside the feasible region of the parameter space and should not be used.

$Q_{\max}$ in step 4 is the maximum allowable value of Q for the application at hand. If the number of sections is sufficient to reduce Q below $Q_{\max}$, the algorithm is terminated.

Example 14.5. The coefficients in Table 14.9 represent an elliptic highpass filter satisfying the following specifications:

Passband ripple: 0.5 dB
Minimum stopband attenuation: 50 dB
Passband edge: 0.75 rad/s

TABLE 14.9
Coefficients of $H_F(z)$ (Example 14.5)

a_{0j}	a_{1j}	a_{2j}	b_{0j}	b_{1j}	b_{2j}
−1.0	1.0	0.0	7.022673×10^{-1}	1.0	0.0
1.0	1.765666×10^{-2}	1.0	6.452156×10^{-1}	1.351877	1.0
1.0	7.880299×10^{-1}	1.0	8.893343×10^{-1}	1.320853	1.0

$H_0 = 1.033262 \times 10^{-2}$

Stopband edge: 0.64 rad/s

Sampling frequency: 2.0 rad/s

Design a delay equalizer that will reduce the Q of the filter-equalizer combination to a value less than 1.0 percent.

Solution. The design was carried out using Algorithm 6 for step 1 and Algorithm 7 for steps 2 and 3, along with the nonuniform variable sampling technique of Sec. 14.6 in each case. In order to achieve the desired degree of flatness in the delay characteristic, it was found necessary to increase the number of equalizer sections to five. The progress of the design is illustrated in Table 14.10. The transfer-function coefficients for the successive equalizers are given in Table 14.11. The delay characteristic of the filter-equalizer combination with zero, two, and five equalizer sections is illustrated in Fig. 14.10.

TABLE 14.10
Progress of design (Example 14.5)

j	(c_{0j}, c_{1j})	$\tilde{\tau}_{FEj}/T$	Q
0	—	11.76	66.38
1	(0.3, 0.3)	16.21	34.41
2	(0.6097, 1.482)	20.16	19.79
3	(0.7582, 1.610)	26.47	8.05
4	(0.7690, 1.579)	32.85	2.72
5	(0.7803, 1.567)	39.08	0.83

TABLE 14.11
Coefficients of $H_E(z)$ (Example 14.5)

Sections	j	c_{0j}	c_{1j}
1	1	6.158622×10^{-1}	1.496936
2	1	7.549257×10^{-1}	1.715040
	2	7.614137×10^{-1}	1.504945
3	1	7.552047×10^{-1}	1.726392
	2	7.826521×10^{-1}	1.431156
	3	7.668681×10^{-1}	1.634637
4	1	7.703755×10^{-1}	1.681226
	2	7.671458×10^{-1}	1.551901
	3	7.945904×10^{-1}	1.391108
	4	7.659007×10^{-1}	1.741710
5	1	7.593030×10^{-1}	1.692920
	2	7.602346×10^{-1}	1.483221
	3	7.985100×10^{-1}	1.365123
	4	7.551977×10^{-1}	1.732325
	5	7.607868×10^{-1}	1.610131

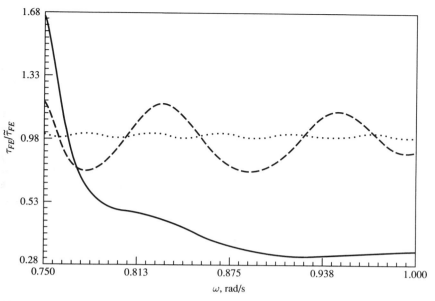

FIGURE 14.10
Delay characteristic of filter-equalizer combination (Example 14.5): ——— no equalization, − − − two equalizer sections, ······ five equalizer sections.

TABLE 14.12
Coefficients of $H_F(z)$ (Example 14.6)

j	a_{0j}	a_{1j}	a_{2j}	b_{0j}	b_{1j}	b_{2j}
1	−1.0	0.0	1.0	7.105797×10^{-1}	-5.558010×10^{-1}	1.0
2	1.0	−1.676442	1.0	8.610875×10^{-1}	-1.312559×10^{-2}	1.0
3	1.0	9.873155×10^{-1}	1.0	8.856595×10^{-1}	−1.099622	1.0

$H_0 = 2.602536 \times 10^{-2}$

Example 14.6. The coefficients in Table 14.12 represent an elliptic bandpass filter satisfying the following specifications:

> Passband ripple: 1.0 dB
> Minimum stopband attenuation: 40 dB
> Low passband edge: 0.3 rad/s
> High passband edge: 0.5 rad/s
> Low stopband edge: 0.2 rad/s
> High stopband edge: 0.7 rad/s
> Sampling frequency: 2.0 rad/s

Design a delay equalizer that will reduce the Q of the filter-equalizer combination to a value less than 2.0 percent.

Solution. The design was carried out as in Example 14.5. In order to achieve the desired degree of flatness in the delay characteristic, it was found necessary to increase the number of equalizer sections to four. The progress of the design is illustrated in Table 14.13. The transfer-function coefficients for the successive equalizers are given in Table 14.14. The delay characteristic of the filter-equalizer combination with zero, two, and four equalizer sections is illustrated in Fig. 14.11.

The mechanism by which Algorithm 8 leads to a series of progressively improved *stable* designs is illustrated in Figs. 14.12 and 14.13. As can be seen in Figs. 14.12a and 14.13a, the error surface for the 1-section equalizer has a well-defined depression in the feasible region of the parameter space which tends to be maintained as the number of equalizer sections is increased; see, for example, the error surface for the 4-section equalizer illustrated in Figs. 14.12b and 14.13b. In effect, a natural barrier is formed around the solution which assures the stability of successive equalizer sections.

TABLE 14.13
Progress of design (Example 14.6)

j	(c_{0j}, c_{1j})	$\tilde{\tau}_{FEj}/T$	Q
0	—	11.95	52.22
1	(0.3, 0.3)	16.38	27.78
2	(0.7332, −0.5297)	23.19	9.36
3	(0.7783, −0.7739)	29.13	3.31
4	(0.7469, −0.1775)	32.44	1.96

TABLE 14.14
Coefficients of $H_E(z)$ (Example 14.6)

Sections	j	c_{0j}	c_{1j}
1	1	7.405829×10^{-1}	-5.245374×10^{-1}
2	1	7.814228×10^{-1}	-7.738830×10^{-1}
	2	7.783450×10^{-1}	-2.892557×10^{-1}
3	1	7.621367×10^{-1}	-1.775013×10^{-1}
	2	7.468771×10^{-1}	-4.845243×10^{-1}
	3	7.925267×10^{-1}	-8.497135×10^{-1}
4	1	7.748554×10^{-1}	-2.501149×10^{-1}
	2	7.393927×10^{-1}	-5.571269×10^{-1}
	3	7.930800×10^{-1}	-8.709362×10^{-1}
	4	5.866017×10^{-1}	1.566691×10^{-1}

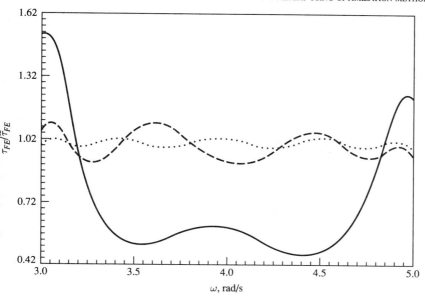

FIGURE 14.11
Delay characteristic of filter-equalizer combination (Example 14.6): ——— no equalization, – – – two equalizer sections, ······ four equalizer sections.

REFERENCES

1. K. Steiglitz, "Computer-Aided Design of Recursive Digital Filters," *IEEE Trans. Audio Electroacoust.*, vol. AU-18, pp. 123–129, June 1970.
2. A. G. Deczky, "Synthesis of Recursive Digital Filters Using the Minimum p-Error Criterion," *IEEE Trans. Audio Electroacoust.*, vol. AU-20, pp. 257–263, October 1972.
3. J. W. Bandler and B. L. Bardakjian, "Least pth Optimization of Recursive Digital Filters," *IEEE Trans. Audio Electroacoust.*, vol. AU-21, pp. 460–470, October 1973.
4. C. Charalambous, "Minimax Design of Recursive Digital Filters," *Computer Aided Design*, vol. 6, pp. 73–81, April 1974.
5. C. Charalambous, "Minimax Optimization of Recursive Digital Filters Using Recent Minimax Results," *IEEE Trans. Acoust., Speech, Signal Process.*, vol. ASSP-23, pp. 333–345, August 1975.
6. R. Fletcher, *Practical Methods of Optimization*, Volume 1, Unconstrained Optimization, Wiley, New York, 1980. (See also R. Fletcher, *Practical Methods of Optimization*, Second Edition, Wiley, New York, 1990.)
7. D. G. Luenberger, *Linear and Nonlinear Programming*, Second Edition, Addison-Wesley, Reading, 1984.
8. P. E. Gill, W. Murray, and M. H. Wright, *Practical Optimization*, Academic Press, London, 1981.
9. D. M. Himmelblau, *Applied Nonlinear Programming*, McGraw-Hill, New York, 1972.
10. B. D. Bunday, *Basic Optimisation Methods*, Edward Arnold, London, 1984.
11. M. Al-Baali and R. Fletcher, "An Efficient Line Search for Nonlinear Least Squares," *J. Opt. Theo. Applns.*, vol. 48, pp. 359–378, 1986.
12. C. Charalambous, "A Unified Review of Optimization," *IEEE Trans. Microwave Theory and Techniques*, vol. MTT-22, pp. 289–300, March 1974.

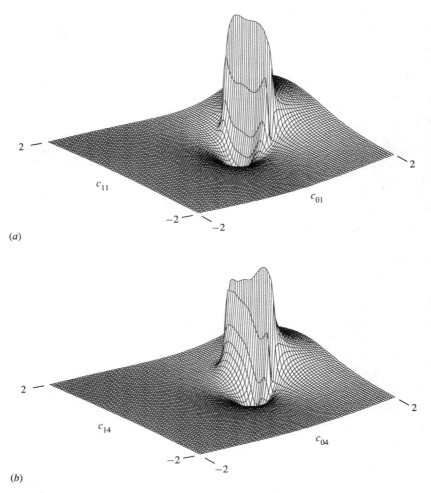

FIGURE 14.12
3-D plots of error function (Example 14.6): (*a*) 1-section equalizer, (*b*) 4-section equalizer (the coefficients of the first three sections have been assumed to have the optimized values achieved in the 3-section equalizer).

13. C. Charalambous and A. Antoniou, "Equalisation of Recursive Digital Filters," *IEE Proc.*, vol. 127, pt. G, pp. 219–225, October 1980.
14. C. Charalambous, "Design of 2-Dimensional Circularly-Symmetric Digital Filters," *IEE Proc.*, vol. 129, pt. G, pp. 47–54, April 1982.
15. C. Charalambous, "Acceleration of the Least pth Algorithm for Minimax Optimization with Engineering Applications," *Mathematical Programming*, vol. 17, pp. 270–297, 1979.
16. A. Antoniou, "Improved Minimax Optimisation Algorithms and Their Application in the Design of Recursive Digital Filters," *IEE Proc.*, vol. 138, pt. G, pp. 724–730, December 1991.

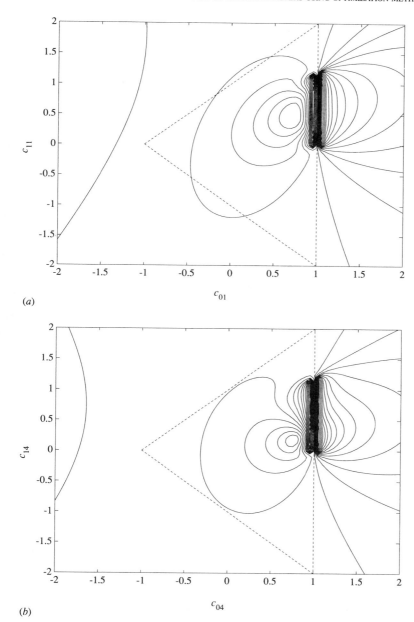

FIGURE 14.13
Contour plots of error function (Example 14.6): (a) 1-section equalizer, (b) 4-section equalizer (the coefficients of the first three sections have been assumed to have the optimized values achieved in the 3-section equalizer).

17. A. Antoniou, M. Degano, and C. Charalambous, "Compensation for the Effects of the D/A Convertor in Recursive Digital Filters," *IEE Proc.*, vol. 129, pt. G, pp. 273–279, December 1982.
18. A. Antoniou and W.-S. Lu, "Design of Two-Dimensional Digital Filters by Using the Singular Value Decomposition," *IEEE Trans. Circuits Syst.*, vol. CAS-34, pp. 1191–1198, October 1987.

ADDITIONAL REFERENCES

Ansari, R., and B. Liu: "A Class of Low-Noise Computationally Efficient Recursive Digital Filters with Applications to Sampling Rate Alterations," *IEEE Trans. Acoust., Speech, Signal Process.*, vol. ASSP-33, pp. 90–97, February 1985.

Bernhardt, P. A.: "Simplified Design of High-Order Recursive Group-Delay Filters," *IEEE Trans. Acoust., Speech, Signal Process.*, vol. ASSP-28, pp. 498–503, October 1980.

Brophy, F., and A. C. Salazar: "Considerations of the Padé Approximant Technique in the Synthesis of Recursive Digital Filters," *IEEE Trans. Audio Electroacoust.*, vol. AU-21, pp. 500–505, December 1973.

―――― and ――――: "Recursive Digital Filter Synthesis in the Time Domain," *IEEE Trans. Acoust., Speech, Signal Process.*, vol. ASSP-22, pp. 45–55, February 1974.

―――― and ――――: "Two Design Techniques for Digital Phase Networks," *Bell System Tech. J.*, vol. 54, pp. 767–781, April 1975.

Charalambous, C.: "A New Approach to Multicriterion Optimization Problem and Its Application to the Design of 1-D Digital Filters," *IEEE Trans. Circuits Syst.*, vol. CAS-36, pp. 773–784, June 1989.

Deczky, A. G.: "Equiripple and Minimax (Chebyshev) Approximations for Recursive Digital Filters," *IEEE Trans. Acoust., Speech, Signal Process.*, vol. ASSP-22, pp. 98–111, April 1974.

Dubois, H., and H. Leich: "On the Approximation Problem for Recursive Filters with Arbitrary Attenuation Curve in the Pass-Band and the Stop-Band," *IEEE Trans. Acoust., Speech, Signal Process.*, vol. ASSP-23, pp. 202–207, April 1975.

Rabiner, L. R., N. Y. Graham, and H. D. Helms: "Linear Programming Design of IIR Digital Filters with Arbitrary Magnitude Function," *IEEE Trans. Acoust., Speech, Signal Process.*, vol. ASSP-22, pp. 117–123, April 1974.

Thajchayapong, P., and P. J. W. Rayner: "Recursive Digital Filter Design by Linear Programming," *IEEE Trans. Audio Electroacoust.*, vol. AU-21, pp. 107–112, April 1973.

Thiran, J.-P.: "Recursive Digital Filters with Maximally Flat Group Delay," *IEEE Trans. Circuit Theory*, vol. CT-18, pp. 659–669, November 1971.

――――: "Equal-Ripple Delay Recursive Digital Filters," *IEEE Trans. Circuit Theory*, vol. CT-18, pp. 664–669, November 1971.

PROBLEMS

14.1. The step response $u(t)$ of a digital filter is required to approximate the ideal step response

$$u_0(t) = \begin{cases} t & \text{for } 0 \leq t < 6 \\ 2 & \text{for } 2 \leq t < 3 \\ -t + 5 & \text{for } 3 \leq t < 4 \\ 1 & \text{for } 4 \leq t < 5 \end{cases}$$

Formulate a least-squares objective function for the solution of the problem.

14.2. The quantity y should in theory be related to parameters x_1 and x_2 by a formula of the form

$$y = \frac{a_0 x_1}{1 + a_1 x_1 + a_2 x_2}$$

TABLE P14.2

x_1	x_2	y_0
1	1	0.1265
2	1	0.2193
1	2	0.0075
2	2	0.1262
0.1	0	0.1859

In a specific experiment, the data in Table P14.2 were collected. Construct an objective function that can be used to find coefficients a_0 to a_2 such that the maximum difference between y and y_0 is minimized.

14.3. Obtain an objective function that can be used to find approximate values of x_1 and x_2 that satisfy the relations

$$x_1 = x_2^2 - 3 \log x_1$$
$$x_2 = (2x_1^2 - 5x_1 + 1)/x_1$$

14.4. The so-called *Rosenbrock function*

$$f(\mathbf{x}) = 100(x_2 - x_1^2)^2 + (1 - x_1)^2$$

represents a highly nonlinear surface in the shape of a narrow curved falling valley. It is often used to test the ability of algorithms to maneuver around curved valleys. Show that $f(\mathbf{x})$ has a minimum at point $[1 \quad 1]^T$.

14.5. Find and classify the stationary points of function

$$f(\mathbf{x}) = x_1^2 - x_2^2 + x_3^2 - 2x_1x_3 - x_2x_3 + 4x_1 + 12$$

14.6. (*a*) Show that the function

$$f(\mathbf{x}) = 2x_1^2 - 2x_1x_2 + x_2^2 + 2x_1 - 2x_2$$

has a minimum.

(*b*) Find the minimum of the function using Algorithm 1 with $\mathbf{x}_0 = [0 \quad 0]^T$ as initial point.

14.7. Repeat Prob. 14.6 for the function

$$f(\mathbf{x}) = x_1^2 + 2x_2^2 + 4x_1 + 4x_2$$

14.8. Show that at point $[1 + \varepsilon_1, 1 + \varepsilon_2]$, where $|\varepsilon_1| \ll 1$ and $|\varepsilon_2| \ll 1$, the Rosenbrock function given in Prob. 14.4 can be approximated by a quadratic function.

14.9. Derive Eq. (14.27).

14.10. Derive Eq. (14.28).

14.11. (*a*) Show that the estimate of α_0 used in step 2*d* of Algorithm 5 (practical quasi-Newton algorithm), can be derived by using Eq. (14.27).

(*b*) Justify the use of $\alpha_0 = 1$, if the estimate in part (*a*) is unreasonable.

14.12. (*a*) Show that the extrapolation in step 6 of Algorithm 5 can yield a negative $\overset{\cup}{\alpha}_0$.

(*b*) How is the problem in part (*a*) avoided.

(c) In the interpolation as well as extrapolation routines of Algorithm 5, the search is aborted if the number of function evaluations exceeds a certain maximum. Why is it advisable to include such termination criteria in optimization algorithms?

(d) Explain the purpose of steps 4c, 4d, and 6c in Algorithm 5.

14.13. Write a computer program for Algorithm 5 using the BFGS updating formula.
  (a) Use the program to obtain a least-squares solution of the problem in Prob. 14.1.
  (b) Repeat part (a) for the problem in Prob. 14.2.
  (c) Repeat part (a) for the problem in Prob. 14.3.

14.14. (a) Use the program in Prob. 14.13 to minimize Rosenbrock's function given in Prob. 14.4
  (b) Repeat part (a) for the function

$$f(\mathbf{x}) = 100[(x_3 - 10\theta)^2 + (r-1)^2] + x_3^2$$

where

$$\theta = \begin{cases} \frac{1}{2\pi} \tan^{-1}\left(\frac{x_2}{x_1}\right) & \text{for } x_1 > 0 \\ 0.25 & \text{for } x_1 = 0 \\ 0.5 + \frac{1}{2\pi} \tan^{-1}\left(\frac{x_2}{x_1}\right) & \text{for } x_1 < 0 \end{cases}$$

and

$$r = \sqrt{(x_1^2 + x_2^2)}$$

14.15. Repeat Prob. 14.14 for the following functions

(a)
$$f(\mathbf{x}) = (x_1 + 10x_2)^2 + 5(x_3 - x_4)^2 + (x_2 - 2x_3)^4 + 100(x_1 - x_4)^4$$

(b)
$$f(\mathbf{x}) = \sum_{i=2}^{5} 100(x_i - 10x_{i-1}^2)^2 + (1 - x_i)^2$$

14.16. Replace the BFGS updating formula in the computer program of Prob. 14.13 by the DFP formula.
  (a) Use the program to obtain a least-squares solution of the problem in Prob. 14.1.
  (b) Repeat part (a) for the problem in Prob. 14.2.
  (c) Repeat part (a) for the problem in Prob. 14.3.

14.17. (a) Run the computer program in Prob. 14.13 (BFGS version) with the function of Prob. 14.4 using 10 different initial points and find the average number of function evaluations. Count one function evaluation for each evaluation of $f(\mathbf{x})$ and one for each partial derivative of $f(\mathbf{x})$. Repeat this process using the program in Prob. 14.16 (DFP version). Compare the results obtained.
  (b) Repeat part (a) with the function in part (b) of Prob. 14.15.

14.18. Replace the line search in the computer program of Prob. 14.13 (BFGS version) by the line search described in [11] (see also second edition of Fletcher [6], pp. 34–35).
  (a) Use the program to obtain a least-squares solution of the problem in Prob. 14.1.
  (b) Repeat part (a) for the problem in Prob. 14.2.
  (c) Repeat part (a) for the problem in Prob. 14.3.

DESIGN OF RECURSIVE FILTERS USING OPTIMIZATION METHODS   **541**

14.19. (a) Run the computer program in Prob. 14.18 (different line search) with the function in Prob. 14.4 using 10 different initial points and find the average number of function evaluations. Count one function evaluation for each evaluation of $f(\mathbf{x})$ and one for each partial derivative of $f(\mathbf{x})$. Repeat this process using the program in Prob. 14.13. Compare the results obtained.
(b) Repeat part (a) with the function in part (b) of Prob. 14.15.

14.20. Write a computer program for Algorithm 6 (least-pth minimax algorithm) using the BFGS updating formula.
(a) Use the program to obtain a minimax solution of the problem in Prob. 14.1.
(b) Repeat part (a) for the problem in Prob. 14.2.
(c) Repeat part (a) for the problem in Prob. 14.3.
(d) Compare the minimax solutions obtained in parts (a) to (c) with the corresponding least-squares solutions obtained in Prob. 14.13 with respect to the minimum error achieved and the amount of computation required.

14.21. Write a computer program for Algorithm 7 (Charalambous minimax algorithm) using the BFGS updating formula.
(a) Use the program to obtain a minimax solution of the problem in Prob. 14.1.
(b) Repeat part (a) for the problem in Prob. 14.2.
(c) Repeat part (a) for the problem in Prob. 14.3.
(d) Compare the minimax solutions obtained with Algorithm 7 with the corresponding solutions obtained with Algorithm 6 in Prob. 14.20.

14.22. (a) Using the computer program in Prob. 14.20, design a fourth-order highpass digital filter with passband and stopband edges of 3.5 and 1.5 rad/s, respectively. The sampling frequency is required to be 10 rad/s.
(b) Repeat part (a) with the program in Prob. 14.21.
(c) Design an elliptic highpass filter satisfying the same specifications as the filter obtained in part (a) (i.e., same maximum passband ripple, minimum stopband attenuation, passband and stopband edges) using the closed-form method of Chap. 8.
(d) Compare the designs obtained in parts (a) and (b), and (a) and (c).

14.23. (a) Using the computer program in Prob. 14.20, design a sixth-order bandpass digital filter with passband edges 900 and 1100 rad/s, and stopband edges 800 and 1200 rad/s. The sampling frequency is required to be 6000 rad/s.
(b) Repeat part (a) with the program in Prob. 14.21.
(c) Design an elliptic bandpass filter satisfying the same specifications as the filter obtained in part (a) (i.e., same maximum passband ripple, minimum stopband attenuation, passband and stopband edges) using the closed-form method of Chap. 8.
(d) Compare the designs obtained in parts (a) and (b), and (a) and (c).

14.24. (a) Modify the computer program in Prob. 14.20 using the nonuniform sampling technique in Sec. 14.6.
(b) Use the program obtained to design a lowpass filter with the idealized piecewise-continuous amplitude response

$$M_0(\omega) = \begin{cases} 0.545455\omega + 1.0 & \text{for } 0 \leq \omega \leq 0.55 \\ 0.75\omega - 0.45 & \text{for } 0.60 \leq \omega \leq 1.0 \end{cases}$$

The frequency range 0.55 to 0.60 rad/s represents a transition band in which the amplitude response is undefined. The sampling frequency is 2.0 rad/s.

14.25. (a) Modify the computer program in Prob. 14.21 using the nonuniform variable sampling technique in Sec. 14.6.

(b) Use the program obtained to design the lowpass filter described in part (b) of Prob. 14.24.

(c) Compare the results with those obtained in part (b) of Prob. 14.24.

14.26. (a) Using the method in Chap. 8, design an elliptic bandpass filter satisfying the following specifications:

Passband ripple: 1.0 dB
Minimum stopband attenuation: 30 dB
Low stopband edge: 0.15 rad/s
Low passband edge: 0.20 rad/s
High passband edge: 0.40 rad/s
High stopband edge: 0.50 rad/s
Sampling frequency: 2.0 rad/s

(b) Design a bandpass filter with the same order and band edges as for the filter in part (a) by using the program in Prob. 14.24. The ideal passband and stopband gains can be assumed to be 1.0 and 0.01, respectively.

(c) The bandpass filter in part (b) is to be used in cascade with a D/A converter. The amplitude response of the D/A converter $\phi(\omega)$ is of the form given in Example 14.3 with $\tau = T$. Using the program in Prob. 14.24, redesign the filter taking into consideration the amplitude response of the D/A converter. Use the coefficients of the filter in part (a) for the initialization of the algorithm.

(d) Repeat part (c) with the program in Prob. 14.25.

14.27. (a) The amplitude response of a recursive filter is specified at 21 frequency points as in Table P14.27 and $\omega_s = 2$ rad/s. Obtain an eighth-order design using the program in Prob. 14.24.

TABLE P14.27
Specified amplitude response

ω	Gain	ω	Gain	ω	Gain
0.00	0.9135	0.35	0.6232	0.70	0.5681
0.05	0.7080	0.40	0.7986	0.75	0.5725
0.10	0.6939	0.45	0.8186	0.80	0.5758
0.15	0.4062	0.50	0.7808	0.85	0.5780
0.20	0.3872	0.55	0.6598	0.90	0.5794
0.25	0.0070	0.60	0.5544	0.95	0.5802
0.30	0.2888	0.65	0.5623	1.00	0.5805

(b) Design the filter in part (a) using the program in Prob. 14.25.

14.28. (a) Using the method in Chap. 8, design an elliptic lowpass filter satisfying the following specifications:

Passband ripple: 1.0 dB
Minimum stopband attenuation: 40 dB
Passband edge: 0.10 rad/s

Stopband edge: 0.15 rad/s
Sampling frequency: 2.0 rad/s

(b) Design a delay equalizer that will reduce the Q of the filter-equalizer combination to a value less than 5.0 percent using the computer program in Prob. 14.24.

(c) Repeat part (b) using the program in Prob. 14.25.

14.29. (a) Using the method in Chap. 8, design an elliptic highpass filter satisfying the following specifications:

Passband ripple: 1.0 dB
Minimum stopband attenuation: 60 dB
Passband edge: 0.90 rad/s
Stopband edge: 0.85 rad/s
Sampling frequency: 2.0 rad/s

(b) Design a delay equalizer that will reduce the Q of the filter-equalizer combination to a value less than 5.0 percent using the computer program in Prob. 14.24.

(c) Repeat part (b) using the program in Prob. 14.25.

14.30. (a) Design a delay equalizer for the filter in Prob. 14.26a that will reduce the Q of the filter-equalizer combination to a value less than 5.0 percent using the computer program in Prob. 14.24.

(b) Repeat part (a) using the program in Prob. 14.25.

CHAPTER 15

DESIGN OF NONRECURSIVE FILTERS USING OPTIMIZATION METHODS

15.1 INTRODUCTION

The methods for the design of nonrecursive filters described so far (see Chap. 9 and Sec. 13.8) are based largely on closed-form solutions. As a result, they are easy to apply and entail a relatively insignificant amount of computation. Unfortunately, they usually lead to suboptimal designs whereby the filter order required to satisfy a set of given specifications is not the lowest that can be achieved. Consequently, the number of arithmetic operations required per output sample is not minimum, and the computational efficiency and speed of operation of the filter are not as high as could be.

This chapter deals with a method for the design of nonrecursive filters known as the *weighted-Chebyshev* method. In this method, an error function is formulated for the desired filter in terms of a linear combination of cosine functions and is then minimized by using a very efficient multivariable optimization algorithm known as the *Remez exchange algorithm*. When convergence is achieved, the error function becomes equiripple, as in other types of Chebyshev solutions (see Sec. 5.4). The amplitude of the error in different frequency bands of interest is controlled by applying weighting to the error function.

The weighted-Chebyshev method is very flexible and can be used to obtain optimal solutions for most types of nonrecursive filters, e.g., digital differentiators, Hilbert transformers, and lowpass, highpass, bandpass, bandstop, and multiband filters with piecewise-constant amplitude responses. Furthermore, like the methods of Chap. 14, it can be used to design filters with arbitrary amplitude responses. In common with other optimization methods, the weighted-Chebyshev method requires a large amount of computation; however, as the cost of computation is becoming progressively cheaper and cheaper with time, this disadvantage is not a very serious one.

The development of the weighted-Chebyshev method began with a paper by Herrmann published in 1970 [1], which was followed soon after by a paper by Hofstetter, Oppenheim, and Siegel [2]. These contributions were followed by a series of papers, during the seventies, by Parks, McClellan, Rabiner, and Herrmann [3–8]. These developments led, in turn, to the well-known McClellan-Parks-Rabiner computer program for the design of nonrecursive filters, documented in [9], which has found widespread applications. The approach to weighted-Chebyshev filters presented in this chapter is based on that reported in [3, 6, 8], and includes several enhancements proposed by the author in [10, 11].

15.2 PROBLEM FORMULATION

Consider a nonrecursive filter characterized by the transfer function

$$H(z) = \sum_{n=0}^{N-1} h(nT)z^{-n} \quad (15.1)$$

and assume that N is odd, the impulse response is symmetrical, and $\omega_s = 2\pi$. Since $T = 2\pi/\omega_s = 1$ s, the frequency response of the filter can be expressed as

$$H(e^{j\omega T}) = e^{-jc\omega} P_c(\omega)$$

where

$$P_c(\omega) = \sum_{k=0}^{c} a_k \cos k\omega \quad (15.2)$$

$$a_0 = h(c)$$

$$a_k = 2h(c-k) \quad \text{for } k = 1, 2, \ldots, c$$

$$c = (N-1)/2$$

(see Table 9.1).

If $e^{-jc\omega} D(\omega)$ is the desired frequency response and $W(\omega)$ is a weighting function, an error function $E(\omega)$ can be constructed as

$$E(\omega) = W(\omega)[D(\omega) - P_c(\omega)] \quad (15.3)$$

If $|E(\omega)|$ is minimized such that

$$|E(\omega)| \leq \delta_p$$

with respect to some compact subset of the frequency interval $[0, \pi]$, say Ω, a filter can be obtained in which

$$|E_0(\omega)| = |D(\omega) - P_c(\omega)| \leq \frac{\delta_p}{|W(\omega)|} \quad \text{for } \omega \in \Omega \quad (15.4)$$

15.2.1 Lowpass and Highpass Filters

The amplitude response of an equiripple lowpass filter is of the form illustrated in Fig. 15.1, where δ_p and δ_a are the amplitudes of the passband and stopband ripples, and ω_p and ω_a are the passband and stopband edges, respectively. Hence, we require

$$D(\omega) = \begin{cases} 1 & \text{for } 0 \leq \omega \leq \omega_p \\ 0 & \text{for } \omega_a \leq \omega \leq \pi \end{cases} \quad (15.5a)$$

with

$$|E_0(\omega)| \leq \begin{cases} \delta_p & \text{for } 0 \leq \omega \leq \omega_p \\ \delta_a & \text{for } \omega_a \leq \omega \leq \pi \end{cases} \quad (15.5b)$$

Therefore, from Eq. (15.4) and (15.5b) we deduce

$$W(\omega) = \begin{cases} 1 & \text{for } 0 \leq \omega \leq \omega_p \\ \delta_p/\delta_a & \text{for } \omega_a \leq \omega \leq \pi \end{cases} \quad (15.6)$$

Similarly, for highpass filters we obtain

$$D(\omega) = \begin{cases} 0 & \text{for } 0 \leq \omega \leq \omega_a \\ 1 & \text{for } \omega_p \leq \omega \leq \pi \end{cases}$$

and

$$W(\omega) = \begin{cases} \delta_p/\delta_a & \text{for } 0 \leq \omega \leq \omega_a \\ 1 & \text{for } \omega_p \leq \omega \leq \pi \end{cases} \quad (15.7)$$

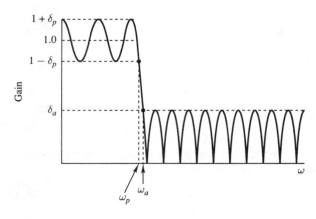

FIGURE 15.1
Amplitude response of equiripple lowpass filter.

15.2.2 Bandpass and Bandstop Filters

The amplitude responses of equiripple bandpass and bandstop filters assume the forms illustrated in Fig. 15.2a and b, respectively, where δ_p and δ_a are the passband and stopband ripples, respectively, ω_{p1} and ω_{p2} are passband edges, and ω_{a1} and ω_{a2} are stopband edges. For bandpass filters

$$D(\omega) = \begin{cases} 0 & \text{for } 0 \leq \omega \leq \omega_{a1} \\ 1 & \text{for } \omega_{p1} \leq \omega \leq \omega_{p2} \\ 0 & \text{for } \omega_{a2} \leq \omega \leq \pi \end{cases}$$

$$W(\omega) = \begin{cases} \delta_p/\delta_a & \text{for } 0 \leq \omega \leq \omega_{a1} \\ 1 & \text{for } \omega_{p1} \leq \omega \leq \omega_{p2} \\ \delta_p/\delta_a & \text{for } \omega_{a2} \leq \omega \leq \pi \end{cases} \quad (15.8)$$

and for bandstop filters

$$D(\omega) = \begin{cases} 1 & \text{for } 0 \leq \omega \leq \omega_{p1} \\ 0 & \text{for } \omega_{a1} \leq \omega \leq \omega_{a2} \\ 1 & \text{for } \omega_{p2} \leq \omega \leq \pi \end{cases}$$

$$W(\omega) = \begin{cases} 1 & \text{for } 0 \leq \omega \leq \omega_{p1} \\ \delta_p/\delta_a & \text{for } \omega_{a1} \leq \omega \leq \omega_{a2} \\ 1 & \text{for } \omega_{p2} \leq \omega \leq \pi \end{cases} \quad (15.9)$$

15.2.3 Alternation Theorem

An effective approach for the solution of the optimization problem at hand is to solve the minimax problem

$$\text{minimize}_\mathbf{x} \left\{ \max_\omega |E(\omega)| \right\} \quad (15.10)$$

where

$$\mathbf{x} = [a_0 \quad a_1 \quad \ldots \quad a_c]^T$$

The solution of this problem exists by virtue of the so-called *alternation theorem* [12] stated below.

Theorem 15.1. If $P_c(\omega)$ is a linear combination of $r = c + 1$ cosine functions of the form

$$P_c(\omega) = \sum_{k=0}^{c} a_k \cos k\omega$$

then a necessary and sufficient condition that $P_c(\omega)$ be the unique, best, weighted-Chebyshev approximation to a continuous function $D(\omega)$ on Ω, where Ω is a compact subset of the frequency interval $[0, \pi]$, is that the weighted error function $E(\omega)$ exhibit at least $r + 1$ *extremal frequencies* in Ω, i.e., there must exist at least $r + 1$ points $\hat{\omega}_i$

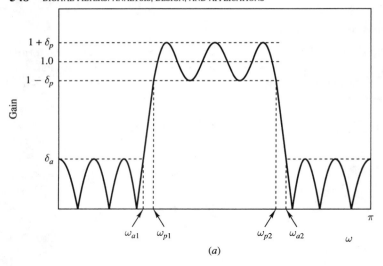

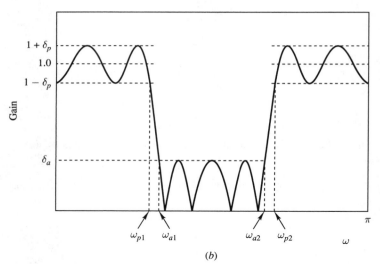

FIGURE 15.2
Amplitude responses of equiripple filters: (*a*) bandpass filter, (*b*) bandstop filter.

in Ω such that

$$\hat{\omega}_0 < \hat{\omega}_1 < \ldots < \hat{\omega}_r$$

$$E(\hat{\omega}_i) = -E(\hat{\omega}_{i+1}) \qquad \text{for } i = 0, 1, \ldots, r-1$$

and

$$|E(\hat{\omega}_i)| = \max_{\omega \in \Omega} |E(\omega)| \qquad \text{for } i = 0, 1, \ldots, r$$

From the alternation theorem and Eq. (15.3), we can write

$$E(\hat{\omega}_i) = W(\hat{\omega}_i)[D(\hat{\omega}_i) - P_c(\hat{\omega}_i)] = (-1)^i \delta \qquad (15.11)$$

for $i = 0, 1, \ldots, r$, where δ is a constant. This system of equations can be put in matrix form as

$$\begin{bmatrix} 1 & \cos\hat{\omega}_0 & \cos 2\hat{\omega}_0 & \cdots & \cos c\hat{\omega}_0 & \frac{1}{W(\hat{\omega}_0)} \\ 1 & \cos\hat{\omega}_1 & \cos 2\hat{\omega}_1 & \cdots & \cos c\hat{\omega}_1 & \frac{-1}{W(\hat{\omega}_1)} \\ \vdots & \vdots & \vdots & & \vdots & \vdots \\ 1 & \cos\hat{\omega}_r & \cos 2\hat{\omega}_r & \cdots & \cos c\hat{\omega}_r & \frac{(-1)^r}{W(\hat{\omega}_r)} \end{bmatrix} \begin{bmatrix} a_0 \\ a_1 \\ \vdots \\ a_c \\ \delta \end{bmatrix} = \begin{bmatrix} D(\hat{\omega}_0) \\ D(\hat{\omega}_1) \\ \vdots \\ D(\hat{\omega}_r) \end{bmatrix} \qquad (15.12)$$

If the extremal frequencies (or extremals for short) were known, coefficients a_k and, in turn, the frequency response of the filter could be computed using Eq. (15.2). The solution of this system exists since the above $(r+1) \times (r+1)$ matrix can be shown to be nonsingular [12].

15.3 REMEZ EXCHANGE ALGORITHM

The Remez exchange algorithm is an *iterative multivariable* algorithm which is naturally suited for the solution of the minimax problem in Eq. (15.10). It is based on the second optimization method of Remez [13] and involves the following basic steps:

Algorithm 1: Basic Remez exchange algorithm

1. Initialize extremals $\hat{\omega}_0, \hat{\omega}_1, \ldots, \hat{\omega}_r$ and ensure that an extremal is assigned at each band edge.
2. Locate the frequencies $\overset{\frown}{\omega}_0, \overset{\frown}{\omega}_1, \ldots, \overset{\frown}{\omega}_\rho$ at which $|E(\omega)|$ is maximum and $|E(\overset{\frown}{\omega}_i)| \geq \delta$. These frequencies are *potential* extremals for the next iteration.
3. Compute the convergence parameter

$$Q = \frac{\max|E(\overset{\frown}{\omega}_i)| - \min|E(\overset{\frown}{\omega}_i)|}{\max|E(\overset{\frown}{\omega}_i)|} \qquad (15.13)$$

   where $i = 0, 1, \ldots, \rho$.
4. Reject $\rho - r$ *superfluous* potential extremals $\overset{\frown}{\omega}_i$ according to an appropriate rejection criterion and renumber the remaining $\overset{\frown}{\omega}_i$ sequentially; then set $\hat{\omega}_i = \overset{\frown}{\omega}_i$ for $i = 0, 1, \ldots, r$.
5. If $Q > \varepsilon$, where ε is a convergence tolerance (say $\varepsilon = 0.01$), repeat from step 2; otherwise continue to step 6.
6. Compute $P_c(\omega)$ using the last set of extremals; then deduce $h(n)$, the impulse response of the required filter, and stop.

The amount of computation required by the algorithm tends to depend quite heavily on the initialization scheme used in step 1, on the search method used for the location of the maxima of the error function in step 2, and on the criterion used to reject superfluous frequencies $\hat{\omega}_i$ in step 4.

15.3.1 Initialization of Extremals

The simplest scheme for the initialization of extremals $\hat{\omega}_i$ for $i = 0, 1, \ldots, r$ is to assume that they are uniformly spaced in the frequency bands of interest. If there are J distinct bands in the required filter of widths $B_1, B_2, \ldots, B_J$ and extremals are to be located at the left-hand and right-hand band edges of each band, the total bandwidth should be divided into $r + 1 - J$ intervals. Under these circumstances, the average interval between adjacent extremals is

$$W_0 = \frac{1}{r+1-J} \sum_{j=1}^{J} B_j$$

Since the quantities B_j/W_0 need not be integers, the use of W_0 for the generation of the extremals will almost always result in a fractional interval in each band. This problem can be avoided by rounding the number of intervals B_j/W_0 to the nearest integer and then readjusting the frequency interval for the corresponding band accordingly. This can be achieved by letting the numbers of intervals in bands j and J be

$$m_j = \text{Int}\left(\frac{B_j}{W_0} + 0.5\right) \quad \text{for } j = 1, 2, \ldots, J-1 \quad (15.14a)$$

and

$$m_J = r - \sum_{j=1}^{J-1}(m_j + 1) \quad (15.14b)$$

respectively, and then recalculating the frequency intervals for the various bands as

$$W_j = \frac{B_j}{m_j} \quad \text{for } j = 1, 2, \ldots, J \quad (15.15)$$

A more sophisticated initialization scheme which was found to give good results is described in [14].

15.3.2 Location of Maxima of the Error Function

The frequencies $\hat{\omega}_i$, which *must include maxima at band edges* if $|E(\hat{\omega}_i)| \geq |\delta|$, can be located by simply evaluating $|E(\omega)|$ over a dense set of frequencies. A reasonable number of frequency points that yields sufficient accuracy in the determination of the frequencies $\hat{\omega}_i$ is $8(N + 1)$. This corresponds to about 16 frequency points per ripple of $|E(\omega)|$. A suitable frequency interval for the jth band is $w_j = W_j/S$ with $S = 16$.

The above *exhaustive* step-by-step search can be implemented in terms of Algorithm 2 below where ω_{Lj} and ω_{Rj} are the left-hand and right-hand edges in band j; W_j is the interval between adjacent extremals and m_j is the number of intervals W_j in band j; w_j is the interval between successive samples of $|E(\omega)|$ in interval W_j

and S is the number of intervals w_j in each interval W_j; N_j is the total number of intervals w_j in band j; and J is the number of bands.

Algorithm 2: Exhaustive step-by-step search

1. Set $N_j = m_j S$, $w_j = B_j/N_j$ and $e = 0$.
2. For each of bands $1, 2, \ldots, j, \ldots, J$ do:
   For each of frequencies $\omega_{1j} = \omega_{Lj}$, $\omega_{2j} = \omega_{Lj} + w_j, \ldots, \omega_{ij} = \omega_{Lj} + (i-1)w_j, \ldots,$ $\omega_{N_{j+1}} = \omega_{Rj}$, set $\hat{\omega}_e = \omega_{ij}$ and $e = e + 1$ provided that $|E(\omega_{ij})| \geq |\delta|$ and one of the following conditions holds:
   (a) Case $\omega_{ij} = \omega_{Lj}$: if $|E(\omega_{ij})|$ is maximum at $\omega_{ij} = \omega_{Lj}$ (i.e., $|E(\omega_{Lj})| > |E(\omega_{Lj} + \varepsilon)|$);
   (b) Case $\omega_{Lj} < \omega_{ij} < \omega_{Rj}$: if $|E(\omega)|$ is maximum at $\omega = \omega_{ij}$ (i.e., $|E(\omega_{ij} - w_j)| < |E(\omega_{ij})| > |E(\omega_{ij} + w_j)|$);
   (c) Case $\omega_{ij} = \omega_{Rj}$: if $|E(\omega_{ij})|$ is maximum at $\omega_{ij} = \omega_{Rj}$ (i.e., $|E(\omega_{Rj})| > |E(\omega_{Rj} - \varepsilon)|$.

The parameter ε in steps 2(a) and 2(c) is a small positive constant and a value $10^{-2} w_j$ was found to yield satisfactory results.

In practice, $|E(\omega)|$ is maximum at an interior left-hand band edge[†] if its first derivative at the band edge is negative, and a mirror-image situation applies at an interior right-hand band edge. In such cases, $|E(\omega)|$ has a zero immediately to the right or left of the band edge and the inequality in step 2(a) or 2(c) may sometimes fail to identify a maximum. However, the problem can be avoided by using the inequality $|E(\omega_{Lj} - \varepsilon)| > |E(\omega_{Lj})|$ in step 2(a) and $|E(\omega_{Rj})| < |E(\omega_{Rj} + \varepsilon)|$ in step 2(c) for interior band edges. An alternative approach to the problem is to use gradient information based on the formulas found in Sec. 15.6.

In rare circumstances, a maximum of $|E(\omega)|$ may occur between a band edge and the first sample point. Such a maximum may be missed by Algorithm 2 but the problem can be easily identified since the number of potential extremals will then be less than the minimum. The remedy is to check the number of potential extremals at the end of each iteration and if it is found to be less than $r + 1$, the density of sample points, i.e., S, is doubled and the iteration is repeated. If the problem persists, the process is repeated until the required number of potential extremals is obtained. If a value of S equal to or less than 256 does not resolve the problem, the loss of potential extremals is most likely due to some other reason.

An important precaution in the implementation of the preceding as well as the subsequent search methods is to ensure that extremals belong to the dense set of frequency points to avoid numerical ill-conditioning in the computation of $E(\omega)$ [see Eqs. (15.11) and (15.17)]. In addition, the condition $|E(\omega_{ij})| \geq |\delta|$ should be replaced by $|E(\omega_{ij})| > |\delta| - \varepsilon_1$ where ε_1 is a small positive constant, say 10^{-6}, to ensure that no maxima are missed owing to roundoff errors.

[†]An interior band edge is one in the range $0 < \omega < \pi$, i.e., not at $\omega = 0$ or π.

The search method is very reliable and its use in Algorithm 1 leads to a *robust* algorithm since the entire frequency axis is searched using a dense set of frequency points. Its disadvantage is that it requires a considerable amount of computation and is, therefore, inefficient. Improved search methods will be considered in Sec. 15.4.

A more efficient version of Algorithm 2 is obtained by maintaining all the interior band edges as extremals throughout the optimization independently of the behavior of the error function at the band edges. However, the algorithm obtained tends to fail more frequently than Algorithm 2.

15.3.3 Computation of $|E(\omega)|$ and $P_c(\omega)$

In steps 2 and 6 of the basic Remez algorithm (Algorithm 1), $|E(\omega)|$ and $P_c(\omega)$ need to be evaluated. This can be done by determining coefficients a_k by inverting the matrix in Eq. (15.12). This approach is inefficient and may be subject to numerical ill-conditioning, in particular, if δ is small and N is large. An alternative and more efficient approach is to deduce δ analytically and then interpolate $P_c(\omega)$ on the r frequency points using the *barycentric* form of the *Lagrange interpolation* formula. The necessary formulation is as follows.

Parameter δ can be deduced as

$$\delta = \frac{\sum_{k=0}^{r} \alpha_k D(\hat{\omega}_k)}{\sum_{k=0}^{r} \frac{(-1)^k \alpha_k}{W(\hat{\omega}_k)}} \tag{15.16}$$

and $P_c(\omega)$ is given by

$$P_c(\omega) = \begin{cases} C_k & \text{for } \omega = \hat{\omega}_0, \hat{\omega}_1, \ldots, \hat{\omega}_{r-1} \\ \dfrac{\sum_{k=0}^{r-1} \frac{\beta_k C_k}{x-x_k}}{\sum_{k=0}^{r-1} \frac{\beta_k}{x-x_k}} & \text{otherwise} \end{cases} \tag{15.17}$$

where

$$\alpha_k = \prod_{i=0,\ i\neq k}^{r} \frac{1}{x_k - x_i} \tag{15.18}$$

$$C_k = D(\hat{\omega}_k) - (-1)^k \frac{\delta}{W(\hat{\omega}_k)} \tag{15.19}$$

$$\beta_k = \prod_{i=0,\ i\neq k}^{r-1} \frac{1}{x_k - x_i} \tag{15.20}$$

with

$$x = \cos\omega \qquad x_i = \cos\hat{\omega}_i \qquad \text{for } i = 0, 1, 2, \ldots, r$$

In step 2 of the Remez algorithm, $|E(\omega)|$ often needs to be evaluated at a frequency which was an extremal in the previous iteration. For these cases, the magnitude of the error function is simply $|\delta|$, according to Eq. (15.11), and need not be evaluated.

15.3.4 Rejection of Superfluous Potential Extremals

The solution of Eq. (15.12) can be obtained only if *precisely* $r+1$ extremals are available. By differentiating $E(\omega)$, one can show that in a filter with one frequency band of interest (e.g., a digital differentiator) the number of maxima in $|E(\omega)|$ (potential extremals in step 2 of Algorithm 1) can be as high as $r+1$. In the weighted-Chebyshev method, band edges at which $|E(\omega)|$ is maximum and $|E(\omega)| \geq |\delta|$ are treated as potential extremals (see Algorithm 2). Therefore, whenever the number of frequency bands is increased by one, the number of potential extremals is increased by 2, i.e., for a filter with J bands there can be as many as $r + 2J - 1$ frequencies $\hat{\omega}_i$ and a maximum of $2J - 2$ superfluous $\hat{\omega}_i$ may occur. This problem is overcome by rejecting $\rho - r$ of the potential extremals $\hat{\omega}_i$, if $\rho > r$, in step 4 of the algorithm.

A simple rejection scheme is to reject the $\rho - r$ frequencies $\hat{\omega}_i$ that yield the lowest $|E(\hat{\omega}_i)|$ and then renumber the remaining $\hat{\omega}_i$ from 0 to r [8]. This strategy is based on the well known fact that the magnitude of the error in a given band is inversely related to the density of extremals in that band, i.e., a low density of extremals results in a large error and a high density results in a small error. Conversely, a low band error is indicative of a high density of extremals, and rejecting superfluous $\hat{\omega}_i$ in such a band is the appropriate course of action.

A problem with the above scheme is that whenever a frequency remains an extremal in two successive iterations, $|E(\omega)|$ assumes the value of $|\delta|$ in the second iteration by virtue of Eq. (15.11). In practice, there are almost always several frequencies that remain extremals from one iteration to the next, and the value of $|E(\omega)|$ at these frequencies will be the same. Consequently, the rejection of potential extremals on the basis of the magnitude of the error can become arbitrary and may lead to the rejection of potential extremals in bands where the density of extremals is low. This tends to increase the number of iterations, and it may even prevent the algorithm from converging on occasion. This problem can to some extent be alleviated by rejecting only potential extremals that are not band edges.

An alternative rejection scheme based on the aforementioned strategy, which was found to give excellent results for 2-band and 3-band filters, involves ranking the frequency bands in the order of lowest average band error, dropping the band with the highest average error from the list, and then rejecting potential extremals, one per band, in a cyclic manner starting with the band with the lowest average error [11]. The steps involved are as follows:

Algorithm 3: Alternative rejection scheme for superfluous potential extremals

1. Compute the average band errors

$$E_j = \frac{1}{\nu_j} \sum_{\hat{\omega}_i \in \Omega_j} |E(\hat{\omega}_i)| \quad \text{for } j = 1, 2, \ldots, J$$

where Ω_j is the set of potential extremals in band j given by

$$\Omega_j = \{\hat{\omega}_i : \omega_{Lj} \leq \hat{\omega}_i \leq \omega_{Rj}\}$$

ν_j is the number of potential extremals in band j, and J is the number of bands.

2. Rank the J bands in the order of lowest average error and let $l_1, l_2, \ldots, l_J$ be the ranked list obtained, i.e., l_1 and l_J are the bands with the lowest and highest average error, respectively.

3. Reject one $\hat{\omega}_i$ in each of bands $l_1, l_2, \ldots, l_{J-1}, l_1, l_2, \ldots$ until $\rho - r$ superfluous $\hat{\omega}_i$ are rejected. In each case, reject the $\hat{\omega}_i$, other than a band edge, that yields the lowest $|E(\hat{\omega}_i)|$ in the band.

For example, if $J = 3$, $\rho - r = 3$, and the average errors for bands 1, 2, and 3 are 0.05, 0.08, and 0.02, then $\hat{\omega}_i$ are rejected in bands 3, 1, and 3. Note that potential extremals are not rejected in band 2, which is the band of highest average error.

15.3.5 Computation of Impulse Response

The impulse response in step 6 of Algorithm 1 can be determined by noting that function $P_c(\omega)$ is the frequency response of a noncausal version of the required filter. The impulse response of this filter, represented by $h_0(n)$ for $-c \leq n \leq c$, can be determined by computing $P_c(k\Omega)$ for $k = 0, 1, 2, \ldots, c$, where $\Omega = 2\pi/N$, and then using the *inverse discrete Fourier transform*. It can be shown that

$$h_0(n) = h_0(-n) = \frac{1}{N}\left[P_c(0) + \sum_{k=1}^{c} 2P_c(k\Omega)\cos\left(\frac{2\pi kn}{N}\right)\right] \quad (15.21)$$

for $n = 0, 1, 2, \ldots, c$ (see Prob. 15.1). Therefore, the impulse response of the required causal filter is given by

$$h(n) = h_0(n - c)$$

for $n = 0, 1, 2, \ldots, N - 1$.

15.4 IMPROVED SEARCH METHODS

For a filter of length N, with the number of intervals w_j in each interval W_j equal to S, the exhaustive step-by-step search of Sec. 15.3.2 (Algorithm 2) requires about $S \times (N+1)/2$ function evaluations, where each function evaluation entails $N-1$ additions, $(N + 1)/2$ multiplications, and $(N + 1)/2$ divisions [see Eq. (15.17)]. A Remez optimization usually requires four to eight iterations for lowpass or highpass filters, six to ten iterations for bandpass filters, and eight to twelve iterations for bandstop filters. Further, if prescribed specifications are to be achieved and the appropriate value of N is unknown, typically two to four Remez optimizations have to be performed (see Sec. 15.7). For example, if $N = 101$, $S = 16$, number of Remez optimizations = 4, iterations per optimization = 6, the design would entail 24 iterations, 19,200 function evaluations, 1.92×10^6 additions, 0.979×10^6 multiplications, and 0.979×10^6 divisions. This is in addition to the computation required for the evaluation of δ and coefficients α_k, C_k, and β_k once per iteration. In effect, the amount of computation required to complete a design is quite substantial. In this section, alternative search techniques are described which reduce the amount of computation to a fraction of that required by the exhaustive search described in the previous section.

15.4.1 Selective Step-by-Step Search

When Eq. (15.12) is solved, the error function $|E(\omega)|$ is forced to satisfy the alternation theorem of Sec. 15.2.3. This theorem can be satisfied in several ways. The most likely possibility is illustrated in Fig. 15.3a, where ω_{Lj} and ω_{Rj} are the left-hand and right-hand edges, respectively, of the jth frequency band. In this case, ω_{Lj} and ω_{Rj} are extremal frequencies and there is strict alternation between maxima and zeros of $|E(\omega)|$. Additional maxima of $|E(\omega)|$ can be introduced under the following circumstances:

1. To the right of $\omega = 0$ (first band), if there is an extremal and $|E(\omega)|$ has a minimum at $\omega = 0$, as depicted in Fig. 15.3b (see properties of $|P_c(\omega)|$ in Sec. 15.6);
2. To the left of $\omega = \pi$ (last band), if there is an extremal and $|E(\omega)|$ has a minimum at $\omega = \pi$, as depicted in Fig. 15.3c (see Sec. 15.6);
3. At $\omega = 0$, if there is no extremal at $\omega = 0$, as depicted in Fig. 15.3d;
4. At $\omega = \pi$, if there is no extremal at $\omega = \pi$, as depicted in Fig. 15.3e;
5. To the right of an interior left-hand edge, as depicted in Fig. 15.3f;
6. To the left of an interior right-hand edge, as depicted in Fig. 15.3g;
7. At $\omega = \omega_{Lj}$, if there is no extremal at $\omega = \omega_{Lj}$, as depicted in Fig. 15.3h;
8. At $\omega = \omega_{Rj}$, if there is no extremal at $\omega = \omega_{Rj}$, as depicted in Fig. 15.3i;
9. Two consecutive *new* maxima at the interior of a band between two adjacent extremals, as depicted in Fig. 15.3j.

The maxima in Fig. 15.3a can be located by searching in the neighborhood of each extremal frequency using gradient information since there is a one-to-one correspondence between extremals and maxima of $|E(\omega)|$. If the first derivative is positive (negative), there is a maximum of $|E(\omega)|$ to the right (left) of the extremal, which can be readily located by increasing (decreasing) the frequency in steps w_j until $|E(\omega)|$ begins to decrease. The maxima in items (1) and (2) in the above list can be found by searching to the right of $\omega = 0$ in the first case or to the left of $\omega = \pi$ in the second case, if the second derivative is positive at $\omega = 0$ or π. Similarly, the maxima in (3) and (4) can be identified by checking whether $|E(\omega)|$ has a maximum and $|E(\omega)| \geq |\delta|$ at $\omega = 0$ in the first case or at $\omega = \pi$ in the second case. The maxima in (5) and (6) can be found by searching to the right of an interior left-hand edge if the first derivative is positive or to the left of a right-hand interior edge if the first derivative is negative. Similarly, the maxima in (7) and (8) can be identified

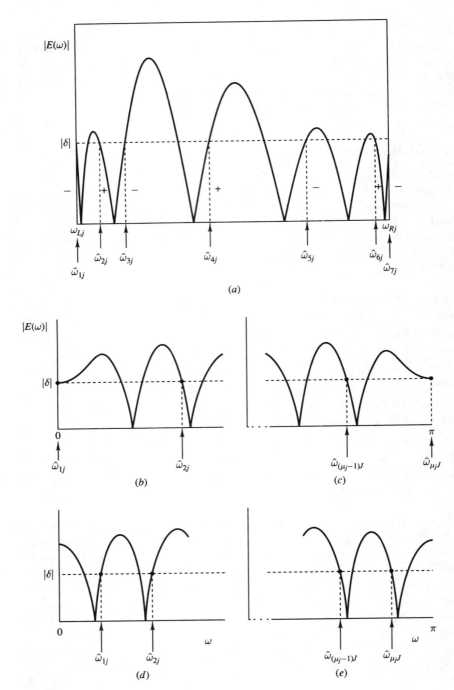

FIGURE 15.3a-e
Types of maxima in $|E(\omega)|$.

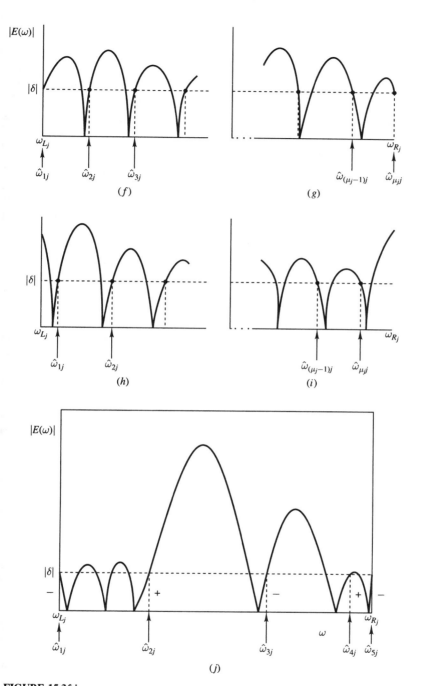

FIGURE 15.3f-j
Types of maxima in $|E(\omega)|$.

by checking whether the first derivative is negative at $\omega = \omega_{Lj}$ in the first case and positive at $\omega = \omega_{Rj}$ in the second case, and $|E(\omega)| \geq |\delta|$ in each of the two cases.

If a *selective* step-by-step search based on the above principles is used in Algorithm 1, then at the start of the optimization the distance between a typical extremal $\hat{\omega}_i$ and the nearby maximum point $\overset{\frown}{\omega}_i$ will be less than half the period of the corresponding ripple of $|E(\omega)|$, owing to the relative symmetry of the ripples of the error function. In effect, in the first iteration only half of the combined width of the different bands needs to be searched. This will reduce the number of function evaluations by more than fifty percent relative to that required by the exhaustive search of Sec. 15.3.2 without degrading the accuracy of the optimization in any way. As the optimization progresses and the solution is approached, extremal $\hat{\omega}_i$ and maximum point $\overset{\frown}{\omega}_i$ tend to coincide and, therefore, the cumulative length of the frequency range that has to be searched is progressively reduced, thereby resulting in further economies in the number of function evaluations. In the last iteration, only two or three function evaluations are needed (including derivatives) per ripple. As a result, the total number of function evaluations can be reduced by 65 to 70 percent relative to that required by the exhaustive search [10].

A selective search of the type just described will miss maxima of the type in item (9) in the above list and the algorithm will fail. However, the problem can be overcome relatively easily. Maxima of the type in (9) can sometimes occur in the stopbands of bandstop filters, and it was found possible to reduce the number of failures by increasing somewhat the density of extremals in the stopband relative to the density of extremals in the passbands [11]. An alternative approach, which was found to give good results, is to check the distance between adjacent potential extremals at the end of the search; if the difference exceeds the initial difference by a significant amount [say if $(\overset{\frown}{\omega}_{(k+1)} - \overset{\frown}{\omega}_k) > RW_j$ for some k, where R is a constant in the range 1.5 to 2.0], then an exhaustive search is undertaken between $\overset{\frown}{\omega}_k$ and $\overset{\frown}{\omega}_{(k+1)}$ in order to locate any missed maxima.

15.4.2 Cubic Interpolation

In this section, an alternative search method is described which can increase further the computational efficiency of the Remez algorithm. The method is based on *cubic interpolation* [11].

Assume that the error function, depicted in Fig. 15.4, can be represented by the third-order polynomial

$$|E(\omega)| = M = a + b\omega + c\omega^2 + d\omega^3 \qquad (15.22)$$

where a, b, c, and d are constants. The first derivative of M with respect to ω is obtained from Eq. (15.22) as

$$\frac{dM}{d\omega} = G = b + 2c\omega + 3d\omega^2$$

DESIGN OF NONRECURSIVE FILTERS USING OPTIMIZATION METHODS **559**

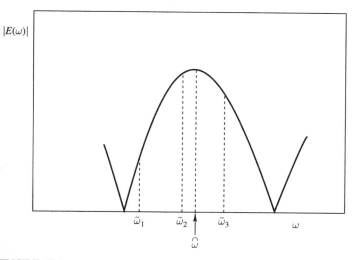

FIGURE 15.4
Frequency points for cubic interpolation.

Hence, the frequencies at which M has stationary points are given by

$$\bar{\omega} = \frac{1}{3d}[-c \pm \sqrt{(c^2 - 3bd)}] \quad (15.23)$$

The stationary point that corresponds to a maximum point, designated as $\overset{\frown}{\omega}$, can be selected by noting that M is maximum when

$$\frac{d^2 M}{d\omega^2} = 2c + 6d\overset{\frown}{\omega} < 0$$

or

$$\overset{\frown}{\omega} < -\frac{c}{3d} \quad (15.24)$$

Evidently, if constants $b, c,$ and d are known, $\overset{\frown}{\omega}$ can be readily determined. If we assume that $\tilde{\omega}_1, \tilde{\omega}_2,$ and $\tilde{\omega}_3$ are distinct frequencies, we can write

$$M\bigg|_{\omega=\tilde{\omega}_k} = M_k = a + b\tilde{\omega}_k + c(\tilde{\omega}_k)^2 + d(\tilde{\omega}_k)^3$$

for $k = 1, 2,$ and 3 and

$$G\bigg|_{\omega=\tilde{\omega}_1} = G_1 = b + 2c\tilde{\omega}_1 + 3d(\tilde{\omega}_1)^2$$

By solving this system of simultaneous equations, we can show that

$$d = \frac{\beta - \gamma}{\theta - \psi} \quad (15.25)$$

$$c = \beta - \theta d \quad (15.26)$$

$$b = G_1 - 2c\tilde{\omega}_1 - 3d(\tilde{\omega}_1)^2 \quad (15.27)$$

where

$$\beta = \frac{(M_2 - M_1) + G_1(\tilde{\omega}_1 - \tilde{\omega}_2)}{(\tilde{\omega}_1 - \tilde{\omega}_2)^2} \quad (15.28)$$

$$\gamma = \frac{(M_3 - M_1) + G_1(\tilde{\omega}_1 - \tilde{\omega}_3)}{(\tilde{\omega}_1 - \tilde{\omega}_3)^2} \quad (15.29)$$

$$\theta = \frac{2(\tilde{\omega}_1)^2 - \tilde{\omega}_2(\tilde{\omega}_1 + \tilde{\omega}_2)}{(\tilde{\omega}_1 - \tilde{\omega}_2)} \quad (15.30)$$

$$\psi = \frac{2(\tilde{\omega}_1)^2 - \tilde{\omega}_3(\tilde{\omega}_1 + \tilde{\omega}_2)}{(\tilde{\omega}_1 - \tilde{\omega}_3)} \quad (15.31)$$

By evaluating constants $\beta, \gamma, \theta, \psi, d, c$, and b and then using Eqs. (15.23) and (15.24), $\overset{\frown}{\omega}$ can be determined.

The search just described entails four function evaluations, three for M_1, M_2, M_3, and one for G_1, per iteration. The method is useful as a possible replacement of the selective search only if it gives a fairly accurate estimate of $\overset{\frown}{\omega}$. To achieve this goal, the cubic interpolation should not be used near band edges where $|E(\omega)|$ is not well behaved (see Fig. 15.3).

If the above cubic interpolation is to be used to find the maximum of $|E(\omega)|$ in the neighborhood of extremal $\hat{\omega}_i$, the most suitable value for frequency $\tilde{\omega}_1$ is the extremal itself, since the value of M is known at this frequency and need not be computed. It is given by Eq. (15.11) as

$$M|_{\omega=\hat{\omega}_i} = |\delta|$$

The frequency $\tilde{\omega}_3$ should bracket the required maximum point $\overset{\frown}{\omega}_i$, but it should not bracket the adjacent zero of the error function. It should be recalled at this point that as the solution is approached, $\overset{\frown}{\omega}_i$ tends to move closer to $\hat{\omega}_i$. Therefore, the accuracy of the cubic interpolation can be improved by reducing the interval $\tilde{\omega}_3 - \tilde{\omega}_1$ as the solution is approached. Such an *adjustable bracket* can be formed by using the convergence parameter Q of Eq. (15.13), which is known to reduce from 1 to 0 as the solution is approached. A formula for $\tilde{\omega}_3$ which was found to give good results [11] is

$$\tilde{\omega}_3 = \begin{cases} \hat{\omega}_i + \dfrac{Q}{2}(\hat{\omega}_{i+1} - \hat{\omega}_i) & \text{for } G_1 > 0 \\ \hat{\omega}_i - \dfrac{Q}{2}(\hat{\omega}_i - \hat{\omega}_{i-1}) & \text{for } G_1 < 0 \end{cases} \quad (15.32)$$

Frequency $\tilde{\omega}_2$ can be placed at the center of the frequency range $\tilde{\omega}_1$ to $\tilde{\omega}_3$, i.e.,

$$\tilde{\omega}_2 = \tfrac{1}{2}(\tilde{\omega}_1 + \tilde{\omega}_3) \quad (15.33)$$

The computational efficiency of the cubic interpolation method described remains constant from iteration to iteration since the number of function evaluations required to perform an interpolation is constant. At the start of the optimization, the cubic-interpolation search is more efficient than the selective step-by-step method.

However, as the solution is approached the number of function evaluations required by the selective search is progressively reduced, as was stated earlier, and at some point the selective search becomes more efficient. A prudent strategy under these circumstances is to use the cubic-interpolation search at the start of the optimization and switch over to the selective step-by-step search when some suitable criterion is satisfied. Extensive experimental results have shown that computational advantage can be gained by using the cubic-interpolation search if $Q > 0.65$, and the selective search otherwise [11]. The use of the cubic interpolation along with the selective step-by-step search of the preceding section can reduce the number of function evaluations by 70 to 75 percent relative to that required by the exhaustive search.

15.4.3 Quadratic Interpolation

An alternative method for the location of the maxima of $|E(\omega)|$ that was found to work well is based on a *two-stage quadratic interpolation* search. However, the computational efficiency that can be achieved with this approach was found to be somewhat inferior relative to the above one-stage cubic interpolation search.

15.4.4 Improved Formulation

In the problem formulation considered so far, the extremals $\hat{\omega}_0, \hat{\omega}_1, \ldots, \hat{\omega}_r$ are treated as a 1-D array and are numbered sequentially from 0 to r. Through the rejection of superfluous extremals, as detailed above, the distribution of extremals can change from iteration to iteration. In order to evaluate δ and coefficients C_k correctly [see Eqs. (15.16) and (15.19)], it is necessary to monitor and track the indices of the first and last extremal of each band throughout the optimization. This tends to complicate the implementation of the Remez algorithm quite significantly. The problem can be eliminated by representing the extremals in terms of a 2-D array of the form

$$\hat{\Omega} = \begin{bmatrix} \hat{\omega}_{11} & \hat{\omega}_{12} & \cdots & \hat{\omega}_{1j} & \cdots & \hat{\omega}_{1J} \\ \hat{\omega}_{21} & \hat{\omega}_{22} & \cdots & \hat{\omega}_{2j} & \cdots & \hat{\omega}_{2J} \\ \vdots & \vdots & \cdots & \vdots & \cdots & \vdots \\ \hat{\omega}_{\mu_1 1} & \hat{\omega}_{\mu_2 2} & \cdots & \hat{\omega}_{\mu_j j} & \cdots & \hat{\omega}_{\mu_J J} \end{bmatrix}$$

where the jth column represents the extremals of the jth band, μ_j is the number of extremals in the jth band, and J is the number of bands. The use of this notation necessitates that the formulas for δ and $P_c(\omega)$ be modified accordingly. From Eqs. (15.16)–(15.20) one can show that (see Probs. 15.2 and 15.3)

$$\delta = \frac{\sum_{\{k,m\} \in \mathbf{K}_r} \alpha_{km} D(\hat{\omega}_{km})}{\sum_{\{k,m\} \in \mathbf{K}_r} \frac{(-1)^q \alpha_{km}}{W(\hat{\omega}_{km})}} \tag{15.34}$$

and

$$P_c(\omega) = \begin{cases} C_{km} & \text{for } \omega \in \hat{\Omega} \\ \dfrac{\sum_{\{k,m\} \in \mathbf{K}_{r-1}} \frac{\beta_{km} C_{km}}{x - x_{km}}}{\sum_{\{k,m\} \in \mathbf{K}_{r-1}} \frac{\beta_{km}}{x - x_{km}}} & \text{otherwise} \end{cases} \tag{15.35}$$

where

$$\beta_{km} = \prod_{\{i,j\} \in \mathbf{I}_{r-1}} \frac{1}{x_{km} - x_{ij}} \tag{15.36}$$

$$\alpha_{km} = \begin{cases} \beta_{km} & \text{if } k = \mu_J \text{ and } m = J \\ \dfrac{\beta_{km}}{x_{km} - x_{\mu_J J}} & \text{otherwise} \end{cases} \tag{15.37}$$

$$C_{km} = D(\hat{\omega}_{km}) - (-1)^q \frac{\delta}{W(\hat{\omega}_{km})} \tag{15.38}$$

with

$$q = \begin{cases} k-1 & \text{if } m = 1 \\ k-1 + \sum_{j=1}^{m-1} \mu_j & \text{if } m \geq 2 \end{cases} \tag{15.39}$$

and

$$x = \cos \omega \qquad x_{ij} = \cos \hat{\omega}_{ij} \qquad \text{for } \{i, j\} \in \mathbf{I}_r$$

In the above formulation $\mathbf{K}_r$, $\mathbf{K}_{r-1}$, $\mathbf{I}_r$, and $\mathbf{I}_{r-1}$ are sets given by

$$\mathbf{K}_r = \{\{k, m\} : (1 \leq k \leq \mu_m) \text{ and } (1 \leq m \leq J)\} \tag{15.40}$$

$$\mathbf{K}_{r-1} = \{\{k, m\} : (1 \leq k \leq l) \text{ and } (1 \leq m \leq J)\} \tag{15.41}$$

$$\mathbf{I}_r = \{\{i, j\} : (1 \leq i \leq \mu_j) \text{ and } (1 \leq j \leq J)\} \tag{15.42}$$

and

$$\mathbf{I}_{r-1} = \{\{i, j\} : (1 \leq i \leq h) \text{ and } (1 \leq j \leq J) \text{ and } (i \neq k \text{ or } j \neq m)\} \tag{15.43}$$

with

$$l = \begin{cases} \mu_J - 1 & \text{for } m = J \\ \mu_m & \text{otherwise} \end{cases}$$

and

$$h = \begin{cases} \mu_J - 1 & \text{for } j = J \\ \mu_j & \text{otherwise} \end{cases}$$

15.5 EFFICIENT REMEZ EXCHANGE ALGORITHM

The above principles will now be used to construct an efficient Remez exchange algorithm. As in Algorithm 2, ω_{Lj} and ω_{Rj} are the left- and right-hand edges in band j; W_j is the interval between adjacent extremals and m_j is the number of intervals

W_j in band j; w_j is the interval between successive samples in interval W_j, and S is the number of intervals w_j in each interval W_j; N_j is the total number of intervals w_j in band j; and J is the number of bands. The frequencies $\hat{\omega}_{1j}, \hat{\omega}_{2j}, \ldots, \hat{\omega}_{\mu_j j}$ are the current extremals and $\overset{\frown}{\hat{\omega}}_{1j}, \overset{\frown}{\hat{\omega}}_{2j}, \ldots, \overset{\frown}{\hat{\omega}}_{\nu_j j}$ are the potential extremals for the next iteration in band j. The magnitude of the error function and its first and second derivatives with respect to ω are denoted as

$$M = |E(\omega)|, \qquad G_1 = \frac{d|E(\omega)|}{d\omega}, \qquad G_2 = \frac{d^2|E(\omega)|}{d\omega^2}$$

The improved algorithm consists of a main part called MAIN which calls routines EXTREMALS, SELECTIVE, and CUBIC. The steps involved are detailed below.

Algorithm 4: Efficient Remez exchange algorithm

MAIN

M1.
  (a) Initialize S, say $S = 16$, and set $Q = 1$.
  (b) For $j = 1, 2, \ldots, J$ do:
    Compute m_j and W_j for $j = 1, 2, \ldots, J$ using Eqs. (15.14) and (15.15), respectively.
    Initialize extremals by letting $\hat{\omega}_{1j} = \omega_{Lj}, \ldots, \hat{\omega}_{ij} = \omega_{Lj} + (i-1)W_j, \ldots, \hat{\omega}_{\mu_j j} = \omega_{Rj} = \omega_{Lj} + m_j W_j$.
    Set $N_j = m_j S$ and $w_j = B_j / N_j$.

M2.
  (a) Compute coefficients β_{kj}, α_{kj}, and C_{kj} for $j = 1, 2, \ldots, J$ using Eqs. (15.36)–(15.38).
  (b) Compute δ using Eq. (15.34).

M3. Call EXTREMALS.

M4.
  (a) Set $\rho = \nu_1 + \nu_2 + \ldots \nu_J$.
  (b) Reject $\rho - r$ superfluous potential extremals using Algorithm 3, renumber the remaining $\overset{\frown}{\hat{\omega}}_{ij}$ sequentially, and update μ_j if necessary.
  (c) Update extremals by letting $\hat{\omega}_{ij} = \overset{\frown}{\hat{\omega}}_{ij}$ for $i = 1, 2, \ldots, \mu_j$ and $j = 1, 2, \ldots, J$.

M5.
  (a) Compute Q using Eq. (15.13).
  (b) If $Q > 0.01$, go to step M2.

M6.
  (a) Compute $P_c(k\Omega)$ for $k = 0, 1, \ldots, r-1$ using Eq. (15.35).
  (b) Compute $h(n)$ using Eq. (15.21).
  (c) Stop.

EXTREMALS

E1. For each of bands $1, 2, \ldots, j, \ldots, J$ do:
  (A) Set $e = 0$.

(B) For each of extremals $\hat{\omega}_{1j}, \hat{\omega}_{2j}, \ldots, \hat{\omega}_{ij}, \ldots, \hat{\omega}_{\mu_j j}$ do:
   (a) Case $\hat{\omega}_{ij} = \hat{\omega}_{1j}$:
      If $\hat{\omega}_{ij} = \omega_{Lj}$, then do:
         Case $j = 1$ (first band):
            If $G_2 < 0$, then set $e = e + 1$ and $\overset{\frown}{\omega}_{ej} = \hat{\omega}_{ij}$; otherwise call SELECTIVE.
         Case $j \neq 1$ (other bands):
            If $G_1 > 0$, then call SELECTIVE; otherwise set $e = e + 1$ and $\overset{\frown}{\omega}_{ej} = \hat{\omega}_{ij}$.
      If $\hat{\omega}_{ij} \neq \omega_{Lj}$, then call SELECTIVE.
   (b) Case $\hat{\omega}_{1j} < \hat{\omega}_{ij} < \hat{\omega}_{\mu_j j}$:
      If $Q < 0.65$, then call SELECTIVE; otherwise call CUBIC.
      If $flag0 = 1$ (CUBIC was unsuccessful in generating a good estimate of the maximum point), then call SELECTIVE.
   (c) Case $\hat{\omega}_{ij} = \hat{\omega}_{\mu_j j}$:
      If $\hat{\omega}_{ij} = \omega_{Rj}$, then do:
         Case $j = J$ (last band):
            If $G_2 < 0$, then set $e = e + 1$ and $\overset{\frown}{\omega}_{ej} = \hat{\omega}_{ij}$; otherwise call SELECTIVE.
         Case $j \neq 1$ (other bands):
            If $G_1 < 0$, then call SELECTIVE; otherwise set $e = e + 1$ and $\overset{\frown}{\omega}_{ej} = \hat{\omega}_{ij}$.
      If $\hat{\omega}_{ij} \neq \omega_{Rj}$ then call SELECTIVE.

(C) Check for an additional potential extremal at the left-hand edge of band j: If $\hat{\omega}_{1j}$ and $\overset{\frown}{\omega}_{1j} \neq \omega_{Lj}$, $|E(\omega_{Lj})| > |E(\omega_{Lj} + w_j)|$, and $|E(\omega_{Lj})| \geq |\delta|$, then set $e = e + 1$ and insert new potential extremal at $\omega = \omega_{Lj}$.

(D) Check for an additional potential extremal at the right-hand edge of band j: If $\hat{\omega}_{\mu_j j}$ and $\overset{\frown}{\omega}_{ej} \neq \omega_{Rj}$, $|E(\omega_{Rj} - w_j)| < |E(\omega_{Rj})|$, and $|E(\omega_{Rj})| \geq |\delta|$, then insert new potential extremal at $\omega = \omega_{Rj}$ and set $e = e + 1$.

(E) Check for additional potential extremals in band j:
   (a) For $k = 1, 2, \ldots, e - 1$ check if
   $$\overset{\frown}{\omega}_{(k+1)j} - \overset{\frown}{\omega}_{kj} > RW_j$$
   For each value of k for which the inequality is satisfied, use an exhaustive search between frequencies $\overset{\frown}{\omega}_{kj}$ and $\overset{\frown}{\omega}_{(k+1)j}$ (see Algorithm 2). For each new maximum of M such that $|E(\omega)| \geq |\delta|$, insert a new potential extremal sequentially between $\overset{\frown}{\omega}_{kj}$ and $\overset{\frown}{\omega}_{(k+1)j}$ and set $e = e + 1$ (R is a constant in the range 1.5 to 2.0).
   (b) If there is a large gap (larger than RW_j) between the left-hand edge and the first potential extremal, check for additional potential extremals in the range $\omega_{Lj} < \omega < \hat{\omega}_{1j}$; for each new maximum such that $|E(\omega)| \geq |\delta|$, insert a new potential extremal sequentially between ω_{Lj} and $\hat{\omega}_{1j}$ and set $e = e + 1$.

(c) If there is a large gap (larger than RW_j) between the last potential extremal and the right-hand edge, check for additional potential extremals in the range $\hat{\omega}_{ej} < \omega < \omega_{Rj}$; for each new maximum such that $|E(\omega)| \geq |\delta|$, insert a new potential extremal sequentially between $\hat{\omega}_{ej}$ and ω_{Rj} and set $e = e + 1$.

(F) Set $v_j = e$.

E2. Return.

SELECTIVE

S1. If ($G_1 \geq 0$ and $\hat{\omega}_{ij} \neq 0$) or ($G_2 > 0$ and $\hat{\omega}_{ij} = 0$), then increase ω in steps w_j until a maximum of M is located. Set $e = e+1$ and assign the frequency of this maximum to $\hat{\omega}_e$. If no maximum is located in the frequency range $\hat{\omega}_{ij} \leq \omega < (\hat{\omega}_{(i+1)j}$ or $\omega_{Rj})$, discontinue the search.

S2. If ($G_1 \leq 0$ and $\hat{\omega}_{ij} \neq \pi$) or ($G_2 > 0$ and $\hat{\omega}_{ij} = \pi$), then decrease ω in steps w_j until a maximum of M is located. Set $e = e + 1$ and assign the frequency of this maximum to $\hat{\omega}_e$. If no maximum is located in the frequency range (ω_{Lj} or $\hat{\omega}_{(i-1)j}) \leq \omega < \hat{\omega}_{ij}$, discontinue the search.

S3. Return.

CUBIC

C1. Set $flag0 = 0$.

C2. Set $\tilde{\omega}_1 = \hat{\omega}_{ij}$ and compute frequencies $\tilde{\omega}_3$ and $\tilde{\omega}_2$ using Eqs. (15.32)–(15.33).

C3. Compute constants β, γ, θ, and ψ using Eqs. (15.28)–(15.31).

C4. Compute constants d, c, b using Eqs. (15.25)–(15.27). If $3bd > c^2$ (third-order polynomial has no maximum), then set $flag0 = 1$ and return.

C5. Compute $\hat{\omega}$ using Eqs. (15.23) and (15.24). If frequency $\hat{\omega}$ is outside the interval $[\tilde{\omega}_1, \tilde{\omega}_3]$ (estimate of the maximum point is unreliable), then set $flag0 = 1$ and return.

C6. Set $\hat{\omega} = w_j \times \text{Int } (\hat{\omega}/w_j + 0.5)$.

C7. Set $e = e + 1$ and $\hat{\omega}_{ej} = \hat{\omega}$.

C8. Return.

Step E1(B)(a) checks for maxima at or near the left-hand edge of each band for the cases illustrated in Fig. 15.3a, b, d, f, and h. Step E1(B)(b) locates the interior maxima in Fig. 15.3a that correspond to extremals $\hat{\omega}_{2j}$ to $\hat{\omega}_{(\mu_j-1)j}$. Step E1(B)(c) checks for maxima at or near the right-hand edge of each band for the cases illustrated in Fig. 15.3a, c, e, g, and i. Step E1(C) checks for a new maximum at left-hand edge ω_{Lj} in the special case where there is no extremal and a maximum has not been picked up already at this frequency by step E1(B)(a). Such a situation can arise as shown in Fig. 15.3d where step E1(B)(a) will pick up the maximum at the right of point $\omega = \hat{\omega}_{1j}$, since $G_1 > 0$, but miss the maximum at $\omega = 0$. A similar situation can arise as illustrated in Fig. 15.3h. Step E1(D) checks for a new maximum at right-hand edge ω_{Rj} for the case where there is no extremal and a maximum has not been picked up already at this frequency by step E1(B)(c). Such a situation can arise as shown in Fig. 15.3e where step E1(B)(c) will pick up the maximum at the left of point $\omega = \hat{\omega}_{\mu_J J}$, since $G_1 < 0$, but miss the maximum at $\omega = \pi$. A similar situation can arise as illustrated in Fig. 15.3i. Steps E1(E)(a) to E1(E)(c) check for any missed maxima, like the maxima between $\hat{\omega}_{1j}$ and $\hat{\omega}_{2j}$ in Fig. 15.3j, in cases where the interval between any two adjacent maxima, between the left-hand edge and the first maximum, or between the last maximum and the right-hand edge is significantly larger than the average interval between adjacent extremals.

When the ripple of the error function is seriously skewed (e.g., near band edges in the first or second iteration) routine CUBIC may yield a poor estimate of the maximum point, and on rare occasions the third-order polynomial may not have a maximum. If either of these cases is detected, CUBIC is aborted and SELECTIVE is called in its place. CUBIC will almost always yield a value of $\hat{\omega}$ between two adjacent sample points. In order to ensure that each potential extremal is a member of the set of sample points, $\hat{\omega}$ is rounded to the nearest sample point in step C6. This makes the estimate produced by CUBIC compatible with that produced by SELECTIVE and prevents numerical ill-conditioning in the evaluation of $E(\omega)$, G_1, and G_2. The CUBIC interpolation routine may be disabled by modifying step E1(B)(b).

Extensive experimentation by the author has shown the above algorithm to be quite robust. It never failed in the design of 81 2-band filters chosen at random, it failed twice in the design of 67 3-band filters, three times in the design of 50 4-band filters, and three times in the design of 33 5-band filters. Lack of convergence is usually brought about by a cyclic pattern of rejected potential extremals, but the problem can be easily overcome by changing one of the specified filter parameters slightly, e.g., a passband or stopband edge or the order of the filter.

15.6 GRADIENT INFORMATION

Routines SELECTIVE and CUBIC in the above algorithm rely heavily on the *first and second derivatives* of $|E(\omega)|$ with respect to ω. From Eq. (15.3), we have

$$\frac{d|E(\omega)|}{d\omega} = \operatorname{sgn}\left[D(\omega) - P_c(\omega)\right] \left[\frac{dD(\omega)}{d\omega} - \frac{dP_c(\omega)}{d\omega}\right] \tag{15.44}$$

and

$$\frac{d^2|E(\omega)|}{d\omega^2} = \operatorname{sgn}\left[D(\omega) - P_c(\omega)\right] \left[\frac{d^2 D(\omega)}{d\omega^2} - \frac{d^2 P_c(\omega)}{d\omega^2}\right] \tag{15.45}$$

where

$$\operatorname{sgn}(x) = \begin{cases} 1 & \text{for } x \geq 0 \\ -1 & \text{for } x < 0 \end{cases}$$

The first and second derivatives of $|E(\omega)|$ under different circumstances can be obtained from the following properties of $|P_c(\omega)|$ [10].

15.6.1 Property 1

For any frequency including extremal $\hat{\omega}_{\mu_J J}$ (last extremal of last band) but excluding all other extremals,

$$\frac{dP_c(\omega)}{d\omega} = \frac{d(\omega)n_1(\omega) - d_1(\omega)n(\omega)}{d^2(\omega)} \qquad (15.46)$$

where

$$n(\omega) = \sum_{\{k,m\} \in \mathbf{K}_{r-1}} \frac{\beta_{km} C_{km}}{x - x_{km}}$$

$$d(\omega) = \sum_{\{k,m\} \in \mathbf{K}_{r-1}} \frac{\beta_{km}}{x - x_{km}}$$

$$n_1(\omega) = \sin \omega \sum_{\{k,m\} \in \mathbf{K}_{r-1}} \frac{\beta_{km} C_{km}}{(x - x_{km})^2}$$

$$d_1(\omega) = \sin \omega \sum_{\{k,m\} \in \mathbf{K}_{r-1}} \frac{\beta_{km}}{(x - x_{km})^2}$$

$\mathbf{K}_{r-1}$ is given by Eq. (15.41).

15.6.2 Property 2

For all extremal frequencies except $\hat{\omega}_{\mu_J J}$,

$$\left.\frac{dP_c(\omega)}{d\omega}\right|_{\omega=\hat{\omega}_{ij}} = \frac{\sin(\hat{\omega}_{ij})}{\beta_{ij}} \sum_{\{k,m\} \in \mathbf{K}_{r-1}^{ij}} \frac{\beta_{km}(C_{ij} - C_{km})}{x_{ij} - x_{km}} \qquad (15.47)$$

where

$$\mathbf{K}_{r-1}^{ij} = \{\{k,m\} : (1 \leq k \leq l) \text{ and } (1 \leq m \leq J) \text{ and } (k \neq i \text{ or } m \neq j)\}$$

with

$$l = \begin{cases} \mu_J - 1 & \text{for } m = J \\ \mu_m & \text{otherwise} \end{cases}$$

15.6.3 Property 3

For $\omega = 0$ or π, it follows from properties 1 and 2 that

$$\frac{dP_c(\omega)}{d\omega} = 0 \qquad (15.48)$$

15.6.4 Property 4

For $\omega = 0$ if no extremal occurs at zero or for $\omega = \pi$ under all circumstances,

$$\frac{d^2 P_c(\omega)}{d\omega^2} = \frac{d(\omega)n_2(\omega) - d_2(\omega)n(\omega)}{d^2(\omega)} \qquad (15.49)$$

where

$$n_2(\omega) = \cos\omega \sum_{\{k,m\} \in \mathbf{K}_{r-1}} \frac{\beta_{km} C_{km}}{(x - x_{km})^2}$$

$$d_2(\omega) = \cos\omega \sum_{\{k,m\} \in \mathbf{K}_{r-1}} \frac{\beta_{km}}{(x - x_{km})^2}$$

15.6.5 Property 5

If there is an extremal at $\omega = 0$, then

$$\left.\frac{d^2 P_c(\omega)}{d\omega^2}\right|_{\omega=0} = \frac{1}{\beta_{11}} \sum_{\{k,m\} \in \mathbf{K}^{11}_{r-1}} \frac{\beta_{km}(C_{11} - C_{km})}{x - x_{km}} \qquad (15.50)$$

where $\mathbf{K}^{11}_{r-1} = \mathbf{K}^{ij}_{r-1}$ with $i = j = 1$.

Example 15.1. Design a nonrecursive equiripple highpass filter using the Remez algorithm (*a*) with the exhaustive search of Sec. 15.3.2, (*b*) with the selective step-by-step search of Sec. 15.4.1, and (*c*) with the selective step-by-step search in conjunction with the cubic interpolation search of Sec. 15.4.2. Compare the results obtained. The required specifications are as follows:

Filter length N: 21
Passband edge ω_p: 2.0 rad/s
Stopband edge ω_a: 1.0 rad/s
Ratio δ_p/δ_a: 18.0
Sampling frequency ω_s: 2π rad/s

Solution. The design in (*a*) was carried out using Algorithm 1 in conjunction with Algorithms 2 and 3, whereas the designs in (*b*) and (*c*) were carried out using Algorithm 4, first without and then with routine CUBIC, respectively. The progress of the design is illustrated in Table 15.1. As can be seen, the exhaustive and selective search methods required four iterations each, whereas the selective search in conjunction with cubic interpolation required five iterations. However, the number of function evaluations [evaluations of $P_c(\omega)$ using Eq. (15.35) plus evaluations of G_1 and G_2 using Eqs. (15.44) and (15.45), respectively] decreased from 616 in the first method to 167 in the second method to 147 in the third method. In the Remez algorithm, approximately 80 to 90 percent of the computational effort involves function evaluations. In effect, relative to that required by the exhaustive search, the use of the selective step-by-step search reduced the amount of computation by about 73 percent, and the use of the selective step-by-step search in

TABLE 15.1
Progress in design of highpass filter (Example 15.1)

Iter. No.	Exhaustive search		Selective search		Sel. with cub. interpolation	
	Q	FE's	Q	FE's	Q	FE's
1	0.9379	154	0.9379	70	0.9379	49
2	0.5792	154	0.5792	48	0.6519	28
3	0.0846	154	0.0846	29	0.0756	28
4	0.0000	154	0.0000	20	0.0309	22
5	—	—	—	—	0.0000	20
Total FE's		616		167		147

conjunction with the cubic interpolation search reduced the amount of computation by about 76 percent.

The three methods resulted in approximately the same impulse responses, as can be seen in Table 15.2, and the passband ripple and minimum stopband attenuation obtained in each case were 0.073 dB and 72.6 dB, respectively. The amplitude response of the filter is illustrated in Fig. 15.5.

Example 15.2. Design a nonrecursive equiripple bandpass filter using the Remez algorithm (*a*) with the exhaustive search, (*b*) with the selective step-by-step search, and (*c*) with the selective step-by-step search in conjunction with the cubic interpolation search. Compare the results obtained. The required specifications are as follows:

Filter length N: 33
Lower passband edge ω_{p1}: 1.00 rad/s

TABLE 15.2
Impulse response of highpass filter (Example 15.1)

n	$h_0(n) = h_0(-n)$	
	Exhaustive or selective search	Selective with cubic interpolation
0	4.976192×10^{-1}	4.976160×10^{-1}
1	-3.120628×10^{-1}	-3.120636×10^{-1}
2	2.462999×10^{-3}	2.466692×10^{-3}
3	8.853907×10^{-2}	8.854032×10^{-2}
4	-2.605336×10^{-3}	-2.609410×10^{-3}
5	-3.790087×10^{-2}	-3.790082×10^{-2}
6	2.553469×10^{-3}	2.555600×10^{-3}
7	1.553835×10^{-2}	1.553849×10^{-2}
8	-2.126568×10^{-3}	-2.127674×10^{-3}
9	-5.222708×10^{-3}	-5.222820×10^{-3}
10	1.898114×10^{-3}	1.898558×10^{-3}

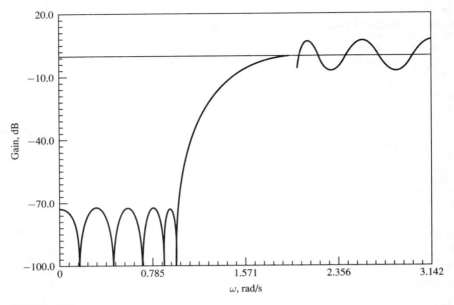

FIGURE 15.5
Amplitude response of equiripple highpass filter (Example 15.1) (the passband gain is multiplied by the factor 200 in order to illustrate the passband ripple).

Upper passband edge ω_{p2}: 2.00 rad/s
Lower stopband edge ω_{a1}: 0.63 rad/s
Upper stopband edge ω_{a2}: 2.40 rad/s
Ratio δ_p/δ_a: 23.0
Sampling frequency ω_s: 2π rad/s

Solution. As in Example 15.1, the design in (*a*) was carried out using Algorithm 1 in conjunction with Algorithms 2 and 3, whereas the designs in (*b*) and (*c*) were carried out using Algorithm 4 first without and then with routine CUBIC. The progress of the design is illustrated in Table 15.3. In this example, each of the three methods required eight iterations, and in each case there was a superfluous maximum at the end of the fourth iteration. The problem was eliminated by rejecting the second maximum of the third band, using the rejection method detailed in Sec. 15.3.4 (see Algorithm 3). As can be seen in Table 15.3, the number of function evaluations decreased from 1848 in the first method to 417 in the second method to 345 in the third method. In effect, the use of the selective step-by-step search reduced the amount of computation by about 77 percent, and the use of the selective step-by-step search in conjunction with the cubic interpolation search reduced the amount of computation by about 81 percent, relative to that required by the exhaustive search.

The three methods resulted in approximately the same impulse responses, as can be seen in Table 15.4. The amplitude response of the filter is illustrated in Fig. 15.6; the passband ripple and minimum stopband attenuation obtained in each case were 0.934 dB and 52.6 dB, respectively.

TABLE 15.3
Progress in design of bandpass filter (Example 15.2)

Iter. No.	Exhaustive search		Selective search		Selective with cubic interpolation	
	Q	FE's	Q	FE's	Q	FE's
1	0.8970	231	0.8970	99	0.8906	42
2	0.6109	231	0.6109	63	0.6098	42
3	0.4556	231	0.4556	54	0.3940	59
4	0.2975	231	0.2975	40	0.2924	41
			$\hat{\omega}_{23}$ rejected			
5	0.6329	231	0.6329	45	0.6342	45
6	0.4035	231	0.4035	44	0.4019	44
7	0.1268	231	0.1268	40	0.1293	40
8	0.0063	231	0.0063	32	0.0078	32
Total FE's		1848		417		345

TABLE 15.4
Impulse response of bandpass filter (Example 15.2)

n	$h_0(n) = h_0(-n)$	
	Exhaustive or selective search	Selective with cubic interpolation
0	4.095939×10^{-1}	4.095922×10^{-1}
1	2.529508×10^{-2}	2.529615×10^{-2}
2	-2.978313×10^{-1}	-2.978309×10^{-1}
3	-3.828648×10^{-2}	-3.828785×10^{-2}
4	7.734350×10^{-2}	7.734476×10^{-2}
5	-1.885007×10^{-3}	-1.885265×10^{-3}
6	5.491991×10^{-2}	5.491921×10^{-2}
7	3.246312×10^{-2}	3.246380×10^{-2}
8	-4.740273×10^{-2}	-4.740238×10^{-2}
9	-1.517104×10^{-2}	-1.517257×10^{-2}
10	-4.296619×10^{-3}	-4.294720×10^{-3}
11	-1.645571×10^{-2}	-1.645695×10^{-2}
12	2.126770×10^{-2}	2.126769×10^{-2}
13	2.137354×10^{-2}	2.137482×10^{-2}
14	-8.758516×10^{-3}	-8.760408×10^{-3}
15	-8.493829×10^{-3}	-8.492478×10^{-3}
16	-3.891420×10^{-5}	-3.931393×10^{-5}

15.7 PRESCRIBED SPECIFICATIONS

Given a filter length N, a set of passband and stopband edges, and a ratio δ_p/δ_a, a nonrecursive filter with approximately piecewise-constant amplitude-response specifications can be readily designed. While the filter obtained will have passband and stopband edges at the correct locations and the ratio δ_p/δ_a will be as required, the

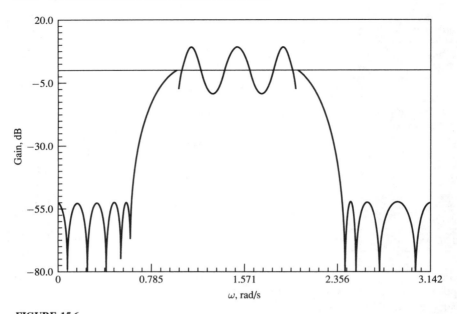

FIGURE 15.6
Amplitude response of equiripple bandpass filter (Example 15.2) (the passband gain is multiplied by the factor 20 in order to illustrate the passband ripple).

amplitudes of the passband and stopband ripples are highly unlikely to be precisely as specified. An acceptable design can be obtained by predicting the value of N on the basis of the required specifications and then designing filters for increasing or decreasing values of N until the lowest value of N that satisfies the specifications is found.

A reasonably accurate *empirical* formula for the prediction of N for the case of lowpass and highpass filters, due to Herrmann, Rabiner, and Chan [15], is

$$N = \text{Int}\left[\frac{(D - FB^2)}{B} + 1.5\right] \quad (15.51)$$

where

$$B = |\omega_a - \omega_p|/2\pi$$
$$D = [0.005309(\log \delta_p)^2 + 0.07114 \log \delta_p - 0.4761]\log \delta_a$$
$$\quad - [0.00266(\log \delta_p)^2 + 0.5941 \log \delta_p + 0.4278]$$
$$F = 0.51244(\log \delta_p - \log \delta_a) + 11.012$$

This formula can also be used to predict the filter length in the design of bandpass, bandstop, and multiband filters in general. In these filters, a value of N is computed for each transition band between a passband and stopband or a stopband and passband using Eq. (15.51) and the largest value of N so obtained is taken to be the predicted filter length. *Prescribed specifications* can be achieved by using the following design algorithm:

Algorithm 5: Design of filters satisfying prescribed specifications

1. Compute N using Eq. (15.51); if N is even, set $N = N + 1$.
2. Design a filter of length N using Algorithm 4 and determine the minimum value of δ, say $\overset{\cup}{\delta}$.
   (A) If $\overset{\cup}{\delta} > \delta_p$, then do:
      (a) Set $N = N + 2$, design a filter of length N using Algorithm 4, and find $\overset{\cup}{\delta}$;
      (b) If $\overset{\cup}{\delta} \leq \delta_p$, then go to step 3; else, go to step 2(A)(a).
   (B) If $\overset{\cup}{\delta} < \delta_p$, then do:
      (a) Set $N = N - 2$, design a filter of length N using Algorithm 4, and find $\overset{\cup}{\delta}$;
      (b) If $\overset{\cup}{\delta} > \delta_p$, then go to step 4; else, go to step 2(B)(a).
3. Use the last set of extremals and the corresponding value of N to obtain the impulse response of the required filter and stop.
4. Use the last but one set of extremals and the corresponding value of N to obtain the impulse response of the required filter and stop.

Example 15.3. In an application, a nonrecursive equiripple bandstop filter is required which should satisfy the following specifications:

Odd filter length
Maximum passband ripple A_p: 0.5 dB
Minimum stopband attenuation A_a: 50.0 dB
Lower passband edge ω_{p1}: 0.8 rad/s
Upper passband edge ω_{p2}: 2.2 rad/s
Lower stopband edge ω_{a1}: 1.2 rad/s
Upper stopband edge ω_{a2}: 1.8 rad/s
Sampling frequency ω_s: 2π rad/s

Design the lowest-order filter that will satisfy the specifications.

Solution. The use of Algorithm 4 in conjunction with Algorithm 5 gave a filter of length 35. The progress of the design is illustrated in Table 15.5. The impulse response of the filter obtained is given in Table 15.6. The corresponding amplitude response is depicted in Fig. 15.7; the passband ripple and minimum stopband attenuation achieved are 0.4342 and 51.23 dB, respectively, and are within the specified limits.

TABLE 15.5
Progress in design of bandstop filter (Example 15.3)

N	Iters.	FE's	A_p, dB	A_a, dB
31	10	582	0.5055	49.91
33	7	376	0.5037	49.94
35	9	545	0.4342	51.23

TABLE 15.6
Impulse response of bandstop filter (Example 15.3)

n	$h_0(n) = h_0(-n)$	n	$h_0(n) = h_0(-n)$
0	6.606345×10^{-1}	9	2.806340×10^{-2}
1	-2.307038×10^{-2}	10	-2.276572×10^{-2}
2	2.711461×10^{-1}	11	-9.924812×10^{-3}
3	4.306831×10^{-2}	12	-1.047638×10^{-3}
4	-1.198723×10^{-1}	13	-1.412229×10^{-2}
5	-1.829974×10^{-2}	14	1.284774×10^{-2}
6	-4.974998×10^{-3}	15	1.096745×10^{-2}
7	-2.016415×10^{-2}	16	8.260758×10^{-4}
8	4.593774×10^{-2}	17	3.482212×10^{-3}

15.8 GENERALIZATION

As was demonstrated in Chap. 9, there are four types of constant-delay nonrecursive filters. The impulse response can be *symmetrical* or *antisymmetrical*, and the filter length can be *odd* or *even*. In the preceding sections, we considered the design of filters with symmetrical impulse response and odd length. In this section, we show that the Remez algorithm can also be applied for the design of the three other types of filters.

15.8.1 Antisymmetrical Impulse Response and Odd Filter Length

Assuming that $\omega_s = 2\pi$, the frequency response of a nonrecursive filter with *antisymmetrical* impulse response and *odd* length can be expressed as

$$H(e^{j\omega T}) = e^{-jc\omega} j P'_c(\omega)$$

where

$$P'_c(\omega) = \sum_{k=1}^{c} a_k \sin k\omega \qquad (15.52)$$

$$a_k = 2h(c-k) \qquad \text{for } k = 1, 2, \ldots, c$$

$$c = (N-1)/2$$

(see Table 9.1).

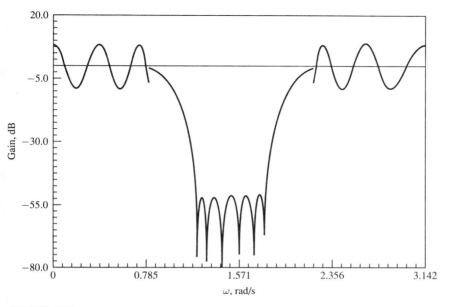

FIGURE 15.7
Amplitude response of equiripple bandstop filter (Example 15.3) (the passband gain is multiplied by the factor 40 in order to illustrate the passband ripple).

A filter with a desired frequency response $e^{-jc\omega} j D(\omega)$ can be designed by constructing the error function

$$E(\omega) = W(\omega)[D(\omega) - P'_c(\omega)] \quad (15.53)$$

and then minimizing $|E(\omega)|$ with respect to some compact subset of the frequency interval $[0, \pi]$. From Eq. (15.52), $P'_c(\omega)$ can be expressed as [6]

$$P'_c(\omega) = \sin \omega \, P_{c-1}(\omega) \quad (15.54)$$

where

$$P_{c-1}(\omega) = \sum_{k=0}^{c-1} \tilde{c}_k \cos k\omega \quad (15.55a)$$

and

$$a_1 = \tilde{c}_0 - \tfrac{1}{2}\tilde{c}_2 \quad (15.55b)$$

$$a_i = \tfrac{1}{2}(\tilde{c}_{i-1} - \tilde{c}_{i+1}) \quad \text{for } i = 2, 3, \ldots, c-2 \quad (15.55c)$$

$$a_{k-1} = \tfrac{1}{2}\tilde{c}_{k-2} \quad (15.55d)$$

$$a_k = \tfrac{1}{2}\tilde{c}_{k-1} \quad (15.55e)$$

Hence Eq. (15.53) can be put in the form

$$E(\omega) = \tilde{W}(\omega)[\tilde{D}(\omega) - \tilde{P}(\omega)] \quad (15.56)$$

where

$$\tilde{W}(\omega) = Q(\omega)W(\omega)$$
$$\tilde{D}(\omega) = D(\omega)/Q(\omega)$$
$$\tilde{P}(\omega) = P_{c-1}(\omega)$$
$$Q(\omega) = \sin \omega$$

Evidently, Eq. (15.56) is of the same form as Eq. (15.3), and on proceeding as in Sec. 15.2 one can obtain the system of equations

$$\begin{bmatrix} 1 & \cos \hat{\omega}_0 & \cos 2\hat{\omega}_0 & \cdots & \cos(c-1)\hat{\omega}_0 & \frac{1}{\tilde{W}(\hat{\omega}_0)} \\ 1 & \cos \hat{\omega}_1 & \cos 2\hat{\omega}_1 & \cdots & \cos(c-1)\hat{\omega}_1 & \frac{-1}{\tilde{W}(\hat{\omega}_1)} \\ \vdots & \vdots & \vdots & & \vdots & \vdots \\ 1 & \cos \hat{\omega}_r & \cos 2\hat{\omega}_r & \cdots & \cos(c-1)\hat{\omega}_r & \frac{(-1)^r}{\tilde{W}(\hat{\omega}_r)} \end{bmatrix} \begin{bmatrix} a_0 \\ a_1 \\ \vdots \\ a_{c-1} \\ \delta \end{bmatrix} = \begin{bmatrix} \tilde{D}(\hat{\omega}_0) \\ \tilde{D}(\hat{\omega}_1) \\ \vdots \\ \tilde{D}(\hat{\omega}_r) \end{bmatrix}$$

where $r = c$ is the number of cosine functions in $P_{c-1}(\omega)$. The above system is the same as that in Eq. (15.12) except that the number of extremals has been reduced from $c+2$ to $c+1$; therefore, the application of the Remez algorithm follows the methodology detailed in Secs. 15.2 and 15.3.

The use of Algorithm 1 or 4 yields the optimum $P_{c-1}(\omega)$ and from Eq. (15.54), the cosine function $P'_c(\omega)$ can be formed. Now $jP'_c(\omega)$ is the frequency response of a noncausal version of the required filter. The impulse response of this filter can be obtained as

$$h_0(n) = -h_0(-n) = -\frac{1}{N}\left[\sum_{k=1}^{c} 2P'_c(k\Omega) \sin\left(\frac{2\pi kn}{N}\right)\right] \quad (15.57)$$

for $n = 0, 1, 2, \ldots, c$, where $\Omega = 2\pi/N$, by using the inverse discrete Fourier transform. The impulse response of the corresponding causal filter is given by

$$h(n) = h_0(n - c)$$

for $n = 0, 1, 2, \ldots, N - 1$.

15.8.2 Even Filter Length

The frequency response of a filter with *symmetrical* impulse response and *even* length is given by

$$H(e^{j\omega T}) = e^{-jc\omega} P_d(\omega)$$

where

$$P_d(\omega) = \sum_{k=1}^{d} b_k \cos\left(k - \frac{1}{2}\right)\omega$$

$$b_k = 2h(d - k) \quad \text{for } k = 1, 2, \ldots, d$$

$$d = N/2$$

(see Table 9.1). $P_d(\omega)$ can be expressed as

$$P_d(\omega) = \cos\frac{\omega}{2} P_{d-1}(\omega)$$

where

$$P_{d-1}(\omega) = \sum_{k=0}^{d-1} \tilde{b}_k \cos k\omega \qquad (15.58a)$$

and

$$b_1 = \tilde{b}_0 + \tfrac{1}{2}\tilde{b}_1 \qquad (15.58b)$$

$$b_i = \frac{1}{2}(\tilde{b}_{i-1} + \tilde{b}_i) \qquad \text{for } i = 2, 3, \ldots, d-1 \qquad (15.58c)$$

$$b_k = \tfrac{1}{2}\tilde{b}_{k-1} \qquad (15.58d)$$

Proceeding as in the case of antisymmetrical impulse response, an error function of the form given in Eq. (15.56) can be constructed with

$$\tilde{P}(\omega) = P_{d-1}(\omega)$$

and

$$Q(\omega) = \cos\frac{\omega}{2}$$

Similarly, if the impulse response is *antisymmetrical* and the filter length is *even*, we have

$$H(e^{j\omega T}) = e^{-jc\omega} j P'_d(\omega)$$

where

$$P'_d(\omega) = \sum_{k=1}^{d} b_k \sin(k - \tfrac{1}{2})\omega$$

$$b_k = 2h(d-k) \qquad \text{for } k = 1, 2, \ldots, d$$

$$d = N/2$$

$P'_d(\omega)$ can now be expressed as

$$P'_d(\omega) = \sin\frac{\omega}{2} P_{d-1}(\omega)$$

where

$$P_{d-1}(\omega) = \sum_{k=0}^{d-1} \tilde{d}_k \cos k\omega \qquad (15.59a)$$

and

$$b_1 = \tilde{d}_0 - \tfrac{1}{2}\tilde{d}_1 \qquad (15.59b)$$

$$b_i = \tfrac{1}{2}(\tilde{d}_{i-1} - \tilde{d}_i) \qquad \text{for } i = 2, 3, \ldots, d-1 \qquad (15.59c)$$

$$b_k = \tfrac{1}{2}\tilde{d}_{k-1} \qquad (15.59d)$$

As in the previous case, an error function of the form given in Eq. (15.56) can be obtained with

$$\tilde{P}(\omega) = P_{d-1}(\omega)$$

and

$$Q(\omega) = \sin\frac{\omega}{2}$$

The various polynomials for the four types of nonrecursive filters are summarized in Table 15.7.

15.9 DIGITAL DIFFERENTIATORS

The Remez algorithm can be easily applied for the design of *equiripple digital differentiators*. The ideal frequency response of a causal differentiator is of the form $e^{-jc\omega} jD(\omega)$ where

$$D(\omega) = \omega \qquad \text{for } 0 < |\omega| < \pi \qquad (15.60)$$

and $c = (N - 1)/2$ (see Sec. 9.5). From Table 15.7, we note that differentiators can be designed in terms of filters with antisymmetrical impulse response of either odd or even length.

15.9.1 Problem Formulation

Assuming odd filter length, Eqs. (15.53) and (15.60) give the error function

$$E(\omega) = W(\omega)[\omega - P'_c(\omega)] \qquad \text{for } 0 < \omega \le \omega_p$$

where ω_p is the required bandwidth. Constant absolute or relative error may be required, depending on the application at hand. Hence $W(\omega)$ can be chosen to be either unity or $1/\omega$. In the latter case, $E(\omega)$ can be expressed as

$$E(\omega) = 1 - \frac{1}{\omega} P'_c(\omega) \qquad \text{for } 0 < \omega \le \omega_p$$

and from Eq. (15.54)

$$E(\omega) = 1 - \frac{\sin\omega}{\omega} P_{c-1}(\omega) \qquad \text{for } 0 < \omega \le \omega_p \qquad (15.61)$$

Therefore, the error function can be expressed as in Eq. (15.56) with

$$\tilde{W}(\omega) = \frac{1}{\tilde{D}(\omega)} = \frac{\sin\omega}{\omega}$$

$$\tilde{P}(\omega) = P_{c-1}(\omega)$$

TABLE 15.7
Functions $H(e^{j\omega T})$, $Q(\omega)$, and $\tilde{P}(\omega)$ for the various types of nonrecursive filters

$h(n)$	N	$H(e^{j\omega T})$	$Q(\omega)$	$\tilde{P}(\omega)$
Symmetrical	odd	$e^{-jc\omega} P_c(\omega)$	1	$P_c(\omega) = \sum_{k=0}^{c} a_k \cos k\omega$
	even	$e^{-jc\omega} P_d(\omega)$ $P_d(\omega) = \sum_{k=1}^{d} b_k \cos(k - \tfrac{1}{2})\omega$	$\cos \tfrac{\omega}{2}$	$P_{d-1}(\omega) = \sum_{k=0}^{d-1} \tilde{b}_k \cos k\omega$
Antisymmetrical	odd	$e^{-jc\omega} j P'_c(\omega)$ $P'_c(\omega) = \sum_{k=1}^{c} a_k \sin k\omega$	$\sin \omega$	$P_{c-1}(\omega) = \sum_{k=0}^{c-1} \tilde{c}_k \cos k\omega$
	even	$e^{-jc\omega} j P'_d(\omega)$ $P'_d(\omega) = \sum_{k=1}^{d} b_k \sin(k - \tfrac{1}{2})\omega$	$\sin \tfrac{\omega}{2}$	$P_{d-1}(\omega) = \sum_{k=0}^{d-1} \tilde{d}_k \cos k\omega$

$a_0 = h(c) \quad a_k = 2h(c-k) \quad c = (N-1)/2$
$b_k = 2h(d-k) \quad d = N/2$

15.9.2 First Derivative

In Algorithm 4, the first derivative of $|E(\omega)|$ with respect to ω is required. From Eq. (15.61), one can show that

$$\frac{d|E(\omega)|}{d\omega} = \text{sgn}\left[1 - \frac{\sin\omega}{\omega}P_{c-1}(\omega)\right] \times \left[\frac{\sin\omega - \omega\cos\omega}{\omega^2}P_{c-1}(\omega) - \frac{\sin\omega}{\omega}\frac{dP_{c-1}(\omega)}{d\omega}\right]$$
(15.62)

The first derivative of $P_{c-1}(\omega)$ can be computed by using the formulas in Sec. 15.6, except that the number of extremals is reduced from $c+2$ to $c+1$. The value of $P_{c-1}(\omega)$ can be computed by using Eq. (15.35) with c replaced by $c-1$. If $\hat{\omega}_i$ is an extremal, then Eq. (15.61) yields

$$P_{c-1}(\hat{\omega}_i) = [1 - (-1)^i \delta]\frac{\hat{\omega}_i}{\sin\hat{\omega}_i}$$

since $E(\hat{\omega}_i) = (-1)^i \delta$.

In Algorithm 4, the second derivative of $|E(\omega)|$ with respect to ω is used to determine whether there is a maximum or minimum at $\omega = 0$. For differentiators, this information is more easily determined by computing the quantity

$$G_2' = |E(w_1)| - |E(0)|$$

where w_1 is the interval between successive samples. Depending on whether G_2' is positive or negative, $|E(\omega)|$ has a minimum or maximum at $\omega = 0$.

15.9.3 Prescribed Specifications

A digital differentiator is fully specified by the constraint

$$|E(\omega)| \leq \delta_p \quad \text{for } 0 < \omega \leq \omega_p$$

where δ_p is the maximum passband error and ω_p is the bandwidth of the differentiator.

The differentiator length N that will just satisfy the required specifications is not normally known a priori and, although it may be determined on a hit and miss basis, a large number of designs may need to be carried out. In filters with approximately piecewise-constant amplitude responses, N can be predicted using the empirical formula of Eq. (15.51). In the case of differentiators, N can be predicted by noting a useful property of digital differentiators. If δ and δ_1 are the maximum passband errors in differentiators of lengths N and N_1, respectively, then the quantity $\ln(\delta/\delta_1)$ is *approximately linear* with respect to $N - N_1$ for a wide range of values of N_1 and ω_p, as illustrated in Fig. 15.8. Assuming linearity, we can show that [16]

$$N = N_1 + \frac{\ln(\delta/\delta_1)}{\ln(\delta_2/\delta_1)}(N_2 - N_1) \qquad (15.63)$$

where δ_2 is the maximum passband error in a differentiator of length N_2.

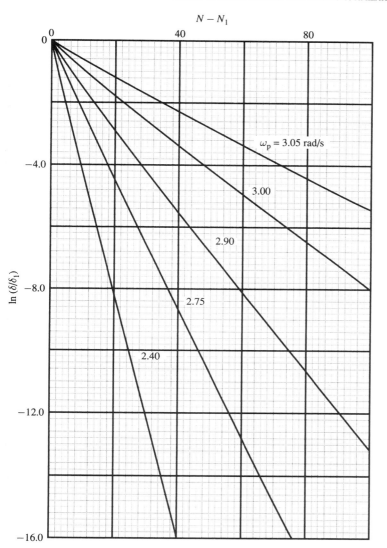

FIGURE 15.8
Variation of $\ln(\delta/\delta_1)$ versus $N - N_1$ for different values of ω_p and $N_1 = 11$.

By designing two low-order differentiators, a fairly accurate prediction of the required value of N can be obtained by using Eq. (15.63). A design algorithm based on this formula is as follows:

Algorithm 6: Design of digital differentiators satisfying prescribed specifications
1. Design a differentiator of length N_1 and find δ_1.
2. Design a differentiator of length $N_2 = N_1 + 2$ and find δ_2.

3. If $\delta_2 \leq \delta_p < \delta_1$, go to step 7.
4. Set $\delta = \delta_p$ and compute N using Eq. (15.63); set $N_3 = \text{Int}(N + 0.5)$; if N_3 is even and a differentiator of odd length is required, then set $N_3 = N_3 + 1$.
5. Design a differentiator of length N_3 and find δ_3.
   (A) If $\delta_3 > \delta_p$, then do:
      (a) Set $N_3 = N_3 + 2$, design a differentiator of length N_3, and find δ_3;
      (b) If $\delta_3 \leq \delta_p$, then go to step 6; else, go to step $5(A)(a)$.
   (B) If $\delta_3 < \delta_p$, then do:
      (a) Set $N_3 = N_3 - 2$, design a differentiator of length N_3, and find δ_3;
      (b) If $\delta_3 > \delta_p$, then go to step 7; else, go to step $5(B)(a)$.
6. Use the last set of extremals and the corresponding value of N to obtain the impulse response of the required differentiator and stop.
7. Use the last but one set of extremals and the corresponding value of N to obtain the impulse response of the required differentiator and stop.

Example 15.4. In an application, a digital differentiator is required which should satisfy the following specifications:

Odd differentiator length
Bandwidth ω_p: 2.5 rad/s
Maximum passband ripple δ_p: 1.0×10^{-6}
Sampling frequency ω_s: 2π rad/s

Design the lowest-order differentiator that will satisfy the specifications.

Solution. The design was carried out using Algorithm 6 in conjunction with Algorithm 4; in Algorithm 4 the relative error of Eq. (15.61) was minimized. The progress of the design is illustrated in Table 15.8. First, differentiators of lengths 21 and 23 were designed and the required N to satisfy the specifications was predicted to be 43 using Eq. (15.63). This differentiator length was found to oversatisfy the specifications, and designs for lengths 41 and 39 were then carried out. The design for $N = 39$ violates the specifications, as can be seen in Table 15.8; therefore, the optimum differentiator length is 41. The impulse response of this differentiator is given in Table 15.9. The amplitude response and passband relative error of the differentiator are plotted in Fig. 15.9a and b.

TABLE 15.8
Progress in design of digital differentiator (Example 15.4)

N	Iters.	FE's	δ_p
21	4	141	7.649×10^{-4}
23	5	187	3.786×10^{-4}
43	5	616	4.078×10^{-7}
41	6	538	8.069×10^{-7}
39	6	500	1.582×10^{-6}

TABLE 15.9
Impulse response of digital differentiator (Example 15.4)

n	$h_0(n) = -h_0(-n)$	n	$h_0(n) = -h_0(-n)$
0	0.0	11	-1.305326×10^{-2}
1	-9.852395×10^{-1}	12	7.955151×10^{-3}
2	4.710789×10^{-1}	13	-4.626299×10^{-3}
3	-2.914014×10^{-1}	14	2.544983×10^{-3}
4	1.966634×10^{-1}	15	-1.309224×10^{-3}
5	-1.371947×10^{-1}	16	6.197315×10^{-4}
6	9.651420×10^{-2}	17	-2.633737×10^{-4}
7	-6.751749×10^{-2}	18	9.638584×10^{-5}
8	4.653727×10^{-2}	19	-2.795288×10^{-5}
9	-3.138375×10^{-2}	20	4.916591×10^{-6}
10	2.058332×10^{-2}		

15.10 ARBITRARY AMPLITUDE RESPONSES

Very frequently nonrecursive filters are required whose amplitude responses cannot be described by analytical functions. For example, in the design of two-dimensional filters (see Sec. 16.6) through the singular-value decomposition [17, 18], the required two-dimensional filter is obtained by designing a set of one-dimensional digital filters whose amplitude responses turn out to have arbitrary shapes. In these applications, the desired amplitude response $D(\omega)$ is specified in terms of a table that lists a prescribed set of frequencies and the corresponding values of the required filter gain. Filters of this class can be readily designed by employing some interpolation scheme which can be used to evaluate $D(\omega)$ and its first derivative with respect to ω at any ω. A suitable scheme is to fit a set of third-order polynomials to the prescribed amplitude response. An interpolation scheme of this type has been used in the design of recursive filters in the previous chapter and is described in detail in Sec. 14.6.

15.11 MULTIBAND FILTERS

The algorithms presented in the previous sections can also be used to design *multiband* filters. While there is no theoretical upper limit on the number of bands, in practice, the design tends to become more and more difficult as the number of bands is increased. The reason is that the difference between the number of possible maxima in the error function and the number of extremals increases linearly with the number of bands, e.g., if the number of bands is 8, then the difference is 14 (see Sec. 15.3.4). As a consequence, the number of potential extremals that need to be rejected is large and the available rejection techniques become somewhat inefficient. The end result is that the number of iterations is increased quite significantly, and convergence is slow and sometimes impossible.

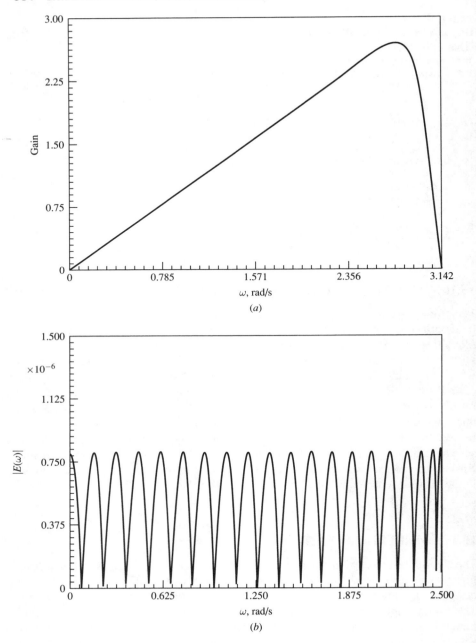

FIGURE 15.9
Design of digital differentiator (Example 15.4): (a) amplitude response, (b) passband relative error.

In mathematical terms, the above difficulty is attributed to the fact that, in the weighted-Chebyshev methods considered in this chapter, the approximating polynomial becomes seriously *underdetermined* if the number of bands exceeds three. The problem can be overcome by using the generalized Remez method described in [14]. This method is based on a different formulation of the design problem and leads to three types of equiripple filters, namely, *maximal-ripple*, *extra-ripple*, and *weighted-Chebyshev* filters. In the case of maximal-ripple filters, the approximating polynomial is fully determined; in the extra-ripple case, it is less underdetermined than the approximating polynomial in the methods described. Therefore, for filters with more than five bands, the method in [14] is preferred.

Example 15.5. In an application, a nonrecursive equiripple 5-band filter is required which should satisfy the specifications in Table 15.10. The sampling frequency is 2π. Design the lowest-order filter that will satisfy the specifications.

Solution. The use of Algorithm 4 in conjunction with Algorithm 5 gave a filter of length 61. The progress of the design is illustrated in Table 15.11. The impulse response of the filter obtained is given in Table 15.12, and the corresponding amplitude response is plotted in Fig. 15.10. As can be seen, the required specifications are satisfied.

The required filter order for multiband filters can be predicted by using the formula in Eq. (15.51), as was stated earlier. A generalized version of this formula which gives improved results can be found in [14].

TABLE 15.10
Specifications of 5-band filter (Example 15.5)

Band:	1	2	3	4	5
$D(\omega)$	1.00	0.00	1.00	0.00	1.00
A_p, dB	0.50	—	0.75	—	1.00
A_a, dB	—	50.00	—	30.00	—
ω_L, rad/s	0.00	0.80	1.50	2.10	2.80
ω_R, rad/s	0.60	1.25	1.90	2.60	π

TABLE 15.11
Progress in design of 5-band filter (Example 15.5)

N	Iters.	FE's	A_{p1}, dB	A_{a2}, dB	A_{p3}, dB	A_{a4}, dB	A_{p5}, dB
61	9	913	0.453	50.46	0.679	30.86	0.905
59	19	2219	0.539	49.35	0.808	29.35	1.077

TABLE 15.12
Impulse response of 5-band filter (Example 15.5)

n	$h_0(n) = h_0(-n)$	n	$h_0(n) = h_0(-n)$
0	5.608208×10^{-1}	16	-8.164458×10^{-4}
1	4.013174×10^{-2}	17	-3.884179×10^{-4}
2	1.006767×10^{-1}	18	2.625242×10^{-3}
3	4.198731×10^{-2}	19	-1.130791×10^{-2}
4	2.414087×10^{-1}	20	9.190432×10^{-3}
5	-1.248415×10^{-1}	21	8.761118×10^{-3}
6	-1.019101×10^{-1}	22	6.476604×10^{-3}
7	6.608448×10^{-3}	23	9.610168×10^{-3}
8	-1.355327×10^{-2}	24	-1.976094×10^{-2}
9	4.780217×10^{-3}	25	-1.075689×10^{-2}
10	-1.549769×10^{-2}	26	3.013727×10^{-3}
11	3.468520×10^{-2}	27	-2.707701×10^{-3}
12	-8.299265×10^{-4}	28	-2.549441×10^{-3}
13	4.694733×10^{-2}	29	-9.605488×10^{-3}
14	2.641761×10^{-3}	30	1.495353×10^{-2}
15	-5.336269×10^{-2}	—	—

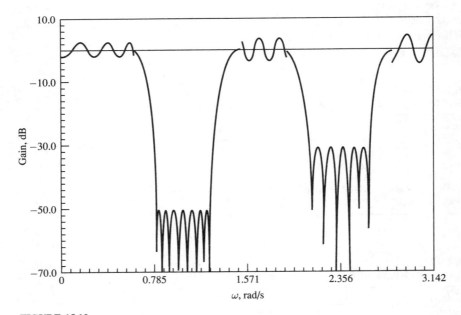

FIGURE 15.10
Amplitude response of equiripple 5-band filter (Example 15.5) (the passband gain is multiplied by the factor 10 in order to illustrate the passband ripple).

REFERENCES

1. O. Herrmann, "Design of Nonrecursive Digital Filters with Linear Phase," *Electron. Lett.*, vol. 6, pp. 182–184, May 1970.
2. E. Hofstetter, A. Oppenheim, and J. Siegel, "A New Technique for the Design of Non-Recursive Digital Filters," *5th Annual Princeton Conf. Information Sciences and Systems*, pp. 64–72, March 1971.
3. T. W. Parks and J. H. McClellan, "Chebyshev Approximation for Nonrecursive Digital Filters with Linear Phase," *IEEE Trans. Circuit Theory*, vol. CT-19, pp. 189–194, March 1972.
4. T. W. Parks and J. H. McClellan, "A Program for the Design of Linear Phase Finite Impulse Response Digital Filters," *IEEE Trans. Audio Electroacoust.*, vol. AU-20, pp. 195–199, August 1972.
5. L. R. Rabiner and O. Herrmann, "On the Design of Optimum FIR Low-Pass Filters with Even Impulse Response Duration," *IEEE Trans. Audio Electroacoust.*, vol. AU-21, pp. 329–336, August 1973.
6. J. H. McClellan and T. W. Parks, "A Unified Approach to the Design of Optimum FIR Linear-Phase Digital Filters," *IEEE Trans. Circuit Theory*, vol. CT-20, pp. 697–701, November 1973.
7. J. H. McClellan, T. W. Parks, and L. R. Rabiner, "A Computer Program for Designing Optimum FIR Linear Phase Digital Filters," *IEEE Trans. Audio Electroacoust.*, vol. AU-21, pp. 506–526, December 1973.
8. L. R. Rabiner, J. H. McClellan, and T. W. Parks, "FIR Digital Filter Design Techniques Using Weighted Chebyshev Approximation," *Proc. IEEE*, vol. 63, pp. 595–610, April 1975.
9. J. H. McClellan, T. W. Parks, and L. R. Rabiner, "FIR Linear Phase Filter Design Program," *Programs for Digital Signal Processing*, IEEE Press, New York, pp. 5.1-1–5.1-13, 1979.
10. A. Antoniou, "Accelerated Procedure for the Design of Equiripple Nonrecursive Digital Filters," *IEE Proc.*, vol. 129, pt. G, pp. 1–10, February 1982 (see p. 107, June 1982 for errata).
11. A. Antoniou, "New Improved Method for the Design of Weighted-Chebyshev, Nonrecursive, Digital Filters," *IEEE Trans. Circuits Syst.*, vol. CAS-30, pp. 740–750, October 1983.
12. E. W. Cheney, *Introduction to Approximation Theory*, McGraw-Hill, 1966, pp. 72–100.
13. E. Ya. Remes, *General Computational Methods for Tchebycheff Approximation*, Kiev, 1957 (Atomic Energy Commission Translation 4491, pp. 1–85).
14. D. J. Shpak and A. Antoniou, "A Generalized Reméz Method for the Design of FIR Digital Filters," *IEEE Trans. Circuits Syst.*, vol. CAS-37, pp. 161–174, February 1990.
15. O. Herrmann, L. R. Rabiner, and D. S. K. Chan, "Practical Design Rules for Optimum Finite Impulse Response Low-Pass Digital Filters," *Bell Syst. Tech. J.*, vol. 52, pp. 769–799, July-August 1973.
16. A. Antoniou and C. Charalambous, "Improved Design Method for Kaiser Differentiators and Comparison with Equiripple Method," *IEE Proc.*, vol. 128, pt. E, pp. 190–196, September 1981.
17. A. Antoniou and W.-S. Lu, "Design of Two-Dimensional Digital Filters by Using the Singular Value Decomposition," *IEEE Trans. Circuits Syst.*, vol. CAS-34, pp. 1191–1198, October 1987.
18. W.-S. Lu, H.-P. Wang, and A. Antoniou, "Design of Two-Dimensional FIR Digital Filters Using the Singular-Value Decomposition," *IEEE Trans. Circuits Syst.*, vol. CAS-37, pp. 35–46, January 1990.

ADDITIONAL REFERENCES

Bonzanigo, F.: "Some Improvements to the Design Programs for Equiripple FIR Filters," *Proc. Int. Conf. Acoust., Speech, Signal Process.*, pp. 274–277, 1982.

Grenez, F.: "Design of Linear or Minimum-Phase FIR Filters by Constrained Chebyshev Approximation," *Signal Processing*, vol. 5, pp. 325–332, 1983.

Herrmann, O.: "Design of Nonrecursive Digital Filters with Linear Phase," *Electron. Lett.*, vol. 6, pp. 328–329, May 1970.

Rabiner, L. R., J. F. Kaiser, and R. W. Schafer: "Some Considerations in the Design of Multiband Finite-Impulse-Response Digital Filters," *IEEE Trans. Acoust., Speech, Signal Process.*, vol. ASSP-22, pp. 462–472, December 1974.

———, and R. W. Schafer: "On the Behavior of Minimax Relative Error FIR Digital Differentiators," *Bell Syst. Tech. J.*, vol. 53, pp. 333–361, February 1974.

Sunder S., W.-S. Lu, A. Antoniou, and Y. Su: "Design of Digital Differentiators Satisfying Prescribed Specifications Using Optimisation Techniques," *IEE Proc.*, vol. 138, pt. G, pp. 315–320, June 1991 (see p. 756, December 1991 for errata).

Vaidyanathan, P. P.: "Optimal Design of Linear-Phase FIR Digital Filters with Very Flat Passbands and Equiripple Stopbands," *IEEE Trans. Circuits Syst.*, vol. CAS-32, pp. 904–917, September 1985.

PROBLEMS

15.1. A noncausal nonrecursive filter has a frequency response $P_c(\omega)$. The filter has a symmetrical impulse response represented by $h_0(n)$ for $-c \leq n \leq c$, where $c = (N-1)/2$. Using the inverse discrete Fourier transform, show that the impulse response of the filter is given by Eq. (15.21).

15.2. Show that δ and $P_c(\omega)$ given by Eqs. (15.16) and (15.17) can be expressed as in Eqs. (15.34) and (15.35), respectively.

15.3. Show that coefficients β_{km}, α_{km}, and C_{km}, which are used to compute δ and $P_c(\omega)$, can be expressed as in Eqs. (15.36) to (15.38).

15.4. Write a computer program based on the Remez algorithm (Algorithm 1) that can be used for the design of filters. Use the exhaustive step-by-step search method in Algorithm 2 in conjunction with the scheme in Algorithm 3 for the rejection of superfluous potential extremals. Then use a routine that will reject the $\rho - r$ superfluous potential extremals $\overset{\frown}{\omega}_i$ on the basis of the lowest error $|E(\overset{\frown}{\omega}_i)|$ (see Sec. 15.3.4) as an alternative rejection scheme and check whether there is a change in the computational efficiency of the program.

15.5. Show that for any frequency including the last extremal of the last band but excluding all other extremals, the first derivative of $P_c(\omega)$ with respect to ω is given by the formula in Eq. (15.46).

15.6. Show that for all extremals other than the last extremal of the last band, the first derivative of $P_c(\omega)$ with respect to ω is given by the formula in Eq. (15.47).

15.7. Show that the first derivative of $P_c(\omega)$ with respect to ω is zero at $\omega = 0$ and $\omega = \pi$ [see Eq. (15.48)]. Hence show that $|E(\omega)|$ has a local maximum or minimum at these frequencies.

15.8. Show that for $\omega = 0$ if no extremal occurs at zero or for $\omega = \pi$ under all circumstances, the second derivative of $P_c(\omega)$ with respect to ω is given by Eq. (15.49).

15.9. Show that if there is an extremal at $\omega = 0$, then the second derivative of $P_c(\omega)$ with respect to ω at $\omega = 0$ is given by Eq. (15.50).

15.10. Write a computer program based on the Remez algorithm that can be used for the design of filters. Use the selective step-by-step search method of Sec. 15.4.1.

15.11. The cubic interpolation search of Sec. 15.4.2 requires the evaluation of constants d, c, b, β, γ, θ, and ψ given by Eqs. (15.25) to (15.31). Derive the formulas for these constants.

15.12. Modify the program of Prob. 15.10 to include the cubic interpolation search of Sec. 15.4.2 (see Algorithm 4).

15.13. Design a nonrecursive equiripple lowpass filter using the Remez algorithm (*a*) with the exhaustive search of Sec. 15.3.2, (*b*) with the selective step-by-step search of Sec. 15.4.1, and (*c*) with the selective step-by-step search in conjunction with the cubic interpolation search of Sec. 15.4.2. Compare the results obtained. The required specifications are as follows:

> Filter length N: 21
> Passband edge ω_p: 1.0 rad/s
> Stopband edge ω_a: 1.5 rad/s
> Ratio δ_p/δ_a: 18.0
> Sampling frequency ω_s: 2π rad/s

15.14. Design a nonrecursive equiripple bandstop filter using the Remez algorithm (*a*) with the exhaustive search, (*b*) with the selective step-by-step search, and (*c*) with the selective step-by-step search in conjunction with the cubic interpolation search. Compare the results obtained. The required specifications are as follows:

> Filter length N: 33
> Lower passband edge ω_{p1}: 0.8 rad/s
> Upper passband edge ω_{p2}: 2.1 rad/s
> Lower stopband edge ω_{a1}: 1.2 rad/s
> Upper stopband edge ω_{a2}: 1.8 rad/s
> Ratio δ_p/δ_a: 23.0
> Sampling frequency ω_s: 2π rad/s

15.15. Modify the program in Prob. 15.10 to include an option for the design of filters satisfying prescribed specifications. Use Algorithm 5.

15.16. In an application, a nonrecursive equiripple highpass filter is required which should satisfy the following specifications:

> Odd filter length
> Maximum passband ripple A_p: 0.1 dB
> Minimum stopband attenuation A_a: 50.0 dB
> Lower passband edge ω_p: 1.8 rad/s
> Lower stopband edge ω_a: 1.0 rad/s
> Sampling frequency ω_s: 2π rad/s

Design the lowest-order filter that will satisfy the specifications.

15.17. In an application, a nonrecursive equiripple bandpass filter is required which should satisfy the following specifications:

> Odd filter length
> Maximum passband ripple A_p: 0.1 dB
> Minimum stopband attenuation A_a: 60.0 dB
> Lower passband edge ω_{p1}: 1.0 rad/s
> Upper passband edge ω_{p2}: 1.6 rad/s

Lower stopband edge ω_{a1}: 0.6 rad/s
Upper stopband edge ω_{a2}: 2.0 rad/s
Sampling frequency ω_s: 2π rad/s

Design the lowest-order filter that will satisfy the specifications.

15.18. Show that the sine polynomial $P'_c(\omega)$ of Eq. (15.52) can be expressed as in Eq. (15.54) where $P_{c-1}(\omega)$ is given by Eq. (15.55a).

15.19. A noncausal nonrecursive filter has a frequency response $jP_c(\omega)$. The filter has an antisymmetrical impulse response represented by $h_0(n)$ for $-c \leq n \leq c$ where $c = (N-1)/2$. Using the inverse discrete Fourier transform, show that the impulse response of the filter is given by Eq. (15.57).

15.20. The relative error in the design of digital differentiators is given by Eq. (15.61). Show that the first derivative of $|E(\omega)|$ with respect to ω is given by Eq. (15.62).

15.21. Write a computer program based on the Remez algorithm that can be used for the design of digital differentiators. Use the selective step-by-step search method in conjunction with the cubic interpolation search.

15.22. Using the program in Prob. 15.21, design a digital differentiator of length $N = 41$ and bandwidth $\omega_p = 3.0$ rad/s. The sampling frequency is 2π rad/s.

15.23. If δ and δ_1 are the maximum passband errors in digital differentiators of lengths N and N_1, respectively, then the quantity $\ln(\delta/\delta_1)$ is approximately linear with respect to $N - N_1$, as can be seen in Fig. 15.8. Assuming linearity, derive the prediction formula of Eq. (15.63).

15.24. Modify the program in Prob. 15.21 to include an option for the design of digital differentiators satisfying prescribed specifications. Use Algorithm 6.

15.25. In an application, a digital differentiator is required which should satisfy the following specifications:

Odd differentiator length
Bandwidth ω_p: 2.75 rad/s
Maximum passband ripple δ_p: 1.0×10^{-4}
Sampling frequency ω_s: 2π rad/s

Design the lowest-order differentiator that will satisfy the specifications.

15.26. In an application, a nonrecursive equiripple 4-band filter is required which should satisfy the specifications in Table P15.26. The sampling frequency is 2π. Design the lowest-order filter that will satisfy the specifications.

15.27. In an application, a nonrecursive equiripple 5-band filter is required which should satisfy the specifications in Table P15.27. The sampling frequency is $\omega_s = 2\pi$. Design the lowest-order filter that will satisfy the specifications.

TABLE P15.26

Band:	1	2	3	4
$D(\omega)$	0.0	1.0	0.0	1.0
A_p, dB	—	0.1	—	0.4
A_a, dB	50.0	—	55.0	—
ω_L, rad/s	0.0	1.2	2.0	2.8
ω_R, rad/s	0.8	1.6	2.4	π

TABLE P15.27

Band:	1	2	3	4	5
$D(\omega)$	1.0	0.0	1.0	0.0	1.0
A_p, dB	0.8	—	0.4	—	1.0
A_a, dB	—	50.0	—	30.0	—
ω_L, rad/s	0.0	0.8	1.6	2.2	2.9
ω_R, rad/s	0.4	1.2	1.9	2.6	π

CHAPTER 16

DIGITAL SIGNAL PROCESSING APPLICATIONS

16.1 INTRODUCTION

In the past several chapters, some sophisticated design methods have been described in detail. In this chapter, we consider a number of extensions of these methods and some of their numerous applications to digital signal processing.

The chapter begins with the underlying principles involved when the sampling frequency is changed from one value to another, and devices that can be used for the conversion, known as *decimators* and *interpolators*, are described [1]. The application of decimators and interpolators in the design of *filter banks* is then considered. These subsystems find widespread applications in communication systems, spectrum analyzers, and speech synthesis. The principles involved are examined and a specific type of filter bank, the so-called *quadrature mirror-image filter bank*, is examined in some detail. Another useful device for communications applications is the *Hilbert transformer*. Its theoretical basis, design principles, and applications to the sampling of bandpassed signals and to single-sideband modulation are described in Sec. 16.4.

Another topic that is receiving considerable attention these days is the design and applications of *adaptive filters*. These are filters that have an adaptation mechanism which allows them to change their characteristics with time in order to satisfy some performance requirement. The topic is a field in its own right, and several textbooks have been written on the theory and design of these filters [2–4]. In Sec. 16.5, we examine some of the fundamentals involved and consider some of the algorithms that are being used as adaptation algorithms.

In many applications, the signal to be processed is a function of two discrete variables. Signals of this type can be processed by *two-dimensional* (2-D) digital filters. The methods involved are usually extensions of their 1-D counterparts, but once in a while there are marked differences. As in the case of adaptive filters, the topic of 2-D digital filters is emerging as an important field of study, and some textbooks have already been written on the subject [5–7]. For the sake of completeness, we conclude this chapter by examining some of the basic principles involved in the analysis and design of these filters.

16.2 SAMPLING-FREQUENCY CONVERSION

For various technical reasons, it may be preferable to store or record a discrete-time signal using one sampling frequency but process or transmit it using another sampling frequency. The conversion of the sampling frequency from one value to another can be accomplished in an indirect way by converting a discrete-time signal back to a continuous-time one and then sampling the latter at a different rate using the techniques described in Sec. 6.7. However, each such conversion introduces quantization noise and $\sin \omega / \omega$ distortion, as well as other problems. An alternative and more satisfactory approach is to carry out the conversion in the discrete-time domain using a class of digital filters known as *decimators* and *interpolators*.

Decimators can be used to reduce the sampling frequency, whereas interpolators can be used to increase it. This section deals with the underlying principles involved in connection with sampling-frequency conversion in the discrete-time domain and highlights some design aspects concerning decimators and interpolators.

16.2.1 Decimators

Let us first examine the situation where the sampling frequency of a discrete-time signal $x(nT)$ is to be reduced.

If $x(nT)$ is deemed to have been obtained by sampling a continuous-time signal $x(t)$ using a sampling frequency ω_s, then the frequency spectrum of $x(nT)$ can be readily obtained from Eq. (6.36) as[†]

$$X(e^{j\omega T}) = \frac{1}{T} \sum_{n=-\infty}^{\infty} X_c[j(\omega - n\omega_s)] \qquad (16.1)$$

where

$$X_c(j\omega) = \mathcal{F}x(t) \quad \text{and} \quad X(z) = \mathcal{Z}x(nT)$$

[†] The notation has been changed somewhat for the sake of convenience.

Let the new sampling frequency be $\omega'_s = \omega_s/M$, where M is an integer greater than unity. Since the sampling period is increased from T to $T' = MT$, a signal $x_d(nT')$ is obtained which is related to $x(nT)$ by the equation

$$x_d(nT') \equiv x(nT') = x(nMT)$$

that is, $x_d(nT')$ is obtained by retaining only samples $\ldots, -2M, -M, 0, M, 2M, \ldots$ of discrete-time signal $x(nT)$ as shown in Fig. 16.1 for the case where $M = 3$. This process has been referred to in the past as *downsampling, sampling-frequency compression*, or *discrete-time sampling* [1]. A device that can perform this operation is said to be a *downsampler, sampling-frequency compressor*, or simply *compressor* and is often represented by the symbol depicted in Fig. 16.2a.

By analogy with Eq. (16.1), the spectrum of signal $x_d(nT')$ can be readily expressed as

$$X_d(e^{j\omega T'}) = \frac{1}{T'} \sum_{n=-\infty}^{\infty} X_c[j(\omega - n\omega'_s)]$$

$$= \frac{1}{MT} \sum_{n=-\infty}^{\infty} X_c\left[j(\omega - n\frac{\omega_s}{M})\right] \quad (16.2)$$

i.e., *the spectrum of $x_d(nT')$ can be regarded as an infinite sum of shifted copies of the spectrum of $x(t)$ divided by MT*, where each copy is shifted by a multiple of ω_s/M. Alternatively, Eq. (16.2) can be expressed as

$$X_d(e^{j\omega T'}) = \frac{1}{MT} \sum_{n=-\infty}^{\infty} \sum_{m=0}^{M-1} X_c\left[j\left(\omega - n\omega_s - m\frac{\omega_s}{M}\right)\right]$$

since $X_d(e^{j\omega T'})$ can be considered to be periodic with period ω_s; hence

$$X_d(e^{j\omega T'}) = \frac{1}{M} \sum_{m=0}^{M-1} \left\{ \frac{1}{T} \sum_{n=-\infty}^{\infty} X_c\left[j\left(\omega - n\omega_s - m\frac{\omega_s}{M}\right)\right] \right\}$$

$$= \frac{1}{M} \sum_{m=0}^{M-1} X(e^{j(\omega - m\omega_s/M)T}) \quad (16.3)$$

According to this representation, *the spectrum of $x_d(nT)$ may be deemed to be the sum of M copies of the spectrum of $x(nT)$ divided by M*, where each copy is shifted by a multiple of ω_s/M.

The above conversion process is illustrated in Fig. 16.3 for the case where $M = 3$ and $X_c(j\omega)$ is real. As can be seen, if $X_c(j\omega) = 0$ for $|\omega| \geq \omega_s/2M$, then the copies of the spectrum of $x(nT)$ produced by downsampling do not overlap, and hence $x(t)$ can be recovered by using an ideal lowpass filter with cutoff frequency $\omega_s/2M$. If the aforementioned requirement is violated, then aliasing will be introduced as detailed in Sec. 6.7.1. To avoid this possibility, a bandlimiting filter with cutoff frequency $\omega_c = \omega_s/2M$ is usually used at the input of the downsampler, as depicted in Fig. 16.2b. The configuration obtained is said to be a *decimator*.

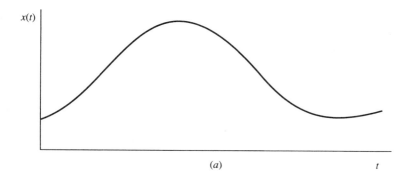
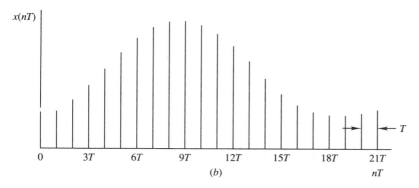
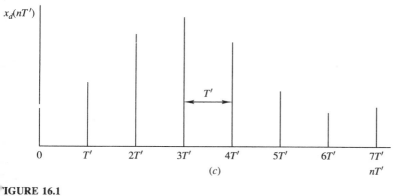

FIGURE 16.1
The process of downsampling.

16.2.2 Interpolators

The preceding section has shown that sampling-frequency reduction can be achieved by applying downsampling which is analogous to the sampling of a continuous-time signal. We now show that a sampling-frequency increase can be achieved through a

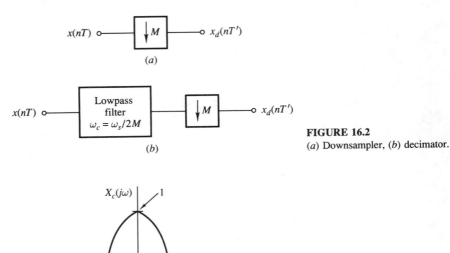

FIGURE 16.2
(a) Downsampler, (b) decimator.

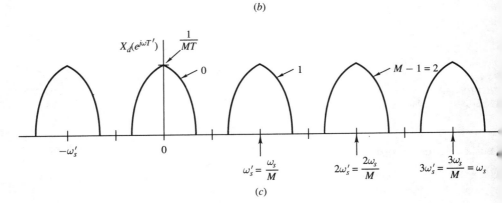

FIGURE 16.3
Operation of downsampler.

process which is analogous to the process of converting a discrete-time signal into a continuous-time signal.

A sampling-frequency increase from ω_s to $L\omega_s$, where L is an integer, can be achieved by using a device known as an *upsampler*, *sampling-frequency expander*, or simply *expander* in conjunction with a lowpass filter [1]. An upsampler is usually represented by the symbol in Fig. 16.4a, and its response to an excitation $x(nT)$, designated as $x_u(nT)$, can be expressed as

$$x_u(nT') = \begin{cases} x(nT/L) & \text{for } n = 0, \pm L, \pm 2L, \ldots \\ 0 & \text{otherwise} \end{cases} \quad (16.4)$$

where $T' = T/L$. This process is illustrated in Fig. 16.5 for the case where $L = 3$. Eq. (16.4) can be written as

$$x_u(nT') = \sum_{k=-\infty}^{\infty} x(kT)\delta(nT - kLT)$$

and by applying the z transform, we obtain

$$X_u(e^{j\omega T'}) = \sum_{n=-\infty}^{\infty} \left[\sum_{k=-\infty}^{\infty} x(kT)\delta(nT - kLT) \right] e^{-j\omega nT'}$$

$$= \sum_{k=-\infty}^{\infty} \sum_{n=-\infty}^{\infty} x(kT)[\delta(nT - kLT)e^{-j\omega nT/L}]$$

$$= \sum_{k=-\infty}^{\infty} x(kT)e^{-j\omega kT} = X(e^{j\omega T}) \quad (16.5)$$

In effect, *the frequency spectra of $x_u(nT')$ and $x(nT)$ are identical*, as illustrated in Figs. 16.6a and b for the case where $L = 3$. Since $T' = T/L$, or $\omega'_s = L\omega_s$, there are L images of the signal in the baseband $-\omega'_s/2 \le \omega \le \omega'_s/2$, i.e., upsampling will simply change the location of the sampling frequency as shown.

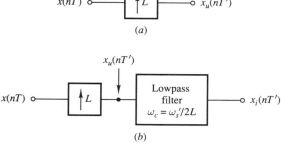

FIGURE 16.4
(a) Upsampler, (b) interpolator.

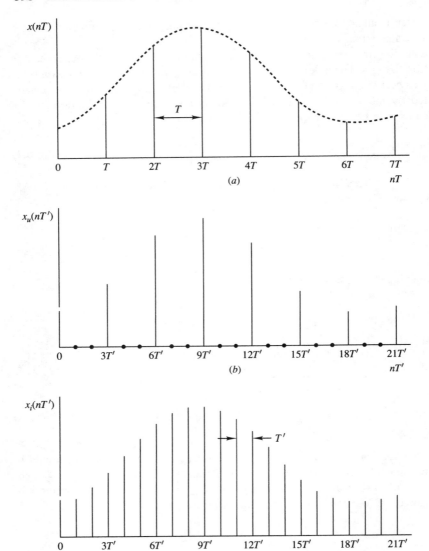

FIGURE 16.5
The process of upsampling.

A discrete-time version of $x(nT)$, sampled at the increased rate of $\omega'_s = L\omega_s$ can now be generated by filtering signal $x_u(nT')$ using an ideal lowpass digital filter with a frequency response

$$H(e^{j\omega T'}) = \begin{cases} 1 & \text{for } 0 < |\omega| < \omega_c \\ 0 & \text{otherwise} \end{cases}$$

DIGITAL SIGNAL PROCESSING APPLICATIONS **599**

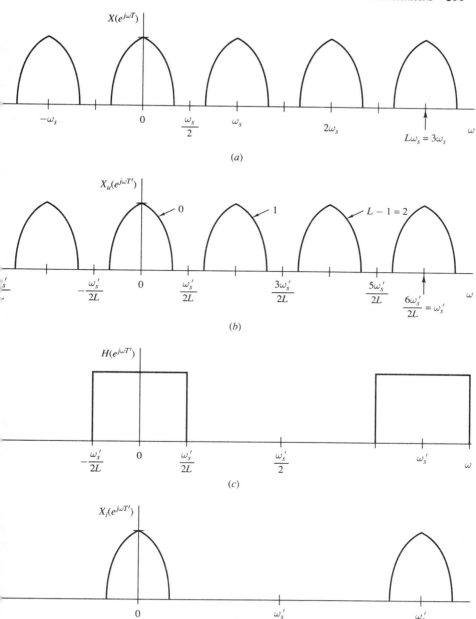

FIGURE 16.6
Operation of upsampler.

where $\omega_c = \omega_s'/2L$, as illustrated in Fig. 16.4b. This filter will reject unnecessary images and yield a discrete-time signal $x_i(nT') = x(nT')$, as illustrated in Fig. 16.6c and d. The configuration of Fig. 16.4b is said to be an *interpolator*.

In order to demonstrate that the output in Fig. 16.4b is indeed an interpolated version of $x(nT)$, we can write

$$X_i(z) = H(z)X_u(z) \quad (16.6)$$

where

$$X_u(z) = \mathcal{Z} x_u(nT')$$

$$= \left. \sum_{k=-\infty}^{\infty} x(kT)e^{-j\omega kLT'} \right|_{e^{j\omega T'}=z}$$

$$= \sum_{k=-\infty}^{\infty} x(kT)z^{-kL}$$

according to Eq. (16.5), since $T = LT'$; hence Eq. (16.6) yields

$$x_i(nT') = \mathcal{Z}^{-1}[H(z)X_u(z)]$$

$$= \mathcal{Z}^{-1}\left[H(z) \sum_{k=-\infty}^{\infty} x(kT)z^{-kL}\right]$$

$$= \sum_{k=-\infty}^{\infty} x(kT)\mathcal{Z}^{-1}[H(z)z^{-kL}] \quad (16.7)$$

Now it can be easily shown that

$$h(nT') = \mathcal{Z}^{-1}H(z) = \frac{\sin(\omega_s nT'/2L)}{\omega_s nT'/2L}$$

[see Eq. (9.25)] and from Eq. (16.7) and Theorem 2.6, $x_i(nT')$ can be expressed as

$$x_i(nT') = \sum_{k=-\infty}^{\infty} x(kT)h[(n-kL)T']$$

$$= \sum_{k=-\infty}^{\infty} x(kT) \frac{\sin[\omega_s(n-kL)T'/2L]}{\omega_s(n-kL)T'/2L} \quad (16.8)$$

Since $h(nT') = 1$ for $n = 0$ and 0 for $\pm kL$, we have $x_i(nT') = x(nT')$ for $n = 0, \pm L, \pm 2L$, etc., as required. The relation in Eq. (16.8) is entirely analogous to that in Eq. (6.44).

Example 16.1. Show that the use of a digital filter characterized by the impulse response

$$h(nT') = \begin{cases} 1 - |n|/L & \text{for } |n| < L \\ 0 & \text{otherwise} \end{cases} \quad (16.9)$$

in the scheme of Fig. 16.4b will result in a system that can perform linear interpolation.

Solution. The response of the filter at nT' to an excitation $x_u(nT')$ is given by the convolution summation as

$$x_i(n) = \sum_{k=-\infty}^{\infty} x_u(k) h(n-k)$$

where period T' is dropped for the sake of brevity. The response at $(mL + \lambda)T'$, where λ is an integer in the range $0 \leq \lambda \leq L$, is given by

$$\begin{aligned} x_i(mL + \lambda) &= \sum_{k=-\infty}^{\infty} x_u(k) h(mL + \lambda - k) \\ &= \ldots + x_u(mL - 1) h(1 + \lambda) + x_u(mL) h(\lambda) \\ &\quad + x_u(mL + 1) h(\lambda - 1) + \ldots \\ &\quad + x_u[(m+1)L - 1] h[\lambda - (L-1)] \\ &\quad + x_u[(m+1)L] h(\lambda - L) \\ &\quad + x_u[(m+1)L + 1] h[\lambda - (L+1)] + \ldots \end{aligned} \quad (16.10)$$

From Eq. (16.4), $x_u(n) = 0$ if n is not a multiple of L; therefore, for $0 < \lambda < L$, Eqs. (16.9) and (16.10) yield

$$\begin{aligned} x_i(mL + \lambda) &= x_u(mL) h(\lambda) + x_u[(m+1)L] h(\lambda - L) \\ &= \left(1 - \frac{\lambda}{L}\right) x_u(mL) + \frac{\lambda}{L} x_u[(m+1)L] \end{aligned}$$

which is the linear interpolation between samples mL and $(m+1)L$.

16.2.3 Sampling-Frequency Conversion by a Noninteger Factor

Through the use of an interpolator, the sampling frequency can be increased from ω_s to $\omega'_s = L\omega_s$. On the other hand, through the use of a decimator, the sampling frequency can be decreased from ω'_s to $\omega''_s = \omega'_s/M$. By cascading an interpolator and a decimator, as depicted in Fig. 16.7, a sampling frequency $\omega''_s = L\omega_s/M$ can be obtained, where L/M is a ratio of integers. By this means, arbitrary conversion factors can be achieved. In Fig. 16.7, the interpolator necessitates a lowpass filter with a cutoff frequency $\omega'_s/2L$, while in the decimator a lowpass filter with a cutoff frequency $\omega'_s/2M$ is needed. One of the two filters, namely, the one with the higher cutoff frequency, is obviously redundant and can be eliminated.

16.2.4 Design Considerations

The design of interpolators and decimators is straightforward, and perhaps the most complicated part of the design involves the design of the required filters. Depending on the system requirements, a recursive or nonrecursive filter may be more appropriate, and any one of the methods described in Chaps. 8, 9, 14, and 15 can be used. Furthermore, a large variety of filter structures can be employed. A comparison between recursive and nonrecursive designs can be found in Sec. 9.6. The merits and demerits of the various types of structures can be found in Sec. 12.13.

16.3 QUADRATURE-MIRROR-FILTER BANKS

Filter banks find applications in many areas of science and engineering, and are used for time-division to frequency-division multiplex translation, subband speech coding, bandwidth compression, and many other types of signal processing [1, 8, 9]. In these systems, a given signal is decomposed into consecutive subbands that are processed independently. The subbands obtained are then used to synthetize a processed version of the signal.

The motivation behind this roundabout way of processing a signal is usually to achieve some economical advantage, e.g., increased utilization of equipment or improved signal-to-noise ratio, but on occasion a filter bank is used to carry out some type of processing that cannot be carried out by any other means. Time-division to frequency-division multiplex translation is used quite extensively in communication systems in order to achieve increased channel capacity [10]. In subband speech coding, the signal is decomposed into several bands that are coded individually, taking advantage of certain perceptive properties of the human ear. In this way, improved signal quality can be achieved without increasing the bit rate [11].

In its most general form, a filter bank consists of a *decomposition* or *analysis* section and a *reconstruction* or *synthesis* section, as depicted in Fig. 16.8. Depending on the application, (1) the spectra of $x_0(nT), x_1(nT), \ldots, x_K(nT)$ should not overlap, may overlap somewhat, or can overlap heavily; (2) the processed signal $y(nT)$ must be a faithful reproduction of $x(nT)$ or it may be a transformed version that bears little resemblance; and (3) the subband widths may be uniform, irregular, or fixed by some logarithmic relation.

In this section, we consider a specific type of filter bank, the so-called *quadrature mirror-filter* (QMF) bank.

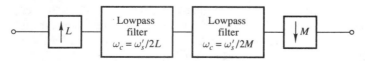

FIGURE 16.7
Sampling-frequency conversion by noninteger factor.

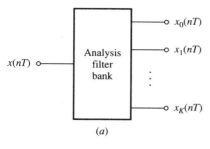

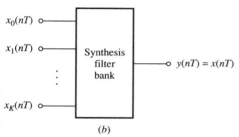

FIGURE 16.8
Filter bank: (a) analysis section, (b) synthesis section.

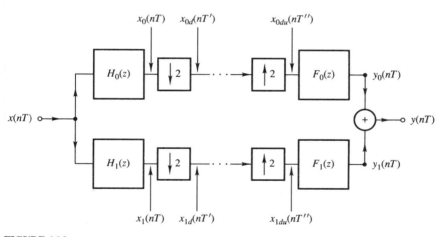

FIGURE 16.9
2-band QMF bank.

16.3.1 Operation

A 2-band QMF bank is illustrated in Fig. 16.9. The analysis section of the bank consists of a lowpass and a highpass filter which decompose the input signal $x(nT)$ into two components $x_0(nT)$ and $x_1(nT)$. Ideally, the lowpass and highpass filters should have the frequency responses

$$H_0(e^{j\omega T}) = \begin{cases} 1 & \text{for } 0 < |\omega| < \omega_s/4 \\ 0 & \text{otherwise} \end{cases} \tag{16.11}$$

and

$$H_1(e^{j\omega T}) = \begin{cases} 1 & \text{for } \omega_s/4 < |\omega| < \omega_s/2 \\ 0 & \text{otherwise} \end{cases} \quad (16.12)$$

respectively. The operation of the filter bank is illustrated in Figs. 16.10 and 16.11. Downsampling by a factor $M = 2$ will produce shifted copies of the lowpassed and highpassed spectra, as depicted in Fig. 16.10g and h and upsampling by a factor $L = 2$ will change the location of the sampling frequency, as shown in Fig. 16.11a and b (see Sec. 16.2); that is, one image of each of the lowpassed and highpassed spectra will be introduced in the baseband, as in Figs. 16.11a and b by analogy with Fig. 16.3. At the output, the synthesis lowpass and highpass filters, which should in principle be identical to the corresponding analysis filters, will regenerate $x_0(nT)$ and $x_1(nT)$, respectively, as shown in Fig. 16.11e and f. Hence the output of the adder will be the required signal, as depicted in Fig. 16.11g.

In practical filters, the passband gain is not unity, the stopband loss is not infinite, and transitions between passbands and stopbands are gradual. Hence amplitude and phase distortion as well as aliasing will be introduced, as can be seen in Fig. 16.12. While the effects of these imperfections can be reduced by designing filters of better

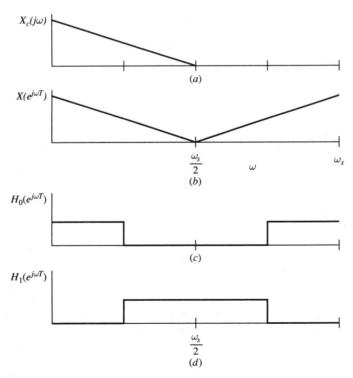

FIGURE 16.10a–d
Operation of analysis section of 2-band QMF bank.

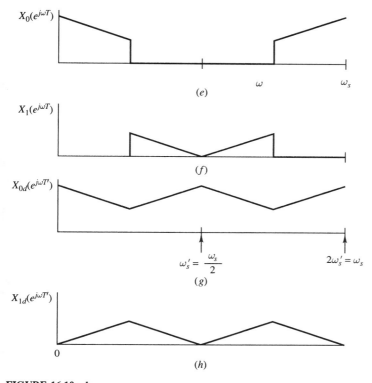

FIGURE 16.10e–h
Operation of analysis section of 2-band QMF bank.

quality, these problems can largely be eliminated in a much simpler way by using lowpass and highpass filters whose frequency responses have certain symmetry properties. This possibility will now be examined.

16.3.2 Elimination of Aliasing Errors

Let $T' = 2T$ and $T'' = T'/2 = T$. From Eq. (16.3), the output of the top downsampler in Fig. 16.9 can be expressed as

$$X_{0d}(e^{j\omega T'}) = \tfrac{1}{2}[X(e^{j\omega T'/2})H_0(e^{j\omega T'/2}) + X(e^{j(\omega T'/2-\pi)})H_0(e^{j(\omega T'/2-\pi)})] \quad (16.13)$$

On the other hand, the output of the top upsampler is given by Eq. (16.5) as

$$X_{0du}(e^{j\omega T''}) = X_{0du}(e^{j\omega T'/2}) = X_{0d}(e^{j\omega T'}) \quad (16.14)$$

and from Fig. 16.9

$$Y_0(e^{j\omega T''}) = X_{0du}(e^{j\omega T''})F_0(e^{j\omega T''}) \quad (16.15)$$

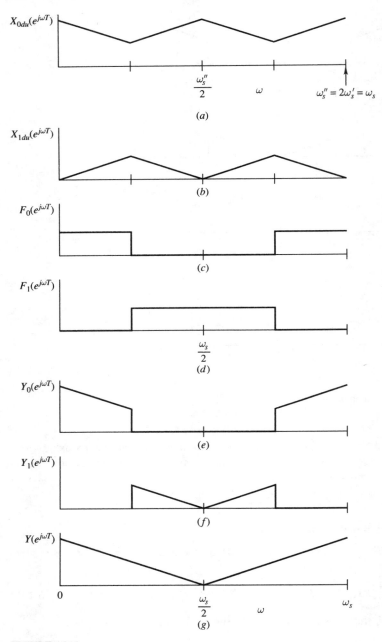

FIGURE 16.11
Operation of synthesis section of 2-band QMF bank.

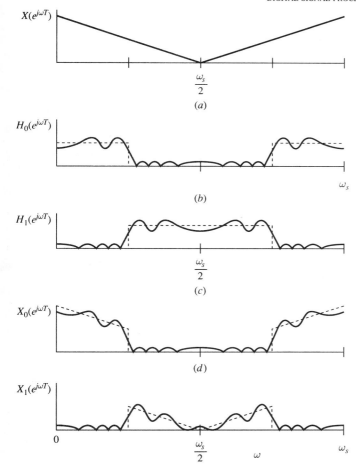

FIGURE 16.12
Effects of filter imperfections in 2-band QMF bank.

Now Eqs. (16.13) to (16.15) give

$$Y_0(e^{j\omega T''}) = \tfrac{1}{2}[X(e^{j\omega T'/2})H_0(e^{j\omega T'/2}) \\ + X(e^{j(\omega T'/2-\pi)})H_0(e^{j(\omega T'/2-\pi)})]F_0(e^{j\omega T''})$$

and on eliminating T' and T'', we have

$$Y_0(e^{j\omega T}) = \tfrac{1}{2}[X(e^{j\omega T})H_0(e^{j\omega T}) + X(e^{j(\omega T-\pi)})H_0(e^{j(\omega T-\pi)})]F_0(e^{j\omega T})$$

Similarly,

$$Y_1(e^{j\omega T}) = \tfrac{1}{2}[X(e^{j\omega T})H_1(e^{j\omega T}) + X(e^{j(\omega T-\pi)})H_1(e^{j(\omega T-\pi)})]F_1(e^{j\omega T})$$

and, therefore,

$$Y(e^{j\omega T}) = Y_0(e^{j\omega T}) + Y_1(e^{j\omega T})$$
$$= \tfrac{1}{2}[H_0(e^{j\omega T})F_0(e^{j\omega T}) + H_1(e^{j\omega T})F_1(e^{j\omega T})]X(e^{j\omega T})$$
$$+ \tfrac{1}{2}[H_0(e^{j(\omega T-\pi)})F_0(e^{j\omega T})$$
$$+ H_1(e^{j(\omega T-\pi)})F_1(e^{j\omega T})]X(e^{j(\omega T-\pi)}) \qquad (16.16)$$

The first term in the above relation represents the required signal, whereas the second term represents spurious components due to aliasing. If the filters are designed such that

$$H_0(e^{j(\omega T-\pi)})F_0(e^{j\omega T}) + H_1(e^{j(\omega T-\pi)})F_1(e^{j\omega T}) = 0 \qquad (16.17)$$

then the aliasing produced by the lowpass filters will be canceled exactly by the aliasing produced by the highpass filters even if the quality of these filters is not particularly good. This relation can be satisfied by designing the analysis and synthesis filters such that

$$F_0(e^{j\omega T}) = 2H_1(e^{j(\omega T-\pi)})$$

and

$$F_1(e^{j\omega T}) = -2H_0(e^{j(\omega T-\pi)})$$

The factor 2 is required to compensate for the factor 1/2 introduced by the downsampling (see Fig. 16.3c).

With Eq. (16.17) satisfied, the output of a 2-band QMF bank is given by Eq. (16.16) as

$$Y(e^{j\omega T}) = T(e^{j\omega T})X(e^{j\omega T})$$

where

$$T(e^{j\omega T}) = H_0(e^{j\omega T})H_1(e^{j(\omega T-\pi)}) - H_1(e^{j\omega T})H_0(e^{j(\omega T-\pi)}) \qquad (16.18)$$

is the overall frequency response of the QMF bank. If we assume that the lowpass and highpass filters have frequency responses that are mirror-image symmetric with respect to frequency $\omega = \omega_s/4$, then

$$H_1(e^{j\omega T}) = H_0(e^{j(\omega T-\pi)}) \qquad (16.19a)$$

or

$$H_1(e^{j(\omega T-\pi)}) = H_0(e^{j\omega T}) \qquad (16.19b)$$

Hence Eq. (16.18) can be expressed as

$$T(e^{j\omega T}) = H_0^2(e^{j\omega T}) - H_0^2(e^{j(\omega T-\pi)}) \qquad (16.20)$$
$$= M(\omega)e^{j\theta(\omega)}$$

where $M(\omega)$ and $\theta(\omega)$ are the overall amplitude and phase responses of the QMF bank. In practice, $M(\omega)$ should as far as possible be constant and $\theta(\omega)$ should be a linear function of ω in order to avoid amplitude and phase distortion in the reconstruction process (see Sec. 3.6).

A transfer function that characterizes the QMF bank can be formed as

$$T(z) = H_0(z)H_1(-z) - H_1(z)H_0(-z) \tag{16.21}$$

by letting $e^{j\omega T} = z$ in Eq. (16.18).

16.3.3 Design Considerations

QMF banks can be designed in terms of nonrecursive or recursive filters. Let us consider the first case.

On assuming that the impulse response of the analysis lowpass filter is symmetrical, its frequency response can be expressed as

$$H_0(e^{j\omega T}) = e^{j\omega(N-1)T/2}|H_0(e^{j\omega T})| \tag{16.22}$$

(see Table 9.1), where N is the filter length. From Eqs. (16.20) and (16.22), we can write

$$T(e^{j\omega T}) = [|H_0(e^{j\omega T})|^2 \pm |H_0(e^{j(\omega T - \pi)})|^2]e^{j\omega(N-1)T}$$

where the plus sign applies in the case where N is even and the minus sign applies in the case where N is odd. In effect, *linear phase response is achieved*, i.e., the design has zero phase distortion. If the required mirror-image symmetry is assumed, the overall frequency response becomes zero at $\omega = \omega_s/4$ if N is odd. Since $T(e^{j\omega T})$ is required to be an allpass function, we conclude that N *must be even*.

A nonrecursive design can now be obtained by solving the approximation problem in Eq. (16.11) such that the constraint

$$|H_0(e^{j\omega T})|^2 + |H_0(e^{j(\omega T - \pi)})|^2 = 1 \quad \text{for } 0 \leq \omega \leq \omega_s/2$$

is satisfied. Designs of this type can be obtained by using the window technique [12].

Nonrecursive QMF banks can also be designed by using optimization methods, as described in [13, 14]. One of many possibilities is to construct an error function of the form

$$E(\mathbf{x}) = E_1(\mathbf{x}) + \alpha E_2(\mathbf{x})$$

where $\mathbf{x}$ is a column vector whose elements are the transfer function coefficients and

$$E_1(\mathbf{x}) = \int_{\omega_s/4}^{\omega_s/2} |H_0(e^{j\omega T})|^2 d\omega$$

$$E_2(\mathbf{x}) = \int_{\omega_s/4}^{\omega_s/2} [|H_0(e^{j\omega T})|^2 + |H_0(e^{j(\omega T - \pi)})|^2 - 1]^2 d\omega$$

Minimizing $E(\mathbf{x})$ with respect to the transfer-function coefficients using one of the quasi-Newton algorithms described in Chap. 14 will minimize the stopband error in the lowpass filter and force the overall amplitude response to approach unity in a least-squares sense. Parameter α can be used to emphasize or de-emphasize the error in the overall amplitude response relative to the stopband error of the lowpass filter.

In certain types of recursive filters, it is quite easy to design a pair of complementary filters whose amplitude responses assume the form

$$H_0(e^{j\omega T}) = \tfrac{1}{2}[A_0(e^{j\omega T}) + A_1(e^{j\omega T})] \tag{16.23}$$

and

$$H_1(e^{j\omega T}) = \tfrac{1}{2}[A_0(e^{j\omega T}) - A_1(e^{j\omega T})] \tag{16.24}$$

where $A_0(e^{j\omega T})$ and $A_1(e^{j\omega T})$ are allpass functions. With these methods, arbitrary amplitude response specifications can be obtained and the required symmetry about the frequency $\omega_s/4$ can be easily achieved. From Eqs. (16.18)–(16.19), the overall frequency response can be put in the form

$$T(e^{j\omega T}) = H_0^2(e^{j\omega T}) - H_1^2(e^{j\omega T})$$

and from Eqs. (16.23)–(16.24)

$$T(e^{j\omega T}) = A_0(e^{j\omega T}) A_1(e^{j\omega T})$$

In effect, *the overall frequency response of the QMF bank is an allpass function*; that is, the design obtained has zero amplitude distortion. However, a certain amount of phase distortion will be present, which may or may not be objectionable depending on the application. Such filters can be designed as wave lattice filters, as described by Gazsi and Vaidyanathan, Mitra, and Neuvo (see Sec. 12.5 and [10] and [24] of Chap. 12).

Example 16.2. Design a 2-band QMF bank as a wave lattice filter using the method described in Sec. 12.5 along with an elliptic approximation.[†] The specifications of the lowpass filter are as follows:

$A_p = 1.0$ dB $A_a = 45.0$ dB $\tilde{\Omega}_p = 4,500$ rad/s

$\tilde{\Omega}_a = 5,500$ rad/s $\omega_s = 2.0 \times 10^4$ rad/s

Solution. Using the formulas in Tables 8.2 and 8.6, the required selectivity factor is found to be $k = 0.729454$. The minimum filter order that will satisfy the required specifications is $n = 5$, according to Eqs. (5.45)–(5.49). The value of λ in the lowpass-to-lowpass transformation (see Table 8.1) that will compensate for the warping effect introduced by the bilinear transformation can be determined as $\lambda = 1.570796 \times 10^{-4}$.

[†]See Sec. VI of [9] for a related design example.

On using the formulas in Eqs. (5.50)–(5.58) and then applying the lowpass-to-lowpass transformation, the denormalized transfer function is obtained as

$$H_A(s) = \frac{1}{s+b_{01}} \prod_{j=2}^{3} \frac{s^2 + a_{0j}}{s^2 + b_{1j}s + b_{0j}}$$

where coefficients a_{ij} and b_{ij} are given in Table 16.1.

Now from Eqs. (12.25a) and (12.25b), polynomials $d_A(s)$ and $d_B(s)$ can be deduced as

$$d_A(s) = (s+b_{01})(s^2 + b_{13}s + b_{03}) \tag{16.25}$$

and

$$d_B(s) = s^2 + b_{12}s + b_{02} \tag{16.26}$$

Digital realizations for networks N_A and N_B shown in Fig. 12.14 can be obtained as depicted in Fig. 16.13 by using the approach described in Sec. 12.5. The multiplier constants of the adaptors can be calculated from Eqs. (12.29)–(12.30) as follows: $\mu_{11} = 0.536886$, $\mu_{12} = 0.432520$, $\mu_{22} = -0.566457$, $\mu_{13} = 0.158183$, and $\mu_{23} = -0.886611$.

The configuration obtained will operate as a halfband lowpass filter with respect to output B_2 and as a halfband highpass filter with respect to output B_1 by virtue of the Feldkeller equation (see Sec. 12.5.2, p. 410).

16.3.4 Perfect Reconstruction

In the preceding section, it has been shown that designs with zero phase distortion can be easily obtained by using linear-phase nonrecursive filters whereas zero amplitude distortion can be achieved by using recursive wave digital filters. In the case of nonrecursive filters, amplitude distortion can be rendered insignificant by using a high-order filter to achieve small passband and stopband ripples. On the other hand, in the case of recursive filters, phase distortion can be rendered insignificant through the use of phase equalization (see Sec. 14.8). In this section, a scheme due to Smith and Barnwell [8, 15] is described by which zero phase distortion as well as zero amplitude distortion can be achieved simultaneously, independently of the quality of the filters used.

TABLE 16.1
Coefficients of $H_A(s)$ (Example 16.2)

j	a_{0j}	b_{0j}	b_{1j}
1	—	1.918343×10^3	—
2	1.279580×10^8	1.605501×10^7	2.459931×10^3
3	5.951444×10^7	2.945786×10^7	6.607256×10^2

$H_0 = 1.191374 \times 10^2$

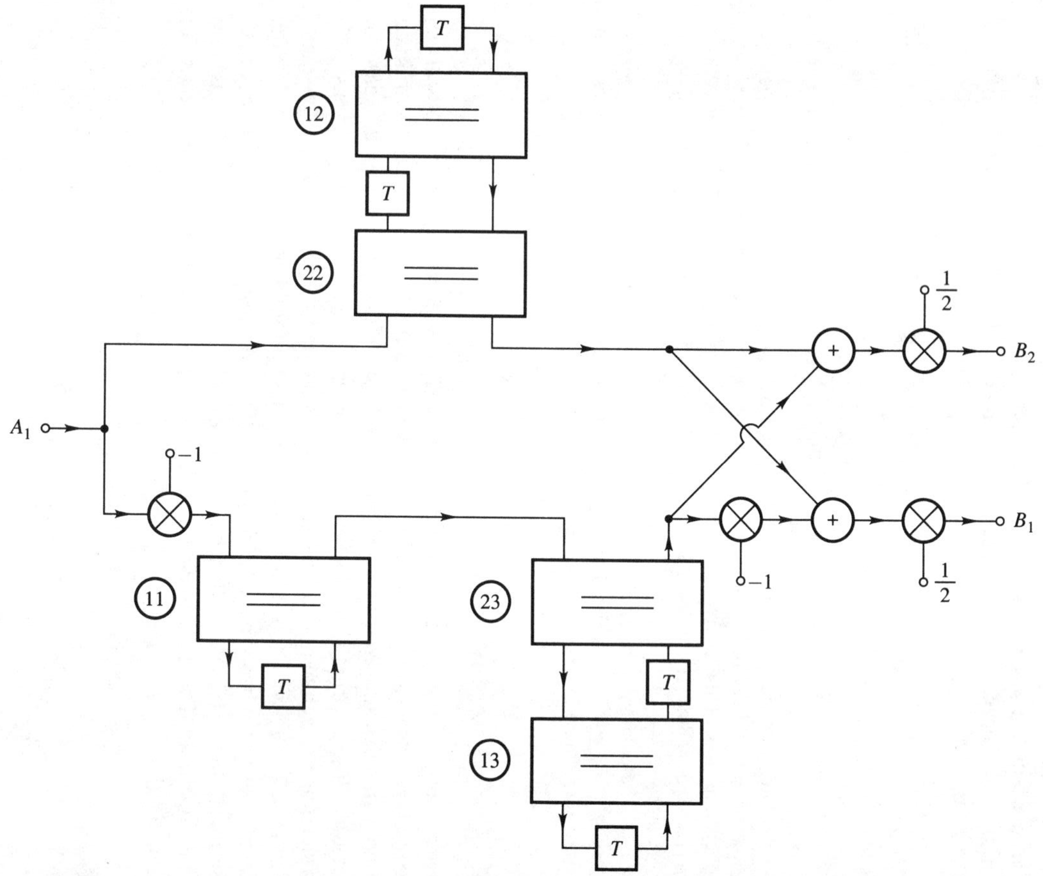

FIGURE 16.13
Lattice wave digital filter realizing a 2-band QMF bank (Example 16.2).

Let $H(z)$ be a linear-phase, lowpass, nonrecursive transfer function and assume that the passband and stopband errors satisfy the symmetry property

$$|1 - H(e^{j\omega T})| = |H(e^{j(\omega_s/2-\omega)T})| \leq \delta \qquad (16.27)$$

as illustrated in Fig. 16.14a. Using $H(z)$, a modified transfer function

$$H_0(z) = H(z) + \delta z^{-(N-1)/2}$$

can be formed, where N is the length of the filter. From the above symmetry property, we can write

$$|H_0(e^{j\omega T})|^2 + |H_0(e^{j(\omega T - \pi)})|^2 = G \qquad (16.28)$$

where G is a constant, i.e., changes in the two functions cancel each other out exactly, as demonstrated in Fig. 16.14a to c.

A linear-phase transfer function with the amplitude response given by Eq. (16.28) can now be constructed as

$$T(z) = H_0(z)H_0(z^{-1})z^{-(N-1)} - [-H_0(-z^{-1})(-z)^{-(N-1)}]H_0(-z) \qquad (16.29)$$

(see Prob. 16.6), where N is even. On comparing Eqs. (16.21) and (16.29), the assignment

$$H_1(z) = z^{-(N-1)}H_0(-z^{-1})$$

can be made and if we let

$$F_0(z) = 2H_1(-z) \quad \text{and} \quad F_1(z) = -2H_0(-z)$$

a realization of the QMF bank is obtained in which both the amplitude and phase distortions are zero at the same time. Realizations of this type are said to have the *perfect reconstruction property*.

Halfband filters with the symmetry property of Eq. (16.27) can be designed by using the methods of Chap. 15.

The above principles can be extended to the design of QMF banks with multiple bands.

16.4 HILBERT TRANSFORMERS

In certain digital signal processing applications, it is necessary to form a special version of a given signal $x(nT)$, designated as $\tilde{x}(nT)$, with the special property that its frequency spectrum is equal to that of $x(nT)$ for the positive Nyquist interval and zero for the negative Nyquist interval, i.e.,[†]

$$\tilde{X}(e^{j\omega T}) = \begin{cases} X(e^{j\omega T}) & \text{for} \quad 0 < \omega < \omega_s/2 \\ 0 & \text{for} \quad -\omega_s/2 \leq \omega < 0 \end{cases} \qquad (16.30)$$

[†]Alternatively, the spectrum of $\tilde{x}(nT)$ may be taken to be equal to that of $x(nT)$ for the negative Nyquist interval and zero for the positive Nyquist interval.

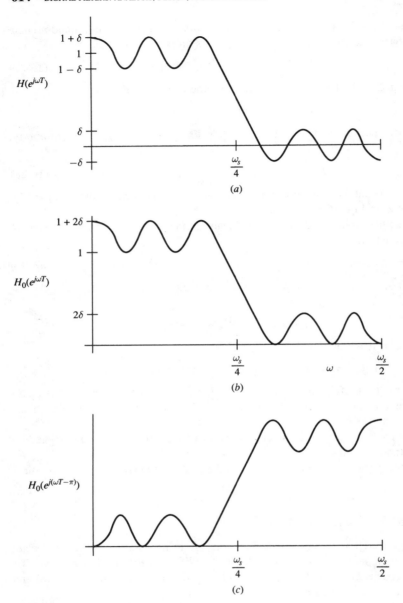

FIGURE 16.14
Required amplitude-response symmetry for perfect reconstruction.

Signals with this property have been referred to as *analytic signals* in the past [16, 17] and are useful in a number of applications, for example, in single-sideband modulation in frequency-division multiplex systems. The motivation for eliminating the spectrum of a signal over the negative Nyquist interval is that it is the mirror image of the spectrum over the positive Nyquist interval, i.e., it contains the same information, and its elimination reduces the required bandwidth for the processing and/or transmission of the signal by half.

An essential property of real continuous-time signals is that their amplitude spectra are even, and their phase spectra are odd functions of ω (see Prob. 6.3). Since analytic signals violate these requirements, they must necessarily be complex of the form

$$\tilde{x}(nT) = x_r(nT) + jx_i(nT) \tag{16.31}$$

where $x_r(nT)$ and $x_i(nT)$ are real sequences. If the spectrum of an analytic signal $\tilde{x}(nT)$ is known, the spectra of $x_r(nT)$ and $x_i(nT)$ can be readily deduced as

$$X_r(e^{j\omega T}) = \tfrac{1}{2}[\tilde{X}(e^{j\omega T}) + \tilde{X}^*(e^{-j\omega T})] \tag{16.32}$$

and

$$jX_i(e^{j\omega T}) = \tfrac{1}{2}[\tilde{X}(e^{j\omega T}) - \tilde{X}^*(e^{-j\omega T})] \tag{16.33}$$

where $\tilde{X}^*$ is the complex conjugate of $\tilde{X}$. These relations are illustrated in Fig. 16.15. From Eqs. (16.32) and (16.33), we obtain

$$\tilde{X}(e^{j\omega T}) = 2X_r(e^{j\omega T}) - \tilde{X}^*(e^{-j\omega T}) \tag{16.34}$$

and

$$\tilde{X}(e^{j\omega T}) = 2jX_i(e^{j\omega T}) + \tilde{X}^*(e^{-j\omega T}) \tag{16.35}$$

and since $\tilde{X}^*(e^{-j\omega T}) = 0$ for $0 < \omega < \omega_s/2$ (see Fig. 16.15b), Eqs. (16.30), (16.34), and (16.35) give

$$\tilde{X}(e^{j\omega T}) = \begin{cases} 2X_r(e^{j\omega T}) & \text{for } 0 < \omega < \omega_s/2 \\ 0 & \text{for } -\omega_s/2 \leq \omega < 0 \end{cases}$$

and

$$\tilde{X}(e^{j\omega T}) = \begin{cases} 2jX_i(e^{j\omega T}) & \text{for } 0 < \omega < \omega_s/2 \\ 0 & \text{for } -\omega_s/2 \leq \omega < 0 \end{cases}$$

Thus

$$X_i(e^{j\omega T}) = -jX_r(e^{j\omega T}) \qquad \text{for } 0 < \omega < \omega_s/2 \tag{16.36}$$

On the other hand, from Eq. (16.31)

$$\tilde{X}(e^{j\omega T}) = X_r(e^{j\omega T}) + jX_i(e^{j\omega T})$$

and since $\tilde{X}(e^{j\omega T}) = 0$ for $-\omega_s/2 \leq \omega < 0$, we have

$$X_i(e^{j\omega T}) = jX_r(e^{j\omega T}) \qquad \text{for } -\omega_s/2 \leq \omega < 0 \tag{16.37}$$

Therefore, Eqs. (16.36) and (16.37) can be expressed as

$$X_i(e^{j\omega T}) = \begin{cases} -jX_r(e^{j\omega T}) & \text{for } 0 < \omega < \omega_s/2 \\ jX_r(e^{j\omega T}) & \text{for } -\omega_s/2 \leq \omega < 0 \end{cases}$$

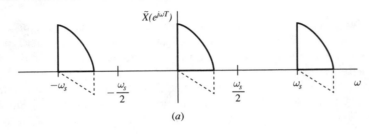
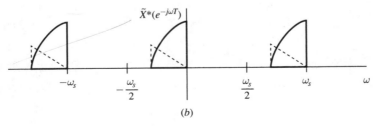
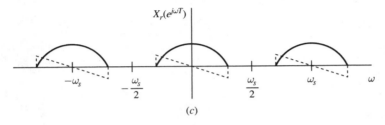
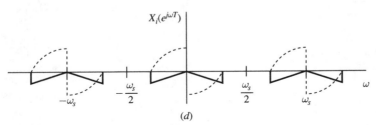

FIGURE 16.15
Derivation of $x_r(nT)$ and $x_i(nT)$ from $\tilde{x}(nT)$ (solid curves represent real parts and dashed curves represent imaginary parts).

or

$$X_i(e^{j\omega T}) = H(e^{j\omega T})X_r(e^{j\omega T}) \quad (16.38)$$

where

$$H(e^{j\omega T}) = \begin{cases} -j & \text{for} \quad 0 < \omega < \omega_s/2 \\ j & \text{for} \quad -\omega_s/2 < \omega < 0 \end{cases} \quad (16.39)$$

These results show that the real and imaginary parts of an analytic signal are interrelated and the imaginary part can be obtained from the real part by using Eq. (16.38). This relation may be deemed to represent a filter with input $x_r(nT)$, output $x_i(nT)$, and frequency response $H(e^{j\omega T})$. A filter of this type is commonly referred to as a *Hilbert transformer*. Its output $x_i(nT)$ is said to be the *Hilbert transform* of $x_r(nT)$.

On the basis of these principles, given a real sequence $x_r(nT)$, a corresponding analytic signal $\tilde{x}(nT)$ can be synthesized by using the configuration depicted in Fig. 16.16. The operation of this scheme is illustrated in Fig. 16.17. As can be seen, if $x_r(nT)$ is a real sequence and $x_i(nT)$ is generated by using a Hilbert transformer, then $x_r(nT) + jx_i(nT)$ is an analytic signal, as shown in Fig. 16.17d.

16.4.1 Design of Hilbert Transformers

Hilbert transformers can be designed either in terms of nonrecursive or recursive filters. In the former case, either the Fourier series method of Chap. 9 or the weighted-Chebyshev method of Chap. 15 can be used.

Using the Fourier series method of Sec. 9.3, the impulse response of a Hilbert transformer can be obtained as

$$h(nT) = \frac{T}{2\pi} \left(\int_{-\omega_s/2}^{0} je^{j\omega nT} d\omega - \int_{0}^{-\omega_s/2} je^{j\omega nT} d\omega \right)$$

$$= \begin{cases} \dfrac{2}{n\pi} \sin^2 \dfrac{n\pi}{2} & \text{for } n \neq 0 \\ 0 & \text{for } n = 0 \end{cases} \quad (16.40)$$

Evidently, a Hilbert transformer, like a digital differentiator, has an antisymmetrical impulse response and can be designed either with odd or even filter length N using the window technique described in Chap. 9. From Table 9.1, we note that for odd N, the amplitude response is zero for $\omega = 0$ and $\omega = \omega_s/2$. Hence the useful bandwidth that can be achieved is restricted to some range $0 < \omega_L \leq \omega \leq \omega_H < \omega_s/2$ where ω_L and ω_H can be made to approach 0 and $\omega_s/2$, respectively, as closely as desired

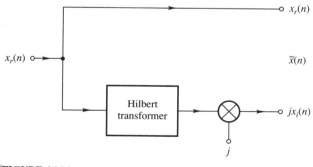

FIGURE 16.16
Synthesis of an analytic signal using a Hilbert transformer.

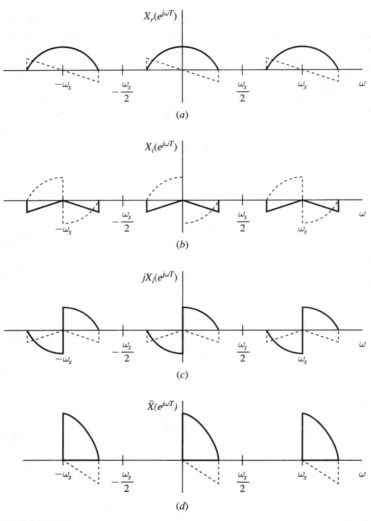

FIGURE 16.17
Operation of configuration in Fig. 16.16 (solid curves represent real parts and dashed curves represent imaginary parts).

by increasing N. On the other hand, for even N, the amplitude response need not be zero at $\omega = \omega_s/2$, as can be seen in Table 9.1, and, in this case, Hilbert transformers can be designed in which $\omega_H = \omega_s/2$.

The ideal response of a causal Hilbert transformer can be expressed in the form $e^{-j\omega c} j D(\omega)$ where

$$D(\omega) = \begin{cases} -1 & \text{for} \quad 0 < \omega < \omega_s/2 \\ 1 & \text{for} \quad -\omega_s/2 < \omega < 0 \end{cases}$$

and $c = (N - 1)/2$. From Table 15.7, we conclude that a Hilbert transformer can be designed using the weighted-Chebyshev algorithm. The problem formulation is similar to that used for the design of digital differentiators in Sec. 15.9 and is left as an exercise to the reader (see Prob. 16.19).

Example 16.3. (*a*) Design a Hilbert transformer of length $N = 21$ using the Kaiser window with $\alpha = 3.0$, assuming a sampling frequency of 10 rad/s. (*b*) Repeat part (*a*) with $N = 22$ and compare the results obtained in the two cases.

Solution. (*a*) Using Eq. (16.40), the impulse response of the Hilbert transformer, $h(n)$, can be computed as shown in column 2 of Table 16.2a. The modified impulse response assumes the form

$$h_w(n) = w_K(n)h(n) \qquad (16.41)$$

where $w_K(n)$ is the Kaiser window. With $\alpha = 3.0$, Eqs. (9.28) and (16.41) give the impulse response in column 3 of Table 16.2a. The amplitude response of the Hilbert transformer is depicted in Fig. 16.18; the maximum passband error is 1.44 percent of the passband gain.

(*b*) For even N, the impulse response must be antisymmetrical about the midpoint between samples $(N-2)/2$ and $N/2$, i.e., samples 10 and 11 for $N = 22$ (see Sec. 9.2.1). Consequently, in a noncausal Hilbert transformer, the impulse response is defined at $n' = \pm(n - 0.5)$ with $n = 1, 2, \ldots, N/2$. Using this transformation in Eq. (16.40), we obtain the required impulse response as

$$h(n - 0.5) = -h(-n + 0.5)$$
$$= \frac{2}{(n - 0.5)\pi} \sin^2\left[\frac{(n - 0.5)\pi}{2}\right] \qquad \text{for } n = 1, 2, \ldots, N/2 \qquad (16.42)$$

TABLE 16.2a
Impulse response of Hilbert transformer [Example 16.3, part (*a*)]

n	$h(n) = -h(-n)$	$h_w(n)$
0	0.0	0.0
1	6.366198×10^{-1}	6.289178×10^{-1}
2	0.0	0.0
3	2.122066×10^{-1}	1.898748×10^{-1}
4	0.0	0.0
5	1.273240×10^{-1}	9.255503×10^{-2}
6	0.0	0.0
7	9.094568×10^{-2}	4.698992×10^{-2}
8	0.0	0.0
9	7.073553×10^{-2}	2.138270×10^{-2}
10	0.0	0.0

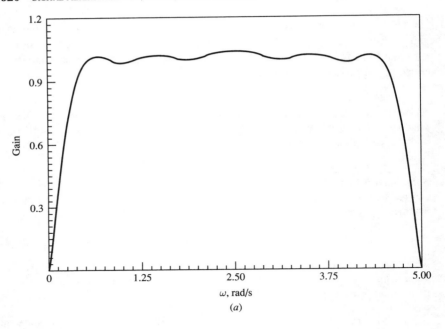

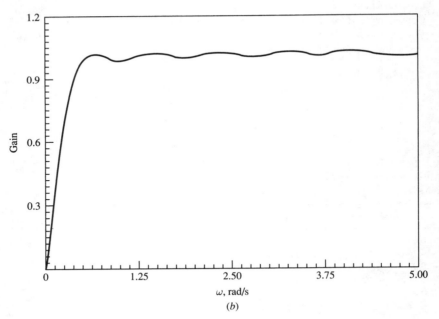

FIGURE 16.18
Amplitude response of Hilbert transformer (Example 16.3): (*a*) $N = 21$, (*b*) $N = 22$.

TABLE 16.2b
Impulse response of Hilbert transformer [Example 16.3, part (b)]

n	$h(n) = -h(-n)$	$h_w(n)$
0.5	6.366198×10^{-1}	6.302499×10^{-1}
1.5	2.122066×10^{-1}	2.038026×10^{-1}
2.5	1.273240×10^{-1}	1.161765×10^{-1}
3.5	9.094568×10^{-2}	7.713458×10^{-2}
4.5	7.073553×10^{-2}	5.447434×10^{-2}
5.5	5.787452×10^{-2}	3.944281×10^{-2}
6.5	4.897075×10^{-2}	2.869059×10^{-2}
7.5	4.244132×10^{-2}	2.066188×10^{-2}
8.5	3.744822×10^{-2}	1.453254×10^{-2}
9.5	3.350630×10^{-2}	9.818996×10^{-3}
10.5	3.031523×10^{-2}	6.211128×10^{-3}

From Eqs. (9.28), (16.41), and (16.42), the design in Table 16.2b is obtained. The amplitude response for this case is depicted in Fig. 16.18b; the maximum passband error in this case is 0.36 percent of the passband gain. We note that for odd N, several values of the impulse response are zero, which renders the design more economical. However, the design for even N provides a wider bandwidth and the approximation error is much smaller.

A feature of the latter design, which may be a problem in certain applications, is that the impulse response must be shifted by a noninteger multiple of the sampling period in order to achieve a causal design.

The frequency response of a Hilbert transformer can also be expressed as

$$H(e^{j\omega T}) = M(\omega)e^{j\theta(\omega)}$$

where

$$M(\omega) = 1 \quad \text{and} \quad \theta(\omega) = \begin{cases} -\pi/2 & \text{for} \quad 0 < \omega < \omega_s/2 \\ \pi/2 & \text{for} \quad -\omega_s/2 < \omega < 0 \end{cases}$$

according to Eq. (16.39). Hence Hilbert transformers can also be designed as recursive filters by assuming a set of cascaded allpass sections, as in Sec. 14.8, and then forcing the overall phase response to approach $-90°$ for the range $0 < \omega < \omega_s/2$ and $90°$ for the range $-\omega_s/2 < \omega < 0$ to within a prescribed tolerance. This can be done by using the optimization methods described in Chap. 14.

16.4.2 Single-Sideband Modulation

One of the important applications of Hilbert transformers is concerned with *single-sideband modulation*. This process involves two steps: first, an analytic version of a real signal is generated; second, it is used to modulate a sinusoidal carrier of frequency

ω_c. Single-sideband modulation can be carried out by using the scheme depicted in Fig. 16.19, as will now be demonstrated.

The signals at nodes A and B in Fig. 16.19a constitute an analytic signal given by

$$\tilde{x}(nT) = x_r(nT) + jx_i(nT)$$
$$= A(nT)e^{j\phi(nT)} \qquad (16.43)$$

where

$$A(nT) = [x_r^2(nT) + x_i^2(nT)]^{1/2} \quad \text{and} \quad \phi(nT) = \tan^{-1}\frac{x_i(nT)}{x_r(nT)}$$

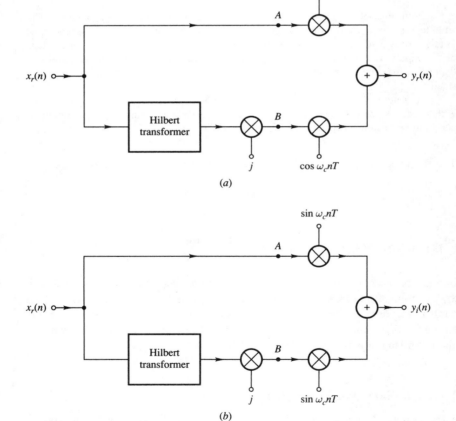

FIGURE 16.19
Single-sideband modulation: (a) generation of real sequence, (b) generation of imaginary sequence.

Hence the outputs of the top and bottom structures are given by

$$y_r(n) = A(nT) \cos[\omega_c nT + \phi(nT)]$$

and

$$y_i(n) = A(nT) \sin[\omega_c nT + \phi(nT)]$$

respectively. Therefore,

$$\tilde{y}(nT) = y_r(n) + j y_i(n)$$
$$= A(nT) e^{[j\omega_c nT + \phi(nT)]}$$

and, if the spectrum of $\tilde{x}(nT)$ [see Eq. (16.30)] is assumed to be zero for $\omega_m \leq \omega \leq \omega_s/2$ and $\omega_c + \omega_m < \omega_s/2$, then the spectrum of $\tilde{y}(nT)$ is given by

$$\tilde{Y}(e^{j\omega T}) = \tilde{X}(e^{j(\omega-\omega_c)T}) \qquad \text{for } 0 < \omega < \omega_s/2 \tag{16.44}$$

(see Prob. 16.20), that is, $\tilde{y}(nT)$ represents a carrier modulated by the upper sideband of signal $x_r(nT)$. The operation of the modulator is illustrated in Fig. 16.20. A simplified realization of the modulator can be easily obtained from Fig. 16.19, as depicted in Fig. 16.21.

16.4.3 Sampling of Bandpassed Signals

In certain applications, it is necessary to sample a signal $x(t)$ whose spectrum occupies some frequency interval $\omega_L \leq \omega \leq \omega_H$. Signals of this type are often generated

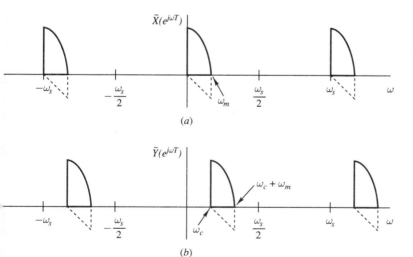

FIGURE 16.20
Operation of single-sideband modulator (solid curves represent real parts and dashed curves represent imaginary parts).

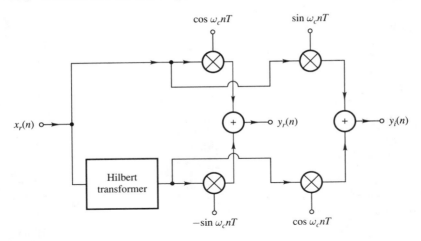

FIGURE 16.21
Simplified single-sideband modulator.

through the use of bandpass filters and can be referred to as *bandpassed*[†] signals. The processing of such signals would necessitate a minimum sampling frequency $\omega_s = 2\omega_H$, according to the sampling theorem. However, through the use of a Hilbert transformer and a pair of decimators, it is possible to generate a downsampled version of the signal that can be processed at a much lower sampling frequency. Such a scheme is shown in Fig. 16.22a, and its mode of operation is illustrated by the frequency spectra of Fig. 16.23.

The signal components at nodes A and B in Fig. 16.22a represent an analytic signal $\tilde{x}(nT)$ with the frequency spectrum depicted in Fig. 16.23b, as demonstrated earlier. Now if $M = \text{Int}(\omega_s/B)$ where $B = (\omega_H - \omega_L)$, the downsamplers will produce exactly M copies of the spectrum of $\tilde{x}(nT)$ in the interval $-\omega_s/2 \leq \omega \leq \omega_s/2$, as illustrated in Fig. 16.23c for the case $M = 8$. As can be seen, the information content of the signal now occupies the interval $-B/2 \leq \omega \leq B/2$ and the signal can be processed using a sampling frequency B. For a narrowband signal, we have $B \ll 2\omega_H$ and, therefore, the necessary speed of operation of the hardware is significantly reduced.

The processed bandpassed signal can be recovered by reversing the above procedure using the configuration of Fig. 16.22b, where the bandpass filter is a complex bandpass filter with a frequency response

$$H(e^{j\omega T}) = \begin{cases} M & \text{for } \omega_L \leq \omega \leq \omega_H \\ 0 & \text{otherwise} \end{cases}$$

The operation of this scheme is illustrated by the frequency spectra in Fig. 16.24. Upsampling will produce $M - 1$ images of the spectrum of $\tilde{x}(nT)$, as shown in

---

[†]Other authors refer to such signals as *bandpass* signals.

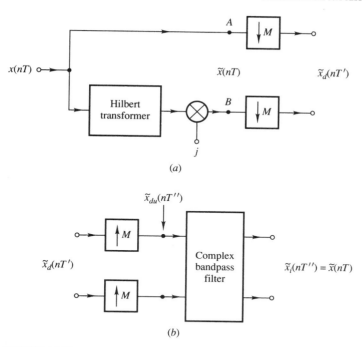

FIGURE 16.22
Processing of bandpassed signals: (*a*) sampling, (*b*) reconstruction.

Fig. 16.24*b* and the desired image can be selected by the bandpass filter, as depicted in Fig. 16.24*d*.

16.5 ADAPTIVE DIGITAL FILTERS

In many applications, time-variable filters are required whose characteristics can be varied with time. Filters of this type can be obtained by using multipliers with time-variable coefficients. A time-variable filter that incorporates some adaptation mechanism by which the multiplier coefficients can be adjusted on line so as to optimize some performance criterion is said to be an *adaptive filter* [2–4]. The adaptation mechanism usually incorporates an optimization algorithm that evaluates the instantaneous values of the multiplier coefficients such that some norm of an error function of the form

$$e(n) = d(n) - y(n) \tag{16.45}$$

is minimized, where $d(n)$ is some desired reference signal and $y(n)$ is the filter output. A typical adaptive-filter configuration is illustrated in Fig. 16.25.

The design of adaptive filters involves the choice of filter structure, the specific error norm to be used as objective function, and the type of adaptation algorithm [18–24]. The structure can be nonrecursive or recursive; the objective function may involve the expected amplitude or square of the error and possibly the expected value

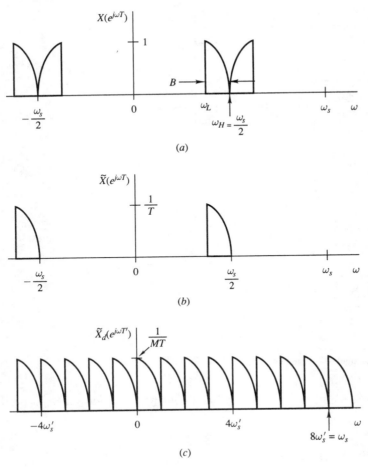

FIGURE 16.23
Operation of scheme in Fig. 16.22a.

of some higher power or the maximum of the error; similarly, the adaptation algorithm can be one of several possibilities. Like other filters, adaptive filters are required to be economical, fast, and insensitive to finite word-length effects; in addition, they should adapt in a short period of time, and the residual error after adaptation should be as small as possible. Hence, the performance criteria for these filters are the simplicity and properties of the structure, and the flexibility, reliability, computational complexity, and convergence properties of the adaptation algorithm employed.

Usually, the most well-behaved of the possible objective functions involves the square of the error (e.g., the mean-squared error), which can be minimized very quickly using some relatively simple optimization algorithms. As a consequence, this objective function is preferred, although a mean-square solution may not be the most appropriate in certain applications.

DIGITAL SIGNAL PROCESSING APPLICATIONS **627**

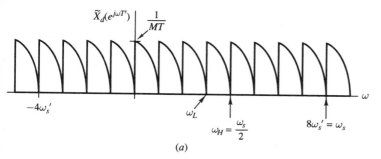

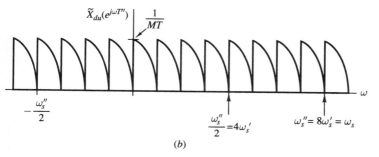

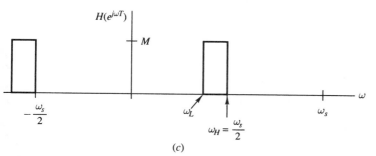

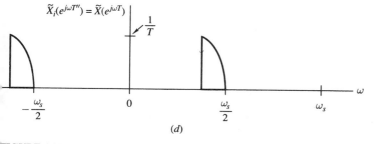

FIGURE 16.24
Operation of scheme in Fig. 16.22b.

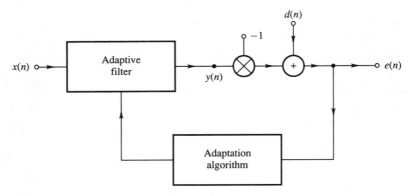

FIGURE 16.25
Typical adaptive-filter configuration.

Below we examine an optimal class of digital filters which are designed by choosing the transfer-function coefficients such that the mean-square value of an error function of the type given in Eq. (16.45) is minimized for some desired response $d(n)$. Filters so designed are commonly referred to as *Wiener filters*, and optimization algorithms that can be used for their design can often be implemented as adaptation algorithms.

16.5.1 Wiener Filters

The simplest structure that can be used for adaptive filters is the direct nonrecursive structure of Fig. 16.26. The output of this configuration is given by

$$y(n) = \sum_{i=0}^{N-1} a_i(n) x(n-i) = \mathbf{a}_n^T \mathbf{x}_n \qquad (16.46)$$

where

$$\mathbf{x}_n = [x(n) \quad x(n-1) \quad \ldots \quad x[n-(N-1)]]^T$$

and

$$\mathbf{a}_n = [a_0(n) \quad a_1(n) \quad \ldots \quad a_{N-1}(n)]^T$$

are the input-signal and coefficient vectors, respectively, at instant nT.

The *mean-square error* (MSE) is defined as

$$\mathbf{\Psi}(\mathbf{a}_n) = E[e^2(n)] \qquad (16.47)$$

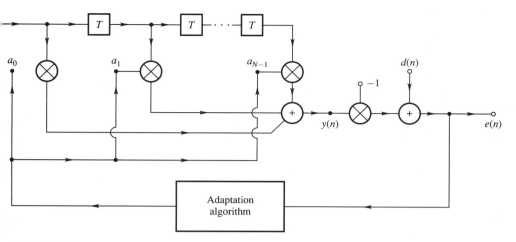

FIGURE 16.26
Nonrecursive adaptive-filter configuration.

where $E[\cdot]$ is the *expected* value of $[\cdot]$. From Eqs. (16.47), (16.45), and (16.46), we can write

$$\Psi(\mathbf{a}_n) = E[d^2(n) - 2d(n)y(n) + y^2(n)]$$
$$= E[d^2(n) - 2d(n)\mathbf{a}_n^T \mathbf{x}_n + \mathbf{a}_n^T \mathbf{x}_n \mathbf{x}_n^T \mathbf{a}_n]$$
$$= E[d^2(n)] - 2E[d(n)\mathbf{a}_n^T \mathbf{x}_n] + E[\mathbf{a}_n^T \mathbf{x}_n \mathbf{x}_n^T \mathbf{a}_n]$$

For a filter with fixed coefficients, the MSE function is given by

$$\Psi(\mathbf{a}_n) = E[d^2(n)] - 2\mathbf{a}_n^T \mathbf{p}_n + \mathbf{a}_n^T \mathbf{R}_n \mathbf{a}_n \tag{16.48}$$

where

$$\mathbf{p}_n = E[d(n)\mathbf{x}_n]$$

and

$$\mathbf{R}_n = E[\mathbf{x}_n \mathbf{x}_n^T]$$

are the *cross correlation* between the desired and input signals and the correlation matrix of the input signal, respectively, at instant nT. In effect, *the objective function in Eq. (16.47) is a quadratic function of the filter coefficients $a_0(n), a_1(n), \ldots, a_{N-1}(n)$.*

The *gradient* vector of the MSE function can be readily expressed as

$$\mathbf{g}_n = \nabla \Psi(\mathbf{a}_n) = \left[\frac{\partial \Psi(\mathbf{a}_n)}{\partial a_0(n)} \quad \frac{\partial \Psi(\mathbf{a}_n)}{\partial a_1(n)} \quad \cdots \quad \frac{\partial \Psi(\mathbf{a}_n)}{\partial a_{N-1}(n)} \right]^T \tag{16.49}$$

and from Eqs. (16.48) and (16.49)

$$\mathbf{g}_n = -2\mathbf{p}_n + 2\mathbf{R}_n \mathbf{a}_n \tag{16.50}$$

Now, on equating the elements of the gradient vector to zero, the coefficient vector that minimizes the MSE function, say $\overset{\cup}{\mathbf{a}}$, can be deduced as

$$\overset{\cup}{\mathbf{a}} = \mathbf{R}_n^{-1}\mathbf{p}_n$$

Evidently, if $\mathbf{p}_n$ and $\mathbf{R}_n$ are known the Wiener solution can be readily obtained. In practice, accurate estimates of $\mathbf{p}_n$ and $\mathbf{R}_n$ are not always available, but *time averages* may be used for their estimation if $d(n)$ and $x(n)$ are stationary and ergodic signals. In such applications, $\mathbf{p}_n$ and $\mathbf{R}_n$ represent a constant vector and a constant matrix, respectively, and the subscript n can be dropped.

Many of the available adaptation algorithms are practical algorithms that lead to the Wiener solution and are borrowed from the field of optimization. Commonly used algorithms are based on the Newton and steepest-descent algorithms.

16.5.2 Newton Algorithm

The Hessian matrix of $\boldsymbol{\Psi}(\mathbf{a}_n)$ can be obtained from Eq. (16.50) as $\mathbf{H}_n = 2\mathbf{R}_n$ and, therefore, the Newton direction (see Sec. 14.3) can be determined as $\mathbf{R}_n^{-1}\mathbf{g}_n/2$. The Wiener solution can, therefore, be approached by obtaining estimates of $\mathbf{a}_{n+1}$ such that

$$\mathbf{a}_{n+1} = \mathbf{a}_n - \alpha\, \mathbf{R}_n^{-1}\tilde{\mathbf{g}}_n \qquad (16.51)$$

for $n = 0, 1, \ldots$, where α is a constant, $\tilde{\mathbf{g}}_n$ is an estimate of $\mathbf{g}_n$, and $\mathbf{a}_0$ is an initial estimate of the transfer-function coefficients. The algorithm is terminated when some convergence criterion is satisfied. This algorithm is essentially the basic Newton algorithm described in detail in Sec. 14.3 except that no line search is used. The constant α, which is sometimes called the *convergence factor*, is chosen to achieve fast convergence for the type of application under consideration. Line searches have not been used in the past owing to the additional amount of computation required, but certain related techniques are likely to be employed in the future.

If $\tilde{\mathbf{g}}_n$ in Eq. (16.51) is the exact gradient, matrix $\mathbf{R}_n$ is a well-behaved positive-definite matrix, and $\alpha = 1/2$, then the Newton algorithm gives the required solution in one iteration, as was demonstrated in Sec. 14.3. However, if $\tilde{\mathbf{g}}_n$ is an approximate estimate of the gradient, a number of iterations is required.

In practice, the Newton algorithm is characterized by a very small number of iterations, but the amount of computation required per iteration is quite large since the inversion of matrix $\mathbf{R}_n$ is required. If $\mathbf{R}_n$ is nearly nonsingular or ill-conditioned, the algorithm can become quite inefficient.

16.5.3 Steepest-Descent Algorithm

If matrix $\mathbf{R}_n$ is assumed to be the $N \times N$ unity matrix, then the updating formula in Eq. (16.51) assumes the form

$$\mathbf{a}_{n+1} = \mathbf{a}_n - \alpha\, \tilde{\mathbf{g}}_n \qquad (16.52)$$

If the negative of the gradient vector, namely, $-\mathbf{g}_n$, is drawn through point $\mathbf{a}_n$, it points in the direction of steepest descent, as can be easily shown, and for this reason the use of Eq. (16.52) as updating formula leads to the so-called *steepest-descent* algorithm.

This algorithm is very simple to implement since the Hessian matrix and its inversion are not required. At the start of the adaptation, the algorithm leads to a large reduction in the error function per iteration. However, as the solution is approached, the elements of the gradient become smaller and smaller and progress in the adaptation process tends to slow down considerably; in particular, if the minimum point is located in the middle of a relatively flat valley. Overall, the steepest-descent algorithm is usually less efficient than the Newton algorithm, but the amount of computation per iteration is much smaller. This makes the steepest-descent algorithm more suitable for real-time applications.

If the ratio of the largest to the smallest eigenvalue of matrix $\mathbf{R}_n$ is large, then the solution point tends to follow a zig-zag trajectory in the parameter space and the performance of the algorithm tends to deteriorate. Under certain circumstances, the algorithm can actually become unstable, as demonstrated by the following example.

Example 16.4. (*a*) Show that the steepest-descent algorithm can be treated as an N-input, N-output first-order digital filter. (*b*) Using the filter obtained, find a necessary and sufficient condition for the stability of the algorithm.

Solution. (*a*) From Eqs. (16.52) and (16.50)

$$\mathbf{a}_{n+1} = \tilde{\mathbf{R}}_n \mathbf{a}_n + 2\alpha\, \mathbf{p}_n \tag{16.53}$$

where

$$\tilde{\mathbf{R}}_n = \mathbf{I} - 2\alpha\, \mathbf{R}_n$$

Matrix $\tilde{\mathbf{R}}_n$ can be expressed as

$$\tilde{\mathbf{R}}_n = \mathbf{Q}\tilde{\mathbf{\Lambda}}\mathbf{Q}^T$$

where $\mathbf{Q}$ is a unitary matrix whose columns comprise an orthogonal set of eigenvectors associated with the eigenvalues of $\tilde{\mathbf{R}}_n$ and $\tilde{\mathbf{\Lambda}}$ is a diagonal matrix whose diagonal elements are the eigenvalues of $\tilde{\mathbf{R}}_n$. Hence Eq. (16.53) can be put in the form

$$\mathbf{a}_{n+1} = \mathbf{Q}\tilde{\mathbf{\Lambda}}\mathbf{Q}^T \mathbf{a}_n + 2\alpha\, \mathbf{p}_n$$

and on premultiplying both sides by $\mathbf{Q}^T$ and letting

$$\mathbf{a}'_{n+1} = \mathbf{Q}^T \mathbf{a}_{n+1} \quad\text{and}\quad \mathbf{p}'_n = 2\alpha\, \mathbf{Q}^T \mathbf{p}_n$$

we obtain

$$\mathbf{a}'_{n+1} = \tilde{\mathbf{\Lambda}} \mathbf{a}'_n + \mathbf{p}'_n \tag{16.54}$$

This equation represents an N-input, N-output, first-order digital filter with input $\mathbf{x}_n = \mathbf{p}'_n$ and output $\mathbf{y}_n = \mathbf{a}'_{n+1}$ as shown in Fig. 16.27.

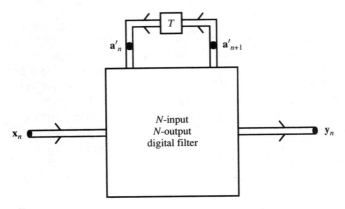

FIGURE 16.27
N-input, N-output, first-order digital filter (Example 16.4).

(b) The algorithm is stable if and only if each of the elements of the impulse response vector $\mathbf{h}_n$ of the filter in Fig. 16.27 is absolutely summable (see Sec. 1.8). From Eq. (16.54)

$$\mathbf{h}_0 = \mathbf{a}'_1 = \tilde{\mathbf{\Lambda}} \mathbf{a}'_0 + \boldsymbol{\delta}_0$$

$$\mathbf{h}_1 = \mathbf{a}'_2 = \tilde{\mathbf{\Lambda}} \mathbf{a}'_1 + \mathbf{0} = \tilde{\mathbf{\Lambda}}^2 \mathbf{a}'_0 + \tilde{\mathbf{\Lambda}} \boldsymbol{\delta}_0$$

$$\mathbf{h}_2 = \mathbf{a}'_3 = \tilde{\mathbf{\Lambda}} \mathbf{a}'_2 + \mathbf{0} = \tilde{\mathbf{\Lambda}}^3 \mathbf{a}'_0 + \tilde{\mathbf{\Lambda}}^2 \boldsymbol{\delta}_0$$

$$\cdots\cdots\cdots\cdots\cdots\cdots\cdots\cdots\cdots\cdots$$

$$\mathbf{h}_n = \mathbf{a}'_{n+1} = \tilde{\mathbf{\Lambda}}^{n+1} \mathbf{a}'_0 + \tilde{\mathbf{\Lambda}}^n \boldsymbol{\delta}_0$$

Without loss of generality, we can assume that $\mathbf{a}_0 = 0$ or $\mathbf{a}'_0 = 0$ and hence

$$\mathbf{h}_n = \tilde{\mathbf{\Lambda}}^n \boldsymbol{\delta}_0$$

Thus, the ith element of the impulse response vector $\mathbf{h}_n$ is obtained as

$$h_i(n) = \tilde{\lambda}_i^n \tag{16.55}$$

The filter is stable if and only if

$$\sum_{n=0}^{\infty} |h_i(n)| < \infty \qquad \text{for } i = 1, 2, \ldots, N \tag{16.56}$$

and, therefore, a necessary and sufficient condition for stability can be obtained from Eqs. (16.55) and (16.56) as

$$|\tilde{\lambda}_i| < 1 \qquad \text{for } i = 1, 2, \ldots, N \tag{16.57}$$

Now the eigenvalues of $\tilde{\mathbf{R}}_n$ are related to those of $\mathbf{R}_n$ by the equation

$$\tilde{\lambda}_i = 1 - 2\alpha \lambda_i$$

(see Prob. 16.21) and hence the inequality in Eq. (16.57) can be expressed as

$$|1 - 2\alpha \lambda_i| < 1 \qquad \text{for } i = 1, 2, \ldots, N$$

Since the eigenvalues of $\mathbf{R}_n$ are real and positive, the steepest-descent algorithm is stable if and only if

$$0 < \alpha < 1/\hat{\lambda}$$

where $\hat{\lambda}$ is the largest eigenvalue of $\mathbf{R}_n$.

Improved performance can often be achieved, in practice, by starting the adaptation process with the steepest-descent algorithm and, when certain progress has been achieved, switching over to the Newton algorithm, which is more efficient at points in the neighborhood of the solution point.

16.5.4 Least-Mean-Square Algorithm

As can be seen in Eq. (16.50), the gradient depends on vector $\mathbf{p}_n$ and matrix $\mathbf{R}_n$, which are not in general available. Nevertheless, an estimate of the gradient, $\tilde{\mathbf{g}}_n$, can be deduced by letting

$$\tilde{\mathbf{p}}_n = d(n)\mathbf{x}_n \tag{16.58}$$

and

$$\tilde{\mathbf{R}}_n = \mathbf{x}_n \mathbf{x}_n^T \tag{16.59}$$

be estimates of $\mathbf{p}$ and $\mathbf{R}$, respectively. From Eqs. (16.50), (16.58), (16.59), (16.45), and (16.46), we can write

$$\begin{aligned}
\tilde{\mathbf{g}}_n &= -2d(n)\mathbf{x}_n + 2\mathbf{x}_n \mathbf{x}_n^T \mathbf{a}_n \\
&= -2[d(n) - \mathbf{x}_n^T \mathbf{a}_n]\mathbf{x}_n \\
&= -2e(n)\mathbf{x}_n
\end{aligned} \tag{16.60}$$

On using this estimate of the gradient in the updating formula of the steepest-descent algorithm given in Eq. (16.52), we obtain

$$\mathbf{a}_{n+1} = \mathbf{a}_n + 2\alpha\, e(n)\mathbf{x}_n \tag{16.61}$$

The use of this formula yields the so-called *least-mean-square* (LMS) algorithm. The convergence factor α is chosen in the range

$$0 < \alpha < \frac{1}{NE[x^2(n)]} \tag{16.62}$$

in order to guarantee the convergence of the algorithm (see Prob. 16.23). The quantity $E[x^2(n)]$ represents the average input power and is usually easy to estimate.

Note that

$$\nabla e^2(n) = 2e(n) \left[\frac{\partial e(n)}{\partial a_0(n)} \quad \frac{\partial e(n)}{\partial a_1(n)} \quad \cdots \quad \frac{\partial e(n)}{\partial a_{N-1}(n)} \right]^T$$

and from Eqs. (16.45), (16.46), and (16.60)

$$\nabla e^2(n) = -2e(n)\mathbf{x}_n = \tilde{\mathbf{g}}_n$$

that is, $\tilde{\mathbf{g}}_n$ is the exact gradient of $e^2(n)$ and thus the LMS algorithm minimizes the instantaneous power of the error signal.

Further details on the above algorithms as well as some others can be found in [2–4].

16.5.5 Recursive Filters

In applications where high selectivity is required, a nonrecursive design would necessitate a transfer function of high order, which may entail a high computational complexity. For such applications, a recursive design may be the only possible solution. Unfortunately, however, the use of recursive structures introduces several new problems. First, the coefficients of the denominator polynomial of the transfer function may assume values in the unstable region of the coefficient space and, if this happens, the adaptive filter will become unstable. Second, the objective function becomes highly nonlinear and may have several local minima, some of them quite shallow. Hence the adaptation algorithm may easily converge to some unsatisfactory solution.

The problem of instability can be overcome by incorporating checks in the adaptation algorithm that can detect an unstable solution and restore stability by making a suitable adjustment to the current coefficient values. For example, the coefficients can be adjusted in such a way as to replace any poles outside the unit circle of the z plane by their reciprocals (see Sec. 7.4, p. 226). The problem of several local minima, which is quite difficult to solve, can be eased to some extent by selecting a well-behaved objective function and by using a good estimate of the solution for the initialization of $\mathbf{a}_0$. For most practical problems, the objective function is well behaved in the neighborhood of the optimum solution, and an initial point in this domain will cause the adaptation algorithm to converge to the optimum solution.

The design of adaptive filters based on recursive structures has been studied by a number of researchers in recent years [23, 24] and more work is anticipated in the future.

16.5.6 Applications

The applications of adaptive filters are numerous and include system identification, channel equalization, signal enhancement, and signal prediction.

In a *system identification* application, a broadband signal, usually white noise, is applied simultaneously at the inputs of an unknown system and an adaptive filter, and an error signal is formed by subtracting the output of the adaptive filter from that of the unknown system, as depicted in Fig. 16.28. If the error signal obtained is minimized, the adaptive filter becomes a model for the unknown system.

The transmission of a signal through an imperfect channel entails amplitude and phase distortion, as was shown in Sec. 3.6. If the frequency response of the channel is known, it can be equalized using a fixed filter, as described in Secs. 8.5.1 and 8.6. However, if the channel response is variable, equalization by means of an adaptive filter is more appropriate. A variable channel response can occur in telephony where

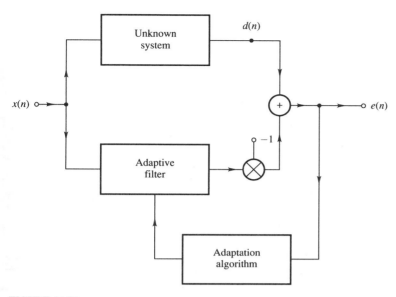

FIGURE 16.28
Use of adaptive filters for system identification.

the physical wire path between any two subscribers depends not only on the locations of the two subscribers but also on the time of the call. *Channel equalization* can be achieved by connecting an adaptive filter in cascade with the channel and comparing the output of the cascade arrangement with a delayed version of the input signal, as illustrated in Fig. 16.29. In this application, the reference signal is a training signal and is known at the receiving end, i.e., it need not be transmitted.

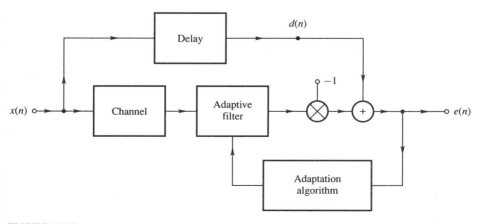

FIGURE 16.29
Use of adaptive filters for channel equalization.

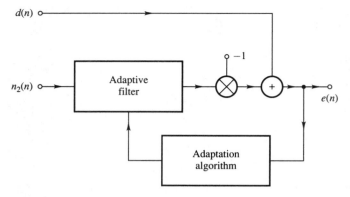

FIGURE 16.30
Use of adaptive filters for signal enhancement.

In *signal enhancement* applications, a signal $x(n)$ corrupted by a noise component $n_1(n)$, namely,

$$d(n) = x(n) + n_1(n)$$

is used as a reference signal and a signal $n_2(n)$ which is correlated to the noise component $n_1(n)$ is applied to the input of an adaptive filter, as depicted in Fig. 16.30. After adaptation, the error signal $e(n)$ will represent an enhanced version of the signal $x(n)$ in which a significant amount of noise has been removed.

In *signal prediction* applications, the signal of interest is used as the reference signal and a delayed version is applied to the input of an adaptive filter. The error signal is generated by subtracting the output of the adaptive filter from the reference signal, as shown in Fig. 16.31. After convergence, the filter coefficients are adjusted in response to the past signal values and can, therefore, be used to reconstruct or

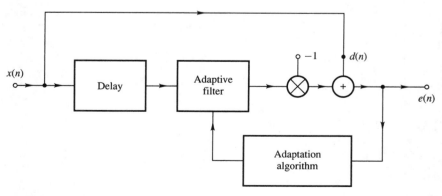

FIGURE 16.31
Use of adaptive filters for signal prediction.

extrapolate the input signal. Signal prediction is used in speech coding, where a model of the signal rather than the signal itself is encoded and transmitted.

16.6 TWO-DIMENSIONAL DIGITAL FILTERS

In many applications, continuous signals are encountered that are inherently two-dimensional (2-D). Signals of this type can be represented by functions of the form $x(t_1, t_2)$, where t_1 and t_2 are independent variables. Each of the two variables can represent an arbitrary physical quantity such as time, length, velocity, acceleration, temperature, and so on. An example of a 2-D continuous signal is the light intensity in an image as a function of the x and y coordinates.

Two-dimensional continuous signals, like their 1-D counterparts, can be represented by frequency spectra. Furthermore, they can be sampled at discrete points $(t_1, t_2) = (n_1 T_1, n_2 T_2)$ to yield discrete[†] signals $x(n_1 T_1, n_2 T_2)$. Signals of this type have frequency spectra that are periodic with respect to frequencies ω_1 and ω_2 with a 2-D period $(\omega_{s1}, \omega_{s2})$, where $\omega_{s1} = 2\pi/T_1$ and $\omega_{s2} = 2\pi/T_2$.

Two-dimensional discrete signals can be processed by *2-D digital filters* whose operation and properties are analogous to those of their 1-D counterparts, that is, they can be time-invariant or time-dependent, causal or noncausal, and linear or nonlinear. Since neither of the two independent variables needs to be time, causality does not have the usual physical interpretation.

The theory, analysis and design methods, and applications of 2-D digital filters are quite extensive but are largely beyond the scope of this book. Below, we present some of the basic principles involved and some of the straightforward extensions of 1-D methods and techniques for the sake of completeness. The reader with more than a casual interest in the analysis and design of 2-D digital filters is referred to the more specialized books on the subject cited earlier.

A 2-D causal recursive (or IIR) digital filter with excitation $x(n_1, n_2)$ and response $y(n_1, n_2)$ can be represented by a difference equation in two variables of the form

$$y(n_1, n_2) = \sum_{i=0}^{N_1} \sum_{j=0}^{N_2} a_{ij} x(n_1 - i, n_2 - j) - \sum_{i=0}^{N_1} \sum_{j=0}^{N_2} b_{ij} y(n_1 - i, n_2 - j)$$

where $b_{00} = 0$. If $b_{ij} = 0$ for $0 \leq i \leq N_1$ and $0 \leq j \leq N_2$, the representation of a 2-D nonrecursive (or FIR) filter is obtained. The pair (N_1, N_2) is the *order* of the filter.

16.6.1 Two-Dimensional Convolution

If the impulse response of a 2-D filter $h(n_1, n_2)$ is known, then its response $y(n_1, n_2)$ to an arbitrary excitation $x(n_1, n_2)$ can be determined by using the *2-D convolution*.

[†]Discrete-time is changed to discrete since neither of the two variables needs to be time.

If the 2-D filter is causal, i.e., its impulse response is zero for $n_1 < 0$ or $n_2 < 0$, then

$$y(n_1, n_2) = \sum_{i=0}^{n_1} \sum_{j=0}^{n_2} x(i,j) h(n_1 - i, n_2 - j)$$

$$= \sum_{i=0}^{n_1} \sum_{j=0}^{n_2} h(i,j) x(n_1 - i, n_2 - j)$$

This formula can be derived by following the approach of Sec. 1.7.

16.6.2 Two-Dimensional z Transform

The most important mathematical tool for the analysis and design of 2-D digital filters is the *2-D z transform*, which is a straightforward extension of its 1-D counterpart. The 2-D z transform of a function $f(n_1, n_2)$ is defined as

$$F(z_1, z_2) = \sum_{n_1=-\infty}^{\infty} \sum_{n_2=-\infty}^{\infty} f(n_1, n_2) z_1^{-n_1} z_2^{-n_2}$$

for all (z_1, z_2) for which the double summation converges. Function $f(n_1, n_2)$ is the 2-D inverse z transform of $F(z_1, z_2)$ and is given by

$$f(n_1, n_2) = \frac{1}{(2\pi j)^2} \oint_{\Gamma_2} \oint_{\Gamma_1} F(z_1, z_2) z_1^{n_1-1} z_2^{n_2-1} dz_1 dz_2 \qquad (16.63)$$

where the two integrals are evaluated in the counterclockwise sense over contours Γ_1 and Γ_2 that are in the region of convergence of $F(z_1, z_2)$.

16.6.3 Two-Dimensional Transfer Function

The transfer function of a 2-D digital filter is the z transform of the impulse response, as can be shown by applying the z transform to the convolution summation. It can be expressed as

$$H(z_1, z_2) = \frac{N(z_1, z_2)}{D(z_1, z_2)} = \frac{\sum_{i=0}^{N_1} \sum_{j=0}^{N_2} a_{ij} z_1^{N_1-i} z_2^{N_2-j}}{z_1^{N_1} z_2^{N_2} + \sum_{i=0}^{N_1} \sum_{j=0}^{N_2} b_{ij} z_1^{N_1-i} z_2^{N_2-j}}$$

where $b_{00} = 0$. The transfer function can be used to find the response of the filter to an arbitrary excitation and its frequency-domain response; furthermore, it contains all the necessary information to determine whether the filter is stable or unstable.

16.6.4 Stability

The *stability* of a 2-D digital filter is closely linked with the singularities of the transfer function, as in 1-D filters. Unfortunately, in 2-D digital filters the singularities are not in general isolated and, as a result, stability analysis is much more complicated. A sufficient condition for the stability of a 2-D filter due to Shanks [25, 26] is that all the singularities of the transfer function are located on the open unit bidisc defined by the set

$$U^2 = \{(z_1, z_2) : |z_1| < 1, |z_2| < 1\}$$

i.e.,

$$D(z_1, z_2) \neq 0 \quad \text{for } (z_1, z_2) \notin U^2 \tag{16.64}$$

where

$$D(z_1, z_2) = z_1^{N_1} z_2^{N_2} + \sum_{i=0}^{N_1} \sum_{j=0}^{N_2} b_{ij} z_1^{N_1-i} z_2^{N_2-j}$$

with $b_{00} = 0$.

Example 16.5. A 2-D digital filter is characterized by the transfer function

$$H(z_1, z_2) = \frac{N(z_1, z_2)}{D(z_1, z_2)}$$

where

$$N(z_1, z_2) = 512(z_1 + 1)^2 (z_2 + 1)^2$$

and

$$D(z_1, z_2) = 512 z_1^2 z_2^2 - 128 z_1 z_2^2 + 256 z_1^2 z_2 - 192 z_2^2$$
$$- 40 z_1^2 - 64 z_1 z_2 + 10 z_1 - 96 z_2 + 15$$

Check the stability of the filter.

Solution. The transfer function can be expressed as

$$H(z_1, z_2) = \frac{N'(z_1, z_2)}{D'(z_1, z_2)}$$

where

$$N'(z_1, z_2) = (z_1 + 1)^2 (z_2 + 1)^2$$

and

$$D'(z_1, z_2) = z_1^2 z_2^2 - \tfrac{1}{4} z_1 z_2^2 + \tfrac{1}{2} z_1^2 z_2 - \tfrac{3}{8} z_2^2$$
$$- \tfrac{5}{64} z_1^2 - \tfrac{1}{8} z_1 z_2 + \tfrac{5}{256} z_1 - \tfrac{3}{16} z_2 + \tfrac{15}{512}$$

The denominator polynomial $D'(z_1, z_2)$ can now be put in the form

$$D'(z_1, z_2) = \left(z_2^2 + \tfrac{1}{2} z_2 - \tfrac{5}{64}\right) z_1^2 - \left(\tfrac{1}{4} z_2^2 + \tfrac{1}{8} z_2 - \tfrac{5}{256}\right) z_1$$
$$- \left(\tfrac{3}{8} z_2^2 + \tfrac{3}{16} z_2 - \tfrac{15}{512}\right)$$
$$= \left(z_1^2 - \tfrac{1}{4} z_1 - \tfrac{3}{8}\right) \left(z_2^2 + \tfrac{1}{2} z_2 - \tfrac{5}{64}\right)$$
$$= \left(z_1 + \tfrac{1}{2}\right) \left(z_1 - \tfrac{3}{4}\right) \left(z_2 + \tfrac{5}{8}\right) \left(z_2 - \tfrac{1}{8}\right)$$

Hence the transfer function $H(z_1, z_2)$ is singular only at points

$$(z_1, z_2) = \begin{cases} \left(-\tfrac{1}{2}, z_2\right) & \left(z_1, -\tfrac{5}{8}\right) \\ \left(\tfrac{3}{4}, z_2\right) & \left(z_1, \tfrac{1}{8}\right) \end{cases}$$

Therefore, $D(z_1, z_2)$ satisfies Eq. (16.64) and as a consequence the filter is stable.

If the denominator of the transfer function can be factored into a product of polynomials of the form $D_1(z_1)D_2(z_2)$ where $D_1(z_1)$ and $D_2(z_2)$ are polynomials in z_1 and z_2, respectively, as in the above example, the stability of the filter can be easily checked by using the techniques of Sec. 3.3, e.g., by applying the Jury-Marden stability criterion. However, if $D(z_1, z_2)$ cannot be factored, the stability analysis can be quite involved (see Chap. 5 of [7]).

16.6.5 Frequency-Domain Analysis

The *frequency response* of a 2-D filter is given by

$$H(e^{j\omega_1 T_1}, e^{j\omega_2 T_2}) = M(\omega_1, \omega_2)e^{j\theta(\omega_1, \omega_2)}$$

where

$$M(\omega_1, \omega_2) = |H(e^{j\omega_1 T_1}, e^{j\omega_2 T_2})| \qquad (16.65)$$

and

$$\theta(\omega_1, \omega_2) = \arg H(e^{j\omega_1 T_1}, e^{j\omega_2 T_2}) \qquad (16.66)$$

are the amplitude and phase response, respectively. A pair of parameters that are sometimes of interest in 2-D digital filters are the *group delays*. These are defined as

$$\tau_1 = -\frac{\partial \theta(\omega_1, \omega_2)}{\partial \omega_1} \quad \text{and} \quad \tau_2 = -\frac{\partial \theta(\omega_1, \omega_2)}{\partial \omega_2} \qquad (16.67)$$

Example 16.6. A 2-D nonrecursive digital filter designed by using the method in [32] has the transfer function

$$H(z_1, z_2) = \sum_{i=1}^{4} (-1)^{i+1} H_i(z_1) H_i(z_2)$$

where

$$H_i(z_k) = \sum_{j=0}^{24} a_{ij} z_k^{-j} \qquad \text{for } k = 1, 2$$

and
$$a_{ij} = a_{i(24-j)} \quad \text{for } i = 1, 2, 3, 4 \tag{16.68}$$

The coefficients a_{ij} are given in Table 16.3. Obtain the amplitude response of the filter.

Solution. The amplitude response can be obtained as shown in Fig. 16.32 by using Eq. (16.65). As can be seen, the given transfer function represents a lowpass filter with a circular passband. Equation (16.68) amounts to symmetrical impulse responses in the 1-D filters, represented by $H_1(z_1)$ and $H_2(z_2)$, i.e., these filters have a linear phase response (see Sec. 9.2). Therefore, the phase response of the 2-D filter, given by Eq. (16.66), is linear with respect to both ω_1 and ω_2, and the group delays in Eq. (16.67) are constant throughout the baseband. It should be mentioned in passing that a linear phase response is highly desirable in image processing applications.

16.6.6 Types of 2-D Filters

As in the case of 1-D filters, different types of 2-D filters can be identified on the basis of their amplitude responses, e.g., lowpass, highpass, etc. Passbands and stopbands are now subareas of the (ω_1, ω_2) plane and can be rectangular or circular. A 2-D lowpass filter has an amplitude response of the form

$$M(\omega_1, \omega_2) \approx \begin{cases} 1 & \text{for } (\omega_1, \omega_2) \in R_1 \\ 0 & \text{for } (\omega_1, \omega_2) \in R_2 \end{cases}$$

where
$$R_1 = \{(\omega_1, \omega_2): |\omega_1| \leq \omega_{p1} \text{ and } |\omega_2| \leq \omega_{p2}\}$$
and
$$R_2 = \{(\omega_1, \omega_2): |\omega_1| \geq \omega_{a1} \text{ or } |\omega_2| \geq \omega_{a2}\}$$

TABLE 16.3
Coefficients of 2-D transfer function (Example 16.6)

j	a_{1j}	a_{2j}	a_{3j}	a_{4j}
0	0.0007	0.0012	0.0031	0.0027
1	−0.0004	−0.0009	0.0006	0.0034
2	−0.0013	−0.0033	−0.0065	−0.0063
3	0.0022	0.0043	−0.0009	−0.0099
4	0.0025	0.0105	0.0173	0.0155
5	−0.0086	−0.0119	−0.0019	0.0301
6	−0.0036	−0.0206	−0.0474	−0.0109
7	0.0241	0.0374	−0.0154	−0.0460
8	−0.0007	0.0549	0.0746	−0.0214
9	−0.0688	−0.0713	0.0744	−0.0007
10	0.0189	−0.1635	0.0015	−0.0137
11	0.3145	−0.0519	−0.0028	−0.0049
12	0.4935	0.0544	0.0250	0.0124

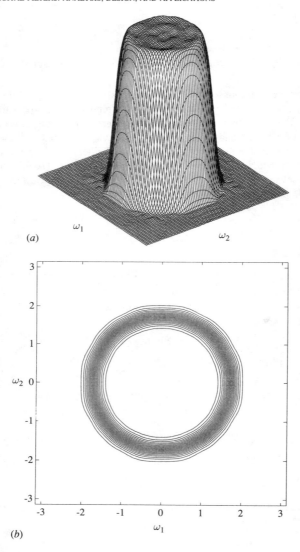

FIGURE 16.32
Amplitude response of 2-D digital filter (Example 16.5): (*a*) 3-D plot, (*b*) contour plot.

in a filter with rectangular band boundaries or

$$R_1 = \left\{ (\omega_1, \omega_2): \sqrt{\omega_1^2 + \omega_2^2} < \omega_p \right\}$$

and

$$R_2 = \left\{ (\omega_1, \omega_2): \sqrt{\omega_1^2 + \omega_2^2} > \omega_a \right\}$$

in a filter with circular band boundaries. A 2-D highpass filter, on the other hand, has an amplitude response

$$M(\omega_1, \omega_2) \approx \begin{cases} 0 & \text{for } (\omega_1, \omega_2) \in R_1 \\ 1 & \text{for } (\omega_1, \omega_2) \in R_2 \end{cases}$$

where R_1 and R_2 are as above.

Another type of filter that has no counterpart in the 1-D domain is the so called *fan filter*. A fan filter has an amplitude response

$$M_1(\omega_1, \omega_2) \approx \begin{cases} 1 & \text{for } (\omega_1, \omega_2) \in S_1 \\ 0 & \text{for } (\omega_1, \omega_2) \in S_2 \end{cases}$$

where

$$S_1 = \left\{ (\omega_1, \omega_2) : \left| \tan^{-1} \frac{\omega_2}{\omega_1} \right| < \theta_1 \text{ or } \left| \tan^{-1} \frac{\omega_2}{\omega_1} \right| > \pi - \theta_1 \right\}$$

and

$$S_2 = \left\{ (\omega_1, \omega_2) : \left| \tan^{-1} \frac{\omega_2}{\omega_1} \right| > \theta_2 \text{ or } \left| \tan^{-1} \frac{\omega_2}{\omega_1} \right| < \pi - \theta_2 \right\}$$

with $\theta_2 > \theta_1$.

16.6.7 Approximations

The most difficult task in the design of 2-D digital filters is the solution of the *approximation* problem, which entails the derivation of a stable transfer function such that prescribed amplitude and/or phase response specifications are achieved. As in 1-D filters, the approximation problem can be solved by using direct or indirect methods in terms of closed-form or iterative solutions.

Nonrecursive filters can be designed by using the 2-D Fourier series in conjunction with 2-D window functions [27] (see Secs. 9.3 and 9.4) or by using a transformation due to McClellan [28, 29]. Recursive filters, on the other hand, can be designed by applying transformations to 1-D filters [30, 31]. Nonrecursive as well as recursive filters can be designed by using the singular-value decomposition [32, 33] or through the application of optimization methods [34-37].

If the numerator and denominator of the transfer function can be factored into products $N_1(z_1)N_2(z_2)$ and $D_1(z_1)D_2(z_2)$, then the transfer function is said to be *separable* and can be expressed as

$$H(z_1, z_2) = H_1(z_1, z_2) H_2(z_1, z_2)$$

where

$$H_1(z_1, z_2) = \frac{N_1(z_1, z_2)}{D_1(z_1, z_2)} \quad \text{and} \quad H_2(z_1, z_2) = \frac{N_2(z_1, z_2)}{D_2(z_1, z_2)}$$

Filters of this class can be readily designed using the approximation techniques for 1-D digital filters described in the previous chapters, and they are suitable for applications where rectangular band boundaries are acceptable. However, if the transfer function is not separable, as may be the case in filters with circular band boundaries, the design is much more involved.

16.6.8 Applications

Two-dimensional digital filters are useful in several areas. Lowpass filters can be used for the *reduction of noise* in images for the same reasons as their 1-D counterparts. Use is made of the fact that the information content of the 2-D signal is often concentrated at low frequencies, whereas noise tends to be distributed throughout the baseband. Highpass filters are sometimes used for the *enhancement of edges in images*; their application is based on the fact that abrupt changes in an image tend to increase the high-frequency content of an image, and its amplification by a highpass filter tends to exaggerate edges or outlines. Edge enhancement finds applications in pattern recognition, surveying, and computer vision. Fan filters have been found very useful for the *processing of geophysical signals*; for example, they can enhance the quality of seismic signals by eliminating signal components that are not associated with the subsurface ground formations. Seismic signals are indispensable for oil prospecting and other geological applications [7].

REFERENCES

1. R. E. Crochiere and L. R. Rabiner, *Multirate Digital Signal Processing*, Prentice-Hall, Englewood Cliffs, N.J., 1983.
2. B. Widrow and S. D. Stearns, *Adaptive Signal Processing*, Prentice-Hall, Englewood Cliffs, N.J., 1985.
3. M. G. Bellanger, *Adaptive Digital Filters and Signal Analysis*, Marcel Dekker, New York, 1987.
4. S. Haykin, *Adaptive Filter Theory*, Second Edition, Prentice-Hall, Englewood Cliffs, N.J., 1991.
5. D. E. Dudgeon and R. M. Mersereau, *Multidimensional Digital Signal Processing*, Prentice-Hall, Englewood Cliffs, N.J., 1984.
6. J. S. Lim, *Two-Dimensional Signal and Image Processing*, Prentice Hall, Englewood Cliffs, N.J., 1990
7. W.-S. Lu and A. Antoniou, *Two-Dimensional Digital Filters*, Marcel Dekker, New York, 1992.
8. P. P. Vaidyanathan, "Quadrature Mirror Filter Banks, M-Band Extensions and Perfect-Reconstruction Techniques," *IEEE ASSP Magazine*, vol. 4, pp. 4–20, July 1987.
9. P. P. Vaidyanathan, "Multirate Digital Filters, Filter Banks, Polyphase Networks, and Applications: A Tutorial," *Proc. IEEE*, vol. 78, pp. 56–93, January 1990.
10. H. Scheuermann and H. Gockler, "A Comprehensive Survey of Digital Transmultiplexing Methods," *Proc. IEEE*, vol. 69, pp. 1419–1450, November 1981.
11. N. S. Jayant and P. Noll, *Digital Coding of Waveforms*, Prentice-Hall, Englewood Cliffs, N.J., 1984.
12. J. D. Johnson and R. E. Crochiere, "An All-Digital Commentary Grade Sub-Band Coder," *J. Audio Eng. Soc.*, vol. 27, pp. 855–865, November 1979.
13. J. D. Johnson, "A Filter Family Designed for Use in Quadrature Mirror Filter Banks," *Proc. IEEE Int. Conf. Acoust., Speech, Signal Process.*, pp. 291–294, April 1980.
14. V. K. Jain and R. E. Crochiere, "Quadrature Mirror Filter Design in the Time Domain," *IEEE Trans. Acoust., Speech, Signal Process.*, vol. ASSP-32, pp. 353–361, April 1984.

15. M. J. T. Smith and T. P. Barnwell, III, "Exact Reconstruction Techniques for Tree-Structured Subband Coders," *IEEE Trans. Acoust., Speech, Signal Process.*, vol. ASSP-34, pp. 434–441, June 1986.
16. B. Gold, A. V. Oppenheim, and C. M. Rader, "Theory and Implementation of the Discrete Hilbert Transform," *Proc. Symp. Computer Process. in Comm.*, vol. 19, pp. 235–250, Polytechnic Press, New York, 1970. (See also *Digital Signal Processing*, edited by L. R. Rabiner and C. M. Rader, IEEE Press, pp. 94–109, 1972.)
17. A. V. Oppenheim and R. W. Schafer, *Discrete-Time Signal Processing*, Prentice-Hall, Englewood Cliffs, N.J., 1989.
18. B. Widrow, J. M. McCool, M. G. Larimore, and C. R. Johnson, Jr., "Stationary and Nonstationary Learning Characteristics of the LMS Adaptive Filter," *Proc. IEEE*, vol. 64, pp. 1151–1162, August 1976.
19. B. Friedlander, "Lattice Methods for Spectral Estimation," *Proc. IEEE*, vol. 70, pp. 990–1017, September 1982.
20. G. Carayannis, D. G. Manolakis, and N. Kalouptsidis, "A Fast Sequential Algorithm for Least-Squares Filtering and Prediction," *IEEE Trans. Acoust., Speech, Signal Process.*, vol. ASSP-31, pp. 1394–1402, December 1983.
21. J. M. Cioffi and T. Kailath, "Fast, Recursive-Least-Squares Transversal Filters for Adaptive Filtering," *IEEE Trans. Acoust., Speech, Signal Process.*, vol. ASSP-32, pp. 304–337, April 1984.
22. M. G. Bellanger, "FLS-QR Algorithm for Adaptive Filtering," *Signal Processing*, vol. 17, pp. 291–304, 1989.
23. J. J. Shynk, "Adaptive IIR Filtering," *IEEE ASSP Magazine*, vol. 6, pp. 4–21, April 1989.
24. M. Nayeri and W. K. Jenkins, "Alternate Realizations to Adaptive IIR Filters and Properties of Their Performance Surfaces," *IEEE Trans. Circuits Syst.*, vol. CAS-36, pp. 485–496, April 1989.
25. J. L. Shanks, *Two-Dimensional Recursive Filters*, SWIEECO Rec., pp. 19E1–19E8, 1969.
26. J. L. Shanks, S. Treitel, and J. H. Justice, "Stability and Synthesis of Two-Dimensional Recursive Filters," *IEEE Trans. Audio Electroacoust.*, vol. AU-20, pp. 115–128, June 1972.
27. T. S. Huang, "Two-Dimensional Windows," *IEEE Trans. Audio Electroacoust.*, vol. AU-20, pp. 88–89, March 1972.
28. J. H. McClellan, "The Design of Two-Dimensional Digital Filters by Transformations," *Proc. 7th Annual Princeton Conf. Information Sciences and Systems*, pp. 247–251, 1973.
29. R. M. Mersereau, W. F. G. Mecklenbräuker, and T. F. Quatieri, Jr., "McClellan Transformations for Two-Dimensional Digital Filtering: I—Design," *IEEE Trans. Circuits Syst.*, vol. CAS-23, pp. 405–413, July 1976.
30. J. M. Costa and A. N. Venetsanopoulos, "Design of Circularly Symmetric Two-Dimensional Recursive Filters," *IEEE Trans. Acoust., Speech, Signal Process.*, vol. ASSP-22, pp. 432–443, December 1974.
31. D. M. Goodman, "A Design Technique for Circularly Symmetric Low-Pass Filters," *IEEE Trans. Acoust., Speech, Signal Process.*, vol. ASSP-26, pp. 290–304, August 1978.
32. A. Antoniou and W.-S. Lu, "Design of Two-Dimensional Digital Filters by Using the Singular Value Decomposition," *IEEE Trans. Circuits Syst.*, vol. CAS-34, pp. 1191–1198, October 1987.
33. W.-S. Lu, H.-P. Wang, and A. Antoniou, "Design of Two-Dimensional FIR Digital Filters by Using the Singular Value Decomposition," *IEEE Trans. Circuits Syst.*, vol. CAS-37, pp. 35–46, January 1990.
34. G. A. Maria and M. M. Fahmy, "An l_p Design Technique for Two-Dimensional Digital Recursive Filters," *IEEE Trans. Acoust., Speech, Signal Process.*, vol. ASSP- 22, pp. 15-21, February 1974.
35. P. A. Ramamoorthy and L. T. Bruton, "Design of Stable Two-Dimensional Analogue and Digital Filters with Applications in Image Processing," *Int. J. Circuit Theory and Applications*, vol. 7, pp. 229–245, 1979.
36. C. Charalambous, "The Performance of an Algorithm for Minimax Design of Two-Dimensional Linear Phase FIR Digital Filters," *IEEE Trans. Circuits Syst.*, vol. CAS-32, pp. 1016–1028, October 1985.
37. C. Charalambous, "Design of 2-Dimensional Circularly-Symmetric Digital Filters," *IEE Proc.*, vol. 129, pt. G, pp. 47–54, April 1982.

ADDITIONAL REFERENCES

Alexander, S. T.: "Fast Adaptive Filters: A Geometrical Approach," *IEEE ASSP Magazine*, vol. 3, pp. 18–28, October 1986.

Friedlander, B.: "Lattice Filters for Adaptive Processing," *Proc. IEEE*, vol. 70, pp. 829–867, August 1982.

Honig, M. L. and D. G. Messerschmitt: "Convergence Properties of Adaptive Digital Lattice Filter," *IEEE Trans. Circuits Syst.*, vol. CAS-28, pp. 482–493, June 1981.

Johns, D. A., W. M. Snelgrove, and A. S. Sedra: "Adaptive Recursive State-Space Filters Using a Gradient-Based Algorithm," *IEEE Trans. Circuits Syst.*, vol. CAS-37, pp. 673–683, June 1990.

Johnson, Jr., C. R., M. G. Larimore, J. R. Treichler, and B. D. O. Anderson: "SHARF Convergence Properties," *IEEE Trans. Circuits Syst.*, vol. CAS-28, pp. 499–509, June 1981.

Marshall, D. F., W. K. Jenkins, and J. J. Murphy: "The Use of Orthogonal Transforms for Improving Performance of Adaptive Filters," *IEEE Trans. Circuits Syst.*, vol. CAS-36, pp. 474–483, April 1989.

Shynk, J. J.: "Adaptive IIR Filtering Using Parallel-Form Realizations," *IEEE Trans. Acoust., Speech, Signal Process.*, vol. ASSP-37, pp. 519–533, April 1989.

PROBLEMS

16.1. The input signal $x(nT)$ in the downsampler of Fig. 16.2a has the real frequency spectrum depicted in Fig. P16.1 and $\omega_s = 20$ rad/s.
  (a) Sketch the frequency spectrum of $x_d(nT')$ if $M = 2$.
  (b) Repeat part (a) if $M = 4$.
  (c) Comment on the answers obtained in parts (a) and (b).

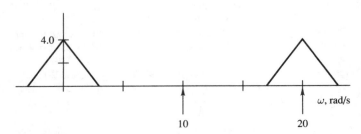

FIGURE P16.1

16.2. Repeat Prob. 16.1 if the spectrum of $x(nT)$ is given by

$$X(e^{j\omega T}) = \text{Re } X(e^{j\omega T}) + j\text{Im } X(e^{j\omega T})$$

where

$$\text{Re } X(e^{j\omega T}) = \begin{cases} 1 - |\omega| & \text{for } -1 < \omega < 1 \\ 0 & \text{for } 1 \leq |\omega| \leq 3 \end{cases}$$

and
$$\text{Im } X(e^{j\omega T}) = \begin{cases} -\omega & \text{for } -1 < \omega < 1 \\ 0 & \text{for } 1 \le |\omega| \le 3 \end{cases}$$
The sampling frequency is the same as in Prob. 16.1.

16.3. The spectrum of signal $x(nT)$ in the downsampler of Fig. 16.2a is given by
$$X(e^{j\omega T}) = e^{-|\omega|} \qquad \text{for } 0 \le |\omega| < 12$$
and $\omega_s = 24$ rad/s. Find the maximum value of M that will limit the aliasing error to a value less than 1 percent relative to the spectrum of the signal at $\omega = 1$ rad/s.

16.4. In an application, the sampling frequency needs to be increased by a factor of 10.
(a) Design a nonrecursive filter that can be used along with an upsampler to construct an interpolator. Linear interpolation is acceptable.
(b) Plot the amplitude response of the filter.

16.5. A signal $x(nT)$ is applied at the input of the configuration depicted in Fig. P16.5a. The frequency spectrum of $x_c(t)$, namely, $X_c(j\omega)$, is zero for $|\omega| \ge \omega_c$, as illustrated in Fig. P16.5b. The filter shown is a nonrecursive filter of length N with a frequency response
$$H(j\omega) = M(\omega)e^{j\theta(\omega)}$$
where
$$M(\omega) = \begin{cases} 3 & \text{for } |\omega| < \omega_c \\ 0 & \text{for } \omega_c \le |\omega| \le \omega_s'/2 \end{cases}$$
and
$$\theta(\omega) = (N-1)\omega T'/2$$

(a) Sketch the frequency spectra at points $A, B, C,$ and D.
(b) Write expressions for the signals and their frequency spectra at points $A, B, C,$ and D.

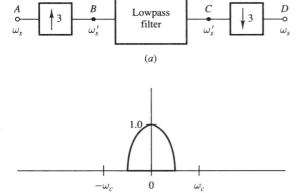

FIGURE P16.5

16.6. Demonstrate the validity of Eq. (16.29).

16.7. The signal $x(nT)$ in a 4-band QMF bank has the triangular frequency spectrum shown in Fig. 16.10b.
(a) Sketch the frequency spectra at the various nodes of the analysis section.
(b) Repeat part (a) for the synthesis section.

16.8. Time-division to frequency-division multiplex translation can be carried out by using the scheme depicted in Fig. P16.8. Signals $x_{ck}(t)$ for $k = 0, 1, \ldots, K-1$ are bandlimited such that $X_{ck}(j\omega) = 0$ for $|\omega| \geq \omega_m$. The lowpass filters shown are identical and each has a cutoff frequency $\omega_c = \omega_m$. On the other hand, the highpass filters have distinct cutoff frequencies $\omega_0, \omega_1, \ldots, \omega_{K-1}$. For correct operation, $\omega_s \geq 2\omega_m$, $\omega'_s > 2(\omega_{LO} + K\omega_m)$, and $\omega_k \geq \omega_{k-1} + \omega_m$ for $k = 1, 2, \ldots, K-1$.
(a) Sketch the frequency spectra at points $A_k, B_k, \ldots, F_k$, and G for the case where $K = 3$.
(b) Explain the role of the lowpass and highpass filters.

16.9. Find the maximum number of channels in the scheme of Fig. P16.8 if $\omega_m = 4$ kHz, $\omega_{LO} = 60$ kHz, $\omega_s = 8$ kHz, and $\omega'_s = 216$ kHz.

16.10. Frequency-division to time-division multiplex translation can be carried out by using the scheme depicted in Fig. P16.10 where the bandpass filters have passbands $\omega_k \leq \omega \leq \omega_k + \omega_m$ for $k = 1, 2, \ldots, K-1$ and each of the lowpass filters has a cutoff frequency $\omega_c \geq \omega_m$. Sketch the frequency spectra of the signals at nodes A, B_k, C_k, D_k, and E_k for the case where $K = 3$.

16.11. Chapter 9 describes the Fourier series method for the design of nonrecursive filters for the case where the filter length N is odd. Derive the impulse response for a lowpass filter with cutoff frequency ω_c for the case where N is even.

16.12. (a) Using the formula for the impulse response obtained in Prob. 16.11 along with the von Hann window design a halfband lowpass filter. Assume that $N = 32$ and $\omega_s = 16$ rad/s.
(b) Design a corresponding halfband highpass filter.
(c) The filters in parts (a) and (b) are used in a QMF bank. Plot the amplitude response of the two filters in cascade.

16.13. Redesign the filters in Prob. 16.12 using the Kaiser window with $\alpha = 3.0$. Compare the results with those obtained using the von Hann window.

16.14. Let the numerator polynomial of transfer function $H_A(s)$ in Example 16.2 be $N(s)$. Demonstrate that $N(s)$ and polynomials $d_A(s)$ and $d_B(s)$ in Eqs. (16.25) and (16.26) satisfy the relation

$$\tfrac{1}{2}[d_A(s)d_B(-s) - d_A(-s)d_B(s)] = N(s)$$

(see Sec. 12.5).

16.15. (a) Redesign the filter in Example 16.2 using a Butterworth approximation.
(b) Demonstrate that the formula in Prob. 16.14 applies.
(c) Two copies of the filter obtained will be used as the analysis and synthesis banks in a transmission system. Plot the overall group delay characteristic of the system.

16.16. (a) Redesign the filter of Example 16.2 using a Chebyshev approximation.
(b) Determine the amplitude response of the lowpass filter by applying the bilinear transformation to the analog transfer function.
(c) Determine the amplitude response of the lowpass filter by analyzing the lattice structure obtained (see Sec. 12.8).

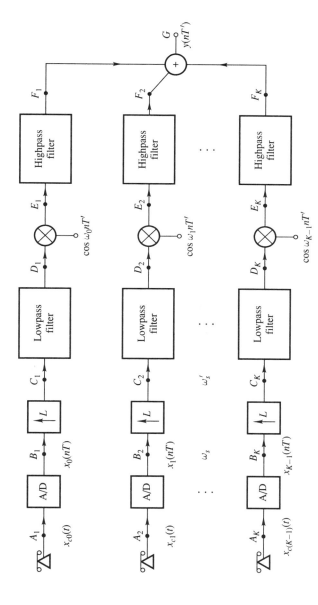

FIGURE P16.8

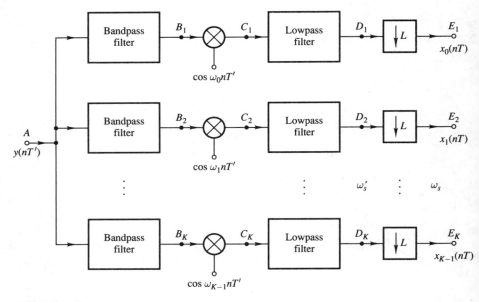

FIGURE P16.10

16.17. The filter obtained in Prob. 16.16 is to be used both for the analysis and synthesis banks in the scheme of Fig. 16.9. Find the overall phase response of the system.

16.18. (a) Design a Hilbert transformer of length $N = 31$ using the Kaiser window with $\alpha = 4.0$, assuming a sampling frequency of 100 rad/s.
  (b) Repeat part (a) with $N = 32$.
  (c) Compare the results obtained in the two cases.

16.19. Formulate the error function and obtain the necessary derivatives to enable the design of Hilbert transformers using the Remez exchange algorithm (say Algorithm 4 in Chap. 15).

16.20. Demonstrate the validity of Eq. (16.44).

16.21. The eigenvalues of an $N \times N$ matrix $\mathbf{R}_n$ are $\lambda_1, \lambda_2, \ldots, \lambda_N$. Show that the eigenvalues of $\tilde{\mathbf{R}}_n = \mathbf{I}_n - 2\alpha \mathbf{R}_n$ are given by $\tilde{\lambda}_i = 1 - 2\alpha \lambda_i$.

16.22. The input and desired signals in an adaptive filter are given by

$$x(n) = e^{-j\omega n/N}$$

and

$$d(n) = e^{-j(\omega n/N + \phi)} + n_1(n)$$

respectively, where $n_1(n)$ is a white noise source with variance σ_n^2.
  (a) Calculate $\mathbf{p}_n$ and $\mathbf{R}_n$ for the case where a nonrecursive filter of length $N = 2$ is employed.
  (b) Obtain the Wiener solution as well as the minimum MSE at the output.

16.23. Show that the inequality in Eq. (16.62) is a sufficient condition for the stability of the LMS algorithm.

16.24. Three variations of the standard LMS updating formula given in Eq. (16.61) are

$$\mathbf{a}_{n+1} = \mathbf{a}_n + 2\alpha \, \text{sgn}\,[e(n)]\mathbf{x}_n$$

$$\mathbf{a}_{n+1} = \mathbf{a}_n + 2\alpha e(n) \, \text{sgn}\,(\mathbf{x}_n)$$

and

$$\mathbf{a}_{n+1} = \mathbf{a}_n + 2\alpha \, \text{sgn}\,[e(n)] \, \text{sgn}\,(\mathbf{x}_n)$$

where

$$\text{sgn}\,(x) = \begin{cases} 1 & \text{for } x \geq 0 \\ -1 & \text{for } x < 0 \end{cases}$$

and

$$\text{sgn}\,(\mathbf{x}) = [\text{sgn}\,(x_1) \quad \text{sgn}\,(x_2) \quad \ldots \quad \text{sgn}\,(x_N)]^T$$

Constant 2α is usually chosen to be a power of two for the sake of computational efficiency. Discuss the effects of these simplifications on the gradient direction, convergence, and the residual error.

16.25. Apply the LMS algorithm and each of the variations described in Prob. 16.24 for the identification of a system characterized by

$$H(z) = \sum_{i=0}^{4} z^{-i}$$

using the initial coefficient vector $\mathbf{a}_0 = [0 \quad 0 \quad 0 \quad 0 \quad 0]^T$. Discuss the results obtained.

16.26. If matrix $\mathbf{R}_n$ is approximated by a diagonal matrix whose diagonal elements are all equal to

$$\|\mathbf{x}_n\|^2 = \mathbf{x}_n^T \mathbf{x}_n$$

the so-called *normalized-LMS* algorithm is obtained.
(a) Show that in this algorithm, the updating formula assumes the form

$$\mathbf{a}_{n+1} = \mathbf{a}_n + \frac{2\alpha \, e(n)\mathbf{x}_n}{\gamma + \mathbf{x}_n^T \mathbf{x}_n}$$

where γ is a small constant.
(b) Explain the purpose of constant γ.

16.27. A transmission channel can be represented by the transfer function

$$H(z) = \sum_{i=0}^{8} (i-4)z^{-i}$$

Equalize the channel by using first the LMS algorithm and then the normalized-LMS algorithm, and compare the results obtained.

16.28. In real-time applications an estimate for $\mathbf{R}_n$, designated by $\tilde{\mathbf{R}}_n$, can be generated as

$$\tilde{\mathbf{R}}_n = (1-\mu)\tilde{\mathbf{R}}_{n-1} + \mu \mathbf{x}_n \mathbf{x}_n^T$$

where μ is a constant. On the other hand, if $\mathbf{A}, \mathbf{B}, \mathbf{C}$ and $\mathbf{D}$ are matrices of appropriate dimensions, then they are interrelated in terms of the so-called *matrix inversion lemma*

which states that
$$(A + BCD)^{-1} = A^{-1} - A^{-1}B(C^{-1} + DA^{-1}B)^{-1}DA^{-1}$$
Using the above formulas, derive a recursive formula for $\tilde{R}_n^{-1}$.

16.29. Algorithms using the gradient estimate given in Eq. (16.60) along with some estimate for $\tilde{R}_n^{-1}$ are referred to as *LMS-Newton* adaptation algorithms.
   (a) Construct such an algorithm using the estimate for $\tilde{R}_n^{-1}$ obtained in Prob. 16.28.
   (b) Apply this algorithm to the system identification problem described in Prob. 16.25.

16.30. A 2-D digital filter has the transfer function
$$H(z_1, z_2) = \frac{N(z_1, z_2)}{D(z_1, z_2)} = \frac{2z_1 z_2}{z_1 z_2 - 0.5z_1 - 0.5z_2 + 0.25}$$
Find its impulse response. The 2-D impulse function is defined as
$$\delta(n_1, n_2) = \begin{cases} 1 & \text{for } n_1 = n_2 = 0 \\ 0 & \text{otherwise} \end{cases}$$

16.31. Repeat Prob. 16.30 if the transfer function is given by
$$H(z_1, z_2) = \frac{N(z_1, z_2)}{D(z_1, z_2)} = \frac{z_1 z_2}{2z_1 z_2 - 1}$$

16.32. Plot the amplitude and phase response of the filter described in Prob. 16.30.

16.33. Check the stability of the filters described in Probs. 16.30 and 16.31.

16.34. A 2-D digital filter is characterized by the transfer function
$$H(z_1, z_2) = \frac{N(z_1, z_2)}{D(z_1, z_2)}$$
where
$$N(z_1, z_2) = 64(z_1 - 1)^2 (z_2 - 1)^2$$
and
$$D(z_1, z_2) = 64z_1^2 z_2^2 - 32z_1 z_2^2 + 48z_1^2 z_2 + 8z_2^2 + 8z_1^2$$
$$- 24z_1 z_2 - 4z_1 + 6z_2 + 1$$
Check its stability.

16.35. A 2-D lowpass digital filter comprises two cascaded 1-D lowpass filters with passband edges ω_{pi} rad/s, stopband edges ω_{ai} rad/s, passband ripples A_{pi} dB, and minimum stopband attenuations A_{ai} rad/s for $i = 1$ and 2. Find the passband and stopband edges, passband ripple, and minimum stopband attenuation of the 2-D filter.

16.36. Using the formulas obtained in Prob. 16.35 design a 2-D lowpass filter satisfying the following specifications:
$$\omega_{p1} = 2.0 \text{ rad/s} \qquad \omega_{p2} = 3.0 \text{ rad/s}$$
$$\omega_{a1} = 2.4 \text{ rad/s} \qquad \omega_{a2} = 3.6 \text{ rad/s} \qquad \omega_{s1} = \omega_{s2} = 10 \text{ rad/s}$$
$$A_p = 1.0 \text{ dB} \qquad A_a \geq 40.0 \text{ dB}$$

APPENDIX A

ELLIPTIC FUNCTIONS

A.1 INTRODUCTION

The Jacobian elliptic functions are derived by employing the Legendre elliptic integral of the first kind. Their theory is quite extensive and is discussed in detail by Bowman [1] and Hancock [2, 3]. We provide here a brief, but for our purposes adequate, treatment of this theory [4].

A.2 ELLIPTIC INTEGRAL OF THE FIRST KIND

The *elliptic integral of the first kind* can be expressed as

$$u \equiv u(\phi, k) = \int_0^\phi \frac{d\theta}{\sqrt{1 - k^2 \sin^2 \theta}} \tag{A.1}$$

where $0 \leq k < 1$. The parameter k is called the *modulus*, and the upper limit of integration ϕ is called the *amplitude* of the integral. Evidently, for a real value of ϕ, $u(\phi, k)$ is real and represents the area bounded by the curve

$$I = \frac{1}{\sqrt{1 - k^2 \sin^2 \theta}}$$

and the vertical lines $\theta = 0$ and $\theta = \phi$. Plots of I and $u(\phi, k)$ for $k = 0.995$ are shown in Fig. A.1. The integrand I has minima equal to unity at $\theta = 0, \pi, 2\pi \ldots$ and maxima equal to $1/\sqrt{1-k^2}$ at $\theta = \pi/2, 3\pi/2, \ldots$. In effect, I is a periodic function of θ with a period π. The area bounded by lines $\theta = n\pi/2$ and $\theta = (n+1)\pi/2$ is constant for any n because of the symmetry of I and is equal to the area bounded by lines $\theta = 0$ and $\theta = \pi/2$. This area is referred to as the *complete* elliptic integral of the first kind and is given by

$$u\left(\frac{\pi}{2}, k\right) = K = \int_0^{\pi/2} \frac{d\theta}{\sqrt{1 - k^2 \sin^2 \theta}} \tag{A.2}$$

(see Fig. A.1).

As a consequence of the periodicity and symmetry of I we can write

$$u(n\pi + \phi_1, k) = 2nK + u(\phi_1, k) \quad \text{and} \quad u\left(\frac{\pi}{2} + \phi_1, k\right) = 2K - u\left(\frac{\pi}{2} - \phi_1, k\right)$$

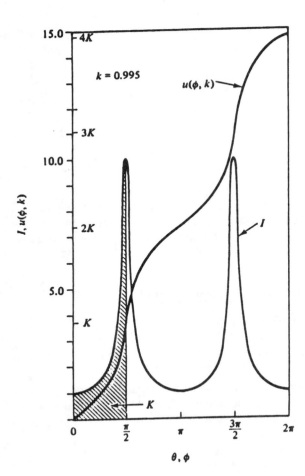

FIGURE A.1
Plots of I versus θ and $u(\phi, k)$ versus ϕ.

where $0 \leq \phi_1 < \pi/2$. That is, the elliptic integral for a given k and any real ϕ can be determined from a table giving the values of the integral in the interval $0 \leq \phi < \pi/2$.

If $k = 0$, Eq. (A.1) gives

$$u(\phi, 0) = \int_0^\phi d\theta = \phi$$

and if $k = 1$,

$$u(\phi, 1) = \int_0^\phi \frac{d\theta}{\cos\theta} = \ln\left[\tan\left(\frac{\pi}{4} + \frac{\phi}{2}\right)\right]$$

according to standard integral tables. Hence $u(\phi, 0)$ increases linearly with ϕ, whereas $u(\phi, 1)$ is discontinuous at $\phi = \pi/2$. For $0 \leq \phi < \pi/2$

$$u(\phi, 0) \leq u(\phi, k) \leq u(\phi, 1)$$

as can be seen in Fig. A.2.

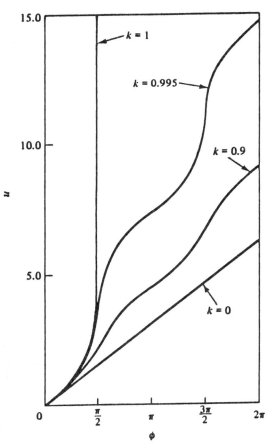

FIGURE A.2
Plots of u versus ϕ for various values of k.

A.3 ELLIPTIC FUNCTIONS

Figure A.2 demonstrates a one-to-one correspondence between u and ϕ. Thus for a given pair of values (u, k) there corresponds a unique amplitude ϕ such that

$$\phi = f(u, k)$$

The *Jacobian elliptic functions* are defined as

$$\text{sn}\,(u, k) = \sin \phi \tag{A.3}$$

$$\text{cn}\,(u, k) = \cos \phi \tag{A.4}$$

$$\text{dn}\,(u, k) = \sqrt{1 - k^2 \sin^2 \phi} \tag{A.5}$$

Many of the properties of elliptic functions follow directly from the properties of trigonometric functions. For example, we can write

$$\text{sn}^2\,(u, k) + \text{cn}^2\,(u, k) = 1 \tag{A.6}$$

and

$$k^2 \,\text{sn}^2\,(u, k) + \text{dn}^2\,(u, k) = 1 \tag{A.7}$$

and so forth.

Plots of the elliptic functions versus u can be constructed as in Fig. A.3. As can be seen, sn (u, k), cn (u, k), and dn (u, k) are periodic functions u with periods $4K, 4K$, and $2K$, respectively, i.e.,

$$\text{sn}\,(u + 4mK, k) = \text{sn}\,(u, k) \tag{A.8}$$

$$\text{cn}\,(u + 4mK, k) = \text{cn}\,(u, k) \tag{A.9}$$

$$\text{dn}\,(u + 2mK, k) = \text{dn}\,(u, k) \tag{A.10}$$

Variations in k tend to change the shape and period of the elliptic functions, as illustrated in Fig. A.4.

If $k = 0$,

$$u(\phi, 0) = \phi$$

and so $\quad$ sn $(u, 0) =$ sn $(\phi, 0) = \sin \phi \quad\quad$ cn $(u, 0) =$ cn $(\phi, 0) = \cos \phi$

i.e., *the elliptic sine and cosine are generalizations of the conventional sine and cosine*, respectively.

A.4 IMAGINARY ARGUMENT

Thus far the argument of the elliptic functions, namely, u, has been assumed to be a real quantity. By performing the integration of Eq. (A.1) over an appropriate path in a complex plane, the elliptic integral can assume complex values. Let us consider the case of imaginary value whereby

$$jv = \int_0^\psi \frac{d\theta}{\sqrt{1 - k^2 \sin^2 \theta}} \tag{A.11}$$

ELLIPTIC FUNCTIONS **657**

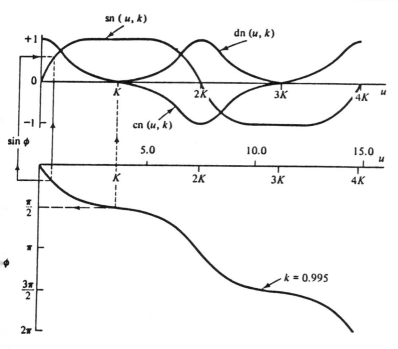

FIGURE A.3
Plots of sn (u, k), cn (u, k), and dn (u, k) versus u.

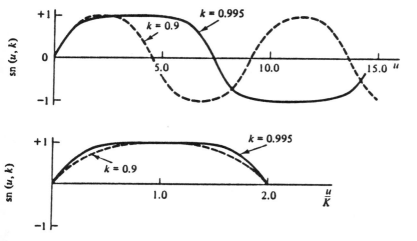

FIGURE A.4
Effect of variations in k on the elliptic sine.

As in Sec. A.3, we can define

$$\operatorname{sn}(jv, k) = \sin \psi \tag{A.12}$$

$$\operatorname{cn}(jv, k) = \cos \psi \tag{A.13}$$

$$\operatorname{dn}(jv, k) = \sqrt{1 - k^2 \sin^2 \psi} \tag{A.14}$$

These functions can be expressed in terms of elliptic functions that have real arguments, as we now show.

By applying the transformations

$$\sin \theta = j \tan \theta' \qquad \sin \psi = j \tan \psi' \tag{A.15}$$

in Eq. (A.11), we have

$$jv = \int_0^{\psi'} \frac{j \, d\theta'}{\sqrt{1 - \sin^2 \theta' + k^2 \sin^2 \theta'}}$$

Alternatively

$$v = \int_0^{\psi'} \frac{d\theta'}{\sqrt{1 - (k')^2 \sin^2 \theta'}}$$

where k', given by

$$k' = \sqrt{1 - k^2}$$

is called the *complementary* modulus. Now from Sec. A.3

$$\operatorname{sn}(v, k') = \sin \psi' \tag{A.16}$$

$$\operatorname{cn}(v, k') = \cos \psi' \tag{A.17}$$

$$\operatorname{dn}(v, k') = \sqrt{1 - (k')^2 \sin^2 \psi'} \tag{A.18}$$

and, therefore, from Eqs. (A.12) to (A.18),

$$\operatorname{sn}(jv, k) = j \tan \psi' = j \frac{\sin \psi'}{\cos \psi'} = \frac{j \operatorname{sn}(v, k')}{\operatorname{cn}(v, k')} \tag{A.19}$$

$$\operatorname{cn}(jv, k) = \frac{1}{\operatorname{cn}(v, k')} \tag{A.20}$$

$$\operatorname{dn}(jv, k) = \frac{\operatorname{dn}(v, k')}{\operatorname{cn}(v, k')} \tag{A.21}$$

By analogy with Eq. (A.2), the *complementary* complete integral of the first kind is given by

$$K' = \int_0^{\pi/2} \frac{d\theta}{\sqrt{1 - (k')^2 \sin^2 \theta}}$$

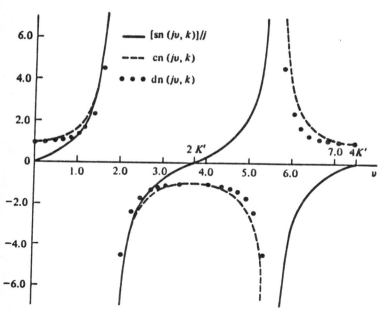

FIGURE A.5
Plots of [sn (jv, k)]/j, cn (jv, k), and dn (jv, k) versus v.

This has a similar interpretation as K; that is, it is the quarter period of sn (v, k') and cn (v, k') or the half period of dn (v, k').

The functions sn (jv, k), cn (jv, k), and dn (jv, k) are periodic functions of jv, as can be seen in Fig. A.5, with periods $j2K'$, $j4K'$, and $j4K'$, respectively, i.e.,

$$\text{sn } (jv + j2nK', k) = \text{sn } (jv, k)$$

$$\text{cn } (jv + j4nK', k) = \text{cn } (jv, k)$$

$$\text{dn } (jv + j4nK', k) = \text{dn } (jv, k)$$

A.5 FORMULAS

Elliptic functions, like trigonometric functions, are interrelated by many useful formulas. The most basic one is the *addition formula*, which is of the form

$$\text{sn } (z_1 + z_2, k) = \frac{\text{sn } (z_1, k) \text{ cn } (z_2, k) \text{ dn } (z_2, k) + \text{cn } (z_1, k) \text{ sn } (z_2, k) \text{ dn } (z_1, k)}{D}$$

(A.22)

where

$$D = 1 - k^2 \text{ sn}^2 (z_1, k) \text{ sn}^2 (z_2, k)$$

The variables z_1 and z_2 can assume real or complex values. By using the above formula and Eqs. (A.6) and (A.7) we can show that

$$\operatorname{cn}(z_1+z_2,k) = \frac{\operatorname{cn}(z_1,k)\operatorname{cn}(z_2,k) - \operatorname{sn}(z_1,k)\operatorname{sn}(z_2,k)\operatorname{dn}(z_1,k)\operatorname{dn}(z_2,k)}{D}$$

(A.23)

$$\operatorname{dn}(z_1+z_2,k) = \frac{\operatorname{dn}(z_1,k)\operatorname{dn}(z_2,k) - k^2 \operatorname{sn}(z_1,k)\operatorname{sn}(z_2,k)\operatorname{cn}(z_1,k)\operatorname{cn}(z_2,k)}{D}$$

(A.24)

Another formula of interest is

$$\operatorname{dn}^2\left(\frac{z}{2},k\right) = \frac{\operatorname{dn}(z,k) + \operatorname{cn}(z,k)}{1 + \operatorname{cn}(z,k)}$$

(A.25)

A.6 PERIODICITY

In the preceding sections we have demonstrated that $\operatorname{sn}(z,k)$, where $z = u + jv$, has a real period of $4K$ if $v = 0$ and an imaginary period of $2K'$ if $u = 0$. In fact these are general properties for any value of v or u as can be easily shown. From the addition formula

$\operatorname{sn}(z + 4mK, k)$

$$= \frac{\operatorname{sn}(z,k)\operatorname{cn}(4mK,k)\operatorname{dn}(4mK,k) + \operatorname{cn}(z,k)\operatorname{sn}(4mK,k)\operatorname{dn}(z,k)}{1 - k^2 \operatorname{sn}^2(z,k)\operatorname{sn}^2(4mK,k)}$$

and since

$$\operatorname{sn}(4mK,k) = \operatorname{sn}(0,k) = 0$$

$$\operatorname{cn}(4mK,k) = \operatorname{cn}(0,k) = 1$$

$$\operatorname{dn}(4mK,k) = \operatorname{dn}(0,k) = 1$$

according to Eqs. (A.8) to (A.10), it follows that

$$\operatorname{sn}(z + 4mK, k) = \operatorname{sn}(z,k)$$

(A.26)

Similarly

$\operatorname{sn}(z + j2nK', k)$

$$= \frac{\operatorname{sn}(z,k)\operatorname{cn}(j2nK',k)\operatorname{dn}(j2nK',k) + \operatorname{cn}(z,k)\operatorname{sn}(j2nK',k)\operatorname{dn}(z,k)}{1 - k^2 \operatorname{sn}^2(z,k)\operatorname{sn}^2(j2nK',k)}$$

and from Eqs. (A.19) to (A.21)

$$\operatorname{sn}(j2nK', k) = \frac{j\operatorname{sn}(2nK', k')}{\operatorname{cn}(2nK', k')} = 0$$

$$\operatorname{cn}(j2nK', k) = \frac{1}{\operatorname{cn}(2nK', k')} = (-1)^n$$

$$\operatorname{dn}(j2nK', k) = \frac{\operatorname{dn}(2nK', k')}{\operatorname{cn}(2nK', k')} = (-1)^n$$

Hence we have

$$\operatorname{sn}(z + j2nK', k) = \operatorname{sn}(z, k) \tag{A.27}$$

Therefore, by combining Eqs. (A.26) and (A.27) we obtain

$$\operatorname{sn}(z + 4mK + j2nK', k) = \operatorname{sn}(z, k)$$

that is, $\operatorname{sn}(z, k)$ is a *doubly periodic* function of z with a real period of $4K$ and an imaginary period of $2K'$.

The z plane can be subdivided into *period parallelograms* by means of lines

$$u = 4mK \quad \text{and} \quad jv = j2nK'$$

as illustrated in Fig. A.6. The specific parallelogram defined by vertices $(0, 0)$, $(4K, 0)$, $(4K, j2K')$, and $(0, j2K')$ is called the *fundamental period parallelogram*. If the value of $\operatorname{sn}(z, k)$ is known for each and every value of z within this parallelogram and along any two adjacent sides, the function is known over the entire z plane.

Similarly the functions $\operatorname{cn}(z, k)$ and $\operatorname{dn}(z, k)$ can be shown to be doubly periodic. The first has a real period of $4K$ and an imaginary period of $4K'$, whereas the second has a real period of $2K$ and an imaginary period of $4K'$.

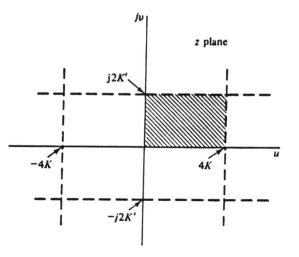

FIGURE A.6
Period parallelograms of $\operatorname{sn}(z, k)$.

A.7 TRANSFORMATION

The equation
$$\omega = \sqrt{k} \, \text{sn} \, (z, k) \tag{A.28}$$
is essentially a variable transformation that maps points in the z plane onto corresponding points in the ω plane. Let us examine the mapping properties of this transformation. These are required in the derivation of $F(\bar{\omega})$ in Sec. 5.5.1.

A point z_p as well as all points
$$z = z_p + 4mK + j2nK'$$
map onto a single point in the ω plane by virtue of the periodicity of sn (z, k). Hence only points in the fundamental period parallelogram need be considered. Three domains of $\sqrt{k}$ sn (z, k) are of interest as follows:

Domain 1: $z = u$ with $0 \le u \le K$

Domain 2: $z = K + jv$ with $0 \le v \le K'$

Domain 3: $z = u + jK'$ with $0 \le u \le K$

In domain 1
$$\omega = \sqrt{k} \, \text{sn} \, (u, k)$$
If $u = 0$,
$$\omega = \sqrt{k} \, \text{sn} \, (0, k) = 0$$
and if $u = K$,
$$\omega = \sqrt{k} \, \text{sn} \, (K, k) = \sqrt{k}$$
i.e., Eq. (A.28) maps points on the real axis of the z plane between 0 and K onto points on the real axis of the ω plane between 0 and $\sqrt{k}$.

In domain 2 we have
$$\omega = \sqrt{k} \, \text{sn} \, (K + jv, k)$$
From the addition formula
$$\omega = \frac{\sqrt{k} \, \text{cn} \, (jv, k) \, \text{dn} \, (jv, k)}{1 - k^2 \, \text{sn}^2 \, (jv, k)} \tag{A.29}$$
since cn $(K, k) = 0$, and from Eqs. (A.19) to (A.21)
$$\omega = \frac{\sqrt{k} \, \text{dn} \, (v, k')}{\text{cn}^2 \, (v, k') + k^2 \, \text{sn}^2 \, (v, k')}$$
Now from Eqs. (A.6) and (A.7)
$$\text{cn}^2 \, (v, k') + k^2 \, \text{sn}^2 \, (v, k') = 1 - \text{sn}^2 \, (v, k') + k^2 \, \text{sn}^2 \, (v, k')$$
$$= 1 - (k')^2 \, \text{sn}^2 \, (v, k') = \text{dn}^2 \, (v, k')$$

Therefore, Eq. (A.29) simplifies to

$$\omega = \frac{\sqrt{k}}{\text{dn }(v, k')}$$

If $v = 0$,

$$\omega = \frac{\sqrt{k}}{\text{dn }(0, k')} = \sqrt{k}$$

and if $v = K'$, we have

$$\omega = \frac{\sqrt{k}}{\text{dn }(K', k')} = \frac{1}{\sqrt{k}}$$

For $v = K'/2$, the use of Eq. (A.25) yields

$$\omega = \frac{\sqrt{k}}{\text{dn }(K'/2, k')} = \sqrt{k} \left[\frac{1 + \text{cn }(K', k')}{\text{dn }(K', k') + \text{cn }(K', k')} \right]^{1/2} = 1$$

Thus Eq. (A.28) maps points on the line $z = K + jv$ for v between 0 and K' onto points on the real axis of the ω plane between $\sqrt{k}$ and $1/\sqrt{k}$; in particular, point $z = K + jK'/2$ maps onto point $\omega = 1$.

In domain 3 Eq. (A.28) becomes

$$\omega = \sqrt{k}\ \text{sn }(u + jK', k)$$

and, as above, Eq. (A.22) yields

$$\omega = \frac{1}{\sqrt{k}\ \text{sn }(u, k)}$$

If $u = 0$,

$$\omega = \frac{1}{\sqrt{k}\ \text{sn }(0, k)} = \infty$$

and if $u = K$,

$$\omega = \frac{1}{\sqrt{k}\ \text{sn }(K, k)} = \frac{1}{\sqrt{k}}$$

i.e., points on line $z = u + jK'$ with u between 0 and K map onto the real axis of the ω plane between ∞ and $1/\sqrt{k}$.

By considering mirror-image points to those considered so far, the mapping depicted in Fig. A.7 can be completed, where points $A, B, \ldots$ map onto points $A', B', \ldots$.

A.8 SERIES REPRESENTATION

Elliptic functions, like other functions, can be represented in terms of series. From [3] or [4]

$$\text{sn }(z, k) = \frac{1}{\sqrt{k}} \frac{\theta_1(z/2K, q)}{\theta_0(z/2K, q)} \qquad (A.30)$$

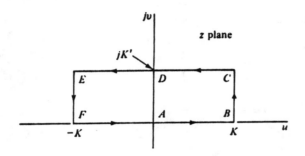

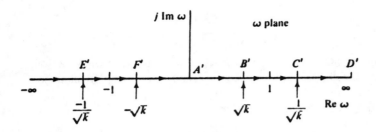

FIGURE A.7
Mapping properties of transformation $\omega = \sqrt{k}\ \text{sn}\ (z, k)$.

$$\text{cn}\ (z, k) = \sqrt{\frac{k'}{k}} \frac{\theta_2(z/2K, q)}{\theta_0(z/2K, q)} \quad \text{(A.31)}$$

$$\text{dn}\ (z, k) = \sqrt{k'} \frac{\theta_3(z/2K, q)}{\theta_0(z/2K, q)} \quad \text{(A.32)}$$

The parameter q is known as the *modular constant* and is given by

$$q = e^{-\pi K'/K}$$

The functions $\theta_0(z/2K, q)$ to $\theta_3(z/2K, q)$ are called *theta functions* and are given by

$$\theta_0\left(\frac{z}{2K}, q\right) = 1 + 2\sum_{m=1}^{\infty} (-1)^m q^{m^2} \cos\left(2m \frac{\pi z}{2K}\right)$$

$$\theta_1\left(\frac{z}{2K}, q\right) = 2q^{1/4} \sum_{m=0}^{\infty} (-1)^m q^{m(m+1)} \sin\left[(2m+1)\frac{\pi z}{2K}\right]$$

$$\theta_2\left(\frac{z}{2K}, q\right) = 2q^{1/4} \sum_{m=0}^{\infty} q^{m(m+1)} \cos\left[(2m+1)\frac{\pi z}{2K}\right]$$

$$\theta_3\left(\frac{z}{2K}, q\right) = 1 + 2\sum_{m=1}^{\infty} q^{m^2} \cos\left(2m \frac{\pi z}{2K}\right)$$

The above series converge rapidly and can be used to evaluate the elliptic functions to any desired degree of accuracy.

REFERENCES

1. F. Bowman, *Introduction to Elliptic Functions with Applications*, Dover, New York, 1961.
2. H. Hancock, *Elliptic Integrals*, Dover, New York.
3. H. Hancock, *Lectures on the Theory of Elliptic Functions*, Dover, New York, 1958.
4. A. J. Grossman, "Synthesis of Tchebyscheff Parameter Symmetrical Filters," *Proc. IRE*, vol. 45, pp. 454–473, April 1957.

INDEX

Absolute convergence of z transform, 48
A/D converter (*see* Analog-to-digital converter)
Adaptive filters:
   adaptation algorithms, 625–626
   for channel equalization, 634–635
   convergence factor, 630
   cross correlation, 629
   gradient vector of the MSE function, 629
   introduction, 592
   least-mean-square algorithm, 633–634
   mean-square error, 628
   Newton algorithm, 630
   recursive, 634
   for signal enhancement, 636
   for signal prediction, 636–637
   steepest-descent algorithm, 630–633
      stability, 631–633
   for system identification, 634
   Wiener filters, 628–630
Adaptors:
   1-multiplier parallel, 397
   1-multiplier series, 395
   2-multiplier parallel, 397
   2-multiplier series, 395
   2-port, 397–398
   unconstrained, 397
Adder, 9
Addition:
   floating-point, 337
   formula for elliptic functions, 659
   one's-complement, 336
   overflow in one's- or two's-complement addition, 336
   signed-magnitude, 336
   two's-complement, 336
Additivity property, 7
Adjoint flow graph, 124
Adjustable bracket, 560
Admittance conversion function, 426
Agarwal, R. C., 345
Al-Baali, M., 505
Algorithms:
   alternative Newton algorithm, 498
   alternative rejection scheme for superfluous potential extremals, 553–554
   basic Newton algorithm, 495
   basic quasi-Newton algorithm, 499
   basic Remez exchange algorithm, 549
   Charalambous minimax algorithm, 512–513
   decimation-in-frequency FFT algorithm, 472–476
      $N = 8$, 476
   decimation-in-time FFT algorithm, 466–472
      $N = 8$, 471–472
   design of digital differentiators satisfying prescribed specifications, 582
   of equalizers, 529
   of nonrecursive filters satisfying prescribed specifications, 573
   efficient Remez exchange algorithm, 563–566
   exhaustive step-by-step search, 551
   Fletcher inexact line search, 504–505
   least-mean-square algorithm, 633–634
   least-pth minimax algorithm, 510
   steepest-descent algorithm, 630–633
Aliasing:
   errors in QMF banks, 605–609
   frequency-domain, 205

667

Aliasing (*cont.*)
  time-domain, 451–453
Allpass CGIC section, 432
Allpass filter, 94
Allpass transfer function:
  continuous-time, 175
  discrete-time, 269
Alternation theorem, 547–548
Amplitude distortion, 87
Amplitude equalization, 270–271
Amplitude of elliptic integral, 653
Amplitude response:
  in analog filters, 80
  in digital filters, 21, 81
  graphical method, 81–82
  influence of warping effect, 235–236
  in 2-dimensional digital filters, 640
Amplitude spectrum, continuous Fourier transform, 178
  discrete Fourier transform, 445
Analog filters:
  amplitude response, 80
  approximations (*see* Approximations)
  attenuation, 140
  bandpass, 141
  bandstop, 141
  basic concepts, 139–140
  Cauer forms, 408
  cutoff frequency, 141
  delay characteristic, 140
  denormalized approximations, 143
  Feldkeller equation, 410
  Foster forms, 408
  frequency-domain characterization, 179
  gain, 80
  group delay, 140
  highpass, 141
  ideal, 140
  implementation using digital filters, 207–208
    using sampled-data filters, 207–208
  insertion loss, 389
  loss, 140
  loss characteristic, 140
  loss function, 140
  lowpass, 140
  maximum output power, 389
  maximum passband loss, 141
  minimum stopband loss, 141
  normalized approximations, 143
  passband, 141
  passband edge, 141
  phase characteristic, 140
  phase response, 80
  phase shift, 80, 140
  practical, 141–142
  sensitivity of passband loss, 389
  specifications, 172
  stopband, 141
  stopband edge, 141
  terminology, 139–140
  transfer function, 140
  transformations (*see* Transformations)
Analog G-CGIC configuration, 426
Analog integrator, 231
Analog-to-digital converter:
  encoder, 210
  ideal, 208
  practical, 210
  quantization error, 212
  sample-and-hold device, 210
Analysis:
  analog lattice network, 406–407
  complex (*see* Complex analysis)
  frequency-domain, 80–85
    in wave digital filters, 419–422
  of networks, 10–13
  section in filter banks, 602
  sensitivity, 125–129
  of signal flow-graphs, 13–15
  steady-state sinusoidal, 21
  time-domain:
    introduction, 1
    using induction, 17–21
    using state-space characterization, 29–31
    using z transform, 78–79
Analytic functions, 45
Analytic signals, 613–617
Analyticity, 45
Anderson, B. D. O., 76
Antisymmetrical impulse response, 277
Antoniou, A., 346, 378
Applications:
  of adaptive filters, 634–637
  of Constantinides transformations, 243
  of Hilbert transformers, 621–625
  of state-space method, 31
  of 2-dimensional digital filters, 644
  use of adaptive filters for signal enhancement, 636
    in channel equalization, 634–635
    in frequency-division to time-division translation, 648
    for signal prediction, 636–637
    in system identification, 634

INDEX **669**

Applications (*cont.*)
  use of sampling frequency conversion in time-division to frequency-division translation, 648
Approximations:
  for analog filters:
    basic concepts, 139–140
    Bessel, 168–169
      (*see also* Bessel approximation)
    Butterworth, 143–145
      (*see also* Butterworth approximation)
    Chebyshev, 146–152
      (*see also* Chebyshev approximation)
    definition, 141
    denormalized, 143
    elliptic, 153–168
      (*see also* Elliptic approximation)
    introduction, 138–139
    inverse-Chebyshev, 153
    maximally flat, 143
    normalized, 143
    realizability constraints, 141–143
    using transformations, 169–172
  the approximation step in the design process, 87–88
  closed-form, 88
  direct, 88
  indirect, 88
  iterative, 88
  for nonrecursive filters (*see* Nonrecursive filters)
  for recursive filters (*see* Recursive filters)
  terminology, 139–140
  for 2-dimensional digital filters, 643–644
Arithmetic:
  addition (*see* Addition)
  binary system, 331
  end-around carry, 336
  fixed-point, 334–336
  floating-point, 337
  multiplication (*see* Multiplication)
  one's-complement, 336
  overflow in one's- or two's-complement addition, 336
  signed-magnitude, 336
  two's-complement, 336
Attenuation, 140
Autocorrelation function:
  continuous-time, 318
  discrete-time, 324
Avenhaus, E., 341

Bandpass filters:
  CGIC section, 431
  ideal, 141
  nonrecursive, 298–299
  practical, 141–142
  problem formulation for nonrecursive filters, 547
  recursive, 253–257
Bandpassed signals, 623–625
Bandstop filters:
  ideal, 141
  nonrecursive, 298–299
  practical, 141–142
  problem formulation for nonrecursive filters, 547
  recursive, 257
Barnwell, III, T. P., 611
Bartlett window function, 307
Baseband, 83
Basic quasi-Newton algorithm, 496–500
Bessel approximation, 168–169
  gamma function, 169
  normalized transfer function, 168
  properties, 169
Bessel functions, 168
  zeroth-order modified, 292
Bessel polynomial, 168
BFGS updating formula for inverse Hessian, 500
BIBO stability, 26
Bilinear transformation:
  application, 250–251, 262
  derivation, 231–233
  design method, 231–238
  design procedure, 250–251, 262
  mapping properties, 233–234
  prewarping of analog filters, 235
  warping effect, 235–238
    influence on amplitude response, 235–236
    influence on phase response, 237
    in wave digital filters, 416
Binary point, 332
Binary system, 331
Binomial series, 56
Biquads, 434
Bits, 332
Blackman window function, 290
  main-lobe width, 290
Block-optimal structures, 362
Bounded-input bounded-output stability, 26
Bowman, F., 653
Broyden-Fletcher-Goldfarb-Shanno updating formula for inverse Hessian, 500
Bruton, L. T., 416, 425
Burrus, C. S., 345
Butterfly flow graph, 467
Butterworth approximation, 143–145
  derivation, 143

670 INDEX

Butterworth approximation (*cont.*)
   loss, 144
   normalized transfer function, 144–145
   zero-pole plots, 145
Butterworth recursive filters, 257–259

Canonic realization, 102–104
Cascade realization, 113–114
   ordering of sections, 360
   using CGICs, 429–432
      output noise, 433–434
      signal scaling, 432–433
Cauchy-Riemann equations, 45
Cauer forms, 408
Causality, 5
Central moment of random variable, 313
CFT (*see* Fourier transform)
CGIC (Current-conversion generalized-immittance converter), 426
   admittance conversion function, 426
CGIC realization, 429–432
   output noise, 433–434
   scaling, 432–433
Chain matrix, 401
Chan, D. S. K., 572
Channel equalization by adaptive filters, 634–635
Characteristic impedance, 401
Characterization:
   of analog filters in frequency domain, 179
   by difference equation, 8–9
      nonrecursive filters, 9
      recursive filters, 9
   in state-space domain, 26–29
   wave, 390
Charalambous, C., 511
Charalambous minimax algorithm, 512–513
Chebyshev approximation, 146–152
   derivation, 146–149
   fourth-order, 146–148
   inverse, 153
   loss, 146
   normalized transfer function, 149–151
   nth-order, 148–149
   zero-pole plot, 152
Chebyshev polynomial, 292
Chebyshev recursive filters, 259–260
Choice of structure, 434–436
Circulators, wave realization of, 402–403
Claasen, T. A. C. M., 379
Coefficient quantization, 330, 339–344
   errors, 330
   optimum word length, 340
   sensitivity to, 343

   statistical word length, 342
Common Fourier transforms, 195
Common z transforms, 53
Comparison between recursive and nonrecursive filters, 305–306
Complementary complete elliptic integral, 658
Complementary modulus, 658
Complete elliptic integral of the first kind, 654
Complex analysis:
   analytic functions, 45
   analyticity, 45
   Cauchy-Riemann equations, 45
   complex variable, 44
   continuous functions, 44
   differentiability, 44
   essential singularities, 46
   holomorphic functions, 45
   isolated singularities, 45
   Laurent series, 45–46
      principal part, 46
   limit, 44
   nonisolated singularities, 45
   order of a pole, 46
      of a zero, 45
   pole of a function, 46
   regular functions, 45
   residue:
      definition, 46
      for multiple pole, 54
      for simple pole, 54
      theorem, 46
   singularities, 45
   zero of a function, 45
Complex convolution, 59–61
   application in the design of nonrecursive filters, 282–286
   graphical interpretation, 284
Complex variable, 44
Complex-convolution theorem, 59–61
Complex-differentiation theorem, 51
Complex-scale-change theorem, 50–51
Computability, 133
Computation of inverse discrete Fourier transform, 476–477
Computational complexity:
   in basic Remez exchange algorithm, 554
   in fast Fourier transform, 472
   in implementation of nonrecursive filters, 477
Concurrency, 116
Constant-delay allpass transfer function, 175
   approximation (*see* Bessel approximation)
   nonrecursive filters, 275–277
   recursive filters, 268–269

INDEX **671**

Constantinides, A. G., 239
Constantinides transformations, 238–243
Constant-input limit cycles, 378
Continued-fraction expansion, 107
Continuous Fourier transform (*see* Fourier transform)
Continuous functions, 44
Continuous-time random processes (*see* Random processes)
Continuous-time signals, processing using digital filters, 207–208
Continuous-time transfer function, 80
 (*see also* Transfer function: continuous)
Convergence factor in adaptive filters, 630
Convergence in Fourier transform, 178
 in z transform:
  absolute, 48
  uniform, 50
Conversion from binary to decimal numbers, 332
 from decimal to binary numbers, 332
Convex functions, 493
Convolution:
 complex, 59–61
  application in the design of nonrecursive filters, 282–286
  graphical interpretation, 284
 frequency, 179
 frequency-domain periodic, 464–465
 integral, 60
 real, 51
 summation, derivation, 21–23
  graphical interpretation, 23–24
 time, 179
 time-domain periodic, 464
 2-dimensional, 637–638
Crochiere, R. E., 341
Cross correlation, 629
Cubic-interpolation search, 558–561
Current-conversion generalized-immittance converter (CGIC), 426
 admittance conversion function, 426
 (*see also* CGIC)
Cutoff frequency, 141
Cyclic convolution (*see* Periodic convolution)

D/A converter (*see* Digital-to-analog converter)
Davidon-Fletcher-Powell updating formula for inverse Hessian, 500
Deadband effect (*see* Limit cycles)
Deadband range, 369
Decimation-in-frequency FFT algorithm, 472–476
Decimation-in-time FFT algorithm, 466–472
Decimators, 593–595

Decomposition section in filter banks, 602
Delay (*see* Group delay or Phase delay)
Delay characteristic, 85, 140
Delay distortion, 87
Delay equalization, 268–269
Delay equalizers, 521–535
Delay-free loops, 26, 390
Denormalized approximations, 143
Descent direction, 494
Design:
 approximation step, 87–88
 concurrency, 116
 decimators, 602
 differentiators (*see* Differentiators)
 general considerations, 115–116
 Hilbert transformers, 617–621
 interpolators, 602
 introduction to, 1, 87–89
 latency, 117
 nonrecursive filters (*see* Nonrecursive filters)
 nonrecursive filters by optimization (*see* Nonrecursive filters by optimization)
 pipelining, 116
 processing element, 116
 processing rate, 117
 QMF banks, 609–611
 realization step, 88
 recursive filters (*see* Recursive filters)
 recursive filters by optimization (*see* Recursive filters by optimization)
 wave digital filters (*see* Wave digital filters)
DFP updating formula for inverse Hessian, 500
DFT (*see* Discrete Fourier transform)
Differentiability, 44
Differentiation formulas (numerical), 302–303
Differentiators:
 using Fourier series method, 303–305
 using Kaiser window function, 305
 using numerical-analysis formulas, 301–303
 using Remez exchange algorithm, 578–583
  first derivative of error function, 580
  prescribed specifications, 580–583
  problem formulation, 578–580
Digital filters:
 allpass, 94
 amplitude equalization, 270–271
 amplitude response, 21, 81
 analysis, 1
 approximation step, 87–88
 causality, 5
 characterization (*see* Characterization)
 choice of structure, 434–436
 comparison with recursive filters, 305–306

672 INDEX

Digital filters (cont.)
  delay characteristic, 85
  delay equalization, 268–269
  design of nonrecursive filters (see Nonrecursive filters)
    of recursive filters (see Recursive filters)
    process, 87–89
      using optimization (see Recursive filters by optimization or Nonrecursive filters by optimization)
  effect of interfacing devices, 89
  effects of arithmetic errors, 88
  elementary functions for time-domain analysis, 15–17
  elements, 9
  excitation, 2
  frequency response, 21, 81
  gain, 81
  group delay, 85
  hardware implementation, 88–89
  linearity, 7
  network representation, 9–10
  noncausal, 5
  nonlinear, 8
  nonrecursive (see Nonrecursive filters)
  order, 9
  overflow, 352
  phase response, 21, 81
  phase shift, 81
  phasor diagram, 82
  processing of continuous-time signals, 207–208
  realization (see Realization)
  recursive (see Recursive filters)
  relaxed, 3
  representation (see Representation of digital filters)
  representation by signal flow graphs, 13, 119–121
  response, 2
  rudimentary implementation, 332–333
  signal flow-graph representation, 119–121
  software implementation, 88–89
  structures (see Structures)
  synthesis, 88
  system representation, 2–8
  time-dependent, 3
  time-invariance, 3
  transformations (see Transformations)
  wave (see Wave digital filters)
  zero-phase filters, 269–270
  zero-pole plots, 68
Digital G-CGIC configuration, 426
Digital integrator, 232

Digital-to-analog converter:
  ideal, 208
  practical, 213
Diniz, P. S. R., 346, 378
Direct canonic realization, 102–104
Direct realization, 98
Discrete Fourier transform (DFT):
  amplitude spectrum, 445
  butterfly flow graph, 467
  of complex discrete-time signals, 484
  computation of inverse discrete Fourier transform, 476–477
  definition, 445
  design of nonrecursive filters, 458–462
  fast-Fourier-transform (FFT) algorithms, 465–477
    decimation-in-frequency, 472–476
    decimation-in-time, 466–472
  interrelation with continuous Fourier transform, 454–456
    with Fourier series, 456–458
    with z transform, 449
  inverse, 446
  linearity, 446
  periodic convolution:
    frequency-domain, 464–465
    time-domain, 464
  periodicity, 446
  phase spectrum, 445
  properties, 446–448
  simplified notation, 462–463
  symmetry, 447
Discrete-time random processes, 323–325
  (see also Random processes)
Discrete-time signals (see Signals)
Discrete-time transfer function (see Transfer function: discrete-time)
Discrimination factor, 155
Dolph-Chebyshev window function, 291–292
  main-lobe width, 291–292
  ripple ratio, 291–292
Downsampler, 594
Downsampling, 594

Ebert, P. M., 373
Effect of interfacing devices, 89
Effects of arithmetic errors, 88
Eigenvalues, 72
Elementary functions, 15–17
  exponential, 16
  sinusoid, 16
  unit impulse, 16–17
  unit ramp, 16

INDEX **673**

Elementary functions (*cont.*)
  unit step, 15
Elements for digital filters:
  adder, 9
  multiplier, 9
  unit delay, 9
Elimination of aliasing errors in QMF banks, 605–609
  of limit cycles in wave digital filters, 423–425
  of overflow limit cycles, 378–380
  of quantization limit cycles, 373–378
Elliptic approximation:
  derivation, 154–160
  discrimination factor, 155
  elliptic functions (*see* Elliptic functions)
  elliptic integral (*see* Elliptic integral)
  fifth-order, 154–160
  infinite-loss frequencies, 159
  introduction, 139
  loss function, 161–163
  minimum order, 166
  minimum stopband loss, 165
  modular constant, 163
  normalized transfer function, 166–167
  nth-order:
    even n, 163–164
    odd n, 160–163
  selectivity factor, 155
  specification constraint, 164–166
  zero-loss frequencies, 158
  zeros and poles of loss function, 161–163
Elliptic functions:
  addition formula, 659
  definition, 656
  in filters, 156
  formulas, 659–660
  fundamental period parallelogram, 661
  imaginary argument, 656–658
  imaginary period, 661
  modular constant, 664
  period parallelogram, 661
  periodicity, 660–661
  plots, 657, 659
  series representation, 663–665
  theta functions, 664
  transformation, 662–663
Elliptic integral:
  amplitude, 653
  complementary complete, 658
  complementary modulus, 658
  complete, 654
  definition, 653
  in filters, 156

  of the first kind, 653
  modulus, 653
Elliptic recursive filters, 261–262
Encoder, 210
End-around carry, 336
Energy spectral density, 180
Energy spectrum, 180
Ensemble in random processes, 315
Equivalent sequences, 184
Error function for delay equalizers, 527
  for recursive filters, 490–491
Errors:
  in A/D converter, 212
  coefficient quantization, 330
  input quantization, 331
  product quantization, 331
  roundoff noise, 331
Error-spectrum shaping, 364–367
Essential singularities, 46
Eswaran, C., 434
Euclidean norm, 492
Euclidean space, 494
Excitation, 2
Exhaustive step-by-step search, 551–552
Expander, 597
Expected value of random variable, 312
Exponent, 337
Exponential function, 16
Extrapolation formula for inexact line search, 503–504
Extremal frequencies (or extremals), 547–548
  initialization, 550

Fan filters, 643
Fast-Fourier-transform (FFT) algorithms, 465–477
  butterfly flow graph, 467
  computation of inverse DFT, 476–477
  computational complexity, 472
  decimation-in-frequency, 472–476
  $N = 8$, 476
  decimation-in-time, 466–472
  $N = 8$, 471–472
Fast-Fourier-transform method, introduction, 444
FDNR networks, 416
Feasible region of (c_0, c_1) plane, 527
Feldkeller equation, 410
Fettweis, A., 388, 416, 423
FFT (*see* Fast Fourier transform)
Filtering of discrete-time random processes, 325–326
Finite-duration impulse response, 25
FIR filters, 25
  (*see also* Nonrecursive filters)

First-order statistics, 316
Fixed-point arithmetic, 334–336
Fletcher inexact line search, 504–505
Fletcher, R., 500, 504, 505, 507
Flip-flop, 332
Floating-point arithmetic, 337
Flow-graph representation (*see* Signal flow-graph representation)
Forced response, 379
Formulas for elliptic functions, 659–660
Formulas for frequency response of nonrecursive filters, 278
Foster forms, 408
Fourier series, 197, 280
   design of differentiators, 303–305
   design of nonrecursive filters, 280–282
   interrelation with discrete Fourier transform, 456–458
     with Fourier transform, 197–198
Fourier transform:
   amplitude spectrum, 178
   convergence, 178
   definition, 177
   discontinuous functions, 178
   energy spectral density, 180
   energy spectrum, 180
   frequency convolution, 179
   frequency shifting, 179
   generalized functions, 182–197
   impulse function, 185–191
     alternative sequences, 187–191
     definition, 185
     properties, 185–186
   of infinite-energy signals, 182
   interrelation with discrete Fourier transform, 454–456
     with the Fourier series, 197–198
     with z transform, 200
   inverse, 178
   linearity, 178
   notation, 178
   Parseval's formula, 179
   phase spectrum, 178
   properties, 178–180
   of sampled signals, 200
   symmetry, 179
   table of common Fourier transforms, 195
   theorems, 178–179
   time convolution, 179
   time scaling, 179
   time shifting, 179
   unity function, 185–191
     alternative sequences, 187–191

     definition, 186
Fourier-series kernel, 196
Fourier-series method, 280–282
Frequency convolution, 179
Frequency folding, 205
Frequency response:
   in digital filters, 21, 81
   in nonrecursive filters, 277–278
   formulas, 278
   periodicity, 82–83
   in QMF banks, 610
Frequency shifting in Fourier transform, 179
Frequency spectrum, 85
Frequency-dependent negative resistance (FDNR) networks, 416
Frequency-division to time-division multiplex translation, 648
Frequency-domain aliasing, 205
Frequency-domain analysis, 80–85
   in 2-dimensional digital filters, 640–641
   in wave digital filters, 419–422
Frequency-domain periodic convolution, 464–465
Frequency-domain representation of random processes:
   continuous-time, 319–323
   discrete-time, 324–325
Frequency-domain sampling theorem, 449–451
Frequency-sampling approximation method, 458–462
   antisymmetrical impulse response, 460–462
   formulas, 461
   symmetrical impulse response, 459–460
Fundamental period parallelogram of elliptic functions, 661

Gain:
   in analog filters, 80
   in digital filters, 81
Gain sensitivity, 136
Gamma function, 169
Ganapathy, V., 434
Gaussian probability-density function, 311
Gazsi, L., 610
G-CGIC configuration:
   analog, 426
   digital, 426
Generalized functions:
   definitions, 182–193
   equivalent sequences, 184
   good functions, 182–184
   impulse function, 185–191
   introduction, 182
   limit of a regular sequence, 184

Generalized functions (*cont.*)
  ordinary functions as, 191
  regular sequences, 184
  theorems, 182–185
  unity function, 185–191
Generalized impedance converter (*see* CGIC)
Geometric series, 17
Gibbs' oscillations, 281
Global minimum, 495
Gold, B., 462
Good functions, 182–184
Gradient information for Remez exchange algorithm, 566–568
Gradient vector:
  definition, 493
  for delay equalizers, 528
  of the MSE function, 629
  of Nth-order recursive filters, 518
Gramian:
  observability, 361
  reachability, 361
Granularity limit cycles, 367
Graph determinant, 14
Graphical interpretation of convolution summation, 23–24
Graphical method for amplitude and phase responses, 81–82
Gray, Jr., A. H., 109
Green, B. D., 374
Gregory-Newton interpolation formulas, 301
Grossman, A. J., 154
Group delay:
  in analog filters, 140
  in digital filters, 85
  in nonrecursive filters, 275
  in 2-dimensional digital filters, 640

Hamming window function, 287–290
  main-lobe width, 289
  ripple ratio, 289
  spectrum, 289–290
Hancock, H., 653
Hanning window function (*see* von Hann window function)
Hardware implementation, 88–89, 115
Herrmann, O., 545, 572
Hessian matrix, 493
Higgins, W. E., 367
Higher-order statistics, 316
Highpass filters:
  CGIC section, 430
  ideal, 141
  nonrecursive, 297
  practical, 141–142
  problem formulation for nonrecursive filters, 546
  recursive, 253
  2-dimensional, 643
Hilbert transform, 617
Hilbert transformers:
  analytic signals, 613–617
  application to single-sideband modulation, 621–623
    to the sampling of bandpassed signals, 623–625
  design, 617–621
  Hilbert transform, 617
Hirano, K., 349
Hofstetter, E., 545
Holder inequality, 355
Holomorphic functions, 45
Homogeneity property, 7
Hwang, S. Y., 360

Ideal filters, 140
IDFT (inverse discrete Fourier transform), 446
IIR filters, 25
  (*see also* Recursive filters)
Images in signal interpolation, 597
Imaginary period of elliptic functions, 661
Impedances, wave realization of, 393
Implementation:
  of analog filters using digital filters, 207–208
    using sampled-data filters, 207–208
  hardware, 88–89, 115
  nonreal-time, 89
  nonrecursive filters, 477–482
  real-time, 89
  rudimentary, 332–333
  software, 88–89, 115
  VLSI, 116
Impulse function, 185–191
  alternative sequences, 187–191
  definition, 185
  of first-order filter, 17
  properties, 185–186
Impulse response of nonrecursive filters:
  antisymmetrical response and odd filter length, 576
  of Nth-order filter, 70
  symmetrical response and odd filter length, 554
  using induction, 17
  using state-space characterization, 30
Impulse sampler, 199
Incident wave quantity, 390
Induction method, 17–21
Inexact line searches, 500–505

Infinite-duration impulse response, 25
Infinite-energy signals, 182
Infinite-loss frequencies in elliptic approximation, 159
Input functions in wave digital filters, 420
Input quantization, 331
  errors, 331
Insertion loss, 389
Interface, analog-to-digital, 210–212
Interior band edge, 551
Interpolation:
  cubic, 558–561
  formula for inexact line search, 503–504
  formulas of numerical analysis, 301
  Gregory-Newton formulas, 301
  linear, 601
  Stirling formula, 301
Interpolators, 595–601
Interreciprocal multipole, 123
Interreciprocity, 123–124
Interrelations:
  between continuous-time, sampled, and discrete-time signals, 205–206
  between discrete and continuous Fourier transforms, 454–456
  between discrete Fourier and z transforms, 449
  between discrete Fourier transform and Fourier series, 456–458
  between Fourier and z transforms, 200
  between the Fourier series and Fourier transform, 197–198
Invariant-impulse-response method, 221–223
Invariant-sinusoid-response method, 246
Invariant-unit-step-response method, 245
Inverse discrete Fourier transform (IDFT), 446
  computation of, 476–477
Inverse Fourier transform, 178
Inverse z transform:
  definition, 53
  by equating coefficients, 65
  by using binomial series, 56
  by using convolution theorem, 58–59
  by using long division, 65
  by using partial fractions, 57
Inverse-Chebyshev approximation, 153
  normalized transfer function, 153
Inverse-Chebyshev recursive filters, 260–261
Isolated singularities, 45

Jackson, L. B., 352, 360, 362, 363, 367
Jacobian elliptic functions (*see* Elliptic functions)
Jenkins, W. K., 344
Joint distribution function, 312

Joint probability-density function, 312
Jury, E. I., 48, 52, 76
Jury-Marden array, 77
Jury-Marden stability criterion, 76–78

Kaiser, J. K., 374
Kaiser window function, 292–293
  design of differentiators, 305
  main-lobe width, 293
  ripple ratio, 293
  spectrum, 292–293
Kim, Y., 362

L_1 norm for recursive filters, 492
L_2 norm for recursive filters, 492
L_2 scaling, 356
L_p norm, 354
L_p norm for recursive filters, 491
L_∞ norm for recursive filters, 492
L_∞ scaling, 356
Ladder realization, 104–109
Ladder wave digital filters, 412–414
Lagrange interpolation, 552
Laplace transform of sampled signals, 202
Latency, 117
Lattice realization, 109–113
Lattice wave digital filters:
  alternative analog realization, 407–408
  analog network, 406
  analysis, 406–407
  digital realization, 411–412
Laurent, P. A., 45
Laurent series, 45–47
  principal part, 46
  residues, 54
Laurent theorem, 45–46
Least-mean-square algorithm, 633–634
Least-pth algorithm, 510–511
Least-squares solution, 492
Legendre elliptic integral (*see* Elliptic integral)
Leon, B. J., 344
l'Hôpital's rule, 18
Lighthill, M. J., 178, 182
Limit, 44
Limit cycles:
  bounds, 373
  constant-input, 378
  deadband effect, 367
  deadband range, 369
  elimination of limit cycles in wave digital filters, 423–425
  of overflow limit cycles, 378–380
  of quantization limit cycles, 373–378

Limit cycles (cont.):
  in first-order filter, 367–369
  forced response, 379
  frequency of, 369
  granularity, 367
  overflow, 367, 372–373
  quantization, 367–372
  resonant frequency, 370
  in second-order filter, 369–372
Limit of a regular sequence, 184
Lindgren, A. G., 362
Line searches, 494
Linearity:
  additivity property, 7
  in digital filters, 7
  in discrete Fourier transform, 446
  in Fourier transform, 178
  homogeneity property, 7
  in z transform, 50
Liu, V., 374
LMS algorithm (see Least-mean-square algorithm)
Location of zeros in nonrecursive filters with constant group delay, 279–280
Long, J. L., 374
Loss, 140
Loss characteristic, 140
Loss function, 140
  for elliptic approximation, 161–163
  zero-pole plots for, 140
Lossless-discrete-integrator ladder, 425
Lowpass filters:
  CGIC section, 430
  ideal, 140
  nonrecursive, 293–297
  practical, 141–142
  problem formulation for nonrecursive filters, 546
  recursive, 251–253
Lowpass-to-bandpass transformation:
  for analog filters, 170–172
  for digital filters, 243
Lowpass-to-bandstop transformation:
  for analog filters, 172
  for digital filters, 242–243
Lowpass-to-highpass transformation:
  for analog filters, 172
  for digital filters, 243
Lowpass-to-lowpass transformation:
  for analog filters, 169–170
  for digital filters, 240–242
Low-sensitivity structures, 344–350
Lyapunov equation, 78
Lyapunov stability criterion, 78

Main lobe, 283
Main-lobe width:
  in Blackman window function, 290
  definition, 283
  in Dolph-Chebyshev window function, 291–292
  in Hamming window function, 289
  in Kaiser window function, 293
  in rectangular window function, 287
  in von Hann window function, 289
Maison's gain formula, 13
Mantissa, 337
Mapping properties of bilinear transformation, 233–234
Markel, J. D., 109
Matched-z-transformation method, 228–231
Maxima of error function in Remez exchange algorithm, 550–551
Maximally-flat approximation, 143
Maximum output power, 389
Maximum passband loss, 141
Mazo, J. E., 373
McClellan, J. H., 545, 643
McGoneral, C. A., 462
Mean value, of continuous-time random processes, 318
  of discrete-time random processes, 324
  of random variable, 312
Mean-square error in adaptive filters, 628
Mean-square value of continuous-time random processes, 318
  of discrete-time random processes, 324
Mecklenbraüker, W. F. G., 379
Meerkötter, K., 374, 423
Meerkötter's structure, 377
Mills, W. L., 374
Minimax algorithms, 509–517
Minimax solution, 492
Minimization of output roundoff noise, 360–364
Minimum point, 494
Minimum stopband loss, 141
  in elliptic approximation, 165
Minors, 75
Mirror-image polynomials, 279
Mitra, S. K., 104, 425, 610
Modified invariant-impulse-response method, 224–226
Modular constant of elliptic functions, 163, 664
Modulus of elliptic integral, 653
Moments, 312–314
MSE (see Mean-square error)
Mullis, C. T., 360, 362, 374
Multiband filters using Remez exchange algorithm, 583–586

Multiplication:
  floating-point, 337
  one's-complement, 336
  signed-magnitude, 336
  two's-complement, 336
Multiplier, 9
  noise model, 350
Multipole, 122
  interreciprocal, 123
  reciprocal, 122
Munson, Jr., D. C., 367

Negative-impedance converter, 439
Network analysis, 10
Network representation of digital filters, 9–10
Neuvo, Y., 610
Newton algorithm, 495
  application to adaptive filters, 630
Newton direction, 494
Newton method, 492–496
Nishimura, S., 349
Noise model for a multiplier, 350
Noise source, 350
  power spectral density, 352
Noncausal digital filters, 5
Nonisolated singularities, 45
Nonlinear digital filters, 8
Nonquadratic functions, 493
Nonquantized signals, 1
Nonrecursive filters:
  antisymmetrical impulse response, 277
  approximations, introduction, 274
  characterization by difference equation, 9
  comparison with recursive filters, 305–306
  constant-delay, 275–277
  differentiators (see Differentiators)
  formulas for frequency response, 278
  frequency response, 277–278
  frequency-sampling method, 458–462
    antisymmetrical impulse response, 460–462
    formulas, 461
    symmetrical impulse response, 459–460
  Gibbs' oscillations, 281
  group delay, 275
  implementation, 477–482
    overlap-and-add method, 478–480
    overlap-and-save method, 480–482
  location of zeros in filters with constant group delay, 279–280
  mirror-image polynomials, 279
  by optimization, 544–586
    adjustable bracket, 560
    alternation theorem, 547–548
    antisymmetrical impulse response and even filter length, 577–578
    odd filter length, 574–576
    arbitrary amplitude responses, 583
    basic Remez exchange algorithm, 549
    computational complexity of basic Remez exchange algorithm, 554
    differentiators, 578–583
      first derivative of error function, 580
      prescribed specifications, 580–583
      problem formulation, 578–580
    efficient Remez exchange algorithm, 562–566
    exhaustive step-by-step search, 551–552
    extremals, 547–548
      initialization, 550
    generalization of Remez exchange algorithm, 574–578
    gradient information for Remez exchange algorithm, 566–568
    improved formulation for Remez exchange algorithm, 561–562
    impulse response:
      antisymmetrical and odd filter length, 576
      symmetrical and odd filter length, 554
    interior band edge, 551
    introduction, 544
    Lagrange interpolation, 552
    maxima of error function, 550–551
    multiband filters, 583–586
    potential extremals, 549
    prediction formula for filter length, 572
    prescribed specifications, 571–574
    problem formulation, 545–547
      bandpass filters, 547
      bandstop filters, 547
      highpass filters, 546
      lowpass filters, 546
    rejection of superfluous potential extremals, 553–554
    search based on cubic interpolation, 558–561
      on quadratic interpolation, 561
    selective step-by-step search, 555–558
    symmetrical impulse response and even filter length, 576–577
  phase delay, 275
  prescribed specifications:
    bandpass, 298–299
    bandstop, 298–299
    highpass, 297
    lowpass, 293–297
  stability, 71
  symmetrical impulse response, 276
2-dimensional digital filters, 637

Nonrecursive filters (*cont.*)
  using discrete Fourier transform
    (frequency-sampling method), 458–462
  using Fourier-series method, 280–282
  using numerical-analysis formulas, 302–303
  using window functions, 282–286
  window functions:
    Blackman, 290
    Dolph-Chebyshev, 291–292
    Hamming, 287–290
    Kaiser, 292–293
    rectangular, 287
    von Hann, 287–290
  zero-pole plot, 280
Nonuniform variable sampling technique, 513–517
Normalized approximations, 143
Normalized sensitivity, 344
Notation for random processes, 316
Notch CGIC section, 431
Number quantization, 337–339
  by rounding, 337
  by truncation, 337
Number representation:
  binary point, 332
  binary system, 331
  bits, 332
  conversion from binary to decimal, 332
    from decimal to binary, 332
  exponent, 337
  mantissa, 337
  one's-complement, 335
  radix, 331
  radix point, 331
  signed-magnitude, 334
  two's-complement, 335
  word length, 335
Numerical differentiation formulas, 301–302
Numerical integration formulas, 302
Numerical-analysis formulas, 301
Nyquist frequency, 83

Objective function for Nth-order filters, 517–518
  for recursive filters, 491
Observability gramian, 361
One's-complement:
  addition, 336
  arithmetic, 336
  multiplication, 336
  representation, 335
Operation of QMF banks, 603–605
Oppenheim, A., 545
Optimization:
  nonrecursive filters:

adjustable bracket, 560
alternation theorem, 547–548
computational complexity of basic Remez
  exchange algorithm, 554
cubic-interpolation search, 558–561
differentiators using Remez exchange
  algorithm, 578–583
  first derivative of error function, 580
  prescribed specifications, 580–583
  problem formulation, 578–580
efficient Remez exchange algorithm, 562–566
exhaustive step-by-step search, 551–552
extremal frequencies (or extremals), 547–548
generalization of Remez exchange algorithm,
  574–578
improved formulation for Remez exchange
  algorithm, 561–562
impulse response of nonrecursive filters:
  antisymmetrical and odd filter length, 576
  symmetrical and odd filter length, 554
initialization of extremal frequencies, 550
interior band edge, 551
Lagrange interpolation, 552
maxima of error function in Remez exchange
  algorithm, 550–551
potential extremals, 549
prediction formula for filter length in
  nonrecursive filters, 572
prescribed specifications in nonrecursive
  filters, 571–574
problem formulation for nonrecursive filters,
  545–547
  bandpass filters, 547
  bandstop filters, 547
  highpass filters, 546
  lowpass filters, 546
quadratic-interpolation search, 561
rejection of superfluous potential extremals,
  553–554.
Remez exchange algorithm:
  antisymmetrical impulse response and even
    filter length, 577–578
  antisymmetrical impulse response and odd
    filter length, 574–576
  arbitrary amplitude responses, 583
  basic, 549
  gradient information, 566–568
  multiband filters, 583–586
  symmetrical impulse response and even
    filter length, 576–577
  selective step-by-step search, 555–558
recursive filters:
  alternative Newton algorithm, 498

Optimization: recursive filters (cont.)
   basic Newton algorithm, 495
   basic quasi-Newton algorithm, 496–500
   BFGS updating formula for inverse Hessian, 500
   Charalambous minimax algorithm, 512–513
   convex functions, 493
   definition of gradient vector, 493
   delay equalizers, 521–535
   descent direction, 494
   design of equalizers, 529
      of recursive filters, 489–534
   DFP updating formula for inverse Hessian, 500
   error function for delay equalizers, 527
      for recursive filters, 490–491
   Euclidean norm, 492
   Euclidean space, 494
   extrapolation formula for inexact line search, 503–504
   feasible region of (c_0, c_1) plane, 527
   Fletcher inexact line search, 504–505
   global minimum, 495
   gradient vector for delay equalizers, 528
      for Nth-order recursive filters, 518
   Hessian matrix, 493
   inexact line searches, 500–505
   interpolation formula for inexact line search, 503–504
   L_1 norm for recursive filters, 492
   L_2 norm for recursive filters, 492
   L_p norm for recursive filters, 491
   L_∞ norm for recursive filters, 492
   least-pth algorithm, 510–511
   least-squares solution, 492
   line searches, 494
   minimax algorithms, 509–517
   minimax solution, 492
   minimum point, 494
   Newton algorithm, 495
   Newton direction, 494
   Newton method, 492–496
   nonquadratic functions, 493
   objective function for Nth-order filters, 517–518
      for recursive filters, 491
   practical quasi-Newton algorithm, 505–507
   problem formulation for recursive filters, 490–492
   quadratic functions, 493
   quasi-Newton algorithms, 496–509
   Rosenbrock function, 539
   stationary point, 493
   sum of the squares, 492
   Taylor series, 493
   termination tolerances, 495
   unimodal functions, 501
   virtual sample points, 514
Optimum word length, 340
Order of a digital filter, 9
   of a pole, 46
   of a 2-dimensional digital filter, 637
   of a zero, 45
Ordering of sections in cascade realization, 360
Ordinary functions as generalized functions, 191
Output noise in CGIC structures, 433–434
Overflow in digital filters, 352
   in one's- or two's-complement addition, 336
Overflow limit cycles, 367, 372–373
   elimination of, 378–380
Overlap-and-add method, 478–480
Overlap-and-save method, 480–482

Pal, R. N., 349
Papoulis, A., 178
Parallel adaptors, 397
   1-multiplier, 397
   2-multiplier, 397
Parallel realization, 114–115
Parallel wire interconnections, wave realization of, 395–397
Parks, T. W., 545
Parseval's continuous-time formula, 179
   discrete-time formula, 61
Partial fractions, 57
Passband, 141
Passband edge, 141
Peek, J. B. H., 379
Perfect reconstruction in QMF banks, 611–613
Period parallelogram for elliptic functions, 661
Periodic continuation, 204, 454
Periodic convolutions, 463–465
Periodicity of discrete Fourier transform, 446
   of elliptic functions, 660–661
   of frequency response, 82–83
   of spectrum in sampled signals, 200–201
Phase characteristic, 140
Phase delay, 275
Phase distortion, 87
Phase response:
   in analog filters, 80
   in digital filters, 21, 81
   graphical method, 81–82
   influence of warping effect, 237
   in 2-dimensional digital filters, 640

INDEX **681**

Phase shift:
  in analog filters, 80, 140
  in digital filters, 81
Phase spectrum:
  continuous Fourier transform, 178
  discrete Fourier transform, 445
Phase-shift sensitivity, 136
Phasor, 81
Phasor diagram, 82
Pipelining, 116
Plots of elliptic functions, 657, 659
Poisson's summation formula, 198–199
Pole of a function, 46
Port conductance, 397
Port resistance, 390
Positive definite matrix, 75
Potential extremals, 549
  rejection of superfluous potential extremals, 553–554
Power density spectrum, 320
Power in discrete-time random processes, 325
Power spectral density:
  of continuous-time random processes, 320
  of discrete-time random processes, 324
  of noise source, 352
  of output noise, 352
Practical filters, 141–142
Practical quasi-Newton algorithm, 505–507
Prediction formula for filter length in nonrecursive filters, 572
Prescribed specifications:
  digital differentiators, 580–583
  introduction, 249
  nonrecursive filters, 293–301, 571–574
    bandpass, 298–299
    bandstop, 298–299
    highpass, 297
    lowpass, 293–297
  recursive filters, 249–268
    bandpass, 253–257
    Butterworth, 257–259
    bandstop, 257
    Chebyshev, 259–260
    elliptic, 261–262
    highpass, 253
    inverse-Chebyshev, 260–261
    lowpass, 251–253
  wave digital filters, 416–419
Prewarping of analog filters, 235
Principal minor determinants (minors), 75
Principal part of Laurent series, 46
Probability-density function:
  definition, 311

  gaussian, 311
  Rayleigh, 327
  uniform, 311
Problem formulation for digital differentiators, 578–580
  for nonrecursive filters, 545–547
    bandpass filters, 547
    bandstop filters, 547
    highpass filters, 546
    lowpass filters, 546
  for recursive filters, 490–492
Processing element, 116
Processing of continuous-time signals using digital filters, 207–208
Processing rate, 117
Product quantization, 331, 350–352
  errors, 331
  noise model for a multiplier, 350
  power spectral density of noise source, 352
    of output noise, 352
PSD (*see* Power spectral density)
Pseudopower in wave digital filters, 423

Quadratic form, 75
Quadratic functions, 493
Quadratic-interpolation search, 561
Quadrature-mirror-filter banks:
  analysis section, 602
  decomposition section, 602
  design, 609–611
  elimination of aliasing errors, 605–609
  frequency response, 610
  introduction, 602
  operation, 603–605
  perfect reconstruction, 611–613
  reconstruction section, 602
  synthesis section, 602
Quantization, coefficient, 330, 339–344
Quantization error, 338
  in A/D converter, 212
  introduced by rounding, 339
    by truncation, 338
  in one's-complement representation, 338
  in two's-complement representation, 338
Quantization:
  input, 331
  limit cycles, 367–372
    elimination of, 373–378
  product, 331, 350–352
Quantized signals, 2
Quantizer, 339
Quasi-Newton algorithms, 496–509

Rabiner, L. R., 462, 545, 572

Radius of convergence, 49
Radix, 331
Radix point, 331
Random processes:
  continuous-time:
    autocorrelation function, 318
    definition, 314–316
    ensemble, 315
    first-order statistics, 316
    frequency-domain representation, 319–323
    higher-order statistics, 316
    mean square value, 318
    mean value, 318
    notation, 316
    power spectral density, 320
    power-density spectrum, 320
    sample function, 315
    second-order statistics, 316
    strictly stationary, 319
    white-noise processes, 323
    wide-sense stationary, 319
    Wiener-Khinchine relation, 322
  discrete-time, 323–325
    autocorrelation function, 324
    filtering of, 325–326
    mean value, 324
    power in, 325
    power-spectral density, 324
Random signals (*see* Random processes)
Random variables:
  central moment, 313
  definition, 310–311
  expected value, 312
  joint distribution function, 312
  joint probability-density function, 312
  mean value, 312
  moments, 312–314
  *n*th central moment, 313
  *n*th moment, 313
  probability-density function:
    definition, 311
    gaussian, 311
    Rayleigh, 327
    uniform, 311
  probability-distribution function, 311
  statistical independence, 312
  variance, 313
Rayleigh probability-density function, 327
Reachability gramian, 361
Real convolution, 51
Realizability constraints:
  continuous-time transfer functions, 141–143
  discrete-time transfer functions, 221

wave digital filters, 405
Realization:
  canonic, 102–104
  cascade, 113–114
  continued-fraction expansion, 107
  direct, 88, 98
  direct canonic, 102–104
  indirect, 88
  introduction, 97
  ladder, 104–109
  ladder wave, 412–416
  lattice, 109–113
  lattice wave, 411–412
  Meerkötter, 377
  parallel, 114–115
  state-space, 104
  step, 88
  using CGICs, 429–432
  wave digital filters, 391–393
    circulators, 402–403
    impedances, 393
    parallel wire interconnections, 395–397
    resonant circuits, 404–405
    series wire interconnections, 395
    transformers, 398–400
    unit elements, 401–402
    voltage sources, 393–395
Reciprocal multipole, 122
Reciprocity, 122
Reconstruction section in filter banks, 602
Rectangular window function, 287
  main-lobe width, 287
  ripple ratio, 287
  spectrum, 287
Recursive filters:
  adaptive, 634
  bilinear-transformation method, 231–238
    (*see also* Bilinear transformation)
  Butterworth, 257–259
  characterization by difference equation, 9
  Chebyshev, 259–260
  comparison with nonrecursive filters, 305–306
  constant-delay, 268–269
  design procedure, 262
  elliptic, 261–262
  frequency response, 81
  group delay, 85
  invariant-impulse-response method, 221–223
  invariant-sinusoid-response method, 246
  invariant-unit-step-response method, 245
  inverse-Chebyshev, 260–261
  matched z-transformation method, 228–231

INDEX **683**

Recursive filters (*cont.*)
  modified invariant-impulse-response method, 224–226
  by optimization, 489–534
    algorithm for the design of equalizers, 529
    basic quasi-Newton algorithm, 496–500
    BFGS updating formula for inverse Hessian, 500
    Charalambous minimax algorithm, 512–513
    convex functions, 493
    delay equalizers, 521–535
    descent direction, 494
    DFP updating formula for inverse Hessian, 500
    error function, 490–491
    error function for delay equalizers, 527
    Euclidean norm, 492
    Euclidean space, 494
    extrapolation formula, 503–504
    feasible region of (c_0, c_1) plane, 527
    Fletcher inexact line search, 504–505
    global minimum, 495
    gradient vector:
      definition, 493
      for delay equalizers, 528
      of Nth-order filters, 518
    Hessian matrix, 493
    inexact line searches, 500–505
    interpolation formula, 503–504
    introduction, 489
    L_1 norm, 492
    L_2 norm, 492
    L_p norm, 491
    L_∞ norm, 492
    least-pth algorithm, 510–511
    least-squares solution, 492
    line searches, 494
    minimax algorithms, 509–517
    minimax solution, 492
    minimum point, 494
    Newton algorithm, 495
    Newton direction, 494
    Newton method, 492–496
    nonquadratic functions, 493
    nonuniform variable sampling technique, 513–517
    objective function, 491
    objective function for Nth-order filters, 517–518
    practical quasi-Newton algorithm, 505–507
    problem formulation, 490–492
    quadratic functions, 493
    quasi-Newton algorithms, 496–509
    stationary point, 493
    sum of the squares, 492
    Taylor series, 493
    termination tolerances, 495
    unimodal functions, 501
    virtual sample points, 514
  phase response, 81
  prescribed specifications, 250–267
    bandpass, 253–257
    bandstop, 257
    highpass, 253
    lowpass, 251–253
  prewarping, 235
  realizability constraints, 221
  by transformations, 237–243
  2-dimensional digital filters, 637
  warping effect, 235–238
  wave (*see* Wave digital filters)
Reflected wave quantity, 390
Region of convergence, 49
Register, 332
Regular functions, 45
Regular sequences, 184
Relation between continuous-time and discrete-time impulse functions, 206
Relative output-noise power spectral density, 386
Relatively prime polynomials, 73
Relaxed digital filters, 3
Remez exchange algorithm:
  antisymmetrical impulse response and even filter length, 577–578
  antisymmetrical impulse response and odd filter length, 574–576
  arbitrary amplitude responses, 583
  basic, 549
  efficient, 562–566
  generalization, 574–578
  gradient information, 566–568
  improved formulation, 561–562
  multiband filters, 583–586
  symmetrical impulse response and even filter length, 576–577
Representation of digital filters:
  by difference equation, 8–9
  by matrices, 26–29
  by networks, 9–10
  by signal flow graphs, 13, 119–121
  by zeros and poles, 68
Residue:
  definition, 47
  for multiple pole, 54
  for simple pole, 54
  theorem, 46

Resonant circuits, wave realization of, 404–405
Resonant frequency, 370
Response, 2
Rhodes, J. D., 410
Ripple ratio:
  definition, 287
  in Dolph-Chebyshev window function, 291–292
  in Hamming window function, 289
  in Kaiser window function, 293
  in rectangular window function, 287
  in von Hann window function, 289
Roberts, R. A., 360, 362, 374
Rosenbrock function, 539
Rounding of numbers, 337
Roundoff noise, 331
  minimization, 360–364
RPSD (relative power spectral density), 386
Rudimentary implementation, 332–333

s^2-impedance elements, 416
Saal, R., 417
Sample function in random processes, 315
Sample-and-hold device, 210
Sampled signals, 199–202
  Fourier transform of, 200
  Laplace transform of, 202
  periodicity of spectrum in, 200–201
Sampled-data filter, 207
Sampler, impulse, 199
Sampling frequency, 83
Sampling of bandpassed signals, 623–625
Sampling theorem:
  frequency-domain, 449–451
  time-domain, 203–204
Sampling-frequency conversion:
  compression, 594
  compressor, 594
  decimators, 593–595
  design of decimators, 602
    of interpolators, 602
  downsampler, 594
  downsampling, 594
  expander, 597
  images, 597
  interpolators, 595–601
  linear interpolation, 601
  by a noninteger factor, 601–602
  upsampler, 597
Sampling-frequency expander, 597
Sandberg, I. W., 374
Scaling, application, 358–359
  based on L_2 norm, 356
  based on L_∞ norm, 356
  in CGIC realization, 432–433
  of signals, 352–360
  in wave digital filters, 422–423
Schur polynomials, 75
Schur-Cohn stability criterion, 74–75
Schur-Cohn-Fujiwara stability criterion, 75–76
Schwarz inequality, 355
Second-order statistics, 316
Section-optimal structures, 362
Sedlmeyer, A., 388
Selective step-by-step search, 555–558
Selectivity:
  analysis, 125–129
  to coefficient quantization, 343
  considerations in wave digital filters, 389–390
  factor, 155
  gain, 136, 343
  normalized, 344
  of passband loss, 390
  phase-shift, 136, 343
Series:
  binomial, 56
  Fourier, 197, 280
  Laurent, 45–47
    residues, 54
  representation for elliptic functions, 663–665
  Taylor series, 143, 493
Series adaptors, 395
  1-multiplier, 395
  2-multiplier, 395
Series wire interconnections, wave realization of, 395
Shanks, J. L., 638
Sherwood, R. J., 104
Shift operator:
  definition, 10
  properties, 10–11
Side lobes, 283
Siegel, J., 545
Signal enhancement by adaptive filters, 636
Signal flow-graph analysis, 13–15
  representation of digital filters, 13, 119–121
Signal prediction by adaptive filters, 636–637
Signal scaling, 352–360
Signals:
  analytic, 613–617
  bandpassed, 623–625
  continuous-time:
    definition, 1
    processing using digital filters, 207–208
    unit step, 181
  discrete-time:
    definition, 1

INDEX **685**

gnals: discrete-time (*cont.*)
    exponential, 16
    notation, 1
    sinusoid, 16
    unit impulse, 16–17
    unit ramp, 16
    unit step, 15
  infinite-energy, 182
  interrelations between continuous-time, sampled, and discrete-time signals, 205–206
  nonquantized, 1
  quantized, 2
  random (*see* Random processes)
  sampled, 199–202
    Fourier transform of, 200
    periodicity of spectrum in, 200–201
    Laplace transform of, 202
  2-dimensional, 637
ignal-to-noise ratio, 367
igned-magnitude addition, 336
  arithmetic, 336
  multiplication, 336
  number representation, 334
ingle-sideband modulation using Hilbert transformers, 621–623
ingularities, 45
  essential, 46
  isolated, 45
  nonisolated, 45
ink, 116
inusoidal function, 16
inusoidal response, 19–21
kwirzynski, J. K., 417
mith, M. J. T., 611
oftware implementation, 88–89, 115
  overlap-and-add method, 478–480
  overlap-and-save method, 480–482
ource, 116
pecification constraint in elliptic approximation, 164–166
pecifications in analog filters, 172
pectral density, energy, 180
pectrum:
  amplitude (in continuous Fourier transform), 178
  amplitude (in discrete Fourier transform), 445
  energy (in continuous Fourier transform), 180
  in Hamming window function, 289–290
  in Kaiser window function, 292–293
  phase (in continuous Fourier transform), 178
  phase (in discrete Fourier transform), 445
  power-density, 320
  in rectangular window function, 287
  in von Hann window function, 289–290

Stability:
  BIBO, 26
  bounded-input bounded-output, 26
  constraint on eigenvalues, 72
  on impulse response, 25–26
  on poles, 70–71
  criteria:
    Jury-Marden, 76–78
    Lyapunov, 78
    Schur-Cohn, 74–75
    Schur-Cohn-Fujiwara, 75–76
  definition, 25
  of forced response, 380
  Jury-Marden array, 77
  in nonrecursive filters, 71
  stabilization of recursive filters, 226
  of steepest-descent algorithm, 631–633
  in 2-dimensional digital filters, 638–640
Stabilization technique, 226
State variables, 27
State-space characterization, 26–28
  derivation of discrete-time transfer function, 68–69
  impulse response, 30
State-space realization, 104
Stationary point, 493
Stationary random processes, 319
Statistical independence in random variables, 312
Statistical word length, 342
Steady-state sinusoidal analysis, 21
  first-order filter, 19–21
  Nth-order filter, 80–81
Steepest-descent algorithm, 630–631
  stability, 631–633
Stirling interpolation formula, 301
Stockham Jr., T. G., 213
Stopband, 141
Stopband edge, 141
Stored power in wave digital filters, 423
Strictly-stationary random processes, 319
Structures:
  block-optimal, 362
  canonic, 102–104
  cascade, 113–114
  choice among, 434–436
  direct, 98
  direct canonic, 102–104
  G-CGIC, 429–432
  ladder, 104–109
  ladder wave, 412–416
  lattice, 109–113
  lattice wave, 411–412
  low-sensitivity, 344–350

Structures (cont.)
    Meerkötter, 377
    optimal state-space, 360–364
    parallel, 114–115
    section-optimal, 362
    state-space, 104
Subgraph determinant, 15
Sum of the squares, 492
Symmetrical impulse response, 276
Symmetry in discrete Fourier transform, 447
    in Fourier transform, 179
Synthesis, 88
Synthesis section in filter banks, 602
System identification by adaptive filters, 634
System representation of digital filters, 2–8
Szentirmai, G., 417

Tables:
    common Fourier transforms, 195
    common z transforms, 53
    design formulas for frequency-sampling
        approximation method, 461
        for recursive filters:
            bandpass and bandstop, 258
            Butterworth, 259
            Chebyshev, 260
            elliptic, 260
            lowpass and highpass, 253
        for Remez exchange algorithm, 579
    discrete-time elementary functions, 16
    elements for digital filter, 10
    formulas for frequency response of nonrecursive
        filters, 278
    possible sequences for impulse and unity
        functions, 193
    transformations for analog filters, 172, 251
        for digital filters, 243
    window parameters, 288
Taylor, M. G., 373
Taylor series, 143, 493
Tellegen's theorem, 121–122
Termination tolerances, 495
Test for common factors, 73–74
Theorems:
    alternation theorem, 547–548
    Fourier transform:
        convergence, 178
        frequency convolution, 179
        frequency shifting, 179
        linearity, 178
        Parseval's formula, 179
        symmetry, 179
        time convolution, 179
        time scaling, 179
        time shifting, 179
    Laurent theorem, 45–46
    Parseval's discrete-time formula, 61
    residue theorem, 46
    sampling theorem:
        frequency-domain, 449–451
        time-domain, 203–204
    Taylor's theorem, 493
    Tellegen's theorem, 121–122
    z transform:
        convergence:
            absolute, 48
            uniform, 50
        complex convolution, 59–61
        complex differentiation, 51
        complex scale change, 50–51
        linearity, 50
        real convolution, 50–51
        translation, 50
Theta functions, 664
Time convolution, 179
Time invariance, 3
Time scaling in Fourier transform, 179
Time-dependent digital filters, 3
Time-division to frequency-division multiplex
    translation, 648
Time-domain aliasing, 451–453
Time-domain analysis:
    elementary functions, 15–17
    impulse response, 17
    sinusoidal response, 19–21
    unit-step response, 17–19
    using induction, 17–21
    using state-space characterization, 29–31
    using z transform, 78–79
Time-domain periodic convolution, 464
Time-domain sampling theorem, 203–204
Time-shifting in Fourier transform, 179
Topological properties:
    adjoint flow graph, 124
    computability, 133
    interreciprocal multipole, 123
    interreciprocity, 123–124
    multipole, 122
    reciprocal multipole, 122
    reciprocity, 122
    sensitivity analysis, 125–129
    Tellegen's theorem, 121–122
    transpose flow graph, 124
    transposition, 124–125
Transfer characteristic:
    adder incorporating saturation mechanism, 373

Transfer characteristic (*cont.*)
  one's- or two's-complement adder, 373
  quantizer, 339
Transfer function:
  continuous-time:
    in analog filters, 80, 140
    constant-delay allpass, 175
Transfer function: continuous-time (*cont.*)
  from convolution integral, 179
  normalized, 143
    Bessel, 168
    Butterworth, 144–145
    Chebyshev, 149–151
    elliptic, 166–167
    inverse-Chebyshev, 153
  realizability constraints, 141–143
  discrete-time:
    allpass, 269
    bandlimited, 222
    definition, 67
    derivation from difference equation, 67–68
      from filter network, 68
      from state-space characterization, 68–69
    in G-CGIC configuration, 429
    realizability constraints, 221
    relation with impulse response, 67
    of sampled-data filter, 208
    in 2-dimensional filters, 628
    in wave digital filters, 412, 420–421
Transformations:
  for analog filters, 169–172
    lowpass to bandpass, 170–172
    lowpass to bandstop, 172
    lowpass to highpass, 172
    lowpass to lowpass, 169–170
    table, 172
  for digital filters, 237–243
    application, 243
    bilinear, 233
    general transformation, 239–240
    lowpass to bandpass, 243
    lowpass to bandstop, 242–243
    lowpass to highpass, 243
    lowpass to lowpass, 240–242
    table, 243
  for elliptic functions, 662–663
Transformers, wave realization of, 398–400
Transforms (*see* Discrete Fourier transform, Fast-Fourier transform, Fourier transform, Hilbert transform, Laplace transform, or z transform)
Translation theorem of z transform, 50
Transmission zeros, 94

Transmittance, 13, 116
Transpose flow graph, 124
Transposition, 124–125
Triangular window function, 307
Trick, T. N., 374
Truncation of numbers, 337
Tschebyscheff approximation (*see* Chebyshev approximation)
Turner, L. E., 374, 378
Two-dimensional convolution, 637–638
Two-dimensional digital filters:
  amplitude response, 640
  applications, 644
  approximations, 643–644
  with circular band boundaries, 642
  fan, 643
  frequency-domain analysis, 640–641
  group delays, 640
  highpass, 643
  introduction, 593
  nonrecursive, 637
  order, 637
  phase response, 640
  with rectangular band boundaries, 641
  recursive, 637
  stability, 638–640
Two-dimensional signals, 637
Two-dimensional transfer function, 638
Two-dimensional z transform, 638
Two-port adaptors, 397–398
Two's-complement:
  addition, 336
  arithmetic, 336
  multiplication, 336
  number representation, 335

Unconstrained adaptors, 397
Uniform convergence of z transform, 50
Uniform probability-density function, 311
Unimodal functions, 501
Unit delay, 9
Unit elements, wave realization of, 401–402
Unit-impulse function, 16–17
Unit-ramp function, 16
Unit-step function:
  continuous-time, 181
  discrete-time, 15
Unit-step response:
  using induction, first-order filter, 17–19
  using state-space characterization, 30
Unity function, 185–191
  alternative sequences, 187–191
  definition, 186

Universal CGIC sections, 429–432
Upsampler, 597

Vaidyanathan, P. P., 374, 425, 610
Variance of random variable, 313
Vaughan-Pope, D. A., 425
Verkroost, G., 378
Virtual sample points, 514
VLSI implementation, 116
Voltage sources, wave realization of, 393–395
von Hann window function, 287–290
   main-lobe width, 289
   ripple ratio, 289
   spectrum, 289–290

Warping effect, 235–238
   influence on amplitude response, 235–236
      on phase response, 237
   in wave digital filters, 416
Wave digital filters:
   adaptors:
      parallel, 397
      series, 395
      2-port, 397–398
   cascade realization using CGICs, 429–432
      output noise, 433–434
      signal scaling, 432–433
   chain matrix, 401
   choice of structure, 434–436
   circulators, 402–403
   design procedure for ladder filters, 413
      for lattice filters, 412
   element realization, 391–393
   frequency-dependent negative resistance networks, 416
   frequency-domain analysis, 419–422
   input functions, 420
   introduction, 388
   ladder wave digital filters, 412–416
   lattice:
      alternative analog realization, 407–408
      analog network, 406
      analysis, 406–407
      digital realization, 411–412
   prescribed specifications, 416–419
   pseudopower, 423
   realizability constraint, 405
   realization of impedances, 393
      of parallel wire interconnections, 395–397
      of resonant circuits, 404–405
      of series wire interconnections, 395
      of transformers, 398–400
      of unit elements, 401–402
      of voltage sources, 393–395

   scaling, 422–423
   sensitivity considerations, 389–390
   stored power, 423
   transfer function, 412, 420–421
   unconstrained adaptors, 397
   warping effect, 416
   wave network characterization, 390
   wire interconnections, 391–392
Wave network characterization, 390
   incident wave quantity, 390
   port conductance, 397
   port resistance, 390
   reflected wave quantity, 390
Weighted-Chebyshev method, introduction, 544–545
White-noise processes, 323
   random processes, 319
Wiener filters, 628–630
Wiener-Khinchine relation, 322
Window functions:
   Bartlett, 307
   Blackman, 290
      main-lobe width, 290
   design of nonrecursive filters, 282–286
   Dolph-Chebyshev, 291–292
      main-lobe width, 291–292
      ripple ratio, 291–292
   Hamming, 287–290
      main-lobe width, 289
      ripple ratio, 289
      spectrum, 289–290
   Kaiser, 292–293
      design of differentiators, 305
      main-lobe width, 293
      ripple ratio, 293
      spectrum, 292–293
   main lobe, 283
   main-lobe width, 283
   rectangular, 287
      main-lobe width, 287
      ripple ratio, 287
      spectrum, 287
   ripple ratio, definition, 287
   side lobes, 283
   triangular, 307
   von Hann, 287–290
      main-lobe width, 289
      ripple ratio, 289
      spectrum, 289–290
Wire interconnections, 391–392
Word length, definition, 335
   optimum, 340
   statistical, 342

z transform:
  application to digital filters, 66
  complex convolution, 59–61
  complex differentiation, 51
  complex scale change, 50–51
  convergence:
    absolute, 48
    uniform, 50
  definition, 47
  interrelation with discrete Fourier transform, 449
    with Fourier transform, 200
  introduction, 43
  inverse:
    definition, 53
    by equating coefficients, 65
    by using binomial series, 56
    by using convolution theorem, 58–59
    by using long division, 65
    by using partial fractions, 57
  Laurent theorem, 45–46
  linearity, 50
  radius of convergence, 49
  real convolution, 50–51
  region of convergence, 49
  residue theorem, 46
  table of common z transforms, 53
  translation, 50
  2-dimensional, 638
Zero of a function, 45
Zero-delay filters, 269–270
Zero-loss frequencies in elliptic approximation, 158
Zero-phase filters, 269–270
Zero-pole plots:
  Butterworth approximation, 145
  Chebyshev approximation, 152
  in digital filters, 68
  for loss function, 140
  for nonrecursive filters, 280
Zeros, transmission, 94
Zeroth-order modified Bessel function, 292
Zverev, A. I., 417